Mixing Audio

Mixing Audio
Concepts, Practices and Tools

Roey Izhaki

AMSTERDAM • BOSTON • HEIDELBERG • LONDON
NEW YORK • OXFORD • PARIS • SAN DIEGO
SAN FRANCISCO • SINGAPORE • SYDNEY • TOKYO

Focal Press is an imprint of Elsevier

Focal Press is an imprint of Elsevier
The Boulevard, Langford Lane, Kidlington, Oxford, OX5 1GB, UK
30 Corporate Drive, Suite 400, Burlington, MA 01803, USA

First edition 2008
Reprinted 2008, 2009 (thrice)

British Library Cataloguing in Publication Data
A catalogue record for this book is available from the British Library

Library of Congress Cataloging-in-Publication Data
A catalog record for this book is available from the Library of Congress

ISBN: 978-0-2405-2068-1

For information on all Focal Press publications
visit our website at www.elsevierdirect.com

Printed and bound in *Great Britain*

09 10 11 12   12 11 10 9 8 7 6 5

Working together to grow
libraries in developing countries

www.elsevier.com | www.bookaid.org | www.sabre.org

ELSEVIER   BOOK AID
International   Sabre Foundation

Contents

Acknowledgments

Carlos Lellis Ferreira for reviewing early drafts and his invaluable comments. Paul Frindle for keenly sharing his priceless knowledge of compressors and other mixing tools. Kelly Howard-Garde, Laura Marr and Chris Zane for their essential help during the proposal phase.

Charlotte Dawson from Digidesign UK; Colin McDowell and Aja Sorensen from McDSP; Jim Cooper from MOTU; Antoni Ożyński, Mateusz Woźniak and Hubert Pietrzykowski from PSP; Nathan Eames from Sonnox; Angus Baigent from Steinberg; and Andreas Sundgren from Toontrack.

Catharine Steers, David Bowers, VijaiSarath Parthasarathy and everyone at Focal Press who were involved with this book.

Luca Barassi, Amir Dotan, Preben Hansen, Guy Katsav, Mooncat, Mandy Parnell and Matthias Postel.

My family and friends for their support.

Contributors

Some of the leading companies in the industry supported the production of the audio material that accompanies this book. I would like to thank:

Audioease for Altiverb.
www.audioease.com

Digidesign for their Music Production Toolkit and many other plugins presented in this book.
www.digidesign.com

McDSP for their full range of plugins.
www.mcdsp.com

MOTU for Digital Performer 5.
www.motu.com

PSP for their full range of plugins.
www.PSPaudioware.com

Sonnox for their full range of plugins.
www.sonnoxplugins.com

Steinberg for Cubase 4.
www.steinberg.net

Toontrack for dfh SUPERIOR and dfh EZdrummer with its full expansion libraries.
www.toontrack.com

Universal Audio for their UAD-1e card and their plugins.
www.uaudio.com

Sample mixes

The following artists contributed productions for the sample mixes presented in Part III.

Hero
By AutoZero (Dan Bradley, Lish Lee and Lee Ray Smith).
www.autozero.co.uk

It's Temps Pt. II
By Brendon 'Octave' Harding and Temps.
www.myspace.com/octaveproductions
www.myspace.com/temps14

Donna Pomini
By TheSwine.
www.theswine.co.uk

The Hustle
By Dan 'Samurai' Havers and Tom 'Dash' Petais.
www.dc-breaks.co.uk

Introduction

It's not often a new form of art is conceived; where or when the art of mixing was born is not an easy question to answer. We can look at the instrumentation of orchestral pieces as a very primitive form of mixing – different instruments that played simultaneously could mask one another; composers understood this and took it into account. In the early days of recording, before multitrack recorders came about, a producer would place musicians in a room so that the final recording would make sense in terms of levels and depth. Equalizers, compressors and reverb emulators hadn't yet been invented. There was no such thing as a mixing engineer. But combining various instruments in order to produce a more appealing result was already an ambition shared by many.

Like many other new forms of creative expression that emerged in the twentieth century, mixing was dependent on technology. It was the appearance of the multitrack tape machine during the 1960s that signified the dawn of mixing as we know it today. Yes, there was a time when having the ability to record eight instruments separately was a dream come true. Multitracks allowed us to repeatedly play recorded material before committing sonic treatment to the mix. Equalizers, compressors and reverb emulators soon became familiar residents in studios; audio consoles grew in size to accommodate more tracks and facilities; 8 tracks became 16 and then 16 became 24. We had more sonic control over individual tracks and over the final master. The art of mixing was flourishing. Music sounded better.

The 1990s significantly reshaped much of the way music is made, produced, recorded, mixed and even distributed – computers triumphed. Realtime audio plugins were first introduced with the release of Pro Tools III as far back as 1994 (although back then they could only run on a dedicated DSP card). It was Steinberg and its 1996 release of Cubase VST that pioneered the audio plugins that we now take for granted – a piece of software that can perform realtime audio calculations using the computer's CPU. The term *project studio* was soon coined as computers became more affordable and competent, and the hiring of expensive studios was no longer a requisite for multitracking and mixing. However, the processing power of computers back then could still not compete with the quality and quantity of mixing devices found in a professional studio. Things have changed – running ten quality reverbs simultaneously on a modern DAW has been a reality for some time. The quality and quantity of audio plugins is improving daily, and new technologies – like convolution – could hint at an even brighter future. Professional studios will always, it seems, have an advantage over project studios, if only for their acoustic qualities, however, DAWs offer outstanding value for money, constantly improving quality and increasingly wider possibilities.

So is everything rosy in the realm of mixing? Not quite. It is thanks to computers that mixing has moved from large and expensive studios into bedrooms. More people than ever are mixing music, but only a few can be said to be true mixing engineers. Mixing used to be done by skilled engineers, who were familiar with their studio and the relatively small set of expensive devices within it. Mixing was their occupation and for many – their *raison d'être*. On the contrary, project studio owners generally do much more than just mixing – for many it is just another stage in an independent production chain. So how can these people improve their mixing?

Despite the monumental importance of mixing, resources have always been lacking. The many magazine articles and the handful of books on the topic are generally either less-than-comprehensive or a clutter of technical information that would require a great degree of involvement from those who wish to learn the true concepts and techniques of this fascinating field. This book was conceived to fill the gap.

I would like, in this opening text, to expose the greatest misconception that exists about mixing: it is wrongly assumed by some that mixing is a purely technical service, some even declare that mixing is simply a remedy for imperfect recordings. There is no doubt that mixing entails technical aspects: a problematic level balance, uncontrolled dynamics and deficient frequency response are just a few of the technical issues we encounter. Yet, with the right amount of effort, almost anybody could master the technical aspects of mixing – after compressing 100 vocal tracks, one should be getting the hang of it. Technical skills are advantageous but can be acquired: the true essence of mixing does not lie in these skills. Many mixes are technically great, but nothing more than that; equally, many mixes are not technically perfect, but as a listening experience they are breathtaking. It is for their sheer creativity – not for their technical brilliance – that some mixes are highly acclaimed and their creators deemed sonic visionaries.

The sonic qualities of music are inseparable from the music itself – the Motown sound, the NEVE sound, the *Wallace sound* and so forth. The non-technical side of mixing entails crafting the sonic aspects of music: shaping sounds, crystallizing soundscapes, establishing sonic harmony between instruments and building sonic impact – all rely on the many creative decisions that we make, all are down to the talent and vision of each individual, all have a profound influence on how the music is perceived. It is in the equalization we dial, in the reverb we choose, in the attack we set on the compressor, to name but a few. There simply isn't one correct way of doing things – be it an acoustic guitar, a kick, or any other instrument, it can be mixed in a hundred different ways; all could be considered technically correct, but some would be more remarkable than others. A mix is a sonic portrait of the music. The same as different portraits of a person can each project a unique impression, different mixes can convey the essence of the music in extremely different ways. We are mixing engineers but more importantly: we are *mixing artists*.

By the time you finish reading this book, you should have far more knowledge, a greater understanding and improved auditory skills that will together enable you to craft better mixes. However, I hope that you keep in mind this maxim:

| *Mixing is an art.* |

A Friendly Warning

It would not make sense for wine tasters to sip boiling oil, just as it would not make sense for mixing engineers to stick sharp needles into their eardrums. While I have yet to meet an engineer who fancies needles in his or her eardrums, very loud levels can be equally harmful. Unlike needle-sticking, the hearing damage caused by loud levels is often not immediate, whether involving short or long exposures.

Sparing the medical terminology, with years one might lose the ability to hear high frequencies, and the really unlucky could lose substantial hearing ability. In some circumstances, very loud levels can cause permanent damage to the eardrum and even deafness. Most audio engineers, like myself, have had one or two level-accidents; the majority of us are fine. But hearing a continuous 7 kHz tone is no laughing matter, especially when it lasts for three days.

The allowance, as they say in Italian, is *forte ma non troppo* – loud but not too much. The National Institute for Occupational Safety and Health in the USA recommends that sound exposure to 85 dBSPL should not exceed eight hours per day, halving the time for each 3 dB increase. A quick calculation reveals that it is only safe to listen to 100 dBSPL for 15 minutes. A screaming child a meter away is roughly 85 dBSPL. A subway train one meter away produces roughly 100 dBSPL when cruising at normal speed.

In the DVD accompanying this book, I have done my best to keep relatively consistent levels. Still, some samples had to be louder than others. Please mind your monitoring level when listening to these samples, and remember that too quiet can easily be made louder, but it might be too late to turn down levels once they are too loud.

Why we like loud levels so much is explained in Chapter 2. But if we are all to keep enjoying music, all we have to do is be sensible about the levels at which we mix and listen to music.

| *Levels, like alcohol, are best enjoyed responsibly.* |

Symbols and formats used

Audio samples

Tracks referenced within these boxes are included on the accompanying DVD, organized in a different folder per chapter. Readers are advised to copy the DVD content to their hard drive before playing these tracks. Please mind your monitoring level when playing these tracks.

Recommended listening

These boxes include references to commercial tracks, which are *not* included on the accompanying DVD.

Notes

Tips or other ideas worth remembering.

Part I
Concepts and Practices

1 Music and mixing

Music – An extremely short introduction

You love music. All of us are mixing because music is one of our greatest passions, if not *the* greatest. Whether starting as a songwriter, bedroom producer, performer or a studio tea boy, we were all introduced to mixing through our love of music and the desire to take part in its creation.

Modern technology has pushed aside – or in some cases replaced – some art forms. Reading literature, for example, has been replaced by watching TV and staring at computer monitors. But with music, technology has provided new opportunities, increased reach and improved quality. The invention of the wax cylinder, radio transmission, tapes, CDs and software plugins have all made music more readily accessible, widely consumed and easily created. One of mankind's most influential inventions – the Internet – is perhaps music today's greatest catalyst. Nowadays, a mouse is all one needs to sample and purchase music. Music is universal and all encompassing. It is in our living rooms, in our cars, in malls, on our televisions and in hairdressing salons. Now that most cellphones have an integrated MP3 player, music seems almost impossible to escape from.

There is a strong bond between music and mixing (other than the obvious connection that music is what's being mixed), and to understand it we should start by discussing the not-too-distant past. History teaches us that in the western world, sacred music was prevalent up until the 19th century, with most compositions commissioned for religious purposes. Secular music has evolved throughout the years, but changed drastically with the arrival of Beethoven. At the time, Beethoven was daring and innovative, but it was the way that his music made people *feel* that changed the course of music so dramatically. Ernest Newman once wrote about Beethoven's symphonies:

> The music unfolds itself with perfect freedom; but it is so heart-searching because we know all the time it runs along the quickest nerves of our life, our struggles & aspirations & sufferings & exaltations.[1]

---

[1] Allis, Michael (2004). *Elgar, Lytton, and The Piano Quintet, Op. 84.* Music & Letters, Vol. 85 No. 2, pp. 198–238. Oxford University Press. Originally a Letter from Newman to Elgar, 30 January 1919.

3

We can easily identify with this when we think about modern music – there is no doubt that music can have a huge impact on us. After Beethoven, music became a love affair between two willing individuals, the artist and the listener, fueled by what is today an inseparable part of music – emotion.

Today music rarely fails to produce emotions – all but a few pieces of music have some sort of mental or physical effect on us. *Killing in the Name* by Rage Against the Machine can trigger a sense of rage or rebellious anger. Many find it hard to remain stationary when hearing *Hey Ya!* by OutKast, and for some this tune can turn a bad morning into a good one. Music can also trigger sad or happy memories, and so the same good morning can turn into a more retrospective afternoon after hearing Albinoni's *Adagio for Strings and Organ in G Major.* (Which goes to show that it's not just emotive lyrics that affect us.) In many cases, our response to music is subconscious, but sometimes we deliberately listen to music in order to incite a certain mood – some listen to ABBA as a warm up for a night out, others to Sepultura. Motion-picture directors understand very well how profoundly music can affect us and how it can be used as a device to garner certain emotional responses from the audience. We all know what kind of music to expect when a couple fall in love or when the shark is about to attack. It would be a particular genre of comedy that used *YMCA* during a funeral scene.

As mixing engineers, one of our prime functions, which is actually *our responsibility*, is to help deliver the emotional context of a musical piece. From the general mix plan to the smallest reverb nuances, the tools we use – and the way we use them – can all sharpen or sometimes even create power, aggression, softness, melancholy, psychedelia and many other emotions or moods. It would make little sense to distort the drums on a mellow love song, just as it would not be right to soften the beat of a hip-hop production. When approaching a new mix, we should ask ourselves a few questions:

- What is this song about?
- What emotions are involved?
- What message is the artist trying to convey?
- How can I support and enhance the song's vibe?
- How should the listener respond to this piece of music?

As basic as this idea might seem, it is imperative to comprehend – the mix is dependent on the music, and mixing is not just a set of technical challenges. What's more, the questions above lay the foundation for an ever so important quality of the mixing engineer – a mixing vision.

> *A mix can, and should enhance the music, its mood, the emotions it entails, and the response it should incite.*

The role and importance of the mix

A basic definition of mixing is: a process in which multitrack material – whether recorded, sampled or synthesized – is balanced, treated and combined into a multichannel format, most commonly two-channel stereo. But in addition to that – and more importantly – **a mix is a sonic presentation of emotions, creative ideas and performance**.

Even for the layman, sonic quality does matter. Take the cellphone, for example, people find it annoying when background noise masks the other party. Intelligibility is the most elementary requirement of sonic quality, but it goes far beyond that. Some new cellphone models with integrated speaker are no better than playback systems from the 1950s. There is no wonder that people prefer listening to music via their kitchen's mini-system or the living room hi-fi. What would be the point of more expensive hi-fi systems if the mixes we play on them sound like they are being played through a cellphone speaker?

Sonic quality is also a powerful selling point. It was a major contributor to the rise of the CD and the fall of compact cassettes. Novice classical music listeners often favor new recordings to the older, monophonic ones, regardless of how acclaimed the performance on these early recordings is. Many record companies these days issue digitally remastered versions of classic albums, which allegedly sound better than the originals. The now ubiquitous iPod owes its popularity to the MP3 format – no other lossy compression format has managed to produce audio files so small, that are still of an acceptable sonic quality.

> *The majority of people appreciate sonic quality more than they ever realize.*

So we know that it is our responsibility as mixing engineers to craft the sonic aspects of the final mix but we also control the quality of the individual instruments that constitute the mix. Let us consider for a moment the differences between studio and live recordings: During a live concert, there are no second chances. You are unable to rectify problems such as bad performance or a buzz from a faulty DI box. Both the recording equipment and the environment are inferior compared to the ones found in most studios – it would be unreasonable to place Madonna in front of a U87 and a pop shield during a live gig. Also, when a live recording is mixed on location, a smaller and cheaper arsenal of mixing equipment is used. All of these elements result in different instruments suffering from masking, poor definition, slovenly dynamics and deficient frequency response, to name just a few of the possible problems. Audio terms aside, these can translate into a barely audible bass guitar, honky lead vocals that come and go, a kick that lacks power, and cymbals that lack spark. The combination of all these makes a live recording less appealing. A studio recording is not immune to these problems, but in most cases it provides much better raw material to work with, and in turn better mixes. With all this in mind, the true art of mixing is far more than just making things sound right . . .

Many people are familiar with Kurt Cobain, Dave Grohl and Krist Novoselic as the band members of Nirvana, who back in 1991, changed the face of alternative rock with the release of *Nevermind*. The name Butch Vig might ring a bell for some, but the general public will be unlikely to have heard of Andy Wallace. The front cover of my *Kill Bill* DVD makes it extremely difficult to not notice Tarantino's writer and director credits. But it is seldom that an album cover credits the producer, let alone the mixing engineer. Arguably, the production of Dr Dre can be just as important as the artists he produces, and perhaps *Nevermind* would have never gained such an enormous success had it not been Andy Wallace's consummate mixing. Nevertheless, record labels generally see very little marketing potential in production personnel. Ironically, major record companies do frequently write fat checks in order to have a specific engineer mix an album or a track because they all realize that:

| *The mix plays an enormous role in an album's or track's success.* |

To understand why, one should listen to the four versions of *Smells Like Teen Spirit* indicated below. The link between the sonic quality of a recording and its ability to excite us means that it is fair to assume that in order of appeal – the rehearsal demo would be the least appealing and the album version the most appealing. Having looked at the differences between a live and a studio recording, it should be clear why most people would find both the rehearsal demo and the live recording less satisfactory. Compare Vig's and Wallace's mixes and it will give you a great insight into what mixing is really about, and what a huge difference a mix can make.

> **Smells Like Teen Spirit** (rehearsal demo, track 10 on CD2)
> Nirvana. **With the Lights Out**. Geffen Records, 2004.
>
> **Smells Like Teen Spirit** (live version)
> Nirvana. **From the Muddy Banks of the Wishkah**. Geffen Records, 1996.
>
> **Smells Like Teen Spirit** (Butch Vig mix, track 20 on CD2)
> Nirvana. **With the Lights Out**. Geffen Records, 2004.
>
> **Smells Like Teen Spirit** (Andy Wallace mix, album version)
> Nirvana. **Nevermind**. Geffen Records, 1991.

Both Vig and Wallace used the same raw tracks; yet, their mixes are distinctly different. Vig's mix suffers from an unbalanced frequency spectrum that involves some masking and the absence of spark; a few mixing elements, like the snare reverb, are easily discernible. Wallace's mix is polished and balanced; it exhibits high definition and perfect separation between instruments; the ambiance is present, but like many mixing elements it is fairly transparent. Perhaps the most important difference between the two mixes is that Vig's mix sounds more natural (more like a live performance), while Wallace's mix sounds more artificial. It is not equipment, time spent or magic tricks that made these two mixes so dissimilar – it is simply the different sonic visions of Vig and Wallace. Wallace, in nearly an alchemist fashion, managed to paint every aspect of this powerful song into an extremely appealing portrait of sounds. Like many other listeners, Gary Gersh – Geffen Records, A&R – liked it better.

Straight after recording *Nevermind*, it was Vig that started mixing the album. Tight schedule and some artistic disagreements he had with Cobain left everyone feeling (including Vig) that it would be wise to bring fresh ears to mix the album. From the bottom of prospective engineers list, Cobain chose Wallace, mostly for his Slayer mixing credits. Despite the fact that Nirvana approved the mixes, following *Nevermind*'s extraordinary success, Cobain complained that the overall sound of *Nevermind* was too slick – perhaps suggesting that Wallace's mixes were too listener-friendly for his artistic, somewhat anarchic taste. Artistic disagreements are something engineers come across often, especially if they ignore the musical concept the artist wants to put forth. Yet some suggested that Cobain's retroactive complaint was only a mis-targeted reaction to the massive success and sudden fame the album brought. Not only did *Nevermind* leave its mark on music history, it also left a mark on mixing history – its sonic legacy, a part of what is regarded as the *Wallace Sound*, is still heavily imitated today. As testament to Wallace's skill, *Nevermind has* aged incredibly well and still sounds fresh despite enormous advances in mixing technology.

Seldom do we have the opportunity to compare different mixes of the same song but the 10th anniversary edition of *The Holy Bible* by the Manic Street Preachers lets us compare an entire album of 12 tracks. The package contains two versions of the album, each mixed by a different engineer. The UK release was mixed by Mark Freegard and the US release by Tom Lord Alge. There is some similarity here to the Vig vs. Wallace case, where Freegard's mixes are cruder and drier compared to the live, brighter and more defined mixes of Alge. In the included DVD, the band comments on the differences between the mixes, saying that for most tracks Alge's mixes represented better their artistic thinking. Arguably, neither version features exceptional mixes (possibly due to poor recording quality in a cheap facility), but the comparison between the two is a worthwhile and recommended experience.

The two examples above teach us how a good mix can sharpen the emotional message of a musical piece, make it more appealing to the listener and result in more commercial success. On the contrary, a bad mix can negatively affect a potentially great piece and significantly impair its success. This is not only relevant for commercial releases. The price and quality of today's DAWs enable unsigned artists and bedroom producers – with enough talent and vision – to craft mixes that are of an equal standard to commercial mixes. A&R departments are beginning to receive demos of a respectable quality, and a big part of this has to do with the mix. Just like a studio manager might filter through a pile of CVs and eliminate candidates based on poor presentation, an A&R scout might dismiss a demo based on its poor mix.

Mixing engineers know what a dramatic effect mixing can have on the final product. With the right amount of effort, even the poorest recording can be made appealing. Yet, there are a few things we cannot do, for example, correct a truly bad performance, compensate for very poor production or alter the original concept. If the piece does not have potential to begin with, it will fail to impress the listener, no matter how noteworthy the mix is.

| *A mix is as good as the song.* |

The perfect mix

It doesn't take much experience before the novice mixer can begin to recognize problems in a mix. For instance, we quickly learn to identify vocals that are too quiet or a deficient frequency response. We will soon see that once a mix is problem-free, there are still many things we can do in order to make it better. The key question is: What is better?

Figure 1.1 Excerpt set. This sequence of 20-second excerpts from various productions is used as an important comparison tool between mixes.

At this point, I recommend an exercise called **excerpt set** (Figure 1.1) – an essential experiment for anyone involved in mixing. It takes around half an hour to prepare, but provides a vital mixing lesson. The excerpt set is very similar to a DJ set, except each track plays for around 20 seconds, and you do not have to beat-match. Simply pull around 20 albums from your CD library, pick only one track from each album and import each track into your audio sequencer. Then trim a random excerpt of 20 seconds from each track and arrange the excerpts consecutively. It is important to balance the perceived level of all excerpts, and it is always nice to have cross-fades between them. Now listen to your set, beginning to end, and notice the differences between the mixes. You are very likely to learn that all the mixes differ greatly. You might also learn that mixes you thought were good are not as good when played before or after another mix. While listening, try to note mixes that *you* think overpower others. This exercise will help develop a heightened awareness of what a good mix is, and can later be used as a reference.

Most of us do not have a permanent sonic standard stored in our brains, so a mix is only better or worse than the previously played mix. The very same mix can sound dull compared to one mix, but bright compared to another. (With experience, we develop the ability to critically assess mixes without the need for a reference; although usually only in a familiar listening environment.) In addition, our auditory system has a very quick settle-in time, and it gets used to different sonic qualities as long as these remain constant for a while. In essence, all our senses work that way – a black and white scene in a color movie is more noticeable than the lack of color on a black and white TV. The reason why the excerpt set is such an excellent tool for revealing differences is that it does not give the brain a chance to settle in to a particular style. When mixes are played in quick succession, we can more easily perceive the sonic differences between them.

Different engineers have different ideas and mix in different environments and therefore produce different mixes. Our ears are able to tolerate radical differences as long as mixes are not heard in quick succession. The truth is that it is hard to find two albums that share an identical mix because: different genres are mixed differently – jazz, heavy metal and trance will rarely share the same mixing philosophy, different songs involve different emotions and therefore call for different soundscapes, and the quality and nature of the raw tracks vary between projects. But mostly, it is because a mixing engineer is an artist in his own right, and each has different visions and ideas about what is best for the mix. Asking what is a perfect mix is like asking who is the best writer that ever lived, or who was the greatest basketball player of all times – it is all down to personal opinion.

Muse. *Absolution*. A & E Records, 2003.
Franz Ferdinand. *You Could Have It So Much Better*. Domino Records, 2005.

Mixing engineers will often adjust their style depending on the project. One example is Rich Costey, who mixed Muse's *Absolution*, imbuing it with a very polished feel. He later produced Franz Ferdinand's *You Could Have It So Much Better* and his mixes on this were much more raw and had distinct retro feel. Both mixing approaches work perfectly well in the context of each album.

So, there is no such thing as a perfect mix but, as with many subjective things in life, there is a multitude of critically acclaimed works. Just like generations of readers have acknowledged Dostoevsky's talent, and many people consider Michael Jordan to be one of the greatest sportsman ever, many audiophiles hold in high regard the likes of Andy Wallace or Spike Stent and the inspiring and pioneering mixes they have crafted. There might not be such a thing as a perfect mix, but the list below includes a small selection of noteworthy mixes produced by acclaimed engineers.

There are many albums with truly outstanding mixes. I could have easily listed 50 of my personal favorites, but the rationale behind my selections was to provide a small diverse list. In addition, many of these examples include mixing aspects that we will discuss in later chapters. It is worth noting that apart from superb mixes, all these albums are exceptionally well produced.

Kruder Dorfmeister. *The K&D Session.* **!K7 Records, 1998.**
Mixed by Peter Kruder and Richard Dorfmeister.

Fascinatingly, neither Kruder nor Dorfmeister are mixing engineers by profession. This downbeat electronica remixes album is a master class in almost every aspect of mixing. From the entire list, the mixes on this album are probably the most integral to the overall product, clearly blurring the line between producing and mixing.

Nirvana. *Nevermind.* **Geffen Records, 1991.**
Mixed by Andy Wallace.

Considered by many as the godfather of mixing, Wallace is perhaps the most influential mixing engineer of our time. He had two main breakthroughs in his career: The first was *Walk This Way* by Run D.M.C. and Aerosmith – a landmark track in rap history. The second was the release of Nirvana's *Nevermind*. Wallace's impressive and prolific mixing credits are far too extensive to catalog here. Sufficient to say the majority of his mixes are bordering on immaculate (or just plain immaculate) and provide a great learning source.

Massive Attack. *Mezzanine.* **Virgin Records, 1998.**
Mixed by Mark 'Spike' Stent.

One of Britain's most notable mixing engineers, Spike's career soared after his unique mixes on a seminal album – *The White Room* by KLF. With *Mezzanine*, an album that is bursting with mixing nuances, Spike crafted one of the most enchanting soundscapes in trip-hop history, and mixing history in general.

Muse. *Absolution.* **A & E Records, 2003.**
Mixed by Rich Costey.

During his early career, Costey engineered for Phillip Glass, a fact that I believe is still reflected in his unique mixes. He quickly became one of the most distinctive mixing engineers around. To some he is known as the record breaker when it comes to power and aggression – a result of mixes he did for bands like Rage Against The Machine, Audioslave, Mars Volta and Muse. Each mix on *Absolution* is inspiring, but one specific mix – *Hysteria* – sums up an era, whilst also heralding the beginning of the next.

Radiohead. *OK Computer.* **Parlophone, 1997.**
Mixed by Nigel Godrich.

Nigel Godrich has hinted in the past that mixing was never his strongest point. A celebrated producer, who is known for his immense creativity, perhaps Godrich's strongest mixing skill is his exceptional ability to reflect and deliver the emotional vitality of the music he works with. His mixes are rarely polished to commercial perfection, but never fail to be penetrating and emotive.

The Delgados. *The Great Eastern*. Chemikal Underground Records, 2000.
Mixed by Dave Fridmann.

Had a mixing studio been a playground, Fridmann would probably be the hyperactive kid. He is not a purist; that is, his mixes are infused with many tricks, gimmicks and fresh ideas. The opening track on this album, *The Past That Suits You Most*, is a demonstration of how, arguably, a mix can demand more attention than the song itself. A later track, *Witness*, perhaps represents his finest achievement to date.

Further reading

Cook, Nicholas (1998). *Music: A Very Short Introduction*. Oxford University Press.

2 Some axioms and other gems

Due to the correlative nature of a huge area like mixing and the diversity of readers, it is impossible to structure this book as a cover-to-cover manual. For those with little background, some concepts presented in this section may become clearer after reading Part II. The depth of coverage means that some topics might appeal to some more than to others. For certain readers, specific topics might become more relevant once basic understanding and techniques have been acquired. Whether working on a console in a professional studio or at home on a DAW, both the novice and veteran will find many fascinating ideas within the following chapters.

This chapter covers a few principles that will crop up time and again throughout *Mixing Audio* so I have chosen to introduce them at this early stage.

Louder perceived better

Back in 1933, two researchers at Bell Labs, Harvey Fletcher and W.A. Munson, conducted one of the most significant experiments in psychoacoustics. Their experiment was based on a series of tests taken by a group of listeners. Each test involved playing a test frequency followed by a reference tone of 1 kHz. The listener simply had to choose which of the two was louder. Successive tests involved either a different test frequency, or different levels. Essentially, what Fletcher and Munson tried to conclude is how louder or softer different frequencies had to be in order to be perceived as loud as 1 kHz. They compiled their results and charted a graph known as the *Fletcher–Munson Curves*. A chart based on the original Fletcher–Munson study is shown in Figure 2.1. I am presenting it upside-down, as it bears a resemblance to the familiar frequency-response graphs that we see on some equalizers. A similar experiment was conducted two decades later by Robinson and Dadson (resulting in the *Robinson-Dadson Contours*), and today we use the ISO 226 standard (which was last revised just a few years ago). The formal name for the outcome of these studies is termed *equal-loudness contours*.

Each curve in Figure 2.1 is known as a *phon curve*, named after the level of the 1 kHz reference. To give an example of how to read this graph, we can follow the 20-phon curve and see that if 1 kHz is played at 20 dBSPL, 100 Hz would need to be played at 50 dBSPL in order to appear equally loud. (A 30 dB difference, which is by no means marginal.) The graph also teaches us that our frequency perception has a bump around 3.5 kHz – this is due to the resonant frequency of our ear canal. Interestingly, the center frequency of a baby's cry falls within this bump.

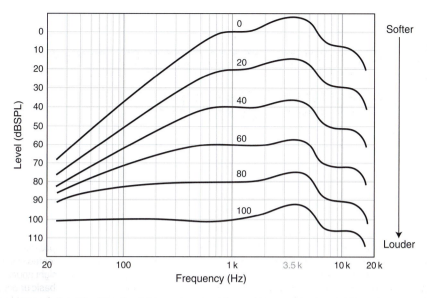

Figure 2.1 The Fletcher-Munson curves (shown here upside-down). Note that on the level axis, soft levels are at the top, loud at the bottom.

One important thing that equal-loudness contours teach us is that we are more sensitive to mid-frequencies – an outcome of the lows and highs roll-off which can be seen on the various curves. Most importantly though, it is evident that at louder levels our frequency perception becomes more even – the 0-phon curve in Figure 2.1 is the least even of all curves, the 100-phon curve is the most even. Another way to look at this is that the louder music is played the louder the lows and highs are perceived. In extremely general terms, we associate lows with power, and highs with definition, clarity and spark. So it is only natural that loud levels make music more appealing – louder perceived better.

This phenomenon explains the ever-rising level syndrome that many experience while mixing – once levels go up, it is no fun bringing them down. The more experienced among us develop the discipline that allows us to defeat this syndrome, or at least slow it down.

> *The louder music is played, the more lows and highs we perceive compared to mids.*

 The latest ISO 226 contours are slightly different than those shown in Figure 2.1; they show additional bump around 12 kHz and a steeper low-frequency roll-off, which also occurs on the louder phon curves.

The fact that our frequency perception alters in relation to levels is a fundamental mixing issue. How are we supposed to craft a balanced mix if the frequency content varies with level? At what level should we mix? And what will happen when the listener plays the track at different levels? The answer is: we check our mix at different levels, and try to make it as level-proof as possible. We know what to expect when we listen at softer levels – less highs and lows. It is possible to equalize the different instruments so that even when the highs and lows are softened, the overall instrument balance hardly changes. For example, if the kick's presence is based solely on low frequencies, it will be heard less at quiet levels, if at all. If we ensure that the kick is also present on the high-mids, it will be heard much better at quiet levels. Many believe that the mids, which vary little with level, are the key to a balanced mix, and if the lows and highs are crafted as an extension to the mids, a mix will exhibit more stable balance at different levels. Also, many agree that if a mix sounds good when played quietly, it is likely to sound good when played loudly; the opposite is not always true. Another point worth remembering is that we can sometimes guess the rough level at which the mix is likely to be played (e.g., dance music is likely to be played louder than ambient), and so we can use that level as the main reference while mixing.

> *Two common beliefs: The mids are the key to a balanced mix at varying levels. A mix that sounds good at quiet levels is likely to sound good at loud levels.*

There is another reason why louder is perceived better. When listening at soft levels, we hear more of the direct sound coming from the speakers and less of the sound reflected from the room boundaries (the room response). Sound energy is being absorbed, mostly as it encounters a surface. The small amount of energy our speakers emit at quiet levels is absorbed by walls to such a degree that only a fraction of it reflects back to our ears. At louder levels, more energy is reflected and we start hearing the room response. As a consequence, the louder music is played, the more we hear the reflections coming from around us, which provides us with the appealing sensation that the music is surrounding us. You can experiment to see this effect, which might be more apparent with eyes shut – play a mix at quiet levels and try to define the spatial boundary of the sound image. Most people will imagine a line, or a very short rectangle between the two speakers. As the music is made louder, the sound image grows, and at some point the two-dimensional rectangle turns into an undefined surrounding sense.

When it comes to making individual instruments louder in the mix, their perception is most often improved as well. The core reason for this is masking – the ability of one sound to cover up another. More specifically, frequency ranges of one instrument mask those of another. One of the principal rules of masking is that louder sounds mask quieter sounds. The higher the level of an instrument in the mix, the more it will tend to mask other instruments, and the more clearly it will be perceived.

Percussives weigh less

It is important to distinguish the different natures of the instruments we are mixing. An important resource in a mix is space. When different instruments are combined, they compete for that space (mostly due to masking). Percussive instruments come and go. For example, a kick has little to no sound between various hits. Percussives fight for space in successive, time-limited periods. On the other hand, sustain instruments play over longer periods, thus constantly fight for space. To give one extreme example, we can think of a rich pad that was produced using sawtooths, involves unison, and played in a legato fashion. Such a pad would fill both the frequency spectrum and the stereo panorama in a way that is most likely to mask many other elements in the mix.

In a practical sense, sustained instruments require somewhat more attention. Whether we are setting the level, panning or equalizing them, our actions will have an effect over a longer period. For example, raising the level of a dense pad is likely to cause more masking problems than raising the level of a kick. If the kick masks the pad it would only do so for short periods – perhaps not such a big deal. But if the pad masks the kick, it would constantly do so – a big deal indeed.

Importance

A scene from Seinfeld: Jerry and Kramer stand in a long line of people at a box office, engaged in a conversation. Being the stars of the show, amongst all the people, the production efforts were focused on Jerry and Kramer. The make-up artist, for example, probably spent quite some time with the two stars, perhaps little time with the extras standing next to them, and most likely no time with any other extras standing further away in the line. In the camera shot, Jerry and Kramer would be clearly seen in the center and extras might be out of focus. The importance of the stars would also be evident in the work of the gaffer, the grips, the boom operator or any other crew member, even the chef.

Equally, different mix elements have varying importance within the mix. The importance of each instrument depends on many factors, including the nature of the production being mixed. In hip-hop, for example, the beat and vocals are generally the most important elements. In jazz, the snare is more important than the kick. Spatial effects are an important part of ambient music. A prominent kick is central to club music, but of far less importance in most folk music. Many more examples can be given. We also have to consider the nature of each instrument and its role in the overall musical context. Vocals, for example, are often of prime importance, but the actual lyrics also play a crucial role. The lyrics of *My Way* are vital to the song's impact, and mixing a vocal part such as this requires that more emphasis be put on it. Arguably, the lyrics to *Give It Away* by Red Hot Chili Peppers are of little importance to the overall climate of the song.

Importance affects how we mix different elements, whether it is levels, frequencies, panning or depth we are working on. We will shortly look at how it can also affect the order in which we mix different instruments and sections. Identifying importance can make the mixing process all the more effective as it minimizes the likeliness of delving into unnecessary or less important tasks. For example, spending a fair amount of time on treating pads that only play for a short period of time at relatively low level. Those of us who mix under time constraints have to prioritize our tasks. In extreme circumstances, you might have as little as one hour in which to mix the drums, just half an hour for the vocals, and so on.

| *A beneficial question in mixing: How important is it?* |

Natural vs. artificial

A specific event that took place back in 1947 changed the course of music production and recording forever. Patti Page, then an unknown singer, arrived at a studio to record a song called *Confess*. The studio was set up in the standard way for that era, with all the performers in the same room, waiting to cut the song live. Problem was, that *Confess* was a duet where two voices overlap, but budget limitations meant that there was no second vocalist. Jack Rael, Page's manager, came up with the unthinkable: Patti would sing the second voice as well, provided the engineer could find a way to *overdub* her. Legend has it that at that point the engineer cried in horror: in real-life, no person can sing two voices at the very same time. It's ridiculous. Unnatural! But to the A&R guy of Mercury Records, this seemed like a great gimmick that might secure them a hit. To achieve this, the engineer recorded from one machine to another while mixing the second voice on top. What then seemed so bizarre is today an integral part of music production.

For our purposes, a 'natural' sound is one that emanates from an instrument that is played in our presence. If there are any deficiencies with the raw recordings (which capture the natural sound), various mixing tools can be employed to make instruments sound 'more natural'. A mix is considered more natural if it presents a realistic sound stage (among other natural characteristics). If natural is our goal, it would not make sense to position the kick upfront and the rest of the drum kit behind it.

However, we have to remember that natural is not always best – natural can also be very ordinary. Early on in film-making and before that, photography, it was discovered that shadows, despite being such a natural part of our daily life, impair visuals. Most advertisements have had tone and color enhancements in order to make them look 'better than life'. The same goes for studio recording. It is not uncommon today to place the kick in front of the drum kit, despite the fact that this creates a very unnatural spatial arrangement.

One of the principal decisions we make when we began a mix is whether we want things to sound natural or artificial. This applies on both the mix and instrument levels. Some mixes call for a more natural approach. Jazz enthusiasts, for example, expect a natural sound stage and natural-sounding instruments, although in recent years, more and more jazz mixes have involved a more unnatural approach. For instance, compressed drums with an emphasized kick and snare. This fresh, contemporary sound has attracted a new audience (and even some connoisseurs), and therefore a wider market for the record companies to exploit. Popular music nowadays tends to be all but natural – the use of heavy compression, distortions, aggressive filtering, artificial reverbs, delays, distorted spatial images and the likes is routine. These enhancements, whilst not natural, increase the potential for creativity and profoundly affect the overall sound. Mixes are sonic illusions. The same way that color enhancement improves visuals, our mixing tools allow us to craft illusions that sound better or just different from real life. People who buy live albums expect a natural sound. Those who buy studio albums expect, to some extent, a sonic illusion, even if they don't always realize it.

Some inexperienced engineers are hesitant to process since they consider the raw recording a natural touchstone. Often they are even cautious about gentle processing, as they
are worried it will be harmful. Listening to a commercial track that was mixed with an
artificial approach will reveal just how extreme mixing treatments can be. Take vocals for
example, their body might be removed, they might be compressed so that there are no
dynamic variations or they might be overtly distorted. We have to remember that radical
mixing is generally unperceived by the majority of listeners, who do not have a trained
ear. For example, here are three sentences my mother has never said and will probably
never say:

- Listen to her voice. It's over-compressed.
- That guitar is missing body.
- The snare is too loud.

The average listener does not think or speak in these terms. For them, it is either exciting
or boring; they either feel it or don't; and most importantly, they either like it or they don't.
This leaves a lot of room for wild and adventurous mixing treatments – we can filter the
hell out of a guitar's bottom end; people will not notice. We can make a snare sound like
a Bruce Lee punch; people will not notice. Just to prove a point here, the verse kick on
Smells Like Teen Spirit reminds me more of a bouncing basketball than any bass drum I
have ever heard playing in front of me. People do not notice.

3 Learning to mix

An analogy can be made between the process of learning a new language and that of learning to mix. At the beginning, nothing seems to make sense. With language, you are unable to understand simple sentences or even separate the words within a sentence. Similarly, if you play a mix to most people they will not be able to hear a reverb or compression as they have haven't focused on these sonic aspects before and definitely haven't used reverbs or compressors. After learning some individual words and how to use them, you find yourself able to identify them in a sentence; in the same way that you start learning how to use compressors and reverbs, and learn to recognize them in mixes. Pronouncing a new word can be challenging, since it is not easy to notice the subtle pronunciation differences in a new language, but after hearing and repeating a word 20 times, you get it right; likewise, after compressing 20 vocal tracks, you will start to identify degrees of compression and quickly evaluate what compression is most suitable. Then, you will begin to learn grammar so that you can begin to connect words together and construct coherent sentences; much like all your mixing techniques help you to craft a mix as a whole. Finally, since conversation involves more than one sentence, the richer your vocabulary is and the stronger your grammar, the more sentences you are able to properly construct. In mixing, the more techniques and tools you learn and the more mixes you craft, the better your mixing becomes. Practice makes perfect.

What makes a great mixing engineer

World-class mixing engineers might earn double the annual minimum wage for a single album. Some mixing engineers also receive points – a percentage from album sale revenue. On both sides of the Atlantic, an accomplished mixing engineer can enjoy a six-digit annual paycheck. These individuals are being remunerated for their knowledge, experience, skill, vision and, generally, an immense amount of talent. Record labels reward them for that, and in exchange enjoy greater sales.

It is clear why mixing is often done by a specialized person. And it is such a vast area that it is no wonder some people devote themselves entirely to it – the amount of *knowledge* and *practice* required to make a great mixing engineer is enough to keep anyone busy.

Primarily, the creative part of mixing revolves around the three steps shown in Figure 3.1. The ability to successfully go through these steps can lead to an outstanding mix. But a great mixing engineer will need a notch more than that, especially if hired. These steps are explained, along with the requisite qualities that make a great mixing engineer, in the following sections.

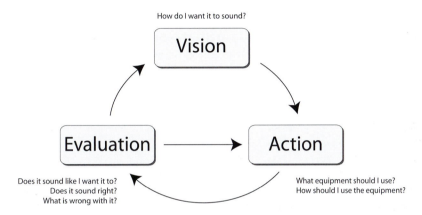

Figure 3.1 The three steps of creative mixing.

Mixing vision

There are different methods of composing. One involves using an instrument, say a piano, then either by means of trial and error or using music theory, coming up with a chord structure and melody lines. Another approach involves imagining or thinking of a specific chord or melody and then playing it, or in some cases, committing it directly to paper. The latter process of 'visualizing' and then playing or writing is favored by many composers and songwriters.

There is an analogy between the two methods of composition and the methods we employ for mixing. If we take the equalization process of a snare, for example, the first approach involves sweeping through the frequencies, then choosing whatever frequency appeals to us most. The second approach involves first imagining the desired sound and then approaching the EQ in order to attain it. Put another way, the first approach might involve thinking such as 'OK, let's try to boost this frequency and see what happens', while the second might sound more like 'I can imagine the snare having less body and sounding more crispy'. Just as some composers can imagine the music before they hear it, a mixing engineer can imagine sounds before attaining it – a big part of mixing vision. Mixing vision is primarily concerned with the fundamental question: **How do I want it to sound?** The response could be soft, powerful, clean etc. But mixing vision cannot be defined by words alone – it is a sonic visualization, which later manifests through the process of mixing.

The selection of tools that we can use to alter and embellish sound is massive – equalizing, compressing, gating, distorting, adding reverb or chorus are just a few. So what type of treatment should we use? There are infinite options available to us (in the analog domain at least) within each category – the frequency, gain and Q controls on a parametric equalizer provide millions of possible combinations. So why should we choose a specific combination and not another? Surely, equalizing something in a way that makes it sound right does not mean that a different equalization would not make it sound better. Mixing vision gives the answer to these questions: 'because this is how I imagined it; this is how I wanted it to sound'.

A novice engineer might lack imagination. The process of mixing for him or her is a trial-and-error affair between acting and evaluating (Figure 3.2). But how can one critically evaluate something without a clear idea of what one wants in the first place? Having no mixing vision can make mixing a very frustrating hit-and-miss process.

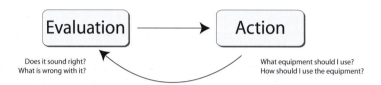

Figure 3.2 The novice approach, without a mixing vision.

Having a mixing vision can make all the difference between the novice and the professional mixing engineer. While the novice shapes the sounds by trial and error, the professional imagines sounds and then achieves them within the mix.

The skill to evaluate sounds

The ability to craft a good mix is based on repeated **evaluations**. One basic question, often asked at the beginning of the mixing process, is **'What's wrong with it?'**. Possible answers might be 'the highs on the cymbals are harsh' or 'the frequency spectrum of the mix is too heavy on the mids'. From the endless amount of treatment possibilities we have, focusing on rectifying the wrongs provides a good starting point. It can also prevent the novice from doing things that aren't actually necessary. For example, equalizing something that didn't really require equalization.

At times, it might be hard to tell what is wrong with the mix, in which case our mixing vision provides the basis for our actions. After applying a specific treatment, the novice might ask, **'Does it sound right?'**. While the veteran might also ask **'Does it sound the way I want it to?'**. Clearly, the veteran has an advantage here since this question is less abstract.

Mastering his or her tools, and knowledge of other common tools

Whether with or without a clear mixing vision we perform many **actions** in order to alter existing sounds. When choosing a reverb for vocals, the novice might tirelessly go through all the available presets on a reverb emulator. There can be upward of 50 of these, and so the entire process can take some time. The veteran, on the other hand, will probably quickly access a specific emulator and choose a familiar preset, a bit of tweaking and the task is done. It takes very little time. Experienced mixing engineers know, or can very quickly find out, which tool would do the best job in a specific situation; they can quickly answer the question: **'What equipment should I use?'**

Professional mixing engineers do not always work in their native environment as they often work in different studios and even though they might take their favorite gear with them, a big part of the mix will be done using in-house equipment. Therefore, professional mixing engineers have to be familiar with the common tools found in a commercial environment.

Mastering the tools at one's disposal does not only stand for the ability to pick the right tool for a specific task, but also having the expertise to employ the equipment in the best way **('How should I use the equipment?')**. Whether to choose high-shelving or high-pass characteristics on an equalizer or knowing that a specific compressor will work well on drums when more than one ratio button is pressed, are a couple of examples.

It is also worth discussing the quantity of tools we have at our disposal. Nowadays, DAW users have a wider selection than those mixing using hardware. Not only are plugins cheaper, they can be used across various tracks, simultaneously, whereas a specific hardware processor cannot. In an analog studio a mixing engineer might have around three favorite compressors to choose from when processing vocals, DAW users might have a choice of ten. Learning each of these compressors – *understanding* each of them – takes time; just reading the manual is time-consuming. However, having an extensive collection of tools can mean that they are not realizing their potential because there is no time to learn and properly experiment with them all. Mixing is a simple process that only requires a pair of trained ears and a few quality tools. Less can be more and more can be less.

|										*Jack of all trades, master of none.*										|

Theoretical knowledge

Four questions:

- When clipping shows on the master track in an audio sequencer, is it the master fader or all of the channel faders that should be brought down?
- For more realistic results, should one or many reverb emulators be used?
- Why and when should stereo linking be engaged on a compressor?
- When should dither be applied?

To say that every mixing engineer knows the answers to these questions would be a lie. So would be saying that one cannot craft an outstanding mix without good theoretical knowledge. There are more than a few highly successful engineers who would be unable to provide answers to many theoretical questions relating to their field. But knowing the answers to these questions is definitely an advantage. Like talent, knowledge is always a blessing and in such a competitive field, can make the difference. Out of two equally talented engineers with different levels of knowledge, it is not difficult to choose.

To acquire knowledge some might undertake an educational program, others learn little by little on the job; but either way, all mixing enthusiasts need to be compulsive learners – if the ratio on a compressor is set to 1:1, a novice will spend hours trying to figure out why no other control has an effect. Learning the difference between shelving and a pass filter is a handy one. The effect that dither has on the final mix quality is good to know. It would seem unreasonable for a mastering engineer not to know when to apply dither but mixing engineers should know too, especially as they are likely to apply it more.

> *It is better to know what you can do, and how to do it, than to understand what you have done.*

Interpersonal skills

Studio producers need an enormous capacity to deal and interact with many people who are known to have different abilities, moods and degrees of dedication. Mixing engineers, tend to work on their own and only occasionally mix in front of the client – whether that be the artist, A&R or the producer. So, although to a lesser extent than a studio producer, like any job that involves interaction with people, mixing also requires good interpersonal skills.

When the band comes to listen to the mix, it should not come as a surprise that each band member insists that his or her instrument is not loud enough. (In their defense, they are used to their instrument appearing louder to them, whether on stage, or through the cans in the live room). Even the old tricks of limiting the mix or blasting the full-range speakers do not always appease them. On many occasions, artists and A&R remark on the work of mixing engineers with the same rational of accountants commenting on the work of graphic designers they hired. While the feedback from fresh ears can sometimes be surprisingly constructive, at other times the mixing engineer is rock-assured that the client's comments are either technically or artistically naïve or inappropriate. Things can easily become personal – mixing engineers, like the artists they mix, can become extremely protective about their work. Interpersonal skills can help avoid or resolve artistic disagreements, and assist with calmly expressing an opinion. But if the artist does not like some aspect of the mix, even if it's technically earth shattering and however much the mixing engineer might disagree, it is he or she who must compromise. The client-always-right law is the same in mixing – after all, a displeased client is a lost client.

The ability to work fast

Learning something new can be tricky and testing – all guitar players experienced some frustration before they could change chords quickly enough or produce a clean sound. It is maddening working on a single verse for a whole day and still being unhappy with the mix. But, as experience accumulates, it takes less time to choose tools and utilize them to achieve the desired sound. Also, our mixing visions become sharper and we can crystallize them more quickly. Altogether, each task takes less time, which leaves more time to elevate the mix or experiment. Needless to say, the ability to work fast is essential for hired mixing engineers, who work under busy schedules and strict deadlines.

Methods of learning

Reading about mixing

Literature is great. Books, magazine articles, even Internet forums can be the source of some extremely valuable theory, concepts, ideas or tips. But reading about mixing will not make a great mixing engineer, the same as reading a cookery book will not make a great chef. Reading about mixing gives us a better chance to understand core concepts and operate our tools, but the one thing it does not do is improve our sonic skills.

Reading manuals is also an important practice, although unfortunately, many people chose to neglect it. The basic aim of a manual is to teach us how to use our equipment, sometimes also how to use it right or how to use it better. In their manuals, many manufacturers will present some practical advice on their products and sometimes on mixing in general. Sometimes the controls of a certain tool are not straightforward and it might take an eternity to understand what their function is without reading the manual.

| *Read the manual.* |

Reading and hearing

This book is an example of this method. An aural demonstration of mixing-related issues provides a chance to develop critical evaluation skills and better understanding of sonic concepts. While this method can contribute to all stages of mixing – vision, action and evaluation – it is a passive way of learning since it does not involve active mixing.

Seeing and hearing

Watching other people mix is another way to learn. Many people want to work in a studio so they can learn from the experienced. Listening to others while they apply their expertise to a mix is a great opportunity and a valuable experience, but comes with two cautions. First, it is impossible to enter other people's minds – while watching them mix it might be possible to understand what they are doing, but not why they are doing it. **Mixing vision and experience are nontransferable**. Second, if we take into account the tricks and tips already published, what is left to learn from these experienced people is mostly their own unique techniques. True, learning the secret techniques of mixing engineers at the top of their game is great, but only if these are used in the right context. There is the belief that the greatest mixing engineers produce incredible mixes because of secret techniques. In practice, these amazing mixes are not down to secret techniques, but to an **extensive understanding of basic techniques and experience using them.** Secret techniques often only add a degree of polish or the individual's idiosyncratic sonic stamp.

Doing it

As discussed, there is more than one approach to learning but actively doing it is the only indispensable method. Without a shadow of a doubt, the best way to learn mixing is simply by doing it. Most of the critical skills and qualities of a great mixing engineer can be acquired through the practice of the art. While mixing, we learn to evaluate sounds and devices, use our equipment in the best way, work faster and articulate our mixing vision quicker. Combined with good theoretical background and enough application, there is very little to stop anyone from becoming a competent mixing engineer. There is a direct link between mixing-miles and the final quality of the mix.

The best way to learn mixing, is to mix.

Mixing analysis

Sometimes, learning the techniques of an art makes it hard to perceive the art as a whole. For example, while watching a movie, film students will analyze camera movements, lighting, edits, lip-sync or acting skills. It can be hard for those students to stop analyzing and just enjoy movies like they did when they were fascinated kids. However, many mixing engineers find it easy to switch in and out from a mixing analysis state – even after many years of mixing, they still find it possible to listen to a musical piece without calculating how long the reverb is, where the trumpet is panned to or questioning the sound of the kick. Others simply cannot help it.

Although it is far less enjoyable to analyze the technical aspects of a movie while watching it, this critical awareness can help make film students into more conscious filmmakers. Sit, watch and learn how the masters did it – simple. The same approach works for mixing as well. Every single mix out there, whether good or bad, is a lesson in mixing. Learning is just a matter of pressing play and actively listening to what has been done before. Although mixing analysis cannot always reveal how things were done, it can reveal much of what was done.

Your CD collection contains hundreds of mixing lessons.

There are endless things to listen for when analyzing others' mixes, and these can cover any and every aspect of the mix. Here are just a few questions you might ask yourself while listening:

- How loud are the instruments in relation to one another?
- How are the instruments panned?
- How do the different instruments appear in the frequency spectrum?
- How far apart are the instruments in the depth field?
- How much compression was applied on the different instruments?
- Can any automation be detected?
- How long are the reverbs?
- How defined are the instruments?
- How do different mix aspects change as the song advances?

A quick demonstration would be appropriate here. The following points provide a partial mixing analysis for the first 30 seconds of *Smells Like Teen Spirit*, the album version:

- The tail of the reverb on the crunchy guitar is audible straight after the first chord (0:01).
- There is extraneous guitar noise coming from the right channel just before the drums are introduced (0:05).
- The crunchy guitar dives in level when the drums are introduced (0:07).
- Along with the power guitars (0:09 –0:25), the kick on downbeats is louder than all other hits. It appears to be the actual performance, but it can also be achieved artificially during mixdown.
- When listening in mono, the power guitars lose some highs (0:09 –0:25).
- The snare reverb changes twice (a particular reverb before 0:09, then no audible reverb until 0:25, then another reverb).
- During the verse, all the kicks have the same timbre (suggesting drum triggers).
- There is kick reverb during the verse.
- It is possible to hear a left/right delay on the hi-hats – especially during open/close hits. This could be the outcome of a spaced microphone technique, but can also occur during mixdown.
- The drums are panned audience-view.

The excerpt set (from Chapter 1) can be a true asset when it comes to mixing analysis, as the quick changes between one mix to another make many aspects more noticeable. Not every aspect of the mix is easily discernible, some are subliminal and are felt rather than heard. To be sure, the more time and practice we put into mixing analysis, the more we discover.

In addition to what we can hear from the plain mix, it is also possible to use different tools in order to reveal extra information. Muting one channel of the mix can disclose additional stereo information (e.g., a mono reverb panned to one extreme). Using a pass filter can help in understanding how things have been equalized. To reveal various stereo effects, one can listen in mono while phase-reversing one channel (this results in a mono version of the difference between the left and right, which tends to make reverbs and room ambiance very obvious).

Reference tracks

Mixing analysis is great, but it is impossible to learn hundreds of mixes thoroughly, and it can be impractical to carry them around just in case we need to refer to them. It is better to focus on a few select mixes, learn them inside out, analyze them scrupulously and have them readily accessible.

Some mixing engineers carry a CD compilation with a few reference tracks (mostly their own past mixes) so they can refer to them. The novice might refer to his reference tracks on a more frequent basis. MP3 players are also used, often with the music stored in a lossless format. When mixing at home or in one's studio, some have a specific folder on the hard drive with their select mixes.

> In addition to reference tracks, including the excerpt set can be great since it enables a quick comparison between many different mixes. It is also possible to include a few raw tracks, which can later be used to evaluate different tools.

Our choice of reference tracks might not be suitable for every mix. If we are working on a mix that includes strings, and none of our reference tracks involve strings, it would be wise to look for a good mix that does. Likewise, if our reference tracks are all heavy metal and we happen to work on a chill-out production, it would be wise to refer to some more appropriate mixes.

Usage of reference tracks

Reference tracks can be employed for different purposes:

- **As a source for imitation** – painting students often go to a museum to copy a familiar painting. While doing so they learn the finest techniques of famous painters. Imitating another's techniques is part of the learning process. Likewise, there is nothing amiss in imitating another's proven mixing techniques – if you like the sound of the kick in a specific mix, why not imitate that sound in your mix? There is no reason why you can't replicate the technique of a specific track that you particularly like. When we are short of a mixing vision we can replace it with the sonic image of an existing mix, try to imitate it or just some aspects of it. Trying to imitate the sound of a known mix is actually a great mixing exercise. However, caution must be exercised. First, because productions can be so diverse – whether in their emotional message, style, arrangement, quality and nature of the raw material etc., what sounds good in another mix might not sound so good in yours. Second, setting a specific sound as an objective can be limiting and mean that nothing better will be achieved. Finally, and most importantly, *imitation is innovation's greatest enemy* – there is little creativity involved in imitation. In fact, it might restrain the development of creative mixing skills.
- **As a source of inspiration** – while imitating a mix requires a constant comparison between the reference track and our own mix, reference tracks can be played before mixing to inspire us in what direction the mix should go and what qualities it should incorporate. For the novice, such a practice can kick-start some mixing vision and set certain sonic objectives.
- **As an escape from a creative dead end** – sometimes we reach a point where we are clearly unhappy with our mix, but cannot tell quite what is wrong with it. We might be simply out of ideas or lacking vision. Learning the difference between our mix and a specific reference mix can trigger new ideas, or suggest problems in our mix.
- **As a reference for a finished mix** – when we finish mixing, we can compare our mix to a specific reference track. Listening to how the professionals do it can help us generate ideas for improvement. The frequency response or relative levels of the two mixes are just two possible aspects that we might compare.
- **To calibrate our ears to different listening environments** – working anywhere but in our usual listening environment reduces our ability to evaluate what we hear. Just before we start to listen critically in an unfamiliar environment, whether mixing or just evaluating our own mixes, playing a mix we know well can help calibrate our ears to unfamiliar monitors, the acoustics or even a different position within the same room.

- **To evaluate monitor models before purchase** – studio monitor retailers usually play customers a popular track, that has an impressive mix and at loud levels. Chances are that the monitors will impress the listener that way. Listening to a mix that you are very familiar with, can improve your judgment.

It is worth remembering that if a reference track has been mastered, it is very likely to contain tighter dynamics, usually in the form of more allied relative levels and heavier compression. In some albums, frequency treatment takes place in order to match the overall sound to that of the worst track. These points are worth bearing in mind when comparing a reference track to a mix in progress – a mastered reference track is an altered version of a mix, usually, for the better.

How to choose a reference track

Choosing some of our own past mixes for reference purposes is always a good idea. Having worked on these mixes we are familiar with the finer details and retrospectively, their faults. Ideally, reference materials should be a combination of both unmastered and commercial tracks. Here are a few of the qualities that reference tracks should have:

- **A good mix** – whilst a matter of opinion, your opinion of what is a good mix is central. It is important to choose a mix you like, not a production you like – despite Elvis Presley's greatness the sonic quality of his original albums is nowhere near today's standards.
- **A contemporary mix** – mixing has always evolved. A good mix from the 1980s is likely to have more profound reverbs than the mix of a similar production from the 1990s. Part of the game is keeping up with the changing trends.
- **Genre related** – clearly, it makes little sense to choose a reference track of a genre that is fundamentally different from the genres you will be working on.
- **A dynamic production** – choosing a dynamic production, which has a dynamic arrangement and a dynamic mix, can be like having three songs in one track. There is more to learn from such a production.

And should not be:

- **A characteristic mix** – the mixing style of some bands, The Strokes for example, is tightly related to the style of music they play. A mix that has a highly distinct character will only serve those distinct productions and bands.
- **Too busy** – it is usually easier to discern mixing aspects in sparse productions.

- **Too simple** – the more there is to learn from a mix the better. An arrangement made of a singer and her acoustic guitar might sound great, but will not teach you how to mix drums.

Witness
The Delgados. *The Great Eastern*. **Chemikal Underground Records, 2000.**

This is one of my reference tracks. It has a rich, dynamic arrangement that includes female and male vocals, acoustic and electronic guitars, keyboards, strings and varying drum sounds between sections. The mix presents a well-balanced frequency spectrum, high definition, beautiful ambiance, musical dynamics and rich stereo panorama – all are retained through quiet and loud passages and changing mix densities.

4 The process of mixing

Mixing and the production chain

Recorded music

There are differences between the production process of recorded music and sequenced music, and these differences affect the mixing process. Figure 4.1 illustrates the common production chain for recorded music. Producers may give input at each stage, but they are mostly concerned with the arrangement and recording stages. Each stage has an impact on the subsequent stage but each of the stages can be carried out by different people. Mixing is largely dependent on both the arrangement and recording stages. For example, an arrangement might involve only one percussion instrument, a shaker for example. If panned center in a busy mix, it is most likely to be masked by other instruments. But panning it to one side can create an imbalanced stereo image. It might be easier for the mixing engineer to have a second percussion instrument, say a tambourine, so the two can be panned left and right. A wrongly placed microphone during the recording stage can result in a lack of body for the acoustic guitar. Recreating this missing body during mixdown is a challenge. Some recording decisions are, to be sure, mixing decisions. For example, the choice of stereo-miking technique for drum overheads determines the localization and depth of the various drums in the final mix. Altering these aspects during mixdown takes effort.

Figure 4.1 Common production chain for recorded music.

Mixing engineers, when a separate entity in the production chain, commonly face arrangement or recording issues mentioned above. There is such a strong link between the arrangement, recordings and the mix that it is actually unreasonable for a producer or

a recording engineer to have no mixing experience whatsoever. A good producer antici-
pates the mix. There is an enormous advantage to having a single person helping with the
arrangement, observing the recording process and mixing the production. This ensures
that the mix is borne in mind throughout the production process.

There is some contradiction between the nature of the recording and mixing stages.
The recording stage is mostly concerned with the capturing of each instrument so that
the sound is as good as it possibly can be. During the mixing stage, different instruments have to be combined, and their individual sound might not work perfectly well
in the context of a mix. For example, the kick and bass might sound unbelievably good
when each is played in isolation, but combined they might mask one another. Filtering
the bass might make it thinner, but will work better in mix context. Much of mixing
involves altering recordings to fit into the mix – no matter how well instruments were
recorded.

Sequenced music

The production process of sequenced music (Figure 4.2) is very different in nature to
that of recorded music. In a way, it is a mixture of songwriting, arranging and mixing –
producing for short. This affects mixing in two principal ways. First, today's DAWs, on
which most sequenced music is produced, make it easy to mix as you go. The mix is
an integral part of the project file, unlike a console mix that is stored separately from the
multitrack. Second, producers commonly select samples or new sounds while the mix
is playing along; unconsciously, they choose sounds based on how well they fit into the
existing mix. A specific bass preset might be dismissed if it lacks definition in the mix, and
a lead synth might be chosen based on the reverb that it brings with it. Some harmonies
and melodies might be transposed so they blend better into the mix. The overall outcome
of this is that sequenced music arrives at the mixing stage partly mixed.

Figure 4.2 Common production chain for sequenced music.

As natural and positive as this practice may seem, it promotes a few mixing problems
which are common to sequenced music. To begin with, synthesizer manufacturers and
sample-library publishers often add reverb (or delay) to presets in order to make them
sound bigger. These reverbs are permanently imprinted into the multitrack submission
and have restricted depth, stereo image and frequency spectrum that might not integrate
well with the mix. Generally speaking, dry synthesized sounds and mono samples offer
more possibilities during mixdown. In addition, producers sometimes get attached to a
specific mixing treatment they have applied, like the limiting of a snare drum, and leave
these treatments intact. Very often the processing is done using inferior plugins, in a
relatively short time, and with very little attention to how the processing affects the overall
mix. Flat dynamics due to over-compression or ear-piercing highs are just two issues that
might have to be rectified during the separate mixing stage.

*Sequenced music often arrives at the mixing stage partly mixed –
which could be more of a hindrance, and less of a help.*

Recording

They say that all you need to get killer drum sounds is a good drum kit in a good room, fresh skins, a good drummer, good microphones, good preamps, some good EQs, nice gates, nicer compressors and a couple of good reverbs. Remove one of these elements and you will probably find it harder to achieve that killer sound; remove three, and you may never achieve it.

The quality of the recorded material has an enormous influence on the mixing stage. A famous saying is 'garbage in; garbage out'. Flawed recordings can be rectified to a certain extent during mixing, but there are limitations. Good recordings leave the final mix quality to the talent of the mixing engineer, and offer greater creative opportunities.

Nevertheless, experienced mixing engineers can testify how drastically the process of mixing can improve poor recordings, and how, even low-budget recordings can turn into an impressive mix. Much of this is thanks to the time, talent and passion of the mixing engineer.

Garbage in; garbage out. Still, a lot can be improved during mixdown.

Arrangement

The arrangement (or instrumentation) largely determines which instruments play, when and how. Mixing-wise, the most relevant factor of the arrangement is its density. A sparse arrangement (Figure 4.3a) will call for a mix that fills various gaps in the frequency, stereo and time domains. An example of this would be an arrangement based solely on an acoustic guitar and one vocal track. The mixing engineer's role in such a case is to create something out of very little. Or at the other extreme, a busy arrangement (Figure 4.3b), where the challenge is to create a space in the mix for each instrument. It is harder to lay

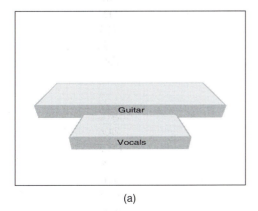
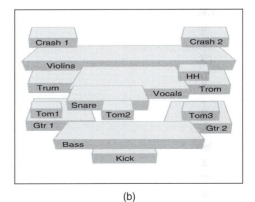

(a)   (b)

Figure 4.3 Sparse vs. dense arrangement.

emphasis on a specific instrument, or emphasize fine details in a busy mix. Technically speaking, masking is the cause.

> **Polly**
> Nirvana. *Nevermind*. Geffen Records, 1991.
> Mixed by Andy Wallace.
>
> **Exit Music (For a Film)**
> Radiohead. *OK Computer*. Parlophone, 1997.
> Mixed by Nigel Godrich.
>
> **Hallelujah**
> Jeff Buckley. *Grace*. Columbia Records, 1994.
> Mixed by Andy Wallace.
>
> Both Andy Wallace and Nigel Godrich faced a sparse arrangement, made of a guitar and vocal only, in sections of *Polly* and *Exit Music*. Each tackled it in a different way – Wallace chose a plain intimate mix, with fairly dry vocal and a subtle stereo enhancement for the guitar. Godrich chose to use very dominant reverbs on both the guitar and vocal. It is interesting to note that Wallace has chosen the latter reverberant approach on his inspiring mix for *Hallelujah* by Jeff Buckley – a nearly seven-minute song with an electric guitar and a single vocal track.

It is not uncommon for the final multitrack to include extra instrumentation along with takes that were recorded as tryouts, or in order to give some choices during mixdown. It is possible, for example, to receive eight power-guitar overdubs for just one song. This is done with the belief that layering eight takes of the same performance will result in an enormous sound. Enormousness aside, properly mixing just two of the eight tracks can sometimes sound much better. There are always opposite situations where the arrangement is so minimalist that it is very hard to produce a rich, dynamic mix. In such cases, nothing should stop the mixing engineer from adding instruments to the mix – as long as time, talent and ability allow this, and the client approves the additions.

| *It is acceptable to remove or add to the arrangement during mixdown.* |

It is worth remembering that the core process of mixing involves both **alteration** and **addition** of sounds – a reverb, for example, is an additional sound that occupies space in the frequency, stereo and time domains. It would therefore be perfectly valid to say that a mix can add to the arrangement. Some producers take this well into account by 'leaving a place for the mix' – the famous vocal echo on Pink Floyd's *Us and Them* is a good example of this. One production philosophy is to keep the arrangements simple, so that greatness can be achieved at the mixing stage.

| *The mix can add sonic elements to the arrangement.* |

Editing

Generally, on projects that are not purely sequenced, editing is the final stage before mixing. Editing is subdivided into two types: selective and corrective. Selective editing is primarily concerned with choosing the right takes, and the practice of comping – combining multiple takes into a composite master take. Corrective editing is done to repair a bad performance. Anyone who has ever engineered or produced in a studio knows that session and professional musicians are a true asset. But as technology is moving forward, enabling more sophisticated performance corrections, poor performance is becoming more accepted – why should we spend money on vocal tuition and studio time, when a plugin can make the singer in tune? (More than a few audio engineers believe that the general public perception of pitch has sharpened in recent years due to the excessive use of pitch-correction.) Drum correction has also become common practice. On big projects a dedicated editor (perhaps the Pro Tools operator from the recording sessions) might work with the producer to do this job. Unfortunately though, sometimes it is the mixing engineer who is expected to perform this function.

Corrective editing can be done to a mechanical extent. Most drums can be quantized to a metronomic precision, and vocals can be made perfectly in tune. Although some pop albums feature such extreme edits, many advocate a more humanized approach, which calls for little more than acceptable performance (perhaps ironically, sequenced music is often humanized to give it feel and swing). Some argue that over-correcting is against all genuine musical values. It is also worth remembering that corrective editing always involves some quality penalty. In addition, audio engineers are much more sensitive to subtle details than most listeners. To give an example, the chorus vocals on Beyoncé's *Crazy in Love* are notoriously late and off-beat, but many listeners don't notice it.

The mix as a composite

Do individual elements constitute the mix, or does the mix consist of individual elements? Those who believe that individual elements constitute the mix might give more attention to how the individual elements sound, but those who think that the mix consists of individual elements care about how the sound of individual elements contribute to the overall mix. It is worth remembering that the mix – as a whole – is the final product. This is not to say that the sound of individual elements is not important, but the overall mix takes priority.

A few examples would be appropriate here. It is extremely common to apply a high-pass filter on a vocal in order to remove muddiness and increase its definition. This type of treatment, which is done to various degrees, can sometimes make the vocals sound utterly unnatural, especially when soloed. However, this unnatural sound often works extremely well in mix context. Another example: Vocals can be compressed while soloed, but the compression can only be perfected when the rest of the mix is playing as well – level variations might become more noticeable with the mix as a reference. Overheads compression should also be evaluated against the general dynamics and intensity of the mix.

It even goes into the realm of psychoacoustics – our brain can separate one sound from a group of sounds. So for example, while equalizing a kick we can isolate it from the rest of the mix in our heads. However, we can just as well listen to the whole mix while equalizing a kick, and by doing so, improve the likelihood of the kick sounding better in mix context. This might seem a bit abstract and unnatural – while we manipulate something it is natural to want to clearly hear the effect. The temptation to focus on the manipulated element always exists, but there's a benefit to listening how the manipulation affects the mix as a whole.

> *It can be beneficial to employ mix-perspective rather than element-perspective, even when treating individual elements.*

Where to start

Preparations

Projects submitted to mixing engineers are usually accompanied by documentation – session notes, track sheets, edit notes etc., all have to be examined before mixing commences. Clients often have their ideas, guidelines, or requirements regarding the mix, which are often discussed at this stage. In addition, there are various technical tasks that might need to be accomplished; these will be discussed shortly.

Auditioning and the rough mix

Unless we were involved in earlier production stages and thus fluent with the raw tracks, we must listen to what is about to be mixed. Auditioning the raw tracks allows us to learn the musical piece; capturing its mood and emotional context; and identify important elements, moments, or problems we will need to rectify. We must study our ingredients before we start cooking.

Often, a rough mix (or a monitor mix) is provided with the raw tracks. It can teach much about the song and inspire a particular mixing direction. Even when a rough mix is submitted, creating our own can be extremely useful – in doing so we familiarize ourselves with the arrangement, structure, quality of recording and, maybe most importantly, how the musical qualities can be conveyed. Our own rough mix is a noncommittal chance to learn much of what we are dealing with, what has to be dealt with, and how. It also helps us to start formulating a mixing plan or vision.

> *Rough mixes, especially our own, are extremely beneficial.*

One issue to be aware of is that it is not unheard of for the artist/s and/or the mixing engineer to unwittingly, become so familiar with the rough mix that they use them as an archetype for the final mix. This unconscious adoption of the rough mix is only natural since it often provides the first opportunity to hear how different elements transform and begin to sound like the real thing. However, this is dangerous territory since rough mixes, by nature, are just that, are done with little attention to small details, and sometimes involve random tryouts that make little technical or artistic sense. Yet, once accustomed to them, we find it hard without them – a point worth remembering.

The plan

Just bringing faders up and doing whatever seems right surely won't be effective – it's like playing a football match without tactics. Every mix is different, and different pieces of music require different approaches. Once we are familiar with the raw material, a plan of action is developed, either in the mind or on paper. Such a plan can help even when the mixing engineer recorded or produced the musical pieces – it resets the mind from any sonic prejudgments and promotes a fresh start to the mixing process. Below is an example of the beginning of a rough plan before mixing commences. As you can see, this plan includes various ideas, from panning positions to the actual equipment to be used; there are virtually no limits to what such a plan might include:

> *I am going to start by mixing the drumbeat at the climax section. In this production the kick should be loud and in-your-face; I'll be compressing it using the Distressor, accent its attack, then add some sub-bass. The pads will be panned roughly halfway to the extremes and should sit behind everything else. I should also try and distort the lead to tuck on some aggression. I would like it to sound roughly like Newman by Vitalic.*

It might be hard to get the feel of the mix at the very early stages. Understandably, it is impossible to write a plan that includes each and every step that should be performed. Moreover, such a detailed plan can limit creativity and chance, which are important aspects of mixing. Therefore, instead of one big plan it can be easier to work using *small plans* – whenever a small plan is finished, a new evaluation of the mix takes place and a new plan is established. Here is a real-life example of a partial task list from a late mixing stage:

- *Kick sounds flat*
- *Snare too far during the chorus – replace with triggers (chorus only)*
- *Stereo imbalance for the violins – amend panning*
- *Solo section: violin reverb is not impressive enough*
- *Haas guitar still not defined*
- *Automate snare reverbs*

Not every mixing engineer approaches mixing with a detailed plan, some do it spontaneously and in their heads. Yet, there is always some methodology, some procedure being followed.

> *Mixing is rarely a case of 'whatever seems right next'. Have a plan, and make sure you identify what's important.*

Technical vs. creative

The mixing process involves both technical and creative tasks. Technical tasks are generally those that do not actually affect the sound, or those that relate to technical problems with the raw material. They usually require little sonic expertise. Here are a few examples:

- **Neutralizing (resetting) the desk** – analog desks should be neutralized at the end of each session, but nature has it that this is not always the case. Line gains and aux sends are the usual suspects that can later cause trouble. Line gains can lead to unbalanced stereo image or unwanted distortion. Aux sends can result in unwanted signals being sent to effect units.

- **Housekeeping** – projects might require some additional care, sometimes simply for our own convenience. For example, file renaming, removing unused files, consolidating edits, etc.
- **Track layout** – organizing the appearance order of the tracks so they are convenient to work with. For example, making all the background vocals, or drum tracks consecutive in appearance, or having the sub-oscillator next to the kick. Sometimes the tracks are organized in the order in which they will be mixed, sometimes the most important tracks are placed in the center of a large console. Different mixing engineers have different layout preferences to which they usually adhere. This enables faster navigation around the mixer, and increased accessibility.
- **Phase check** – recorded material can suffer from various phase issues (these are described in detail in Chapter 11). Phase problems can have subtle to profound effect on sounds, and it is therefore important to deal with them at the beginning of the mixing process.
- **Control/audio grouping** – setting any logical group of instruments as control or audio groups. Common groups are drums, guitars, vocals, etc.
- **Editing** – any editing or performance correction that might be required.
- **Cleaning up** – many recordings require the cleaning up of unwanted sounds, like the buzz from guitar amplifiers. Cleaning up can also include the removal of extraneous sounds like count-ins, musicians talking, pre-singing coughs, etc. These unwanted sounds are often filtered either by gating, a strip-silence process, or region trimming.
- **Restoration** – unfortunately, raw tracks can incorporate noise, hiss, buzz or clicks. These are more common in budget recordings, but can also appear due to degradation issues. It is important to note that clicks might be visible, but not audible (inaudible clicks can become audible if being processed by devices like enhancers or if being played through different D/A converters). Some restoration treatment, de-noising for example, might be applied to solve these problems.

Creative tasks are essentially those that allow us to craft the mix. These might include:

- Using a gate to shape the timbre of a floor-tom.
- Tweaking a reverb preset to sweeten a saxophone.
- Equalizing vocals in order to give them more presence.

While mixing, we have a **flow of thoughts and actions**. The **creative process**, which consists of many creative tasks, requires a high degree of concentration. Any technical task can distract or break the creative flow. If while equalizing a double bass, you find that it is offbeat on the third chorus, you might be tempted to fix the performance straight away (after all, a bad performance can be really disturbing). By the time the editing is done, you might have switched-off from the equalizing process or the creative process altogether. It can take some time to get back in to the creative mood. It is therefore beneficial to go through all the technical tasks first, which clears the path for a distraction-free creative process.

Technical tasks can interrupt the creative flow. Better to complete them first.

Which instruments

Different engineers have a different order in which they mix the different component tracks. There are lots of differences in this business. Some are not committed to one order or another – each production might be mixed in the order they think is most suitable. Here is a summary of common approaches, their possible advantages and disadvantages:

- **The serial approach** – starting with very few tracks, we listen to them in isolation and mix them first, then, gradually more and more tracks are added and mixed. This divide-and-conquer approach enables us to focus well on individual elements (or stems). The danger with this approach is that as more tracks are introduced there is less space in the mix.
  - **Rhythm, harmony, melody** – starting by mixing the rhythm tracks in isolation (drums, beat and bass), then other harmonic instruments (rhythm guitars, pads and keyboards) and finally the melodic tracks (vocals and solo instruments). This method often follows what might have been the overdubbing order. It can also feel a bit odd to work on drums and vocals without any harmonic backing. But arguably, from a mixing point of view, it makes little sense mixing an organ before the lead vocals.
  - **In order of importance** – tracks are brought up and mixed in order of importance. So, for instance, a hip-hop mix might start from the beat, then lead vocals, then additional vocals and then all the other tracks. The advantage here is that important tracks are mixed at early stages when there is still space in the mix and so they can be made bigger. The least important tracks are mixed last into a crowded mix, but there is less of a consequence in making them smaller.
- **Parallel approach** – this involves bringing all the faders up, setting a rough level balance, rough panning and then mixing individual instruments in whatever order one desires. The advantage with such an approach is that nothing is being mixed in isolation. It can work well with small arrangements but can be problematic if many tracks are involved – it can be very hard to focus on individual elements or even make sense of the overall mix at its initial stages. As an analogy, it would be like playing football with eight balls on the pitch.

There are endless variations to each of these approaches. Some, for example, start by mixing the drums (the rhythmical spine), then progress to the vocals (most important), then craft the rest of the mix around these two important elements. Another approach, which is more likely to be taken when an electronic mix makes strong usage of the depth field, involves mixing the front instruments first, then adding the instruments panned to the back of the mix.

There are also different approaches to **drum mixing**. Here are a few things to consider:

- **Overheads** – the overheads are a form of reference for all the other drums. For example, the panning position of the snare might be dictated by its position on the overheads. Changes made to the overheads might affect other drums; so there is some advantage in mixing them first. Nonetheless, many engineers prefer to start from the kick, then the snare, and only then, they might mix the overheads. Sometimes even the overheads are the last drum track to be mixed.

- **Kick** – being the most predominant rhythm element in most productions, the kick is often mixed before any other individual drum and sometimes even before the overheads. Following the kick, the bass might be mixed and only then other drums.
- **Snare** – being the second most important rhythm element in most productions, the snare is often mixed after the kick.
- **Toms** – as they are generally only played occasionally, they are often the least important contributors to the overall sound of the drums. Yet, their individual presence in the mix can be vital.
- **Cymbals** – the hi-hats, ride, crashes or any other cymbals might have a sufficient presence on the overheads. Often in such cases, the cymbals are used to support the overheads, or only mixed at specific sections of the song for interest sake. Sometimes these tracks are not mixed at all.

Which section

With rare exceptions, the process of mixing involves working *separately on the various sections*. Each section is likely to involve different mixing challenges and a slightly different arrangement (choruses are commonly denser than verses). And so, mixing engineers usually loop one section, mix it, then move on to the next section and mix it based on the existing mix. The question is: which section should be first? There are two approaches here:

- **Chronologically** – starting from the first section (intro) and slowly advancing to succeeding sections (verse, chorus). It seems very logical to work this way since this is the order in which music is played and recorded. However, while we might mix the verse perfectly – creating a rich and balanced mix – there will be very little place in the mix for new instruments introduced during the paramount chorus.
- **In order of importance** – the most important section of the song is mixed first, followed by the less important sections. For a recorded production, this section is usually the chorus; for some electronic productions it will be the climax. Very often, the most important sections are also the busiest ones; therefore mixing them first can be beneficial.

Which treatment should be applied first

The standard guideline for treatment order is shown in Figure 4.4. With the exception of faders, that need to be up for sound to be heard, there is no reason to adhere to this order, but it is a useful guideline. Later in the book we will encounter a few techniques that involve a different order. However, there is some logic in using the standard. If we skip panning and mix in mono we lose both the width and depth dimensions. Yet, since masking is most obvious in mono, a few engineers choose to resolve masking by equalization while listening in mono. This might be done before panning, but more frequently done using mono summing. Since processing replaces the original sound, it comes before any effects that add to the sound (modulation, delay or reverb). The assumption is that we would like to have the processed signal sent to the effects, rather than what might be a problematic unprocessed signal. Similarly, it is usually desired to have a modulated sound delayed, rather than having the delays modulated. Finally, since reverb is generally a natural effect, we normally like to have it untreated; treated

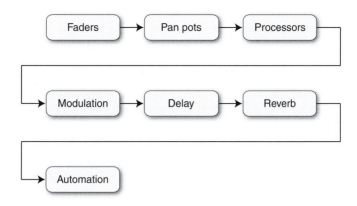

Figure 4.4 The standard treatment order guideline.

reverbs are considered a creative effect. In many cases, automation is the last major stage in a mix.

There is also the **dry–wet approach**, in which all the tracks are first mixed using dry treatment only (faders, pan pots and processors), and only then wet treatment is applied (modulation, delay and reverb). This way, the existing sounds are all dealt with before adding new sounds to the mix. It also leaves any depth or time manipulations (reverbs and delays) for a later stage, which can simplify the mixing process for some. However, some claim that it is very hard to get the real feel and the direction of the mix without depth or ambiance.

Finally, it should be mentioned that the very last stage of the mix, before it is printed, usually involves refinements of panning and levels.

The iterative approach

Back in the days of two-, four- and eight-track tape recorders, mixing was an integral part of the recording process. For example, engineers used to record drums onto six tracks, then mix and bounce them to two tracks and use the previous six tracks for additional overdubs. The only thing that limited the amount of tracks to be recorded was the accumulative noise added in each bounce. Back then, engineers had to commit their mix, time and again, throughout the recording process – once bouncing took place and new material overwrote the previous tracks, there was no way to revert to the original drum tracks. Such a process required enormous forward-planning from the production team – they had to mix whilst all the while considering that what they were doing needed to relate to something that hadn't been recorded yet; diligence, imagination and experience were key.

Today's technology offers hundreds of tracks. Even when submix bouncing is needed (due to channel shortage on a desk or processing shortage on a DAW), the original tracks can be reloaded at later times and a new submix can be created. This practically means that everything can be altered at any stage of the mixing process, and nothing has to be committed before the final mix is printed.

Flexibility to undo mixing decisions at any point in the mixing process is a great asset since **mixing is a correlative process**. The existing mix should normally be retouched to accommodate newly introduced tracks. For example, no matter how good the drums sound when mixed in isolation, the introduction of distorted guitars into the mix might require additional drum treatment (the kick might lose its attack, the cymbals definition and so forth). Plus, treatment in one area might require subsequent treatment elsewhere. For instance, when brightening the vocal by boosting the highs, high frequencies may linger on the reverb tail in an unwanted way; so the damping control on the reverb might need to be adjusted. The equalization might also make the vocal seem louder in the mix and so the fader might need to be adjusted. If the vocal is first equalized and then compressed, the compression settings might need to be altered as well.

Since mixing is such a correlative process, it can benefit from an **iterative coarse-to-fine approach** (Figure 4.5). What this means is, we start with a coarse treatment on which we spend less time, then as the mix progresses we continually refine previous mixing decisions. Most of our attention is given to the late mixing stages where the subtlest changes are made. There is little justification in trying to get everything perfect before these late stages – what is perfect at one point might not be that perfect later.

> *Start with coarse and finish with fine. Only make the mix perfect when it needs to be – at the very end of the mixing process.*

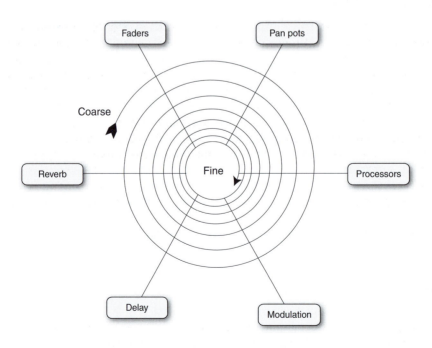

Figure 4.5 The iterative coarse-to-fine mixing approach.

Deadlocks

The evaluation block

This is probably the most common and frustrating deadlock for the novice. It involves listening to the mix, sensing that something is wrong, but not being able to tell what. Reference tracks are a true asset in these situations – they give us the opportunity to compare aspects of our mix with an established work. This comparison can reveal what is wrong with our mix, or at least point us in the right direction.

The circular deadlock

As we mix, we define and execute various tasks – from small tasks like equalizing a shaker to big tasks like solving masking issues or creating the ambiance. The problem is that in the process we naturally tend to remember our most recent actions. While in the moment we might not pay any attention to a specific compression applied two days before, we might be tempted to reconsider an equalization applied an hour previously. Thus, it is possible to enter a frustrating deadlock in which recent actions are repeatedly evaluated – instead of pushing the mix forward the mix goes around in circles. An easy way out of this stalemate situation is often a short break. We might also want to listen to the mix as whole, and re-establish a plan based on what really needs to be done.

| *Be mindful of reassessing recent mixing decisions.* |

The raw tracks factor

Life has it that sometimes the quality of the raw tracks is poor to an extent that makes them impossible to work with. For instance, distortion guitars that exhibit strong comb-filtering will often take blood to fix. A common saying is: What cannot be fixed, can be hidden, trashed or broken even more. For example, if an electric guitar was recorded with a horrible pedal monophonic reverb that just does not fit into the mix, maybe over-equalizing it and ducking it with relation to a delayed version of the drums will yield such an unusual effect that nobody would notice the mono reverb anymore. Clearly there is a limit to how many instruments can receive such dramatic treatment so sometimes re-recording is inevitable. In other cases, it is the arrangement that is to blame, like when the recorded tracks involve a limited frequency content or the absence of a rhythmical backbone. Again, there is nothing to stop the mixing engineer from adding new sounds, or even re-recording instruments; nothing, apart from time availability and ability.

| *What you can't fix, you can hide, trash or break even more.* |

Milestones

The mixing process can have many milestones. On the macro level we can define a few key milestones. The first milestone involves bringing the mix into an adequate state. Once this milestone is reached, the mix is expected to be free of any issues – whether

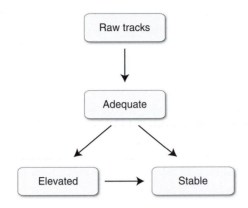

Figure 4.6 Possible milestones in the mixing process.

those existed on the raw tracks or those that were introduced during the actual process of mixing. Such problems can span from basic issues like relative levels (e.g., solo guitar too quiet) to more advanced concepts like untidy ambiance.

Nevertheless, a problem-free mix is not necessarily a good one. The next milestone is making the mix distinctive. The definition of a distinctive mix is abstract and varies between one mix and another, but the general objective is to make the mix notable – memorable – whether by means of interest, power, feel or any other sonic property.

The final milestone, which often only applies to the inexperienced engineer, is stabilizing the mix. This step is discussed in detail in the next section.

Finalizing and stabilizing the mix

All the decisions we make while mixing are based on the evaluation of what we hear at the listening position, otherwise known as the sweet spot. But mixes can sound radically different once one leaves this sweet spot. Here are some prime examples:

- Different **speakers** reproduce mixes differently – each speaker model (sometimes in combination with its amplifier) has its own frequency response, which also affects perceived loudness. While listening on a different set of speakers, the vocals might not sound as loud and the hi-hats might sound harsh. Different monitor positions also affect the overall sound of the mix, mainly with relation to aspects like stereo image and depth.
- Mixes sound different in different **rooms** – each room has its own sonic stamp which affects the music played within it. Frequency response, which is affected by both room modes and combfiltering, is the most notable factor, but the room response (reverberation) also plays a part.
- Mixes sound different in different **points in a room** – both room modes and combfiltering alter our frequency perception as we move around the room. A mix will sound

different when we stand next to the wall and when we are in the middle of the room (this issue is discussed in greater detail in Chapter 7).

- Mixes sound different at different playback levels – as per our discussion on equal-loudness curves.

Many people are familiar with the situation where their mix sounds great in their home studio, but translates badly when played in their car. It would be impossible to have a mix sounding the same in different listening environments for the reasons explained above. This is worth repeating: mixes *do* sound very different in different places. What we can do is ensure that our mixes are problem-free in other listening environments. This is much of the idea behind stabilizing the mix.

The reason that it is usually the novice that has to go through this practice is that the veteran mixing engineer knows his mixing environment so well that he can predict how well the mix will sound in other places. Also, in comparison to a professional studio, the sweet spot in a home studio is actually quite sour. Acoustic problems in home studios can be very profound, while in a professional studio these are rectified by expensive construction and acoustic design, carried out by experts.

The process of stabilizing the mix involves listening to the mix in different locations, different points within those locations and at varying levels. Based on what we learn from this practice we can finalize the mix by fine-tuning some of its aspects. Normally this involves subtle level and equalization adjustments. Here is how and where mixes should be listened to during the stabilizing phase:

- **Quiet levels** – the quieter the level, the less prominent the room becomes and any deficiencies it exhibits. Listening quietly also alters the perceived loudness of the lows and highs, and it is always a good sign if the relative level balance in the mix hardly changes as the mix is played at different levels. Listening at loud levels is also an option, but can be misleading, as mixes usually sound better when played louder.
- **At the room opening** – also known as 'outside the door', many find this listening position highly useful. When listening at its opening, the room becomes the sound source, and instead of listening to highly directional sound coming from the (relatively) small speakers we listen to reflections that come from many points in the room. This reduces many acoustic problems caused by the room itself, for example, combfiltering. The monophonic nature of this position can also be an advantage.
- **Car stereo** – a car is a small, relatively dry listening environment that many people find useful. The same as with the room-opening position, it provides a very different acoustic environment that can reveal problems.
- **Headphones** – with the growing popularity of MP3 players comes the growing importance of checking mixes using headphones. The idea of headphone mix edits for digital downloads seems more and more logical nowadays. Headphones sacrifice some aspects of stereo image and depth for the absence of room modes, acoustic combfiltering and left/right phase issues. We can regard listening in headphones as listening to the two channels in isolation, and it is a known fact that headphones can reveal noises, clicks or other types of problems that would not be as noticeable when the mix is played through speakers. The main issue with headphones is that most of them

have an extremely uneven frequency response, with strong highs and weak lows. Therefore, these are not regarded as a critical listening tool.

- **Specific points in the room** – many home studios present poor acoustic behavior, and very often there are issues at the sweet spot. Moving out of this spot can result in an extended bass response and many other level or frequency revelations. Also, the level balance of the mix changes as we move off axis and out of the monitors' close range. Beyond a certain point (the critical distance) the room's combined reflections become louder than the direct sound – most end-listeners will listen to mixes under such conditions, so it is important to see what happens away from the sweet spot.

Evaluating the mix using any of these methods can be misleading if caution is not exercised. While listening in a specific point in a room, it might seem that the bass disappears and subsequently we might be tempted to boost it in the mix. But there is a chance that it is that point in the room causing the bass deficiency, while the mix itself is fine. How then can we trust what we hear in any of these places? The answer is that we should focus on specific places and learn the characteristics of each. Usually, rather than going to random points, people evaluate mixes in the exact same point of the room every time. In addition, playing a familiar reference track to see how it sounds in each of these places is a great way to learn the sonic nature of each place.

While listening to the mix in different places, it would be wise to **write** down what issues you notice and require further attention, and then go back to the mixing board. In some cases, comments cancel out one another. For example, if while using headphones it seems that the cymbals are too harsh, but when listening at the room opening they sound dull, perhaps nothing should be adjusted.

As some final advice, it should be said that listening in different places is not reserved for final mix stabilization – it can also be beneficial at other stages of the mixing process, especially when decisions seem hard to make.

5 Related issues

How long does it take?

A B-side of a commercial release can be mixed in as little as three hours. An album's lead single might take six days. Generally, for a standard release it would not be abnormal to spend a day or two per song, and albums are very often mixed within about three weeks. Nirvana's *Nevermind,* took about 14 days to mix. The complexity of the track is an obvious factor – a 24-track production that has a simple structure, involving a five-piece rock band should take substantially less time than a 72-track production with many varying sections. The recording quality also plays a part – poor recordings mean that the mixing engineer spends lots of time fixing the recordings rather than creatively crafting the mix.

There is a difference between mixing in a commercial studio and mixing at home. Studio engineers are normally restricted by deadlines. But there is very little to restrict the home producer from mixing one production for a whole month, especially if the production is to become a decisive demo to impress a record label. The question is: what can be done in a month that cannot be done in two weeks? The answer is: a lot. Many engineers agree that mixing is an endless process – there is always something to improve, always a chance to add more excitement or impact. Many say that it is all about jumping from the carousel at the right moment.

It is also important to remember that, unlike the veteran, the novice is walking a long and intense learning path – while the veteran is fluent with his equipment and mixing environment, the novice has an enormous amount of knowledge to acquire. The veteran may choose a compressor and compress lead vocals in a matter of seconds (having done so for years on a daily basis). The novice might have to spend time going through different compressors, trying them in isolation, then with the mix, perhaps while also reading the manual.

But there is some pleasure to be gained from being a novice. If you remember the first time you managed to whistle, you might acknowledge that succeeding in learning can be greatly satisfying. How much satisfaction do you get from whistling today? Discovering mixing and constantly coming up with better mixes can give the novice great satisfaction. For the veteran, mixing is an occupation – a daily job that might involve projects he is not fond of, busy schedules and annoying clients. But the beauty of mixing is that every project differs from the other, and veterans would be lying if they said that there is nothing to learn, even after 30 years of mixing. In fact, even after 30 years there is still lots to learn.

Some people might question how professional engineers can spend eight hours a day mixing. Most of them do not – the music industry is so demanding that it is very common for a mixing engineer to be working in excess of 12 or sometimes even 18 hours a day. This might seem inconceivable for beginners who after just a few hours can find themselves 'unable to hear anything anymore'. Ear-fatigue is common and even when monitoring levels are kept low, the brain can simply get tired after a long period of active and attentive listening. Luckily, with time, our brains learn to handle longer mixing periods, and 18-hour mixing sessions are (relatively) easily handled.

Breaks

Sometimes the process of mixing can be so enthralling that time flies by and we haven't even felt the need to take a break. But it is hard to imagine a continuous 8-hour mixing session being effective. Mixing is a brain- and ear-demanding process that requires attentive listening, which can be hard to perform well for long periods without breaks. Breaks help us to forget our recent mixing actions and therefore provide an opportunity to move the mix forward. After a break, we can bring down the monitoring level in case it was brought up earlier – doing so after a break seems somewhat easier than bringing the level down in the middle of the process when we are used to hearing it at the louder level.

Probably the most important break, for the novice anyway, is the **critical break** – a day or two without listening to the mix after completing it. Having such a long break can clear our mind from individual treatments we have applied and neutralize our brain. This way, the next time we listen to the mix we do so with less sonic prejudice.

Using solos

Solo buttons let us listen in isolation to specific tracks. A useful function but one that can be very easily misused especially by the novice who might spend too much time mixing isolated tracks. We have already established that the mix is a composite and how it can be beneficial to adapt a mix-perspective approach. Using solos result in exactly the opposite since it promotes element-perspective approach. No doubt, soloing a track makes changes easier to discern, but there is also no doubt that it can lead to a treatment that is out of mix context.

One of the problems in using solos is that we lose a reference to the mix. An example already given would be compressing vocals when soloed – they might sound balanced in isolation, but not appear so with the rest of the mix. When soloed, nothing acts as a reference to the loudness of the vocals. Soloing additional tracks, say the acoustic guitar, can give such reference. But as other instruments might affect the perceived loudness of the vocals, only compressing them with the rest of the mix ensures solid compression. On the same basis, panning or depth positioning is pointless unless done in respect to the rest of the mix.

There are situations where using solos is very sensible, like when trying to filter out the buzz from a guitar track. Also, sometimes it is hard to focus on a very specific component of the sound with the whole mix playing along, e.g., the resonant frequency of a snare. While it can be useful to look for such frequency while the snare is soloed, it would be pointless keeping the solo once the frequency has been found.

Nonetheless, solos should be used whenever we manipulate sounds with relation to themselves. For example, while applying EQ automation on vocals in order to balance out frequency changes caused by the proximity effect.

> *Solos should be used with caution since they promote an element-perspective approach.*

Mono listening

Mixes are often checked in mono. It might seem odd that with the increasing popularity of surround systems, mono is still taken into account. We should first consider where music plays in mono:

- **Television** – many televisions are still monophonic.
- **Radio** – standard AM radio broadcast is monophonic, while stereo FM receivers switch to mono when the received signal is weak. Like with televisions, cheap models might only involve a single speaker.
- **Large venues** – it is very hard to maintain a faithful stereo image to large audiences in a big venue. Moreover, in many of these venues there is very little point in distributing a stereo signal, and such a setup would be more expensive. Therefore, in places like malls, sport stadiums and supermarkets, the stereo signal is summed to mono prior to distribution.

The term 'mono-compatible mix' is used in relation to mixes that translate well when summed to mono. Unless mixes are clearly intended to be played in mono (which is fairly rare) the aim is not to create a mix that will present no issues when summed to mono, but to **minimize the side-effects** that this summing might bring about. In practice, there are endless examples of great mixes that present some issues when summed to mono – the loss of high frequencies on *Smells Like Teen Spirit* is just one example. Mixing engineers would sacrifice mono-compatibility for the benefit of a powerful stereo effect, but they will also make sure that the issues caused by mono summing are minimized. A good example of this practice relates to the delay times used for the Haas trick (see Chapter 11) – different time settings result in different combfiltering effects when the stereo material is summed to mono.

> *We usually listen in mono to minimize mono problems, not to eliminate them completely.*

> **Track 1: Hero Stereo**
> The *Hero* production in stereo.
>
> **Track 2: Hero Mono**
> The same production in mono. Most affected are the electric guitars, which can hardly be heard here.

Many studios install a single speaker for the purpose of mono listening. Having the sound only coming out of one speaker is considered *true mono,* as opposed to the slightly blurred *phantom mono* produced by two speakers. As already mentioned, some engineers find it easier to resolve masking when listening in mono. However, it is worth knowing that when we sum to mono the balance of the mix changes. Why exactly this happens is explained in Chapter 13. For now, it is suffice to say that the center remains at the same level while the extremes drop by 3 dB.

Mono listening also helps in **evaluating the stereo aspects of our mix** – by switching between mono and stereo it can be easier to determine the authenticity of panning, various stereo effects and the overall impact of the stereo panorama. A very little change between stereo and mono might suggest that more stereo processing should take place.

Many analog and digital desks offer a mono switch. Very often in the software domain a specific plugin with such functionality has to be inserted on the master bus (Figure 5.1).

Figure 5.1 Logic's Gain plugin enables the summing of a stereo signal to mono.

Bouncing

Bouncing is the process of recording the mix or a submix (the mix of one or more tracks, but not the whole mix). This lets us free up the resources used for the bounced mix. For example, after mixing eight tracks of drums in an analog studio, we can bounce the drum mix onto a stereo track, then play the stereo track instead of the original raw tracks, and use any compressors, equalizers, reverbs or other tools that were used for the drum mix.

With a console-based setup (Figure 5.2) this can be done to free up channels on the desk, for example, in a situation where a digital multitrack recorder is used and there are more tracks available on the recorder than channels on the desk. Sometimes, we want more than one track to be processed using the same piece of outboard gear. For example, we might want to compress both the kick and the snare using the same mono compressor; bouncing the compressed kick frees up the compressor so that it can be used for the snare. Console-bouncing usually involves routing the bounced tracks to a group, which is then sent to the associated track on the multitrack recorder. Whenever bouncing is done on a desk, it is worth remembering to fill in a recall sheet or save the mix using a recall system, so that it can be recalled later if alterations are required.

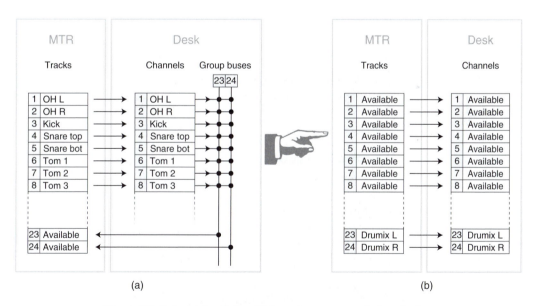

(a)   (b)

Figure 5.2 Before (a) and after (b) bouncing on a console-based setup.

Software bouncing is usually done in order to free up some CPU power so additional plugins can be used. This is done using a dedicated command that brings up an options window (Figure 5.3). In most situations what we have to select is playback range, and solo the tracks we would like to bounce (or mute those we do not want to).

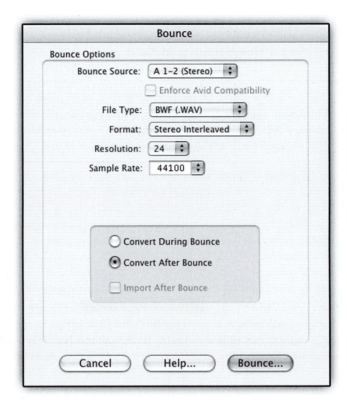

Figure 5.3 Pro Tools' bounce window.

Various applications provide different options in their bounce window. Here are some options worth considering:

- **File type** – WAV files are usually the best option as these are widely supported on both PC and MAC platforms. WAV files can be time-stamped (in which case they are referred to as BWF – Broadcast Wave File), which can be useful when transferring projects between different applications.
- **Bit rate** – most applications' internal architecture uses 32-bit float, which is roughly equivalent in quality to 24-bit integer (more on this in Chapter 10). Therefore, both 32-bit float and 24-bit integer are recommended. If a choice exists between the two, the native option of 32-bit float is recommended, although it will result in a bigger file. Also, 32-bit float files might not be supported at some mastering suites. Selecting 16-bit integer might be useful when disk space is an issue, but is very likely to result in the addition of digital distortion or the dithering noise that rectifies it.
- **Sample rate** – this should always be identical to the project's sample rate.
- **File format** – the two important options are either multiple-mono or stereo-interleaved. Multiple-mono means that a stereo file is stored as two mono files with a respective .L and .R extension (e.g., *Drumix.L.wav* and *Drumix.R.wav*). Stereo-interleaved means that a stereo file is stored as a single stereo file (e.g., *Drumix.wav*). Internally, every application separates a stereo-interleaved file into its discrete left and right channels,

which results in a tiny processing overhead, but makes any additional processing much faster and manageable. Pro Tools, for instance, instead of doing this conversion in realtime every time a stereo-interleaved file is played back, does it once (offline) when the files are imported into the session; this saves the tiny realtime processing overhead, but results in file redundancy that doubles up disk space. It might seem logical to bounce using the multiple-mono format, but doing so involves two issues. First, it results in more file clutter; second, not all applications (some audio editors for example) support this format. The general recommendation here is to use the stereo-interleaved format, unless it results in redundancy such as in the case of Pro Tools.

- **Realtime (online)/offline** – with realtime bouncing our submix is saved to disk as it plays. Thus, if we bounce a six-minute selection, the bouncing process will take six minutes. While being the longer option out of the two, there are a few advantages in realtime bouncing. First, it is usually less prone to timing errors that might occur when we bounce offline. The fact that we listen to what we bounce provides a form of quality control – it can reveal, for instance, clicks that were not audible before the bounced tracks were soloed. It also assures us that we are bouncing exactly the material we want (perhaps we forgot to mute a reverb return that only appears late in the song). All of this is not the case with offline bouncing, where we cannot hear the result until the process is completed. As offline bouncing allocates all available processing resources to the bouncing process, this option is faster than realtime bouncing, which only uses the processing power needed for smooth playback. As a rule in audio engineering we listen to what we commit. Therefore, most professionals bounce in realtime.

One exception to these recommendations involves the bouncing of the final mix. In this case we usually bounce using the destination format. For example, if the file is going to be burnt onto an audio CD it would normally be bounced using 16-bit, 44.1 kHz, stereo interleaved. But if the final mix is to be mastered, **it is always better to leave any bit or sample rate conversions to the mastering stage**, since mastering engineers have better tools for such conversions. A 24-bit, 88.2 kHz project is better bounced using the same bit and sample rates, but with the more standard stereo-interleaved format.

Bouncing issues

What seems like a straightforward process is actually quite scientific and involves one very common mistake that results in quality penalty: bouncing **without checking the levels first**. In other words – using the existing mix levels rather than optimum levels. For example, if we bounce a specific pad track that on a pre-fader meter hits –5 dB and has its fader set to –15 dB, we end up with a bounced version that has its peak at –20 dB. These 20 dB of unused dynamic range result in both the analog and digital domains in smaller signal-to-noise ratio (SNR), which impairs the quality of the signal. Another issue in these circumstances is that the bounced-track fader should be at 0 dB in order for the pad to match the original mixing levels. This means that the fader can only go up by its inherent extra gain (say 10 dB) whereas before it could go up by 25 dB (15 + 10 dB extra gain). You might think that in the digital domain we can normalize the bounced track, but as explained later, in Chapter 10, **normalization is a downgrading process** in which the SNR remains the same, but either distortion or dither noise are introduced. An opposite possible scenario involves bounced signal that is too high in level, in which case clipping distortion could be introduced. As a rule of thumb, we always want our recorded material to be at optimum levels – bounced tracks are no different.

> *Bouncing should be carried out at optimum levels, not mix levels.*

It should be clear by this point that prior to bouncing we must observe the level of the bounced material and, if required, adjust it accordingly. We start by looking at the meters while playing the material we are about to bounce. This helps us in determining the required gain change that will bring about the optimum level. The optimum level is the highest level possible without clipping or distorting. In the analog domain, signals can be pushed beyond the 0 VU, therefore the definition of optimum level is not strict – but is usually accepted as being the level above which undesirable effects become apparent. In the digital domain, optimum level is strictly 0 dBFS, although we often allow a safety margin of 3 dB (i.e., peak at –3 dBFS). The peak-hold feature, which is available on most digital meters, is extremely handy in these situations. It enables us to determine the exact amount of gain change straight after the first listen. When peak-hold is not available, we might have to listen to the material a few times while adjusting the level to optimum.

It does not end here. There are different ways to alter the level of the to-be-bounced material and some are better than others. Here are the possible ways on an analog desk:

- **VCA group** – when available, it provides the quickest way to alter the levels of the original channels. Since VCA grouping does not involve an additional signal path, correct gain structure is maintained.
- **Channel faders** – a correct gain structure will also be maintained if all the channel faders of the original tracks are brought up. If the faders are motorized, there is usually a grouping or linking function that enables simultaneous movement of all the faders by the same amount of decibels. If faders are not motorized, they should be adjusted one at a time, by the same amount of decibels – a slow and imprecise method.
- **Group level** – since all the channels are sent to a group bus anyway, it is possible to bring down (and on many consoles up) the overall group level. While doing so is quick and easy, it does not comply with the correct gain structure concept, and usually results in additional noise (although sometimes unnoticeable).

In the realm of software mixers, the bouncing affair is far more forgiving, as digital audio does not involve noisy components. The ways to alter the bounced material level are:

- **Master fader** – since most master faders in today's software mixers are scaling faders, they provide the quickest and the easiest way to alter the gain of the bounced signal. There is no quality penalty in using them.
- **Subgroup level** – routing all the original tracks to a bus and then altering the bus level. This is a longer method of achieving the same result, but valid just the same.
- **Fader grouping** – this involves grouping a set of track faders, so they all move simultaneously. Here as well, there are no gain structure issues.

There is even more to consider. When we alter the level of the original tracks (like in the case of VCA grouping, channel faders or fader grouping), we should take into account

post-fader aux sends. If a snare is sent to a reverb emulator post-fader, there is no need to bring down the level of the reverb since it will drop down with respect to the snare fader. This is not the case when the send is pre-fader where bringing down the snare level will not alter the reverb level, and so the reverb fader should be brought down as well. To summarize, setting the bounced material level should be done using the master fader on a software mixer, while on a console with bounced mix that uses post-fader sends it might be worth trading the addition of noise with simplicity and time, and using group levels.

On the same principal of bouncing at mix levels, when individual stereo tracks are bounced they often suffer from **limited or shifted stereo width,** as they are bounced with the pan pots positioned anywhere but the extremes. In the case of bounced synthesizer tracks, the same limited stereo width can be the result of a pan-spread control not set to full (like a unison pan-spread set to 50%). Stereo tracks that are not bounced panned to the extremes limit the mixing engineer's ability to control stereo aspects.

Setting the time range prior to bouncing often also requires some attention. It is worth remembering that most time-based effects will require additional time to fade away – if bouncing is limited to original material time range, reverbs and delays might cut abruptly. In addition, plugin delay compensation issues can lead to a bounced track that is out of sync with the original material. Figure 5.4 shows an original snare hit (top) that was bounced with a reverb (bottom). There are three problems with the bounced version: it is lower in level than the original track, it is out of sync and the reverb cuts at the end.

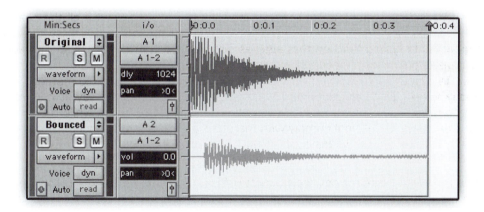

Figure 5.4 Three problems due to bouncing.

Once bouncing is done, it is very difficult to alter aspects of the original raw tracks in the bounced track, for example, the level balance between the kick and the snare. Yet in most cases the original tracks can be retrieved, mixed and bounced again. Nearly every time we bounce there is some quality penalty – noise in the analog domain, digital distortion, dither noise or dynamic range reduction (due to safety margins) can all be introduced during the process. While these unwanted additions might not be audible after the first instance of bouncing, they accumulate and become more profound when material is bounced time and again. Ideally, bouncing should not take place at all.

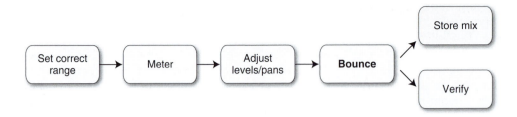

Figure 5.5 Recommended steps in the process of bouncing.

To summarize this section, Figure 5.5 illustrates the recommended steps in the process of bouncing.

Housekeeping

There have been cases in the past where studio assistants lost their jobs for failing to label a tape or for not completing a recall sheet. Some very simple (sometimes boring) tasks can save enormous amounts of time and effort afterward. Housekeeping is one of them. The word 'many' is encountered continually during the process of music creation – we often deal with many clients, many reels, many tracks, many files, many takes, many versions and so forth. These many things are much easier to manage, work with, identify, trace or recall if proper housekeeping is exercised. Take labeling for example, a file labeled *track12_tk9* suggests nothing about its content. A file labeled *Kick_tk3* does. Worse is a file labeled *Kick_tk3* that actually contains the vocal. Going through a project that consists of 200 files (which is not unusual) and encountering meaningless or incorrect file names is every engineer's nightmare (Figure 5.6).

Mix edits

Very often, especially in the commercial industry, more than one version of a mix is required; these versions are referred to as 'edits'. Primarily, this practice is concerned with creating a mix that will conform to the destination playback system. Some of these edits can be produced during the mastering stages. But mixing engineers have more control over the final results, as they have the power to alter individual tracks. Common edit candidates are:

- **Album version** – the mix to be included on the album. Since most albums today are pressed on CD, album mixes are the least restricting ones.
- **Radio edit** – various factors should be considered for a mix that is intended for broadcast on the radio. First, since mixes played on the radio are heavily compressed and limited before transmission, mixing engineers sometimes check their mixes through heavy compression or limiting to see how the mix will translate under such conditions. Second, since it is fair to assume that radio listeners are likely to listen in noisy environments, vocals are commonly pushed up in radio edits. Third, longer songs are less

Figure 5.6 This screenshot shows a partial folder hierarchy for the Hero Project presented in this book. The 'Hero Mix' folder contains the actual mix project. There might be more than one mix per song, so there's a subfolder called 'Hero Mix 01'. The various session files are snapshots of the mix in progress. The 'Hero Mix Ready' contains a project from which a new mix can start; it is a modified version of the edited version, involving the desired track order, groups and so forth. The 'Hero Rough Mix' folder contains the rough mix. The 'In' folder is an archive for incoming material it includes the raw recordings and the editor's submission. The 'Out' folder is an archive folder for material submitted to other entities; in this case, the 'Hero Edit Ready' folder contains the version submitted to editing, which is a modified version of the raw recordings (file names, comments to editor, etc.). Notice the straightforward audio file names, with perhaps the exception of 'gCln 57 t1', which stands for 'guitar clean SM57 take one'.

likely to be played on commercial radio; therefore long album versions are commonly shortened using edits and fades. Fourth, since most radio systems have limited ability to reproduce low frequencies, very often these are filtered or attenuated to some extent. Finally, some lyrical content might require censorship, which is usually done by mutes or 1 kHz tone.

- **Club and LP versions** – it is assumed that both of these will be pressed on a vinyl, which requires centered bass content, and minimum phase differences between the left and right channels, especially at low frequencies. As opposed to radio, club sound systems are expected to provide an extended low-frequency response, so mixing engineers must use full-range monitors to make sure that the low end is properly mixed. Most clubs have a limiter on the signal chain as well.
- **Vocals-up/vocals-down** – the levels of the vocals in mixes is critical. Very often two mixes are bounced, with the vocals varying by around 1 dB between the two. The A&R, producer and artist usually pick their favorite version. If appropriate, it is also possible to record additional variations, like drums-up/drums-down.

As well as these common edits, additional edits or stems might be required. For example, an instrumental mix, a cappella mix, video mix, TV (instrumental and backing vocals), no solo and so on.

Mastering

Mastering engineers have more roles than meet the eye. When assembling an album, they remove extraneous sounds, arrange the tracks in the most compelling order, create smooth fades and natural pauses, and they balance both the frequency spectrum and the level of the various tracks so the album sounds like a coherent piece rather than a collection of unrelated songs. Once a master is completed and approved by the client, they produce a high-quality copy that complies with the requirements of manufacturing plants. Perhaps, their most important role is to bring the sonic aspects of an album to the highest, most appealing state. If the mixes are good, they can make polished diamonds out of gold.

The individual pieces of equipment used in a professional mastering studio usually cost more compared to those found in mixing facilities, and the listening environment is optimized to rectify any possible problems, mostly acoustic ones. It is common, for example, to find mastering studios with nothing that could cause combfiltering (including a desk) between the full- range monitors and the listening position. Theoretically, mastering engineers might have to amend the mixes very little if the mixing engineer did the job right. But rarely mixing engineers have the environment or tools to achieve the critical quality that mastering engineers can.

It should be clear why mastering is so significant – once the finished master leaves the mastering studio, any imperfections will be heard by many and will potentially damage commercial success, sales and most importantly – the joy of listening.

| *Simply put, mastering is an art and science reserved for the experts.* |

Some mixing engineers are tempted to submit mixes that have some stereo treatment, mostly compression. But common sense has it that whatever a mixing engineer can do on a stereo mix, a mastering engineer can do better. Why would you try to fix your company's car, if your company will pay a professional mechanic to do so for you?

Mastering engineers charge a fair amount of money for their valuable job. In cases where such expenditure is not justified (e.g., non-commercial projects), a DIY approach can be taken. This mostly involves the utilization of a limiter or a propriety loudness maximizer, a high-quality equalizer and perhaps a sonic enhancer. These tools are used in the mastering process very similarly – yet very differently – to how they are used in mixing.

Mastering delivery

Historically, mixes used to be submitted to mastering on 1/2" analog tapes. Later, DATs became popular and today it is more and more common to use CD-ROMs (data CDs), data DVD's and external hard drives. Although sometimes the case, it is unprofessional to submit mixes on CD-DAs (audio CDs) as these are prone to errors more than other types of media. The actual media will be accompanied by a specific log, which includes the name of each track and so on.

Having to work on a stereo mix, one of the greatest challenges in mastering is that each instance of processing affects all the elements of the song. For instance, correcting sibilant vocals can reduce the snare's clarity. The high-fidelity tools at the mastering engineer's disposal can fix many issues, but the more the mix needs correction the more distant the hope of perfection becomes. Since nowadays it is possible to find multitrack applications in a mastering studio, it is becoming increasingly common to submit mixes in *stems* (submixes of logical track groups, or even just a single track). If there are any problems in the mix, the mastering engineer might find it easier to process the individual stems rather than the mix as a whole. Common stems are vocals, rhythm, leads, and of course, the residue mix, which consists of everything but what is already included in other stems. In all cases, a full stereo mix should be submitted as well, and should be identical to the mix of all the other stems when their faders are at unity gain (0 dB).

Sometimes, the client or mastering engineer asks for changes to mixes after these have been completed. In a large studio with an analog console, recalling mixes can be time consuming. Saving a mix in stems can be beneficial in such situations – instead of recalling the whole mix (console and outboard gear), we can only recall the mix of the stem that requires alterations, while playing all the other stems untouched.

There are a few additional practices worth considering when submitting mixes to mastering:

- **Use high-quality media** – especially with regard to CD-ROMs where some brands are more reliable than others (usually reflected in the cost). The 650 MB CDs are preferred to the 730 MB ones. Burning speed should always be set to the lowest possible – preferably ×1.
- **Do not fade.** leave any fades at the beginning or end of each track for the mastering engineer. He or she can do more musical fades once the order of the tracks is

determined, and can use any noises at the beginning or end of the track for noise reduction. Make sure to leave the full reverb tail at the end of the track, and as a general guideline, leave 2 seconds of silence before and after the audio of each track.

- **Leave some headroom** – traditionally, 3 dB of headroom was left on tapes for various reasons. For example, mastering engineers could boost on an equalizer without having to attenuate the mix first. Even digital submission would benefit from peaks hitting just below 0 dBFS, since digital audio can still clip during the D/A conversion due to interpolation.
- **Use WAV files** – these are supported on both MAC and PC platforms.
- **Keep the original digital audio quality** – do not perform any sample-rate or bit-depth conversions. These are likely to degrade the quality of the audio, and have no advantage from a mastering point of view – many mastering engineers will convert the mixes to analog before processing, and will use high-quality converters to capture the analog signal and then convert it back to the appropriate digital format.

Further reading

Katz, Bob (2002). *Mastering Audio*. Focal Press.

6 Mixing domains and objectives

There are two main approaches involved with mixing: macromixing and micromixing. **Macromixing** is concerned with the overall mix, for example, its frequency balance. **Micromixing** is concerned with the individual treatment of each instrument, for instance, how natural the vocals sound. When we judge a mix, we evaluate both macromixing and micromixing. When macromixing, there is generally a set of areas and objectives that should be considered. These will also affect micromixing.

The process of mixing can be divided up into five main domains or core aspects: **Time, frequency, level, stereo and depth**. Two of these aspects – stereo and depth – together form a higher domain – **space**. We often talk about the mix as if it exists on an imaginary sound stage, where instruments can be positioned left and right (stereo) or front and back (depth).

> The term *stereo* does not necessarily denote a two-channel system. Any system that is not monophonic can be regarded as stereophonic, for example, Dolby's 4.1 Stereo. For convenience however, throughout this book the term 'stereo' implies a two-channel system.

In many cases, instruments in the mix are fighting for space, like guitars and vocals that mask one another. Each instrument has properties in each domain. We can establish separation between competing mix elements by utilizing various mixing tools to manipulate the sonic aspects of instruments, and their presentation in each domain.

Mixing objectives

There are four principal objectives that apply in most mixes: **Mood, balance, definition and interest**. When evaluating the quality of a mix, we usually start by considering how coherent and appealing each domain is and then assess how well each objective was accomplished within that domain. While such an approach might appear to be extremely technical, it encompasses many important mixing concepts. Also, such an approach will not detract from the creative side of mixing that is down to the individual mixing engineer.

First, we will discuss each of the objectives and then see how they can be achieved in each domain. We will also discuss the key issues and common problems that affect each domain.

Mood

The mood objective is concerned with reflecting the emotional context of the music in the mix. Of all the objectives, this one involves the most creativity and is central to the entire project. It can make the difference between a good mix and a good, pertinent mix. Heavy compression, aggressive equalization, dirty distortion and a loud punchy snare will sound incongruent in a mellow jazz number, and could destroy the songs emotional qualities. Likewise, sweet reverberant vocals, a sympathetic drum mix and quiet guitars will only detract from the raw emotional energy of an angry heavy metal song. Mixing engineers that specialize in one genre can find mixing a different genre like eating soup with a fork – they might try to apply their usual techniques and very familiar sonic vision to a mix that has very different needs.

Balance

We are normally after balance in three domains – frequency balance, stereo image balance, and level balance (which consists of relative and absolute – explained shortly). An example of frequency imbalance is a shortfall of high-mids that can make a mix sound muddy, blurry and distant. However, when it comes to the depth domain, we usually seek coherency, not balance. An in-your-face kick with an in-your-neighbor's-house snare would create a very distorted depth image for drums.

There are generally two things that we will consider trading balance for – a creative effect and interest. For example, in a few sections of The Delgados' *The Past That Suits You Best,* Dave Fridmann chose to pan the drums and vocal to one channel only, creating an imbalanced stereo image but an engaging effect. This interest/imbalance trade off is usually for a very brief period – rolling off some low frequencies during the break section of a dance tune is one of many examples.

Definition

Primarily, definition stands for how distinct and recognizable sounds are. Mostly, we associate definition with instruments, but we can also talk about the definition of a reverb. Not every mix element requires a high degree of definition. In fact, some mix elements are intentionally hidden, like a low-frequency pad that fills a missing frequency range, but does not play any important musical role. A subset of definition also deals with how well each instrument is presented in relation to its timbre – can you hear the plucking on the double bass, or is it pure low energy?

Interest

On an exceptionally cold evening, late in 1806, a certain young man sat at the *Theatre an der Wien* in Vienna, awaiting the premier of Beethoven's *Violin Concerto in D Major.* Shortly after the concert began, a set of four, dissonant D-sharps were played by the first violin section and the orchestra responded with a phrase. Shortly after the same thing happened again; only this time it was the second violin section, accompanied by the violas, playing the same D-sharps. On the repetition the sound appeared to shift to the center of the orchestra and had a slightly different tonality. 'Interesting!' thought the young man.

We have already established that our ears very quickly get used to sounds, and unless there are some changes, we can get bored and lose interest. The verse-chorus-verse structure, double-tracking, arrangement changes and many other production techniques all result in variations that grab the attention of the listener. It is important to understand that even subtle changes give a sense of development, of something happening, even though most listeners are unconscious of it. Even when played in the background, some types of music can distract our attention – most people would find it easier studying for an exam with classical music playing along rather than, say, death metal. Part of mixing is concerned with accommodating inherent interest in productions. For example, we might apply automation to adapt to a new instrument introduced during the chorus.

> *A mix has to retain and increase the intrinsic interest in a production.*

The 45 minutes of Beethoven's violin concerto mentioned above probably contain more musical ideas than all the songs that a commercial radio station plays in a day, but it requires active listening and develops very slowly compared to a pop track. Radio stations usually restrict a track's playing time to around three minutes so more songs (and adverts) can be squeezed in. This way, the listener's attention is held and they are less likely to switch to another station.

Many demo recordings that have not been properly produced include limited ideas and variations, and result in a boring listening experience. While mixing, we have to listen to these productions again and again, and so our boredom and frustration is tenfold. Luckily, we can introduce some interest into boring songs through mixing – we can create a dynamic mix of a static production. Even when a track has been well produced, a dynamic mix can be beneficial. Here are just a few examples of what we can introduce to a recording to increase dynamism:

- Automate levels.
- Have a certain instrument playing in particular sections only – for instance, introducing the trumpets from the second verse.
- Use different EQ settings for the same instrument and toggle them between sections.
- Apply more compression on the drum-mix during the chorus.
- Distort a bass guitar during the chorus.
- Set different snare reverbs in different sections.

> *A mix can add or create interest.*

With all of this in mind, it is worth remembering that not all types of music are meant to force the attention of the listener. Brian Eno said about ambient music that it should be ignorable and interesting at the same time. In addition, some genres, like trance, are based on highly repetitive patterns, which might call for more subtle changes.

Frequency domain

The frequency domain is probably the hardest aspect of mixing to master. Some say that half of the work that goes into a mix is involved with frequency treatment.

The frequency spectrum

Most people are familiar with the bass and treble controls found on many hi-fi systems. Most people also know that the bass control adjusts the low frequencies (lows) and the treble the high frequencies (highs). Some people are also aware that in between the lows and highs are the mid-frequencies (mids).

Our audible frequency range is 20 Hz to 20 kHz (20 000 Hz). In mixing jargon, there are four basic bands in the frequency spectrum: lows, low-mids, high-mids and highs. This division originates from the common four-band equalizers found on many analog desks. There is not a standard definition of where exactly each band begins and ends, and on most equalizers the different bands overlap. Roughly speaking, the crossover points are at 250 Hz, 2 kHz and 6 kHz, as illustrated on the frequency-response graph in Figure 6.1. Out of the four bands, the extreme bands are easiest to recognize since they either open or seal the frequency spectrum. Having the ability to identify the lower or higher midrange can take a bit of practice.

The following tracks demonstrate the isolation of each of the four basic frequency bands. All tracks were produced using a HPF, LPF or a combination of both. The crossover frequencies between the bands were set to 250 Hz, 2 kHz and 6 kHz. The filter slope was set to 24 dB/oct:

Track 6.1: Drums Source
The source, unprocessed track used in the following samples.

Track 6.2: Drums LF Only
Notice the trace of crashes in this track.

Track 6.3: Drums LMF Only
Track 6.4: Drums HMF Only
Track 6.5: Drums HF Only

* * *

And the same set of samples with vocal:

Track 6.6: Vocal Source
Track 6.7: Vocal LF Only
Track 6.8: Vocal LMF Only
Track 6.9: Vocal HMF Only
Track 6.10: Vocal HF Only

Plugin: Digidesign *DigiRack EQ 3*

Frequency balance and common problems

Achieving frequency balance (also referred to as tonal balance) is a prime challenge in most mixes. Here again, it is hard to define what is a good tonal balance and our ears very

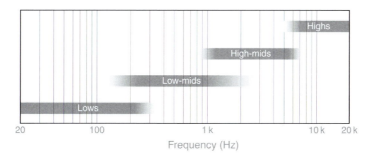

Figure 6.1 The four basic bands of the frequency spectrum.

quickly get used to different mix tonalities. In his book *Mastering Audio*, Bob Katz suggests that the tonal balance of a symphony orchestra can be used as a reference for many genres. Although there might not be an absolute reference, my experience shows that seasoned engineers have an unhesitating conception of frequency balance. Moreover, having a few engineers listening to an unfamiliar mix in the same room, their opinions would be remarkably similar. Although different tonal balances might all be approved, any of the issues presented below are rarely argued on a technical basis (they might be argued on an artistic basis – 'I agree it's a bit dull, but I think this is how this type of music should sound').

The most common problems with frequency balance involve the extremes. A mix is boomy if there is an excess of low-frequency content, and thin if there is a deficiency. A mix is dull if there are not enough highs, and brittle if there are too many. Many novice engineers come up with mixes that present these problems, mainly due to their colored monitors and lack of experience in evaluating these ranges. Generally, the brighter the instruments are, the more defined they become and the more appealing they can sound. There is a dangerous tendency to brighten up everything and end up with very dominant highs. This is not always a product of equalization – enhancers and distortions also add high-frequency content. Either a good night's sleep or a comparison to a reference track would normally help pinpoint these problems. Overemphasized highs are also a problem during mastering, since mastering engineers can use some high-frequency headroom for enhancement purposes. A lightly dull mix can be made brighter with relative ease. A brittle mix can be softened, but this might prevent the mastering engineer from applying sonic enhancements.

Low-frequency issues are also prevalent due to the great variety of playback systems and their limited accuracy in reproducing low frequencies (like in most bedroom studios). For this very reason, low frequencies are usually the hardest to stabilize and it is always worth paying attention to this range when comparing a mix on different playback systems. It can be generalized that it is more common for a mix to have excess of uncontrolled low-end than a deficient one. The next most problematic band is the low-mids, where most instruments have their fundamentals. Separation and definition in the mix is largely dependent on the work done in this busy area, which can very often be cluttered with nonessential content.

Track 6.11: Hero Balanced
Subject to taste, the monitors used and the listening environment, this track presents a relatively balanced frequency spectrum. The following tracks are exaggerated examples of mixes with extreme excesses or deficiencies:

Track 6.12: Hero Lows Excess
Track 6.13: Hero Lows Deficiency
Track 6.14: Hero Highs Excess
Track 6.15: Hero Highs Deficiency

One question always worth asking is: which instrument contributes to which part of the frequency spectrum? Muting the kick and the bass, which provide most of the low-frequency content, will cause most mixes to sound powerless and thin. Based on the 'percussives weigh less' axiom, muting the bass is likely to cause more low-end deficiency. Yet, our ears might not be very discriminating when it comes to short absences of some frequency ranges – a hi-hat can appear to fill the high-frequency range, even when played slowly.

One key aspect of the frequency domain is **separation**. In our perception we want each instrument to have a defined position and size on the frequency spectrum. It is always a good sign if we can say that, for instance, the bass is the lowest, followed by the kick, then the piano, snare, vocals, guitar and then the cymbals. The order matters less – the really important thing is that we can separate one instrument from another. Then we can try and see if there are empty areas: is there a smooth frequency transition between the vocals and the hi-hats, or are they spaced apart with nothing in between them? This type of evaluation, as demonstrated in Figure 6.2, is rather abstract. In practice, instruments can span the majority of the frequency spectrum. But we can still have a good sense of whether instruments overlap, and whether there are empty frequency areas.

The frequency domain and other objectives

Since masking is a frequency affair, **definition** is also bound to the subject of frequency and how the various instruments are crafted into the frequency spectrum. Fighting elements mask one another and have competing content on specific frequency ranges. For example, the bass might mask the kick since both have healthy low-frequency content; the kick's attack might be masked by the snare, as both appear in the high-mids. We equalize various instruments to increase their definition, a practice often done in relation to other masking instruments. Occasionally, we might also want to decrease the definition of instruments that stand out too much.

The frequency content of the mix has some relation to the **mood** we are trying to achieve. Generally, low-frequency emphasis relates to a darker, more mysterious mood, while high-frequency content usually relates to happiness and liveliness. Although power is usually linked to low frequencies, it can be achieved through all the areas of the frequency spectrum. There are many examples of how the equalization of individual instruments can help in conveying one mood or another, but these are discussed later in Chapter 14.

Most DJs are very good at balancing the frequency spectrum. An integral part of their job is making each track in the set similar in tonality to its preceding track. One very

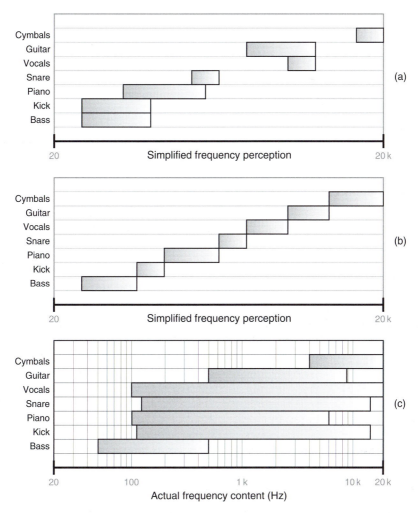

Figure 6.2 An abstraction of instrument distribution on the frequency spectrum. (a) An imbalanced mix with some instruments covering one another (no separation) and a few empty areas. (b) A balanced mix where each instrument has a defined dimension that does not overlap instruments. Altogether, the various instruments constitute a full, continuous frequency response. (c) The actual frequency ranges of the different instruments.

common DJ move involves sweeping up a resonant high-pass filter for a few bars and then switching it off. The momentary loss of low frequencies and the resonant high frequencies creates some tension and makes the re-introduction of the kick and bass very exciting for most clubbers. Mixing engineers do not go to the same extremes as DJs, but attenuating momentarily low frequencies during a break or a transition section can achieve a similar tension. Some frequency **interest** occurs naturally with arrangement changes – choruses might be brighter than verses due to the addition of some instruments. Low frequencies usually remain intact despite these changes, but whether mixing a recorded or an electronic track, it is possible to automate a shelving EQ to add more lows during the exciting sections.

Level domain

High-level signals can cause unwanted distortion, so once we have cleared up that technical issue, our mix evaluation process is much more concerned with the **relative levels** between instruments rather than their absolute level (which varies in relation to the monitoring levels). Put another way, the question is not 'how loud?' but 'how loud compared to other instruments?' The latter question is the basis for most of our level decisions – we usually set the level of an instrument while comparing it to the level of another. There are huge margins for personal taste when it comes to relative levels (as your excerpt set from Chapter 1 should prove), and the only people who are likely to get it truly wrong are the fledglings – achieving a good relative level balance is a task that for most of us comes naturally.

However, as time goes by we learn to appreciate how subtle adjustments can have a dramatic effect on the mix, and subtle level adjustments are no exception. For instance, even a boost of 2 dB on pads can make a mix much more appealing. In contrast to the widespread belief, setting relative levels involves much more than just moving faders; equalizers and compressors are employed to adjust the perceived loudness of instruments in more sophisticated ways – getting an exceptional relative level balance is an art and requires practice.

As we are discussing *relative* levels, the time has arrived to impart one of the most fundamental tips in mixing:

| *To make everything louder in the mix – bring up the monitor level.* |

Levels and balance

What is a good relative level balance is worth discussing. Some mixing engineers are experienced enough to create a mix where all the instruments sound as loud. But only a few, mostly sparse mixes, might benefit from such an approach. All other mixes usually call for some variety of relative levels. In fact, trying to make everything as loud is often a self-defeating habit that novice engineers adopt – it can be both impractical and totally unsuitable. It is worth remembering that the relative level balance of a mastered mix is likely to be tighter – raw mixes usually have greater relative levels variety.

Setting relative balance between the various instruments is usually determined by their importance. For example, the kick in a dance track is more important than any pads. Vocals are usually the most important instrument in rock and pop music. (A very common question is: 'is there anything louder than the vocals?') Maintaining sensible relative levels usually involves gain-rides. As our song progresses, the importance of various instruments might change; like in the case of a lead guitar which is made louder during its solo section.

Track 6.16: Level Balance Original
This track presents a sensible relative level balance between the different instruments. The following tracks demonstrate variations of that balance, arguably for worse:

Track 6.17: Level Balance 1 (Guitars Up)
Track 6.18: Level Balance 2 (Kick Snare Up)
Track 6.19: Level Balance 3 (Vocal Up)
Track 6.20: Level Balance 4 (Bass Up)

Steep level variations of the overall mix might also be a problem. These usually occur due to major arrangement changes that are very common in sections like intros, breaks and outros. If we do not automate levels, the overall mix level can dive or rise in a disturbing way. The level of the crunchy guitar on the first few seconds of *Smells Like Teen Spirit* was ridden exactly for this purpose – having an overall balanced mix level. One commercial example of how disturbing an absence of such balance can be, can be heard on *Everything For Free* by K's Choice. The mix just explodes after the intro and the level burst can easily make you jump off your seat. This track also provides, in my opinion, an example of an occasion when the vocals are not loud enough.

Levels and interest

Although notable level changes of the overall mix are not desired, we want some degree of level variations in order to promote interest and reflect faithfully the intensity of the song. Even if we do not automate any levels, the arrangement of many productions will mean a quieter mix during the verse and a louder one during the chorus. Figure 6.3 shows the waveform of *Witness* by The Delgados, and we can clearly see the level variations between the different sections.

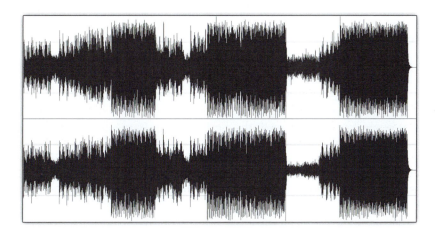

Figure 6.3 The level changes in this waveform clearly reveal the verses, choruses and various breaks.

Very often, however, we spice up these inherent level changes with additional automation. While mastering engineers achieve this by automating the overall mix level, during mixdown we mostly automate individual instruments. The options are endless: the kick, snare or vocals might be brought up during the chorus; the overheads might be brought up on downbeats or whenever a crash hits (crashes often represent a quick intensity burst); we might bring down the level of reverbs to create a tighter, more focused ambiance; we can even automate the drum-mix in a sinusoidal fashion with respect to the rhythm – it has been done before.

> *Level automation is used to preserve overall level balance, but also to break it.*

Levels, mood and definition

If we consider a dance track, the level of the kick is tightly related to how well the music might affect us – a quiet kick will fail to move people in a club. If a loud piano competes with the vocals in a jazz song, we might be unable to comprehend the emotional message of the lyrics or the beauty of the melody. We should always ask ourselves what is the emotional function of each instrument and how can it enhance or damage the mood of a song, then set the levels respectively. In mixing, there are always alternative ways – it is not always the first thing that comes to mind that is the best. For example, a novice way to create power and aggression in a mix is to have very loud, distorted guitars. But the same degree of power and aggression can be achieved using other strategies, which will enable us to bring down the level of the masking guitars, and create a better mix altogether.

Definition and levels are linked, the louder an instrument is the more defined it appears in the mix. However, an instrument with frequency deficiencies might not benefit from a level boost – it might still be undefined, just louder. Remember that bringing up the level of a specific instrument can cause the loss of definition in another.

Dynamic processing

We have discussed so far the overall level of the mix, and the relative levels of the individual instruments it consists of. Another very important aspect of a mix relates to noticeable level changes in the actual performance of each instrument (micromixing). Inexperienced musicians can often produce a single note, or a drum hit, which is either very loud or quiet compared to the rest of the performance. Rarely vocalists produce a level-even performance. These level variations, whether sudden or gradual, break the relative balance of a mix and therefore we need to contain them using gain-riding, compression, or both. We can also encounter the opposite, mostly caused by over-compression, where the perceived level of a performance is too flat, making instruments sound lifeless and unnatural.

Track 6.21: Hero No Vocal Compression
In this track, the vocal compressor was bypassed. Not only is the level fluctuation of the vocal disturbing, but the relative level between the voice and other instruments alters.

Stereo domain

Stereo image criteria

The stereo panorama is the imaginary space we perceive as if existing between the left and right speakers. When we come to talk about stereo image, we either talk about the stereo image of the whole mix or the stereo image of individual instruments within it – a drum kit, for example. One incorrect assumption is that stereo image is only concerned with how far to the left or right instruments are panned. But a stereo image involves concepts slightly more advanced than that, which are based on the following four properties (also illustrated in Figure 6.4):

- **Localization** – concerned with where the sound appears to come from on the left – right axis.
- **Stereo width** – how much of the overall stereo image the sound occupies. A drum kit can appear narrow or wide. The same for a snare reverb.
- **Stereo focus** – how focused the sounds are. A snare can appear to be emanating from a very distinct point in the stereo image, or can be unfocused (smeared) and appear to be emanating from 'somewhere over there'.
- **Stereo spread** – how exactly the various elements are spread across the stereo image. For example, the individual drums on an overhead recording can appear to be coming mostly from left and right, and less from the center.

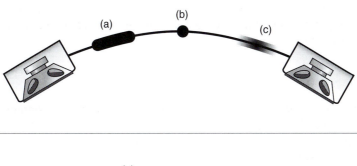

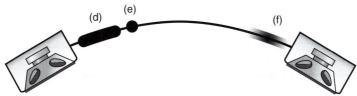

Figure 6.4 Stereo criteria. Localization is concerned with our ability to discern the exact position of an instrument. For example, the position of instrument (b) or (e). Stereo width is concerned with the length of stereo space the instrument occupies. For example, (a) might be a piano wider in image than a vocalist (b). If we can't localize precisely the position of an instrument, its image is said to be smeared (or diffused) like that of (c). The difference between (a, b, c) and (d, e, f) is the way they are spread across the stereo panorama.

To control these aspects of the stereo panorama we use various processors and stereo effects, mainly pan pots and reverbs. As we will learn in Part II, stereo effects like reverbs can fill gaps in the stereo panorama, but can also damage focus and localization.

Stereo balance

Balance is the main objective in the stereo domain. Above all, we are interested in the **balance between the left and right** areas of the mix. Having a stereo image that shifts to one side is unpleasant and can even cause problems if the mix is later cut to vinyl. Mostly we consider the **level balance** between left and right – if one channel has more instruments panned to it, and if these instruments are louder than the ones panned to the other channel, the mix image will converge on one speaker. It is like having a play with most of the action happening on the left side of the stage. Level imbalance between the left and right speakers is not very common in mixes, since we have a natural tendency to balance this aspect of the mix. It is worth remembering that image shifting can also happen at specific points throughout the song, like when a new instrument is introduced. One tool that can help us in identifying image shifting is the L/R swap switch.

> ***Give It Away***
> **Red Hot Chili Peppers. *Blood Sugar Sex Magik*. Warner Bros., 1991.**
>
> ***Useless***
> **Kruder Dorfmeister. *The K&D Session*. !K7 Records, 1998.**
>
> There are more than a few commercial mixes that involve stereo imbalance. One example is *Give It Away* by Red Hot Chili Peppers, where John Frusciante's guitar appears nearly fully to the right, with nothing to balance the left side of the mix. The cymbals panned to the left on Kruder & Dorfmeister's remix for *Useless* is another example.

A slightly more common type of problem is **stereo frequency imbalance**. While the frequency balance between the two speakers is rarely identical, too much variation can result in image shifting. Even two viola parts played an octave apart can cause image shifting if panned symmetrically left and right (the higher octave part will draw more attention). Stereo frequency balance is much to do with arrangement – if there is only one track of an instrument with a very distinguished frequency content, it can cause imbalance when panned. We either pan such an instrument more toward the center (to minimize the resultant image shifting), or we employ a stereo effect to fill the other side of the panorama. One instrument known for causing such problems is the hi-hats, especially in recorded productions where it makes less sense to add delay to it.

While left/right imbalance is the most severe of all stereo problems, **stereo spread imbalance** is the most common. In most mixes, we expect to have the elements spread across the stereo panorama so there are no lacking areas, like a choir is organized on stage. The most obvious type of stereo spread imbalance is a nearly monophonic mix – one that makes very little use of the sides, also know as an *l-mix* (see Figure 6.5). While such a mix can be disturbing, it can sometimes be appropriate. Nearly monophonic mixes are most prevalent in hip-hop – the beat, bass and vocals are gathered around the center,

Figure 6.5 An I-mix. The dark area between the two speakers indicates the intensity area. The white area indicates the empty area. An I-mix has most of its intensity around the center of the stereo panorama, with nothing or very little toward the extremes.

and only a few sounds, like reverbs or backing pads, are sent to the sides. One example of such a mix is *Apocalypse* by Wyclef Jean, although not all hip-hop tracks are mixed that way.

Another type of possible issue with stereo spread involves a mix that has a weak center, and most of its intensity is panned to the extremes (therefore called a *V-mix*, Figure 6.6). V-mixes are usually the outcome of a creative stereo stratagem (or the result of 3-state pan switches like those found on very old consoles) otherwise they are very rare.

Figure 6.6 A V-mix. The term describes a mix that has very little around the center, and most of the intensity is spread to the extremes.

The next type of stereo spread imbalance, and a very common one, is a combination of the previous two, known as the *W-mix* – a mix that has most of its elements panned hard-left, center and hard-right (see Figure 6.7). Many novices produce such mixes because they tend to pan every stereo signal to the extremes. A W-mix is not only unpleasant but in a dense arrangement can be considered a waste of stereo space. Here again, the arrangement plays a role – if there are not many instruments, we need to widen them to fill spaces in the mix, but this is not always appropriate. An example of a W-mix is the verse sections of *Hey Ya!* by Outkast; during the chorus, the stereo panorama becomes more balanced, and the change between the two is an example of how interest can originate from stereo balance variations.

Figure 6.7 A W-mix. This type of mix is common and can cause problems, unless done intentionally. It involves intensity around the extremes and center, but nothing in between.

The last type of a stereo spread imbalance involves a stereo panorama that is lacking in a specific area, say between 13:00 and 14:00. It is like having three adjacent players missing from a row of trumpet players. Figure 6.8 illustrates this. Identifying such a problem is easier with a wide and accurate stereo setup. This type of problem is usually solved with panning adjustments.

Figure 6.8 A mix lacking a specific area on the stereo panorama. The empty area is seen toward the right speaker.

Track 6.22: Full Stereo Spread The full span of the stereo panorama was utilized in this track. The following tracks demonstrate the stereo spread issues discussed above: **Track 6.23: W-Mix** **Track 6.24: I-Mix** **Track 6.25: V-Mix** **Track 6.26: Right Hole** In this track there is an empty area on the right hand side of the mix.

> *I- and W-mixes are the most common types of stereo spread imbalance.*

Stereo image and other objectives

Although it might seem unlikely, stereo image can also promote various **moods**. Soundscape music, chill out, ambient and the likes, tend to have a wider and less-focused stereo image compared to less relaxed genres. The same old rule applies – the less natural, the more powerful. A drum kit as wide as the stereo panorama will sound more natural than a drum kit panned around the center. In *Witness* by The Delgados, Dave Fridmann chose to have a wide drum image for the verse, then a narrow image during the more powerful chorus.

We have seen a few examples of how the stereo image can be manipulated in order to achieve **interest**. Beethoven achieved this by shifting identical phrases across the orchestra, while in *Hey Ya!* the stereo spread changes between the verse and the chorus. It is worth listening to *Smells Like Teen Spirit* and considering the panning strategy and the change in stereo image between the interlude and the chorus. Other techniques include panning automation, auto-pans, stereo delays and so forth. We will see in Chapter 13 how panning can also improve **definition**.

Depth

For a few people, the fact that a mix involves a front/back perspective comes as a surprise. Various mixing tools, notably reverbs, enable us to create a sense of depth in our mix – a vital extension to our sound stage, and our ability to position instruments within it. One magical thing about depth in a mix is that some spatial cue is still preserved even when we listen at random points within a room.

Track 6.27: Depth Demo
This track involves three instruments in a spatial arrangement. The lead is dry and foremost, the congas are positioned behind it, and the flute-like synth is placed way at the back. The depth perception in this track should be maintained whether listening on the central plan between the speakers, at random points in the room, or even outside the door.

Track 6.28: No Depth Demo
This track is the same as the previous arrangement, but excludes the mix elements that contributed to depth perception.

Plugin: Audioease *Altiverb*
Percussion: Toontrack *EZdrummer*

All depth considerations are relative. We never talk about how many meters away – we talk about in front of or behind another instrument. The depth axis of our mix starts with the closest sound and finishes with the farthest. A mix where instruments are very close can be considered tight, and a mix that has an extended depth can be considered spacious. This is used to reflect the **mood** of the musical piece.

The depth field is an outsider when it comes to our standard mixing objectives. We rarely talk about a balanced depth field since we do not aim to have our instruments equally spaced in this domain. **Coherent depth** field is a much more likely objective. Also, in most cases the further an instrument is, the less defined it becomes; crafting the depth field while retaining the definition of the individual instruments can be a challenge. A classical concert with the musicians walking back and forth around the stage would be chaotic. Likewise, depth variations are uncommon, and usually instruments move back and forth in the mix as a creative effect, or in order to promote importance (just like a trumpet player would walk to the front of the stage during his or her solo).

We are used to sonic depth in nature and we want our mixes to recreate this sense. The main objective is to create something natural or otherwise artificial but appealing. Our decisions with regard to instrument placement are largely determined by importance and what we are familiar with in nature. We expect the vocals in most rock productions to be the closest, just like a singer is front-most on stage. But in many electronic dance tracks the kick will be in front and the vocals will appear behind it. Sometimes even way behind it. Then we want all drum-kit components to be relatively close on the depth field because this is how they are organized in real life.

Every two-speaker system has an integral depth, and sounds coming from the sides appear closer than sounds coming from the center. As demonstrated in Figure 6.9, the bending of the front image changes between speaker models, and when we toggle between two sets of speakers our whole mix might appear to shift forward or backward. The sound stage created by a stereo setup has the shape of a skewed rectangle and its limits are set to the angle between our head and the speakers (although we can create an out-of-speaker effect using phase tricks). As illustrated in Figure 6.10, the wider the angle between our head and the speakers, the wider the sound stage will become, especially for far-off instruments.

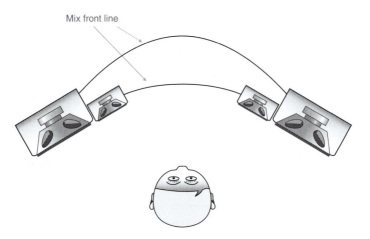

Figure 6.9 The perceived frontline of the mix appears in the stereo image as an arch, the shape of which varies between speaker models.

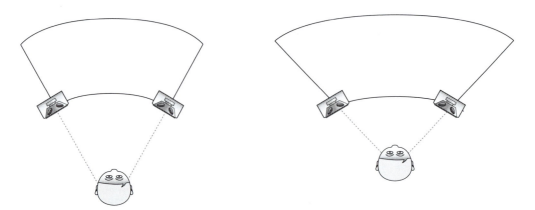

Figure 6.10 The width of the sound stage will change as our head moves forward or backward.

These two tracks demonstrate the integral depth of a two-speaker system. Pink noise and a synth are swept in a cyclic fashion between left and right. When listening on the central plane between properly positioned monitors, sounds should appear to be closest when they reach the extremes, and further away as they approach the center. It should also be possible to imagine the actual depth curve. Listening to these tracks with eyes shut might make the apparent locations easier to discern.

Track 6.29: Swept Pink Noise
Track 6.30: Swept Synth

These tracks were produced using Logic, with -3 dB pan law (which will be explained later).

Part II

Tools

All tools are as good as the person who uses them.

Part II

7 Monitoring

Studio loudspeakers are called monitors. This chapter opens Part II for good reason – an accurate monitoring environment is an absolute mixing requisite, and should be high up in budget planning for every studio, whether home or professional. The main conclusion of this chapter is worth revealing already – the monitors alone do not dictate the overall quality of the monitoring environment; it is the monitors, their position, the listener's position and the acoustic properties of the room that dictate the overall performance of the monitoring environment. Having great monitors badly positioned in a problematic room is like having a Ferrari that only goes up to second gear.

How did we get here?

Sound reproduction

In order to reproduce sound, loudspeaker drivers displace air in response to an incoming voltage that corresponds to a waveform. There is a fundamental difference between the way low and high frequencies are reproduced. Low frequencies call for a rigid, big cone that is capable of displacing a large mass of air. High frequencies, on the other hand, require a light and small diaphragm that can move rapidly. The two requirements obviously conflict. Additionally, low frequencies require a large amount of excursion compared to high frequencies. If a single driver produces both low and high frequencies, the cone displacement caused by low frequencies results in unwanted phase shifts for the high frequencies.

For this reason, one driver cannot faithfully reproduce the full audible range. Therefore, loudspeakers utilize two or more drivers, which are referred to as two-way design, three-way design and so forth. We must make sure that each driver is only fed with the frequencies it specializes in reproducing. A device called a crossover is used to split the incoming signal into different frequency bands (Figure 7.1). A typical two-way loudspeaker would have its crossover around 2 kHz, sending each band to a different driver. It is impossible to build a crossover that simply slices the frequency spectrum in a brick-wall fashion, so there is always some overlapping bandwidth where identical frequencies are sent to both drivers. The insertion of a crossover into the signal path introduces many problems, and manufacturers address them in different ways, but no system is perfect.

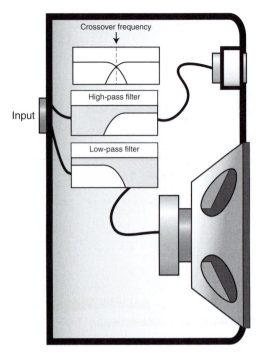

Figure 7.1 Crossover network in a two-way loudspeaker. As the signal enters the speaker, a filter network splits it into two different bands. The low frequencies are sent to the woofer, the high ones to the tweeter.

One issue with multi-way design is that the complete frequency spectrum is produced from different points in space. This might cause unwanted phase interaction when the sounds emitted from the different drivers are summed acoustically. Some manufacturers (like Tannoy) have designs where the high-frequency driver is fitted into the center of the low-frequency driver (where the dust dome is normally present). Such a design is claimed to have, amongst other things, improved stereo imaging. Yet, we never judge the performance of a loudspeaker based on its design.

Two-way studio monitors are able to faithfully reproduce the audible frequency spectrum. A design that involves more than two drivers will bring about better quality only if the problems introduced by the additional drivers and crossovers are addressed. Naturally, a design that addresses these problems involves a higher price tag. A two-way studio monitor, therefore, is more likely to be a better buy than a domestic three-way loudspeaker of the same price.

All ordinary studio monitors can produce frequencies up to 20 kHz. The lower limit of the frequency response is largely determined by the size of the woofer; 6" and 8" are very common diameters. (The size often forms part of the model name, although rarely denotes the exact diameter.) These set the lower frequency limit to around 55 and 45 Hz, respectively. A loudspeaker still produces frequencies below these quoted limits, although these are gradually rolled-off.

Auratones, near-fields and full-range monitors

From the early days of mixing, the assumption was that engineers needed the best monitoring system available. But the main monitors (often referred to as mains) in commercial studios were obviously superior to those in domestic systems, and many mixes did not translate well when listened to on these cheap domestic systems. Mixing engineers soon realized that they needed some speakers that imitated the sound of real-world domestic loudspeakers, and the Auratones 5C, also known as 'the cubes', did the trick (Figure 7.2). These small and not surprisingly cubic, single-driver speakers had a defined midrange, and they got the nickname 'horror-tones' because of their unappealing sound. Nevertheless, engineers soon realized that these small speakers could be used for more than simulating the sound of cheap systems. Being found in many studios even today, the Auratones are very often used for critical level adjustments (notably vocals), and to evaluate definition. (For example, we can tell there's a problem if the kick and bass are lost with these monitors selected.) Having pronounced mids, they also help us to tidy up and equalize this area of the mix where most instruments have their fundamentals and lower-harmonics.

Figure 7.2 The Auratones 5C are the mini-speakers mounted on the meter bridge. Behind them are the Genelec 1031s near-fields. (Courtesy of SAE Institute, London.)

But something was still missing in between the small Auratones and the big mains, and so a new type of compact monitor was released to fill the gap – the near-field monitors (or near-fields). The acoustical term 'near-field' is misleading; a better term would be 'close-field' monitors, which better describes their position within the critical distance. Very often these monitors are placed on top of the console's meter bridge or on stands right behind it. The vast majority of mixes are done using near-fields. Even in professional studios, where mains exist, mixing engineers might use near-fields for the majority of the mixing process, and only occasionally refer to a different set of monitors. The most common type of near-fields nowadays are active, 8", two-way monitors; although budget home studios sometimes may have to compromise with the 6" version.

No book about mixing would be complete without discussing the Yamaha NS10s. In 1978, Yamaha released a bookshelf loudspeaker intended for home-use called the NS10M. It only took a few years before most music studios had a pair installed. Many people tried to explain the immense popularity of the NS10s. Some said that they were the natural successors to the Auratones, providing a compact speaker, better sound but not too flattering. Many engineers testified that the midrange definition of these speakers and their tight low-end was highly beneficial for rock and pop mixes. Others declared that if something sounded right on the NS10s, it would translate well on most consumer systems. In his book *Recording Studio Design*, Philip Newell dedicated more than a few pages to the NS10s and presented detailed research aiming to solve the mystery (including a crane-mounted SSL on which the speakers were then fixed). Whilst some engineers wouldn't touch them with a ten-foot pole, they have become standard in the audio-engineering field, particularly in mixing.

The studio popularity of the NS10s caught Yamaha by surprise. They never designed these speakers for professional use and it soon became apparent that the classic version (NS10M) had some issues when used in studio environments. First, their vertical mounting meant they often obscured the main monitors. Second, they were unable to withstand the abusive studio levels and the tweeters often blew. Finally, and most famously, they had harsh and emphasized highs, which led many people to cover the tweeters with tissue paper. One enthusiastic engineer, Bob Hodas, even researched what type of tissue to use, how many layers and how far they should be from the tweeters. Yamaha themselves used tissue paper as part of their research while developing a redesigned model, and in 1987 the Yamaha *NS10M Studio* was shipped to the market. It solved the issues with the classic version and the direction of the label changed to suggest that the new monitors were designed to be mounted horizontally. Both the classic and studio versions are easily identified by their white cone woofer. As the material used for producing those white cones became unavailable, Yamaha decided to discontinue the production of the NS10s in 2001. Like the Auratones, the NS10s are still found in many studios today, and there is still demand for them in the second-hand market.

While the limited bass response of near-fields might not be an issue for some mixes, it is crucial to have an extended low-frequency response in genres like hip-hop, reggae, dance and others. **Full-range** monitors are so called because they can reproduce the full audible range from 20 Hz to 20 kHz. They are usually large, high-fidelity monitors that provide higher resolution compared to near-fields. In many studios, these monitors are flush mounted, which enhances their acoustic interaction with the room. Where these exist, we refer to them in order to check and stabilize the low end of the mix. When switching to the full-range monitors, we seek the extended low frequencies without losing the level balance. They are also useful when creating separation between the bass and the kick, and due to their high resolution we also refer to them for refinements, like subtle vocal equalization. The high quality of large full-range monitors makes them the favorite choice for mastering, classical music production and many recording situations. But their sound is of such a high quality and far superior to anything that most listeners might have access to, that mixing engineers generally favor the smaller, less impressive, yet accurate near-fields.

Space and budget limitations make full-range monitors a rare breed in home studios, but a dedicated subwoofer provides an excellent alternative. Many professional monitor manufacturers offer a matching subwoofer to their near-fields range, which normally covers the 20-150 Hz range. An optimum configuration for such a setup often involves feeding the stereo mix into the subwoofer first. Most subwoofers have a built-in crossover that splits the frequency spectrum into two bands – the very low frequencies are summed

Figure 7.3 The NEVE VRL studio at SAE Institute, London. The three types of monitors can be seen here: the full-range are the Genelec 1037Bs, the near-fields are the NS10s and the mini-speakers are the AKG LSM50s.

to mono and sent to the subwoofer driver, all other frequencies are sent to the near-fields through a dedicated stereo output at the subwoofer's rear.

Choosing monitors

Active vs. passive

The low-level, low-power line output of our desk or computer needs to be amplified to the powerful speaker-level in order for the mechanical components in a loudspeaker to move. There is always a driving amplifier in the signal chain prior to the loudspeaker drivers, and this amplifier can be either built into the cabinet, or an external unit. A loudspeaker with no integrated amplifier is known as a **passive speaker**, and must be fed with a speaker-level signal that was amplified by an external amplifier. Most multiway speakers contain a crossover within their cabinet. In the case of passive speakers, the crossover is a passive one (there are no active components) and is designed to operate at speaker-level. External amplifiers have a huge influence on the overall sound. The NS10s, for example, can sound distinctively different when driven by different makes with different power ratings. There is some critical interaction between an amplifier and a loudspeaker – it is, essentially, one system – and the amplifier determines many aspects of the overall sound like transient response, low-frequencies reproduction and distortion. Matching an amplifier to a loudspeaker is never an easy affair.

The following should be adhered to when connecting the amplifier to the speakers. Cables must be of a high quality, must be as short as possible and of equal length (for both left and right speaker). If they are not of equal length, there will be stereo imbalance between the two speakers. It is important to make sure that the terminals are correctly connected with plus to plus and minus to minus. If the terminals are crossed on both speakers, the two channels will be in phase, but the overall mix will be phase-inverted. This means that the speaker cones will move backwards instead of forward and vice versa, and might affect the sound. (This is based on the idea that cone suspension does not behave identically between front and back movements. As part of quality control, some mastering engineers check their masters with both speakers inverted.) An easy way to diagnose such incorrect cabling is when a kick pulls the cone rather than pushes it. If only one speaker is cross-connected then the left and right speakers are out of phase with one another, which creates an extremely unfocused, out-of-speakers sonic image that makes accurate mixing an impossible task. Conventionally, the positive (plus) terminals are connected using the red wire (as in the audio phrase: Red is Ring, Right and positive).

Loudspeakers with a built-in amplifier are called **powered speakers**. These have line-level inputs, commonly balanced XLR or 1/4". There are two common designs of powered speaker: the first involves a single amplifier followed by a passive crossover; the second involves an active crossover followed by two or more amplifiers – one for each driver. Technically speaking, a speaker that has an active crossover is called an **active speaker**, and if there is more then one built-in amplifier it is called **bi-amped** or **tri-amped**, depending on the amount of amplifiers. The majority of powered studio monitors are active and multi-amped.

Active monitors are often shielded, which drains magnetic interference that might potentially occur between the speaker and CRT computer screens. Some newer models also have digital inputs. While placing an A/D converter within a loudspeaker might seem to make as much sense as placing a blender arm in a microwave, it prevents any type of analog interference and minimizes the chance of ground-loops.

There are not many cases in mixing where one thing has such a clear advantage over something else, but active speakers provide many advantages over passive ones. In fact, it is hard to come up with a single practical advantage for the passive designs. The fact that a speaker has a built-in amplifier removes the guesswork of matching an amplifier to a loudspeaker and leaves this to the professional designers. Manufacturers can fine-tune the performance of each component for optimal results, and usually the outcome is more cost-effective. Many active speakers include protection circuitry (built-in limiters) that makes them resistant to abuse. If we ignore room acoustics, identical models of active speakers installed in different places are much more consistent in their sound since they are always driven by the same built-in amplifiers. But, for all its advantages, there is no guarantee that an active monitor will perform better than a passive one. Although active speakers gain popularity, passive monitors are still manufactured, some receiving much acclaim.

Enclosure designs and specifications

You may already be aware that some studio monitors have holes on their enclosure (called ports, vents or ducts in professional jargon). Such a design is known as **dipole design** and includes sub-designs like *vented enclosure* (e.g., Genelec 1031A), *bass reflex* (e.g., Dynaudio BM 6A) and *transmission line* (e.g., PMC TB2S-A). Designs with no ports have the air within the enclosure sealed, and are known as **monopole designs**. The most common monopole designs are either *sealed enclosure* (e.g., Yamaha NS10s) or *auxiliary bass radiator*, which is also known as *ABR* or *passive radiator* (e.g., Mackie HR824). While the concept behind the different designs is a fascinating one, it teaches us nothing about the fidelity of the final product – none of the designs ensure a better quality than the other.

There are, however, a few things worth mentioning. Dipole designs are more efficient and provide an extended low-frequency response compared to monopole designs of the same size (although it is claimed that the low-end extension is an imposture and an inaccurate one). On the other hand, monopole designs provide better damping of the woofer cone once the input signal dies abruptly. After a gated kick dies out, momentum will keep the woofer cone moving for a while before it comes to a halt; this extraneous movement generates unwanted low frequencies. While the woofer in monopole designs comes to a halt very quickly, measurements (more specifically waterfall plots) show that some dipole monitors can produce frequencies lower than 100 Hz for more than 100 ms after the input dies. Sound-wise, monopole designs are said to deliver tighter bass response, and it is no wonder that many professional subwoofers employ such a design.

Like many audio devices, the specification sheets for monitors are often inconsistent and can be misleading, teaching us very little about the actual quality of the product. Technical measurements like signal-to-noise ratio, harmonic distortion and maximum output level are often dependent on the system used during measurements, and can easily be manipulated in favor of the manufacturer. The monitors' frequency-response graph, that demonstrates the dips and bumps of various frequencies across the spectrum, has no bearing on the perceived quality or accuracy of the speaker. One speaker with a relatively flat frequency response does not guarantee a better quality than a speaker with noticeable wiggles. Of course, one specification that we do care about is the quoted frequency range of the speaker, specifically its lower limit. There is a meaningful difference between a speaker that rolls-off low frequencies at 70 Hz and one that does so at 50 Hz. For most mixing applications, both specifications will suggest that a subwoofer is needed.

A choice of experience

It should be clear by this point that selecting monitors should not be based on their design or specifications, but on their actual sound. To be sure, one thing we do not want from our monitors is for them to sound flattering – we want them to reveal problems rather than conceal them. Accuracy and detail are the key qualities we are after. 'Because they sound good' is the poorest reason to favor one brand over another. Unfortunately, many buyers fail to comprehend this. Even worse, some manufacturers will sacrifice quality in order to make their monitors more appealing in quick-listening tests. Some retailers have a showroom containing a clutter of many brands and models in an arrangement that bears no resemblance to the actual positioning of monitors in a real studio setup. This supermarket approach does the monitors no justice whatsoever, as often their sound can be greatly influenced by their placement. It can also mean that we are unable to assess any stereo image aspects, or many other critical aspects.

By way of analogy, buying monitors is like buying a bed – no matter how good it looks in the shop or how sophisticated the specifications are, we can only tell how comfortable a bed is after using it for a few days. Mixing engineers will usually become accustomed to their monitors after a while – and having got used to them will remain loyal and depend on them. There truly isn't a 'magic model' – what one praises the other dislikes. Even the NS10s have been the subject of debate. As a guideline: the higher the price tag, *usually*, the higher the quality one should expect.

The room factor

No pair of speakers sounds the same, unless placed in the same room.

It seems unreasonable that sometimes as much money is expended on acoustic treatment as on monitors. The truth is that an expensive set of high-fidelity monitors can perform rather poorly if deficiencies in room acoustics are not treated. We have mentioned the frequency-response graphs and waterfall plots of a loudspeaker. These measurements are taken by manufacturers in anechoic chambers where the room itself is not a variable. This is only fair, as it would be unreasonable for each manufacturer to use a different room – each with its own unique effect. If manufacturers were to record their measurements in domestic rooms, their results would yield variations that can be six times worse than the anechoic measurements. In practice, these untaken worst measurements are what we will hear.

Professional mixing facilities are designed by specialists; therefore this section will only cover the most relevant aspects for smaller project studios. A complete discussion of all the acoustic factors that affect our monitoring environment is far beyond the scope of this book but the most important factors for our purposes are briefly covered below.

Room modes

Room modes are discussed in great detail in many books. A full exploration of room modes is long, technical and requires some background knowledge. Below is a short and simplified explanation of a long and complex topic. See 'Further reading' at the end of this chapter for more information.

Low frequencies propagate in a spherical fashion. For the sake of simplicity, imagine that low frequencies emitted from our monitors travel equally in all directions. Also, whenever low frequencies come in contact with a surface, imagine that a new sound source is created at the point of incident, as if a new speaker is placed there.

The simplest way of describing standing waves involves the use of two parallel surfaces, like walls. Sound emitted from a speaker will hit the left wall, bounce back to the right wall and then bounce back and forth between the two. We can think of the sound as trapped between the two walls. Since every time sound hits a surface some of its energy is being absorbed, after a while the sound dies out. However, continuous monitor output might constantly reinforce waves already trapped.

If the frequency of the trapped waves is mathematically related to the length of the trap, a predictable interaction between all the trapped waves will cause that frequency to be either attenuated or boosted at different points along the trap. For example, halfway between the two walls that frequency might be barely audible, and next to the wall that frequency might be overemphasized. Waves with these characteristics are called **standing waves** and a problematic frequency can be described as a resonant **room**

mode. Acoustic law says that if a specific frequency is trapped in a room, all of its harmonics are also trapped. For example, if 50 Hz is the lowest resonant mode, then 100, 150, 200, 250 Hz and so forth will also be trapped.

If traps could only form between two parallel surfaces, each room would only have a relatively small set of three problematic frequencies (one for each dimension) and their harmonics. However, traps can also form between four or six surfaces, which result in a very complex set of room modes. If two room dimensions are identical, then the problem is twofold since the same frequency is trapped in two dimensions. Cubic rooms are the worst since all three dimensions are identical. Rooms with dimensions that share common multiples (e.g., 3×6×9m) are also very problematic. Despite the common belief, non-parallel walls do not rectify the standing waves problem – they only make the room modes distribution more complex.

The lowest resonant frequencies are also the most profound ones. In small rooms these lower frequencies are well within our audible range. The formula to calculate these fundamental frequencies is quite simple: $f = 172/d$ (d refers to meters). So for example, a dimension of approximately 3m will cause a fundamental room mode at 57 Hz. The bigger the room, the lower the fundamental resonant frequency is. For example, a dimension of 8m will cause a room mode at 21.5 Hz, which is less critical than 57 Hz for mixing applications. Also, as we climb up the frequency scale the effect of room modes becomes less profound, and around 500 Hz we can disregard them. In bigger rooms, problems start lower on the frequency scale, and also end lower – room modes might no longer be an issue above 300 Hz. Big rooms are therefore favored for critical listening applications such as mixing or mastering.

It is crucial to understand that room modes always cause problems at the same frequencies, but the problems are not consistent throughout the room. Eventually, each point in the room has its own frequency response. Room mode calculators are freely available over the Internet – they will output a graph showing problematic frequencies based on given room dimensions that the user enters. However, they do not tell us where exactly in the room we should expect these problems to be noticeable, or in other words – what will the frequency response of each point in the room be.

Luckily, we can quite easily perform a practical test that teaches us just that. It involves playing a sine wave and comparing different frequencies between 20 and 600 Hz. Obviously, we usually listen at our mixing position, but moving around the room would demonstrate how drastically one frequency can be attenuated or boosted at different points across the room. The results of this experiment can be quite shocking for people trying it for the first time – you might learn that a specific frequency (say 72 Hz) is inaudible when you sit, but clearly heard when you stand; you might also learn that one frequency (say 178 Hz) is noticeably louder than a nearby frequency (say 172 Hz).

One more aspect of room modes worth discussing is that they affect the speakers' ability to reproduce the problematic frequencies. If a speaker is positioned in a point where standing waves cause a specific frequency boost, the speaker will be able to produce more of that frequency. The opposite case only applies to monopole speakers – if the speaker is located in a point where a specific frequency is attenuated, the driver will have difficulties producing that frequency.

Included on the DVD are a few 30-second long test tones for readers to experiment with. While listening to each frequency, move around your room to see how at different points that specific frequency is boosted or attenuated. Most readers trying this in a domestic room should recognize at least one frequency that suffers noticeable level variations across the room, which are the consequence of room modes.

Track 7.1: Sine 60 Hz
Track 7.2: Sine 65 Hz
Track 7.3: Sine 70 Hz
Track 7.4: Sine 75 Hz
Track 7.5: Sine 80 Hz
Track 7.6: Sine 90 Hz

Track 7.7: Pink Noise
Pink noise provides equal energy per octave and therefore is commonly used for acoustic measurement. While listening to this noise when seated in the listening position used for mixing, readers are encouraged to move their head back and forth, then sideways. Room modes might cause variation in low frequencies, while early reflections and the directivity of the tweeters might cause variations in high frequencies. One characteristic of a well-tuned room is that moving your head while listening to the music (or pink noise for that matter) hardly alters the perceived frequency content.

Treating room modes

You might instinctively think that you could compensate for room modes effects by connecting a graphic equalizer to the monitors. If 172 Hz is attenuated by 6 dB why not boost that frequency by 6 dB? When it comes to rectifying the acoustic response of the room, monitor equalization is considered futile and in most cases will do more harm than good. There are a few reasons for this: first, the simple fact that any equalization process has its own quality penalty, especially when using less than high-end equalizers (a high-precision graphical EQ can easily exceed the cost of acoustic treatment that will yield better results). Second, the room response varies in relation to different positions within the room. It would make sense to compensate for audible problems at the listening position, but such a treatment can cause greater problems at other positions in the room, including the points where the monitors are situated. Last, there is a difference between the way long sounds and transients excite room modes. Compared to the sound of a kick, room modes would more greatly affect the sustained note of a bass guitar. Equalizing the monitors to make the bass sound good might make the kick sound boomy.

It is worth remembering what exactly we are trying to fix. The overall response at different places in a room is dependent on both the direct sound and the reflected sound. It is room modes caused by reflections that make a perfectly balanced mix unbalanced. But while trying to rectify room modes, monitor equalization also affects the direct sound, which can represent a well-balanced mix. There are situations where monitor equalization is appropriate, but these only happen when the direct sound itself experiences frequency alterations. For example, placing a loudspeaker next to the wall causes low-end emphasis. To compensate for this, many active monitors offer switches for different degrees of bass roll-off. But this type of equalization is not intended to correct room modes, it is merely concerned with correcting coloration of the direct sound.

It should be clear by now that in order to treat room modes we need to treat the reflections. This is achieved by two acoustic concepts: diffusion and absorption. Diffusers scatter sound energy including the low-frequency energy of standing waves. Absorbers soak up sound energy. Diffusers are less welcome in small rooms, partly due to the fact that in close proximity to the listening position they can sometimes impair the overall response rather than enhance it. Absorbers, on the other hand, are a very practical solution. The idea is simple: If we absorb sound as it hits the wall, we damp the reflected energy, and therefore minimize the effect of standing waves. In anechoic chambers there are no reflections and therefore no standing waves, but the unnatural response of these reflection-free spaces makes them highly unsuitable for mixing and some people even find them unbearable. In most mixing situations, we want all reflected frequencies to become inaudible within approximately 500 ms. There is little point covering our walls with excessive absorbent material, since absorbers are most effective at high frequencies – these are readily absorbed by normal materials as well.

And so, in order to minimize the effect of room modes, the key is to target the low frequencies. Both low-frequency diffusers and absorbers are an issue in small rooms since in order to be effective they have to be of considerable depth. For example, in order to absorb 85 Hz, an absorber would have to be around 1m deep. Companies like RPG, RealTraps, Auralex and many others offer affordable, relatively small bass traps that fit most project studios and provide good damping of room modes. It should be pointed out that placing bass traps in a room will not reduce the bass response in any unwanted way. The first reason being that by minimizing the effect of standing waves a smoother frequency response is achieved throughout the room, which in turn means that at various points low frequencies will be heard better. Second, bass traps help in reducing the decay time of reflected low frequencies (even non-resonant ones), which in small domestic rooms is usually longer than desired.

Flutter echo

While room modes are the result of interaction between reflected low-frequency wave-forms, flutter echoes are caused by mid and high frequencies bouncing between two parallel reflective surfaces. In small rooms with reflective surfaces, we can clap our hand in order to produce such an effect. It sounds like quick distinctive echoes with a metallic ringing nature, somewhat similar to the jumping sound effect in cartoons.

The addition of flutter echo to percussive instruments, like snares, is like the addition of nonmusical delay that colors the timbre and adds some metallic tail. For project studios, absorbers are the most practical solution to treat flutter echo. These are placed on the offending walls.

Track 7.8: Flutter Echo
Flutter echo caused by hand claps in a domestic room.

Track 7.9: Snare No Flutter
The source snare track.

Track 7.10: Snare Flutter
An artificial simulation of flutter echo using a 30 ms delay (which roughly corresponds to walls 10m apart) with feedback set to 50%. Although the effect is exaggerated in this track, it demonstrates the timbre coloration that can occur in real life.

Snare: Toontrack *EZdrummer*

Early reflections

Reflections bouncing from nearby surfaces blend with the direct sound. Since they travel a longer distance, they arrive with a specific phase relationship to the direct sound. Mid and high frequencies are more problematic due to their shorter wavelength. Early reflections happening within the first 40 ms cause combfiltering – a set of boosts and cancellations across the perceived frequency response. There is also smearing of the stereo image due to the delayed arrival of the direct sound and its reflections. Early reflections commonly bounce from the desk, sidewalls and ceiling. But sound waves are not limited to specular travel – a sound wave hitting the edge of a computer screen can radiate back to the listening position as well.

Apart from effective positioning, which is discussed next, absorbers are used to suppress these early reflections. Since the mid and high frequencies are the most problematic ones, absorbers do not have to be extremely deep – a 50-mm acoustic foam usually gives the desired results. Prioritizing the placement of these tiles is based on two factors: nearest surfaces are treated first, and in the case of walls, absorbent material is first placed halfway between the sound source and listening position. When possible, nearby objects (like computer screens) are protected by absorptive material, even if the reflections they cause are not specular.

Track 7.11: Left Right 1 kHz
Depending on the monitoring environment, this track might, or might not, demonstrate problems caused by early reflections. A 1 kHz sine toggles between the left and right channels in a two-second interval. Ideally, the test tone should appear to come clearly from the speaker it is emitted from. Also, there should be no difference in level, tone, or stereo width and focus between the two speakers. Being seated in the listening position, moving your head around might reveal differences between the two speakers.

Positioning monitors

Among all the acoustic-related enhancements we can apply, positioning our monitors is the only one that does not cost money. Another thing that does not cost money is reading the manual for our monitors. Nearly all monitor manuals include some practical advice regarding ideal placement and configuration of the speakers.

Where in the room?

Perhaps the most crucial positioning decision is where exactly the listening position is, which is largely determined by the position of the monitors themselves. Room modes affect the frequency response at the listening position and the ability of the speakers to reproduce specific frequencies. Combfiltering also has an effect on what we hear. Since the problems caused by room modes and combfiltering are more profound in small rooms, minor changes to the listening or monitor position in such rooms can have a dramatic effect. This makes the monitor and listening position even more crucial in small mixing environments, like most project studios. Unfortunately, it is in these project studios where space limitations provide little or sometimes no option for monitor placement.

Ideally, we would like to try out different listening and monitor positions in an empty room, usually while playing a familiar mix and test tones. It takes three people moving about – two holding the monitors, the other listening. Despite the time and technical issues involved in connecting the monitors and moving them around, the resultant benefit from correct positioning could be invaluable.

Usually, there is only one dimension in question. Rarely we have a choice as for how high we sit, which consequently determines the height of the monitors. Since left/right symmetry results in more accurate stereo image, the listening position is often halfway on one dimension. Unfortunately, halfway between any two walls is often where profound frequency imbalance is caused by the fundamental room mode. One thing we can do is to experiment to see whether it is the length or the width of a rectangular room that gives us less problems. Since the height is fixed, and on the left/right axis we usually sit in the center, all we have to experiment with is the front/back movement of our listening position and our speakers. It is suggested by some that the ideal listening position might be one-third of the way on the dimension. Nothing, however, beats experimentation.

The equilateral triangle

The most common monitor setup involves an equilateral triangle. The monitors are placed on two vertexes, facing the listener with their rays meeting at a focal point right behind the listener's head (Figure 7.4). Many position the monitors in such a way by using a string to measure an equal distance between all vertexes (the two speakers and the focal point). It is vital that the speakers are angled toward the listener – parallel speakers produce an extremely blurred stereo image with an undefined center. Although the equilateral triangle is a recommended standard, it is worth experimenting with variations involving a wider

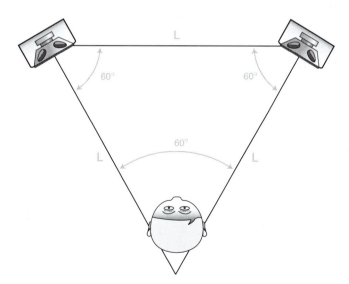

Figure 7.4 An equilateral triangle speaker setup. The angles between the speakers and the focal point behind the listener's head are all 60°. This creates an arrangement where the distance between the two speakers is equal to the distance between each speaker and the focal point.

angle at the focal point. An isosceles with 90° will result in a wider stereo image, but sometimes at the price of an unfocused center.

After the initial positioning of the speakers it is worth playing a familiar mix and moving the head back and forth to find the best listening position. Moving the head backward will narrow the stereo image, while moving it forward will result at some point in a distorted stereo spread and blurred center. There is usually only one point where the stereo image appears to be optimal.

How far?

Once the optimal focal angle has been determined, we can move the speakers closer or further away from the listener while sliding them on the imaginary isosceles sides. While the ear should always be the final judge, a few points are worth considering:

- It takes time for the waves from the different drivers to unite into a cohesive sound. If the speakers are too close to our ears, we can hear the sound as if coming individually from each driver (e.g., highs from the tweeter, lows from the woofer). In such cases, small head movements would result in great changes of the perceived sound, which might render mixing impractical.
- The closer the speakers, the more phase differences between the left and right ears, which results in less solid stereo image.
- The further away the speakers are, the wider the stereo image becomes, which makes panning decisions easier and reverbs somewhat more defined.
- The further away the speakers are, the smaller the direct sound ratio is compared to the reverberant room sound. In a small room with profound resonant room modes, this is not desirable.
- Considering the wall behind the speakers, the further the speakers are from the listener (therefore closer to the wall), the louder the low frequencies will be. This is caused by low frequencies bouncing from the back wall and returning to superimpose on the direct sound. As already mentioned, many active designs feature a bass roll-off switch to compensate for this phenomenon.
- Also, the further away from the listener, the closer the speakers are likely to be to the back and side walls, which can result in more combfiltering.

Horizontal or vertical?

Experts strongly recommend mounting monitors vertically. If the monitors are mounted horizontally, side movements of the head result in individual distance changes from each of the drivers, and thus, unwanted coloration of the frequency response. If for whatever reason monitors are mounted horizontally, the general recommendation is for the tweeters to be on the outside. This ensures a wider stereo image, and some also claim a better bass response.

Very high frequencies are extremely directional, it can easily be demonstrated how a frequency of 18 kHz coming from a single speaker can only be heard if the ear is placed in a very specific point in space. Therefore tweeters should be placed at ear level, or if the monitor is angled, the tweeter's normal should be directed toward the ear.

Damping monitors

Various devices are used to decouple the monitors from the surface they are mounted on. Most widespread are isolation pads made of dense acoustic foam and metal spikes on which the speakers rest. Monitor decouplers function in two principal ways: first, they isolate the monitor from the stand (or desk), ensuring that the speaker operates independently, with no back-vibrations from the stand interfering with the monitor operation. Second, they prevent transmission of vibrations onto the stand, which can generate unwanted resonance. The resonance of a hollow metal stand can easily be demonstrated if you clap your hand next to it. Such stands are designed to be filled with sand, which increases their mass and minimizes resonance. Also, sound generally travels faster through solid matter. It is possible for sound to travel through the stand and floor and reach our body before the sound traveling through air does, possibly confusing our perception of low frequencies. Both foam and spike decouplers are fairly cheap, yet known to have an *extremely positive* effect on the monitor performance. Mostly they yield a tighter, more focused bass response.

Figure 7.5 The Auralex MoPAD monitor isolation pads. Both a Yamaha NS10 and a Genelec 1029 are shown here resting on the Auralex MoPADs.

A/B realm

Virtually every serious mixing studio has more than one pair of monitors. Sometimes, there might even be more than one pair of near-fields. Project studio owners can also benefit from having more than one pair of monitors. Different brands vary in characteristics like frequency response, detail accuracy, depth imaging and stereo imaging. Having more than one brand lets us compare various aspects of the mix using different references, and can help us make important mixing decisions, especially while stabilizing the mix. Different products on the market, often called *control room matrices* or *command centers*,

Figure 7.6 The Mackie Big Knob. Among the features of this studio command system is monitor selection and mono summing.

let us toggle between the different pairs of speakers. The Mackie Big Knob in Figure 7.6 is one of them. Cubase provides internal functionality that achieves the same, provided the audio interface has more than one stereo output.

Further reading

Newell, Philip (2003). *Recording Studio Design*. Focal Press.

8 Meters

A common saying in mixing is 'Listen, don't look'. Being a sonic art, mixing is all about listening. Still, meters are always in sight in mixing environments, and for a good cause. There are various stages and situations where meters can be handy. Of the many types of meters in audio engineering, mixing makes notable use of two: the peak and VU meters. We will also discuss phase meters briefly. It would be hard to discuss metering without giving a little background. The short section below will also become handy in later chapters.

Amplitude vs. level

In general acoustic terms, amplitude describes changes in air pressure compared to the normal atmospheric pressure. A microphone converts changes in air pressure to voltages. An A/D converter converts these voltages into discrete numbers. Changes in air pressure happen above and below the normal atmospheric pressure (the zero reference), resulting in sound amplitude that is bipolar – it has both positive and negative magnitudes. The voltages and numbers used to describe audio signals are also bipolar. An audio system has an equal positive and negative capacity. Professional audio gear, for example, uses −1.23 to +1.23 volts; an audio sequencer uses the −1 to +1 rational range.

One of our main interests is to make sure a signal will not exceed the limits of a system. For example, we do not want the samples within our audio sequencer to rise above +1 or drop below −1. But we do not really care whether the signal exceeded the positive or negative limits, we only care about its absolute magnitude. The term *level* in this book simply denotes the absolute magnitude of signals. Amplitude of +0.5 and −0.5 both denote a sample level of 0.5. Figure 8.1 demonstrates these differences.

Using numbers and voltages to express signal levels would be highly inconvenient. A professional desk telling us that the signal level is 0.30707305603934073775790714109460 V, or an audio sequencer telling us that the sample value is 0.25, is rather overwhelming, especially considering that the two denote exactly the same level only in different units. The decibel system provides an elegant solution – it lets us express levels with more friendly values that mean the same on all systems. The two numbers above are simply −12 dB.

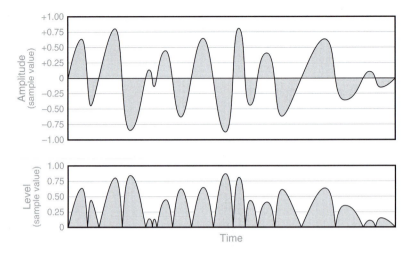

Figure 8.1 Amplitude and level. The top graph shows the amplitude of the waveform, which has both positive and negative magnitudes. The level representation is simply the absolute magnitude of the signal.

The limit, or the standard operating level of a system is denoted by 0 dBr (dB reference). On professional audio equipment 0 dBr is a level of 1.23 V, within an audio sequencer it is a sample level of 1. Since 0 dBr is the high limit of the system, levels are mostly negative. But on analog equipment, signals can go above 0 dBr – they might clip, they might distort, but they can still go there – so we also have positive levels. We call the range above 0 dBr headroom. Even within an audio sequencer, signals can go above 0 dBr, but for a few good reasons we are made to believe that they cannot (more on this later in Chapter 10). For now, we should regard 0 dBr as the absolute limit of a digital system.

For convenience, the 'r' from dBr is omitted henceforth.

Mechanical and bar meters

Mechanical meters involve a magnet and a coil that move a needle. They take up quite some space, and usually only involve a scale of around 24 dB. Bar meters involve either a column of LEDs, a plasma screen or a control on a computer screen. Bar meters might provide some extra indicators in addition to the standard level gauge:

- **Peak hold** – a held line on the meter indicating the highest meter reading. Usually the hold duration can be set to forever, a few seconds or off. This feature tells us how far below 0 dBr the highest peak of the signal hits, which can be useful during recording or bouncing where we use this information to push the signal further up.
- **Peak level** – a numeral display that shows the level of the highest peak.

- **Clip indicator** – an indicator that lights when the signal exceeds the clipping level, which is normally set to 0 dB on a digital system. On analog equipment the clipping level might be set above 0 dB to the level where the signal is expected to distort.

Clear hold and clear clips functions are available on all three facilities.

Clip indicator

Peak hold

Peak hold

Figure 8.2 Cubase track meter facility. In addition to the standard level meter, there is also a graphical and numeral peak hold indicator and a clip indicator.

Peak meters

Peak meters are straightforward – they display the instantaneous level of the signal. Their response to level changes is immediate. Peak meters are mandatory when the signal level must not exceed a predefined limit, like 0 dB on a digital system. In essence, this is their main role.

> The rapid changes in audio signals are too fast for the eye to track, let alone too fast for the refresh rate of screens. In practice, although peak meters are not perfectly instantaneous, in our perception they are close enough to be regarded as such.

On a digital system, the highest level of a peak meter scale is 0 dB. As the lowest level may vary with relation to the bit depth (being −96 dB for 16-bit audio, −144 dB for 24-bit audio and so on), many meters have their bottom level set around −70 dB. If a peak meter is installed on analog equipment, the scale might extend above 0 dB.

Average meters

Our ears perceive loudness with relation to the **average level of sounds, not their peak level** (Figure 8.3 demonstrates the difference between the two). One disadvantage of peak meters is that they teach us little about the loudness of signals. In order to bare any resemblance to loudness, a meter has to incorporate some averaging mechanism. There are various ways to achieve this. A mechanical meter might employ an RC circuit (resistor- capacitor) to slow down the rise and fall of the needle. In the digital domain, the mathematical root mean square (RMS) function might be used. Regardless, the movement of the meter should roughly reflect the loudness we perceive. One example where this can be useful is when compressing vocals – it is the loudness of the vocals we want even.

| *Averaging meters can be considered loudness meters.* |

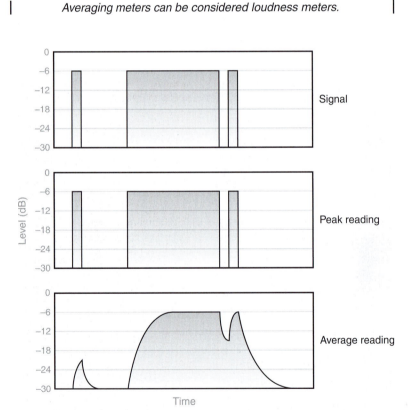

Figure 8.3 Peak vs. average readings. A peak meter tracks the level of the signal instantaneously; therefore its reading is identical to the signal level. An average reading takes time to rise and fall, but the resultant readout is reflective of our perception of loudness. We can say that an average readout seems lazy compared to peak readout.

Of all the averaging meters in mixing, the mechanical VU (Volume Unit) meter is the most popular. Figure 8.4 shows a plugin version of it. The scale spans from −20 to +3 dB, and it is worth noting the distribution of different levels. For example, the right half of the

Figure 8.4 The PSP *VintageMeter*. This free plugin offers averaging metering characteristics similar to those on the mechanical VU meters.

scale covers nearly 6 dB, while the left side covers the remaining 17 dB. Many studio engineers are accustomed to this type of meter, and some even have enough experience to set rough levels of various instruments just by looking at them.

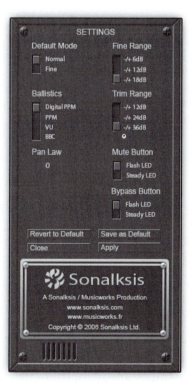

Figure 8.5 The Sonalksis *FreeG* plugin. This free plugin provides extended metering capabilities, among other features. The main meter characteristic is based on one of four ballistics settings (seen on the right panel); these include types of meters not discussed here such as the broadcast BBC type. The narrow bar within the main meter shows RMS levels. There are numeral and graphical (arrows) peak hold indicators for both peak and RMS.

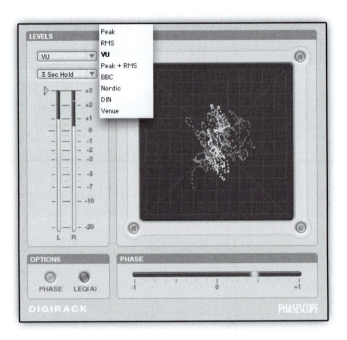

Figure 8.6 The Digidesign *PhaseScope* plugin. The phase meter can be seen at the bottom. Also note the −20 to +3 dB VU scale on the meters, and the amount of possible meter characteristics. The psychedelic graph is actually a very serious tool called the Lissajous Curve. It provides visual imaging of the stereo signal from which we can learn about stereo image shifting, stereo width, phase problems and more.

VU readings can also be displayed on bar meters. This simply involves showing the RMS reading on the standard peak scale (−70 to 0 dB). However, since people are so used to the −20 to +3 range of the mechanical VU meters, a VU bar meter might have its scale covering these 23 dB only.

Phase meters

VU or peak meters are often provided per channel and for the stereo mix. One more type of meter worth briefly mentioning is the phase meter. Phase meters are a common part of large-format consoles. They meter the phase coherency between the left and right channels of the mix. The meter scale ranges from −1 to +1. The +1 position denotes that both the channels are perfectly in phase, i.e., they both output exactly the same signal (which is essentially mono). 0 denotes that each channel plays something completely different (essentially perfect stereo). −1 denotes that the two channels are perfectly phase-inverted. Generally speaking, positive readings tell us that our mix is phase healthy, negative readings suggest that there might be a problem. We want the meter to remain on the positive side of the scale unless we deliberately used an effect that involves phase inversion between the left and the right speakers (like the out-of-speakers effect described later in Chapter 11).

9 Mixing consoles

This chapter reveals the common mixing (not recording) functionality offered by a typical console. It provides vital background information for Chapter 10 on software mixers and the chapters succeeding it. You are advised to read it.

A mixing console (or a *desk* – both terms will be interchanged) is an independent hardware device used alongside a multitrack recorder at the heart of any recording or mixing session in professional studios. Consoles vary in design, features and implementation. Products span from compact 8-channel desks, through 96-channel large-format analog consoles, to large-format digital consoles that can handle more than 500 input signals. In a mixing session, the console's individual channels are fed with the individual tracks from the multitrack recorder. (It is worth noting the terminology – a channel exists on a console; a track exists on a multitrack recorder.) The mixing console then offers three main functionalities:

- **Summing** – combining the audio signals is the heart of the mixing process; most importantly, various channels are summed to stereo via the mix bus.
- **Processing** – most consoles have on-board equalizers, while large-format consoles also offer dynamic processors.
- **Routing** – to enable the use of external processors, effects and grouping, consoles offer routing functionality in the form of insert points, auxiliary sends and routing matrices.

All mixing consoles have two distinctive sections:

- **Channel section** – a collection of channels organized in physical strips. Each channel corresponds to an individual track on the multitrack recorder. The majority of channels on a typical console support a **mono input**. In many cases, all the channels are identical (functionality and layout), yet on some consoles there might be two or more types of channel strips (for example, mono/stereo channels, channels with different input capabilities or channels with different equalizers).
- **Master section** – responsible for central control over the console, and global functionality. For instance, master aux sends, effect returns, control room level and so on.

Large-format analog consoles might also have a **computer section** that provides automation and recall facilities.

Buses

A bus (*or buss* in the UK) is a common signal path to which many signals can be mixed. By way of analogy, a bus is like a highway into which many small roads flow. A summing amplifier is an electronic device used to combine (mix) the multiple sources into one signal, which then becomes the bus signal. Most buses are either mono or stereo, but surround consoles also support multichannel buses, for example, a 7.1 mix bus. Typical buses on a large-format console are:

- Mix bus
- Group buses (or a single record bus on compact desks)
- Auxiliary buses
- Solo bus

Processors vs. effects

The devices used to treat audio signals fall into two categories: processors and effects. It is important to understand the differences between the two.

Processors

A processor is a device, electronic circuit or a software code, used to alter an input signal and *replace* it with the processed, output signal. Processors are used when the original signal is not required after treatment. Processors are therefore connected in series to the signal path (see Figure 9.1). If we take an equalizer for example, it would make no sense to keep the original signal after it had been equalized. Examples of processors include:

- Equalizers
- Dynamic range processors:
  - Compressors
  - Limiters
  - Gates
  - Expanders
  - Duckers
- Distortions

Figure 9.1 Processors are connected in series to the signal path and replace the original signal with the processed one.

- Pitch-correctors
- Faders
- Pan pots

Effects

Effects **add** something to the original sound. Effects take an input signal and generate a new signal based on that original input. There are then different ways to mix the effect output with the original one. Effects are traditionally connected in parallel with the signal path (see Figure 9.2). Most effects are time-based, but pitch-related devices can also fall into this category. Examples of effects include:

- Time-based effects:
  - Reverb
  - Delay
  - Chorus
  - Flanger
- Pitch-related effects:
  - Pitch shifter
  - Harmonizer

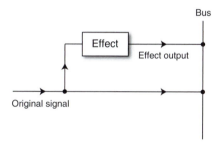

Figure 9.2 In standard operation a copy of the original signal is sent to an effect unit. The original signal and the effect output are then often mixed when summed to the same bus.

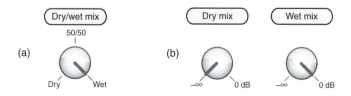

Figure 9.3 Two ways to mix the dry and wet signals within an effect box. (a) A single rotary control with fully dry signal at one extreme, fully wet signal on the other and equal mix of dry and wet in the center. (b) Two independent controls, each determines the level of its respective signal.

Within an effect unit we distinguish between two types of signal. The **dry** signal is simply the unaffected, original input signal. The **wet** signal is the new signal that the effect unit produces. If we take a delay for example, the vocal sent to the delay unit would make the dry signal, and the delays generated by the delay unit constitute the wet signal. While logic has it that effect units should only output the wet signal (and indeed this should be the case when effects are connected in parallel), many effects let us mix internally the dry and wet signals, and output both. The mix between the dry and the wet signal is either determined by a single, percentage-based control, or two separate level controls, one for dry and another for wet (Figure 9.3).

Connecting processors and effects

The standard method for connecting processors is by using an **insert point**, while effects are normally connected using an **auxiliary send**. Both are described in detail in this chapter. If processors and effects are not connected as intended, some problems can arise and bring with them undesired results. Still, exceptions always exist. For example, compressors might be connected in parallel as part of the parallel compression technique. In some situations, it makes sense to connect reverbs using inserts. These exceptions are explored in later chapters.

Basic signal flow

In this section, we will use an imaginary 6-channel console to demonstrate how audio signals flow within a typical console and key concepts in signal routing. We will build the console step-by-step, adding new features each time. Three types of illustrations will be presented in this section: a signal flow diagram, the physical layout of the console and its rear panel (horizontally mirrored for clarity).

The signal flow diagrams introduced here are schematic drawings that help us understand how signals flow, can be routed and where along the signal path, key controls function. By convention, the signal flows from left to right, unless indicated otherwise by arrows. A signal flow diagram is a key part of virtually every console manual. Unfortunately, there isn't a standard notation for these diagrams, so various manufacturers use different symbols. Common to all these diagrams is that identical sections are shown only once. Therefore, a signal flow diagram of a console with 24 identical channels will only show one channel and this channel is often framed to indicate this. Most master section facilities are unique, and any repetitions are noted (for example, if groups 3–4 function exactly like groups 1–2, the diagram will say 'groups 3–4 identical' next to groups 1–2). The signal flow diagrams in this chapter are simplified. For example, they do not include the components converting between balanced and unbalanced.

Step 1: Faders, pan pots, and cut switch

The first step involves the basic console shown in Figure 9.4. Each channel is fed from a respective track on the multitrack recorder. As can be seen in Figure 9.4b each channel has a fader, a pan pot and a cut switch. The signal flow diagram in Figure 9.4a shows us that the audio signal travels from the line input socket through the cut switch, the fader

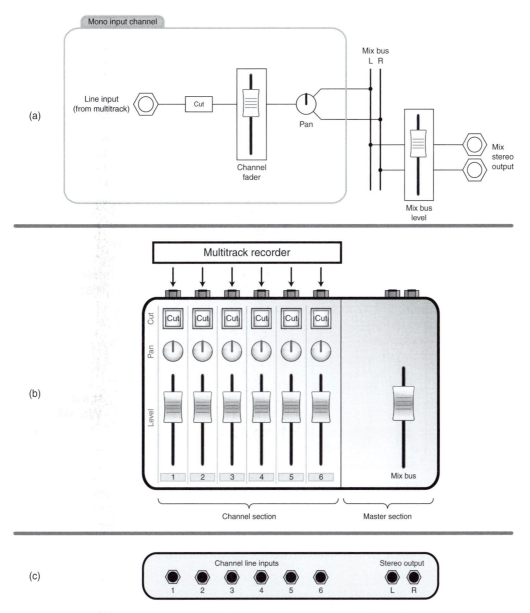

Figure 9.4 The first step in our imaginary six-channel mixer only involves a fader, pan pot and cut switch per channel. Note that the only control to reside in the master section is the mix-bus level.

and then the pan pot. Each pan pot takes in a mono signal and outputs a stereo signal. The stereo signal is then summed to the mix bus (the summing amplifier is omitted from the illustration) and a single fader alters the overall level of the stereo bus signal. Finally, the mix-bus signal travels to a pair of mono outputs, which reside at the rear panel of the console (see Figure 9.4c).

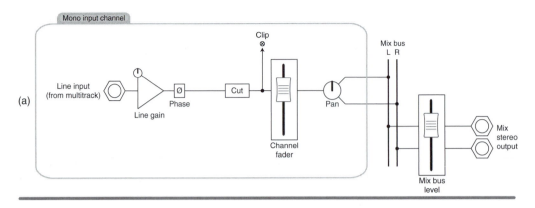

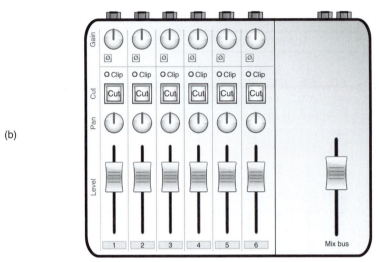

Figure 9.5 In step 2, a line-gain control, a phase-invert switch and a clip indicator have been added per channel strip. The rear panel of the console is the same as in the previous step and not shown.

Step 2: Line gains, phase-invert and clip indicators

Many consoles have a line-gain pot, phase-invert switch and a clip indicator per channel strip. In larger consoles, both the line-gain and phase-invert controls reside in a dedicated input section, which also hosts recording-related controls such as microphone-input selection, phantom power and a pad switch. Figure 9.5 shows the addition of these controls to our console.

The **line-gain** control (or *tape-trim*) lets us boost or attenuate the level of the audio signal before it enters the channel signal path. It serves the mixing engineer in two principal ways. First, it lets us **optimize** the level of the incoming signal. A good recording engineer will make sure that all the tracks on the multitrack recorder are at optimum level, i.e., the highest level possible without clipping (mostly for digital media) or the addition of unwanted distortion (for analog media). If a track level is not optimal, it would be wise to alter it at this early stage. For instance, if the recording level of the vocalist is very low, even the maximum 10 dB extra gain standard faders provide might not be sufficient to make the vocals loud enough in the mix. In such cases, we will need to bring down the rest of the mix, which entails adjusting all other faders. In rare cases, the input signal can be too hot, and might overload the internal channel circuitry. The line-gain pot is used in these situations to bring down the level of the input signal.

Funnily enough, the over-hot input signal is sought after in many situations. An old trick of mixing engineers is to boost the input signal so it intentionally overloads the internal channel circuitry. The reason being that the overloading of the internal analog components is known to produce a **distortion** which adds harmonics that are quite appealing. Needless to say, it has to be done in the right amounts and on the right instrument. Essentially, the line gain in analog consoles adds distortion capabilities per channel. One thing worth keeping in mind is that dynamic range processors, as well as effect sends, succeed the line-gain stage. Therefore any line-gain alterations would also require respective adjustment to, for example, the compressor's threshold or the aux send level. It is therefore wise to set this control during the early stages of the mix, before it can affect other processors or effects in the signal path.

> *The line-gain controls found on analog consoles are used to add distortion.*

A good recording engineer will also ensure that all of the tracks on the multitrack recorder are phase-coherent. In cases where a track was mistakenly recorded with inverted phase, the **phase-invert** control will make it in-phase.

The **clip indicators** on an analog console light up when the signal level overshoots a certain threshold set by the manufacturer. There can be various points along the signal path where the signal is tested. The signal flow in Figure 9.5a tells us that in our console there is only one point before the channel fader. A lit clip indicator on an analog console does not necessarily suggest a sonic impairment, especially if we remember the over-hot signals. If we take some SSL desks, for example, it would not be considered hazardous to have half of the channels clipping when working on a contemporary, powerful mix. The ear should be the sole judge as to whether or not a clipping signal should be dealt with.

Step 3: On-board processors

Most desks offer some on-board processors per channel, from basic tone controls on a compact desk to a filter section, fully featured four-band equalizer, and a dynamics section on large-format consoles. The quality and the amount of these on-board processors dictate much of the console's value, while playing a major role in the console's sound. Figure 9.6b shows the addition of a high-pass filter, an equalizer and a basic compressor to each channel on our console. As can be seen in Figure 9.6a, these processors are located in the signal path between the phase and cut stages, as they would appear on a typical console. Some consoles offer dynamic switching of the various processors. For example, a button would switch the dynamics section to before the equalizer.

Step 4: Insert points

The on-board processors on a large-format console are known for their high quality, yet in many situations we prefer to use an external unit. Sometimes it is simply because a

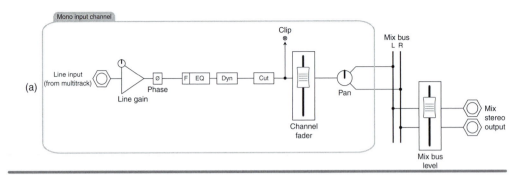

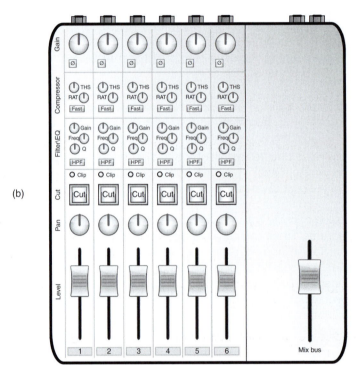

Figure 9.6 As can be seen from the desk layout (b), the additions in step 3 involve a high-pass filter, a single-band fully parametric equalizer and a basic compressor with threshold, release and fast attack controls.

specific processor is not available on-board (duckers, for example, are not very common, even on large-format consoles), and sometimes the sound of an external device is favored over an on-board processor.

As the name suggests, insertion points let us insert an external device into the signal path. They do so by breaking the signal path, diverting the existing signal to an external device and routing the returned, externally processed signal so it **replaces** the original signal. In practice, most consoles always send a copy of the signal to the insert send socket, and the insert button simply determines whether the returned signal is used to

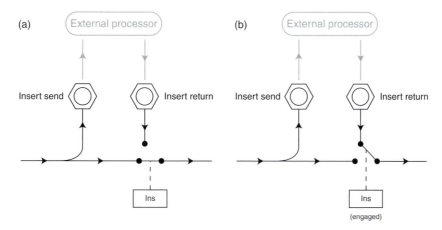

Figure 9.7 A typical insert point. (a) When the insert switch is not engaged, a copy of the signal is still routed to the insert send socket, but it is the original signal that keeps flow in the signal path. (b) When the insert switch is engaged, the original signal is cut and the insert return signal replaces it.

replace the existing signal (Figure 9.7). When insert sends are not used to connect an external device, they are used as a place from which a copy of the signal is taken. For example, when adding sub-bass to a kick, a copy of the kick is needed in order to trigger the opening and closing of the gate (more on this in Chapter 18).

It is important to remember that each external unit can only be connected to one specific channel – when using the channel's insert point it is impossible to send any other channel to the same external unit. However, it is possible to route the output of one external unit into another external unit and return the output of the latter into the console. This way, we can daisy chain external processors, in cases where we want both an external equalizer and a compressor to process the vocal.

Figure 9.8a shows the addition of an insert point into our console's signal flow. Note that the insert point is located after the processors and before the cut stage. Some

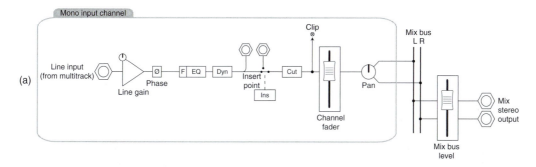

Figure 9.8 In step 4, insert points have been added to the console per channel. The only addition to the layout (b) is the insert switch, while each channel has additional send and return sockets on the rear panel.

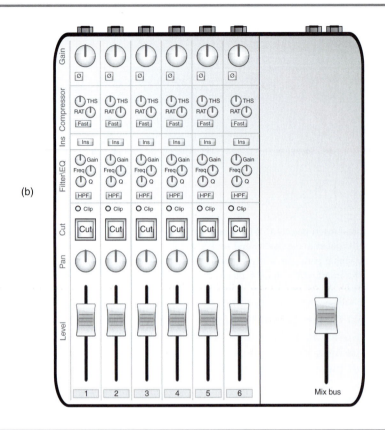

(b)

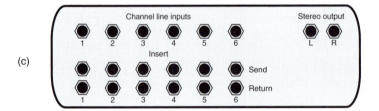

(c)

Figure 9.8 (*Continued*)

consoles enable dynamic switching of the insert point, very often pre or post the on-board processors. The only addition on the layout itself (Figure 9.8b) is the insert-enable button. On the console's rear panel in Figure 9.8c both insert send and return sockets have been added per channel. Two balanced sockets are more common on larger consoles, but some smaller desks utilize one TRS socket that combines both the send (tip) and the return (ring). A Y-lead cable is used in this type of unbalanced connection to connect the external unit.

The importance of signal flow diagrams

The console we have designed so far is very typical – both the order of stages in the signal flow and the layout of the controls on the surface are very similar to those of an ordinary console. One thing that stands out immediately is that the controls on the console's surface are organized for maximum user convenience, and not in the order of processing. Figure 9.9 illustrates how the signal actually travels if we compare it to the channel layout from step 4. We cannot tell just by looking at a desk whether the dynamic processors come before or after the equalizer, whether the insert point comes before or after the processors, or gather other signal-flow related information that can be vital during the mixing process. In many cases the manual will not answer these questions, leaving the signal flow diagram as our only reference.

Step 5: Auxiliary sends

An auxiliary send (*aux send* or *send*) takes a copy of the signal in the channel path and routes it to an internal auxiliary bus. The auxiliary bus is then routed to an output socket, which could then feed an external effect unit. Each channel strip on the console has a set of aux send-related controls. To distinguish these controls from identical auxiliary controls on the master section, we call them *local auxiliary controls*; these may consist of:

- **Level control** – there is always a pot to determine the level of the copy sent to the auxiliary bus.
- **Pre/post-fader switch** – a switch to determine whether the signal is taken before or after the channel fader. Clearly, if the signal is taken post-fader, moving the channel fader would alter the level of the sent signal. When connecting effects like reverbs, we often want the effect level to correspond to the instrument level. Post-fader feed enables this. When the signal is taken pre-fader, its level is independent of the channel fader and will be sent even if the fader is fully down. This is desirable during recording, where the sends determine the level of each instrument in the cue mix – we do not want changes to the monitor mix to affect the artists' headphone mix. If no switch is provided, the signal can be fixed to either pre- or post-fader feed.
- **Pan control** – auxiliary buses can be either mono or stereo. When a stereo auxiliary is used, there will be a pan pot to determine how the mono channel signal is panned to the auxiliary bus.
- **On/off switch** – connects or disconnects the signal from the auxiliary bus. Sometimes simply labeled 'mute'.

It is important to understand that we can send each channel to the same auxiliary bus, while setting the level of each channel individually using the local aux level (see Figure 9.10). For instance, we can send the vocal, acoustic guitar and tambourine to a single auxiliary bus, which is routed to a reverb unit. With the local aux level controls we set the level balance between the three instruments, which results in different amounts of reverb for each instrument. As opposed to insert sends, aux sends allow us to share effects between different channels.

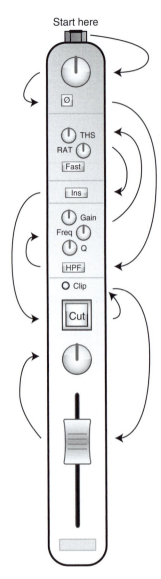

Figure 9.9 This rather overwhelming illustration demonstrates how the audio signal really flows through the channel controls.

Also in Figure 9.10, it is worth noting the controls used to set the overall level of the auxiliary bus or cut it altogether. These **master aux controls** reside in the master section. Master aux controls are often identical to the local ones, with one exception – there is never a pre-/post-fader selection. These are done on a per channel basis and not globally for each auxiliary bus.

Most consoles have more than one auxiliary bus, and thus more than one set of local and master aux controls. On some consoles the amount of auxiliaries determines how many effect units can be used in the mix. For example, in a console with only two auxiliaries,

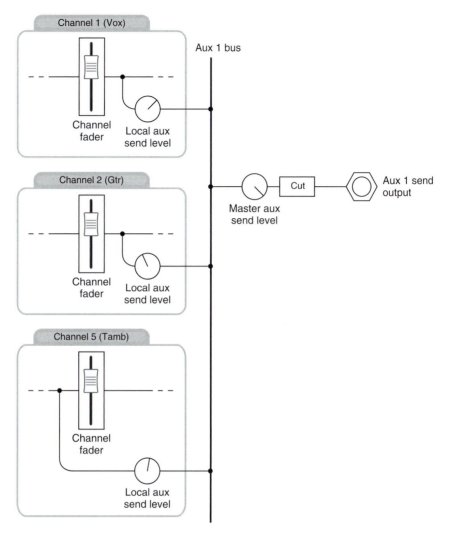

Figure 9.10 In this illustration, the signals from three channels are sent to the same auxiliary bus, each with its own send level settings. Note that for channels 1 and 2 a post-fader feed was selected, while for channel 5 there is a pre-fader feed.

aux 1 might be routed to a reverb unit, while aux 2 to a delay unit. Like with inserts, we can daisy-chain units in order, for example, to have aux 1 sent to a delay, which is then followed by a reverb. The different aux sends are not always identical, and each might involve a slightly different set of controls. In the case of our console, two aux sends were added (Figure 9.11). Aux 1 is stereo, has a pre-/post-switch, pan control and a cut switch. Aux 2 is mono and only features level and cut controls. As can be seen in Figure 9.11a, aux 1 has a fixed post-fader feed. To the desk layout in Figure 9.11b the respective controls for each local aux have been added per channel. Also, the master section now offers both level and cut controls for each master aux. Figure 9.11c shows the addition of output sockets per auxiliary.

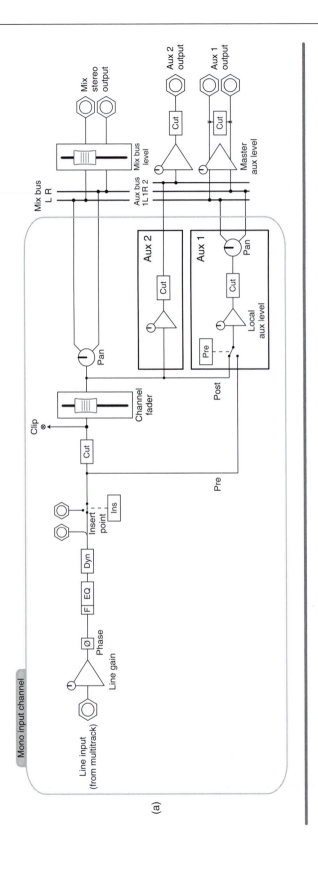

(a)

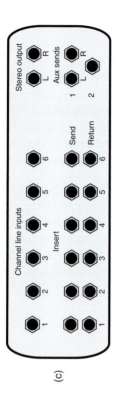

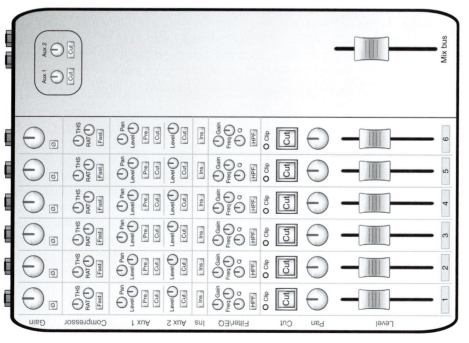

Figure 9.11 In step 5, auxiliary sends were added. The console has two auxiliary buses, one stereo and the other mono. As per channel, the local controls for aux send 1 are level, pre-/post-fader selection, pan and cut. Aux 2 only has local level and cut controls. The master aux controls for both auxiliaries are level and cut.

Step 6: FX returns

So far we have seen how we can send an audio signal to an effect unit. Let us consider now how the effect return can be brought back to the console. FX returns (sometimes called *aux returns*) are dedicated stereo inputs at the back of the desk that can be routed to the mix bus. All but the most basic desks provide some master controls to manipulate the returned effect signal. Level and mute are very common, while large-format desks might also have some basic tone controls. Figure 9.12 shows the addition of two stereo FX returns to our desk. In the layout in Figure 9.12b, a dedicated section provides level and mute controls for each return, and two pairs of additional sockets were added to the rear panel in Figure 9.12c.

FX returns provide a very quick and easy way to blend an effect return into the mix, but even on large-format consoles they offer very limited functionality. For example, they rarely let us pan the returned signal, which might be needed in order to narrow down the width of a stereo reverb return. As an alternative, mixing engineers often return effects into two mono channels, which offer greater variety of processing (individual pan pots, equalizers, compressors, etc.). Such a setup is shown in Figure 9.13.

> *When possible, effects are better returned into channels.*

Groups

Rarely do raw tracks not involve some logical groups. The individual drum tracks are a prime example of a group that exists in most recorded productions; vocals, strings and guitars are also potential groups. While a drum group might consist of 16 individual tracks, some groups might only consist of two – bass-mic and bass-direct for example. During mixdown we often want to change the level of, mute, solo, or process collectively a group of channels. The group facility allows us to do so. There are two types of groups: control and audio.

Control grouping

Control grouping is straightforward. We first allocate a set of channels to a group, so that moving one fader causes all the other faders in the group to move simultaneously. Cutting or soloing a channel also cuts or solos all other group channels. Some consoles define this behavior as *linking*, while grouping denotes a master–slave relationship – only changes to the master channel result in respective changes to the slaves, but changes to slaves do not affect other channels.

VCA grouping

In order for faders to respond to the movement of other faders, they must be motorized. Consoles without motorized faders can achieve something similar using a facility called VCA grouping. With VCA grouping, a set of master VCA group faders are located in the

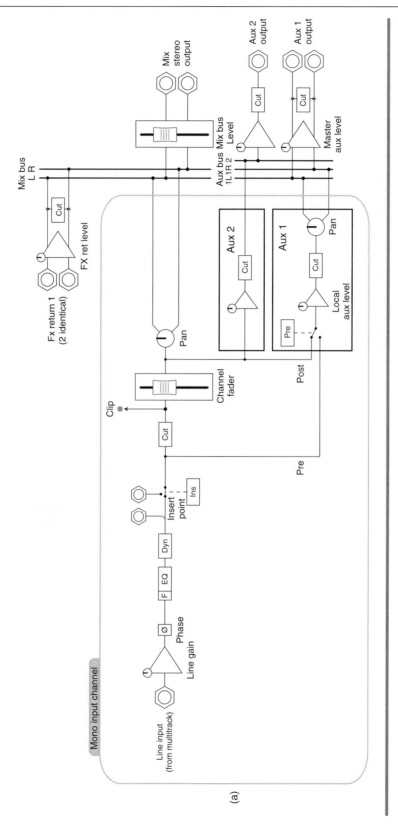

(a)

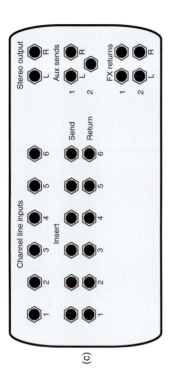

(c)

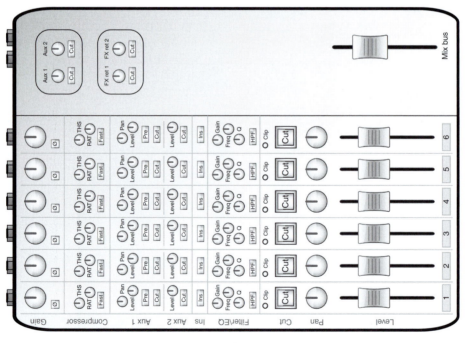

(b)

Figure 9.12 Step 6 involves the addition of two FX returns. There are two pairs of sockets on the rear panel, each FX return has level and cut controls which will affect the returned effect signal before it is summed to the mix bus.

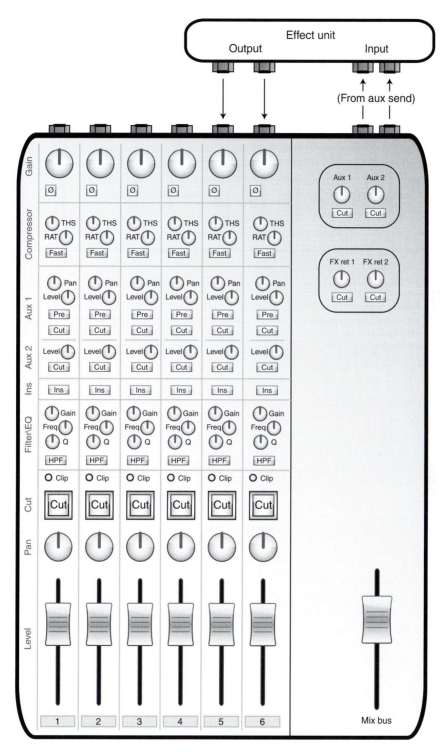

Figure 9.13 When there are free channels on the console, mixing engineers often bring back effect returns into channels rather than using FX returns. Channels offer more processing options compared to FX returns.

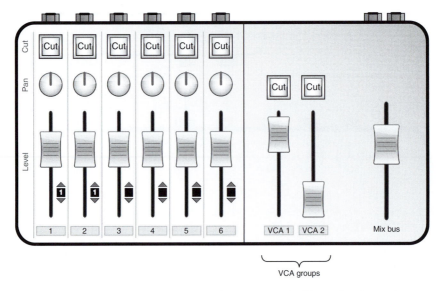

Figure 9.14 A basic console with two VCA groups. In the illustration above, channels 1 and 2 are assigned to VCA group 1. Moving the fader of VCA group 1 will alter the level of channels 1 and 2, although their faders will not move.

master section alongside the cut and solo buttons per group. Individual channel faders can then be assigned to a VCA group using a dedicated control per channel strip. Moving a master VCA group fader alters the level of all channels assigned to the VCA group (obviously, the channel faders will not physically move). Cutting or soloing a VCA group cuts or solos all the assigned channels. Figure 9.14 shows the layout of a small console with VCA groups. Figure 9.14 shows the layout of a small console with VCA groups.

> The concept of VCA faders is explained in Chapter 12.

Control grouping is great when we want to *control* collectively a group of channels. In addition to the standard level, cut and solo controls, digital consoles (and software mixers) sometimes enable the grouping of additional controls such as pan pots, aux sends or the phase-invert switches. However, control grouping does not alter the original signal path, therefore it is impossible to *process* a group of channels collectively. This is often required, for example, in cases where we want to compress the overall drum-mix. When collective processing is needed or if an analog console does not offer control grouping, audio grouping is the solution.

Audio grouping

To handle many signals collectively, a group of channels must first be summed to a group bus – a practice called *subgrouping*. The group signal (essentially a submix) can then be processed and routed to the mix bus. Different consoles provide different amounts of

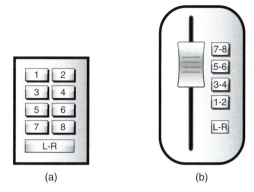

(a)   (b)

Figure 9.15 Routing matrices. (a) A typical arrangement for a routing matrix on a large console. Each group bus has a dedicated button. (b) A typical arrangement for a routing matrix on a smaller desk. Each button assigns the channel signal to a pair of groups, and the channel pan pot can determine to which of the two the signal is sent.

group buses, with multiples of eight being most common. Each group bus is mono, but very often groups are used in stereo pairs made of consecutive odd and even groups – each represents left and right, respectively. For example, in groups 1-2, group 1 is left and 2 is right. The format *Channels:Groups:Mix-buses* is commonly used to describe the amount of these facilities on a console. For example, 16:8:2 denotes 16 channels, 8 group buses and 2 mix buses (i.e., one stereo mix bus).

Each channel can be assigned to different groups using a **routing matrix** that resides on each channel strip. A routing matrix is a collection of buttons that can be situated either vertically next to the fader or in a dedicated area on the channel strip (see Figure 9.15). The latter being more common on larger desks where there can be up to 48 groups. Routing matrices come in two formats – those that have an independent button for each group bus and those that have a button for each pair of groups. A pan pot usually determines how the signal is panned when sent to a pair of groups. For example, if groups 1 and 2 are selected and the pan pot is turned hard-left, the signal will only be sent to group 1.

As far as the signal flow goes, the signal sent to the group is a *copy* of the channel signal. We need a way to disconnect the original signal from the mix bus, since the copy sent to the group will be routed to the mix bus anyway (seldom do we want both the original and the copy together on the mix bus). To achieve this, each routing matrix has a mix assignment switch that determines whether or not the original channel signal feeds the mix bus. The copy is taken *post the channel pan pot* (and thus post the channel fader), and therefore any processing applied to the channel signal (including fader movements) affects the copy sent to the group. It should be made clear that with audio grouping we have full processing control over each individual channel and on the group itself. We can, for example, apply specific compression to the kick, different compression to the snare and different compression still to the drum-mix.

Group facilities fall into two categories: master groups and in-line groups. Consoles that offer **master groups** provide a dedicated strip of controls for each group bus on the

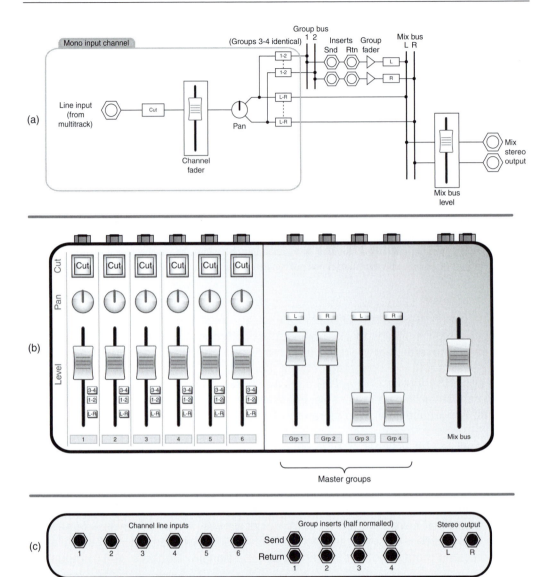

Figure 9.16 A basic console with four master audio groups. In this illustration, channels 1–2 are assigned to groups 1–2; the rest of the channels are assigned to the mix bus and so are groups 1–2.

master section. Each strip contains a fader that sets the overall level of the group, and a few additional controls like a solo button or a mix-bus assignment switch. The latter is required, as during recording or bouncing the group buses are not normally assigned to the mix bus. The mix-bus assignment switch also acts as a mute control since it can disconnect the group signal from the mix bus. To enable the processing of the group signal, each group offers an insertion point.

Figure 9.16a shows the signal flow of a console with master audio groups. Note that the signal to the group bus is a copy taken post-pan. Each group has an insert point (in this case half-normalled, like on a standard patchbay), a fader, and a mix-bus assignment switch. With this console, odd groups can only be assigned to the left, and even groups to the right – effectively forcing the groups to work as stereo pairs. Most consoles also have a button to assign each group to both channels of the mix bus. In Figure 9.16b, we can see the routing matrix next to each channel fader, which involves an assignment button to the mix bus and to each pair of groups. There are also four master group strips, each contains a fader and an assignment switch to either the left or right mix buses. Figure 9.16c shows the addition of the group inserts sockets to the rear of the console.

In-line groups are only found on large-format in-line desks. These desks still have a routing matrix per channel, but they do not offer any master group strips. Instead, each channel strip hosts its associated group (for instance, channel 1 hosts group 1, channel 2 hosts group 2 and so forth). Once hosted, the group signal is affected by all the channel components, including the fader, pan pot, on-board equalizer, compressor and insert point. These provide much more functionality than any master group might. But this functionality comes at a price – each channel can only be used to either handle a track from the multitrack or host its associated group; it cannot do both. In order to use in-line groups, we must have more channels than tracks. On most large consoles this is not an issue – if the multitrack has 24 tracks, they will feed channels 1–24, while channels 25–48 will be available as in-line groups. A possible scenario would involve disconnecting all the drum channels (say, 1–8) from the mix bus and routing them to groups 25–26. Channels 25–26 will then be used to host groups 25–26. To route the two groups to the mix bus, all we need to do is select the mix bus on the routing matrices of the hosting channels (Figure 9.17).

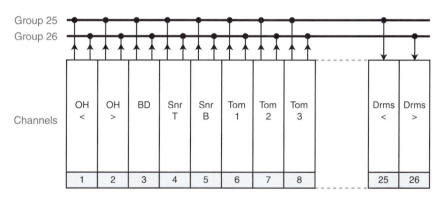

Figure 9.17 Basic schematic of audio groups. Channels 1–8 are routed to groups 25–26, which are then fed into their associated channels.

A group button on the input section of each channel determines what signal feeds the channel path: either the multitrack return signal (the track coming back from the multitrack via the line input) or the associated group signal (which turns the channel into its associated group host). In the case of the drum scenario illustrated above, it will be pressed on channels 25 and 26. Pressing the group button on a channel that has its associated group selected on the routing matrix will cause feedback (the channel will be routed to the same group that feeds it) – and is a recipe for damaged ears.

Figure 9.18 illustrates a hypothetical desk, as it involves in-line groups on a split desk. Yet, it demonstrates well the concept of audio groups. The signal flow in Figure 9.18 shows that pressing the group button will feed the associated group signal into the channel path before the phase button. All the groups in this case are just being sent back to their associated channels. Figure 9.18b reveals a dedicated routing matrix per channel with an assignment button per group (and one button for the mix bus). Also, the group button was added next to the phase button.

Bouncing

The concept of bouncing has been explained in Chapter 5. So far we have only discussed groups in relation to mixing, but groups are an essential part of the recording process as they provide the facility to route incoming microphone signals to the multitrack recorder. Every console that offers group buses also has an output socket per group bus. The group outputs are either hard-wired or can be patched to the track inputs of the multitrack recorder. The process of bouncing is similar to the process of recording, except that instead of recording microphone signals we record a submix. In order to bounce something, all we have to do is route the channels of our submix to a group and make sure that the group output is sent to an available track on the multitrack recorder. Since most submixes are stereo, this process usually involves a pair of groups.

Figure 9.19 shows the console in Figure 9.18 but with the additional group outputs. If you remember, the less-than-optimum level is one of the most common issues with bouncing, and setting the group level is one way to bring the bounced signal to its optimal level. Figure 9.19a shows that the group signal ends up in a group output socket, but only after it has passed a level control. One question that might arise is: where exactly on the desk do we put these group-level controls? We could place them on the master section, but since consoles with in-line groups have as many groups as channels (commonly up to 48 groups) and since these channels and groups are associated, it makes sense to put the group level on each channel strip. As can be seen in Figure 9.19b, the group-level control has been added just below the group button. Figure 9.19c illustrates the addition of the group output sockets, as expected, aligned with the channel inputs.

In-line consoles

The recording aspects

It is impossible to discuss the merits of in-line consoles without talking about recording, but it will be kept brief. During a recording session, the desk accommodates two types of signals: (a) the live performance signals, which are sent via the groups to be recorded on

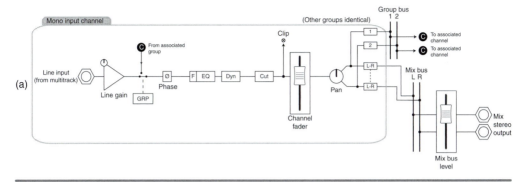

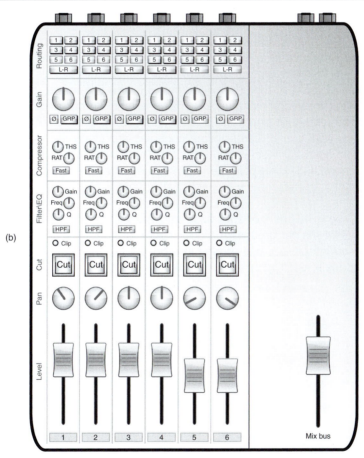

Figure 9.18 A console with in-line group facilities. Channels 1–3 are routed to groups 5–6. The pressed group buttons on channels 5–6 bring the group signal into these channels, which are the only ones assigned to the mix bus.

the multitrack and (b) the multitrack return signals, which include any previously recorded instruments. On a split desk, which we have discussed so far, some channels handle the live signals while others handle the multitrack return signals. But a single channel cannot handle two signals at the same time. Figure 9.20 demonstrates the layout of a split desk during recording.

Split designs can be large. For example, a 24-track recording would require a 48-channel console – 24 channels for live signals and 24 for multitrack returns. If the console also includes 24 master groups, we end up with 72 strips. Size aside, each channel strip costs a good amount of money. However, we rarely require all the expensive features on all channels – we either process the live signals or the multitrack return signals, seldom both.

Going in-line

The way in-line designs solve the size issue is by squeezing 48 channels into 24 I/O modules (referred to as channel strips on a split desk). This is done by introducing additional signal path to the existing channel path on each I/O module. While the channel path is used for live signals, the new path is used for multitrack return signals. Since it is the latter that we monitor (both in the control room and in the live room), the new path is called a *monitor path*. The additional requirements for each I/O module include a line input socket for the multitrack return signal and a fader (or pot on smaller desks) to control the level of the monitor path signal. Large in-line consoles are easily identified by the two faders that reside on each I/O module (Figure 9.21). There will still be only one equalizer, compressor, aux sends and insert point per I/O module but each facility can be switched to either the channel or monitor paths (console permitting).

So we can condense a 48-channel split desk into a 24-channel in-line desk; but what about the additional 24 master groups? If we exchange them for 24 in-line groups (like in Figure 9.18), we can integrate those groups into our 24 I/O modules. We will then end up with 24 I/O modules, each with a channel path, monitor path and a group, all in-line (Figure 9.22).

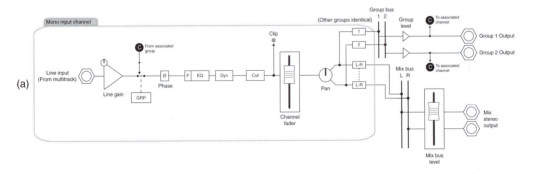

Figure 9.19 A console with physical group outputs, and group-level controls.

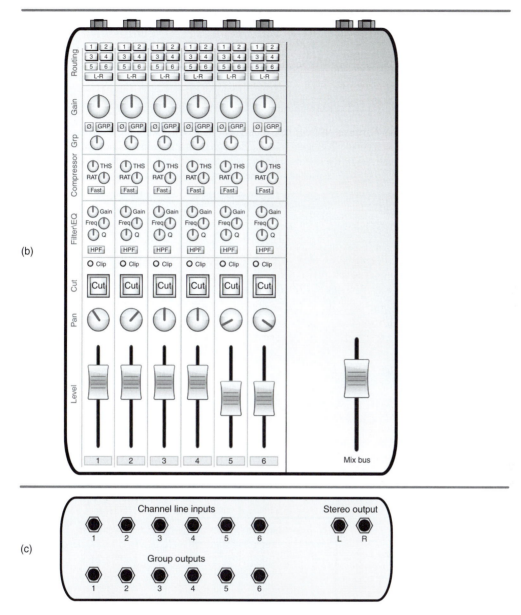

Figure 9.19 (*Continued*)

A 24-track studio might benefit from the in-line concept, but if the studio has a small live room – where only eight microphones may be used simultaneously – it would only need eight groups on its console. A specific design combines a channel and monitor path per I/O module, with master groups instead of in-line ones. Figure 9.23 illustrates this concept, while Figure 9.24 is an example of such a console.

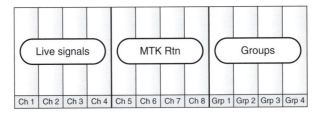

Figure 9.20 An 8-channel, 4-group desk as might be utilized in a recording situation. Channels 1–4 will handle live signals coming through the microphone or line inputs, channels 5–8 will handle the signals returning from the multitrack via the line inputs and the master group will be used to route the incoming signals to the multitrack.

Figure 9.21 The SSL 4000 G+. This photo clearly shows the two faders per I/O module. Depending on the desk configuration, one fader controls the level of the monitor path, the other, the channel path. (Courtesy of SAE Institute, London.)

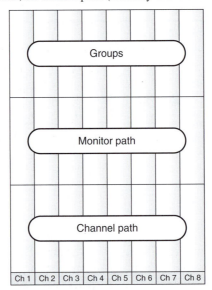

Figure 9.22 An 8-channel in-line console.

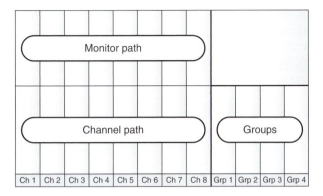

Figure 9.23 An 8-channel, 4-group console.

In-line consoles and mixing

The in-line design makes consoles smaller and cheaper, but as effective for recording. When it comes to mixing, they make things slightly more complex than the split design. The first thing to understand is that a 24-channel console is able to accommodate 48 input signals as each I/O module has both channel and monitor paths. In a 24-track studio this means an addition of 24 free signal paths. As already mentioned, the different processing and routing facilities can only be switched to one path at a time. The catch is that not all facilities can be switched to the monitor path – the insert point, for example, might be

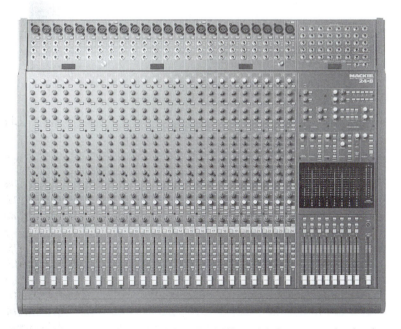

Figure 9.24 The Mackie Analog 8-Bus. Each I/O module has separate level and pan controls per path. In the case of the 8-Bus, the level of one path is controlled using a fader, while the level of the other is controlled using a pot (the white row left of the meters). The eight master group faders can be seen on the right (between the channel faders and master fader).

fixed on the channel path. Generally, the channel path is the 'stronger' path – providing full functionality – while the monitor path is more limited. It is therefore reasonable to use the channel path for the main mix of multitrack return signals and utilize the monitor path for a variety of purposes. Under normal circumstances, the monitor path is used for the following:

- **Effect returns** – effect returns can be easily patched to the monitor path via its line input. We can, for example, send a guitar from the channel path to a delay unit and bring the delay back to the monitor path on the same I/O module.
- **Additional auxiliary sends** – even the most respectable large-format consoles can have a limitation of five auxiliaries, only one of which might be stereo. While this is sufficient for recording, it can restrict effect usage during mixdown – what if we need more than five effects? By having a copy of the channel signal in the monitor path, we can route it to a group, which is then routed to an effect unit. In this scenario, the monitor path level control acts as a local aux send, while the group level acts as master auxiliary level.
- **Signal copies** – in various mixing scenarios a signal duplicate is needed and the monitor path is the native host for these duplicates. One example is the parallel compression technique.

The monitor section

In addition to the global sections we have covered so far, the master section may also contain subsections dealing with global configuration, monitoring, metering, solos, cue mix and talkback. Out of these, the monitor section is most relevant for mixdown. Needless to say, the selection of features is dependent on the actual make. The following section is limited to the most common and useful features.

The monitor output

The imaginary console we have built so far had a stereo mix output all along. The mix output, conventionally, only outputs the mix-bus signal and should be connected to a 2-track recorder. Although most of the time we monitor the mix bus, sometimes we want to listen to an external input from a CD player, a soloed signal, an aux bus or some other signals. To enable this, consoles offer *monitor output* sockets, which, as you might guess, are connected to the monitors. There is also a separate gain control, usually a pot, to control the overall monitoring level.

Figure 9.25 shows the addition of the monitor output facilities to the basic console in step 1. In Figure 9.25a, both the mix bus and an external input (2TK input or 2-track input) are connected to the monitor output. A control circuit cuts all the inactive monitor alternatives. The monitor gain is determined by the *CR Level* pot (control room level). It is worth noting that the mix level will affect the monitor level, but not the other way around. This lets us alter the monitor level without changing the signal level sent to a 2-track recorder via the mix output. If we want to fade our mix, we must use the mix level. Figure 9.25b illustrates the addition of the monitor gain control, labeled 'Monitor Level'. At the rear of the desk in Figure 9.25c, a pair of monitor output sockets have been added, as well as a pair of external input sockets.

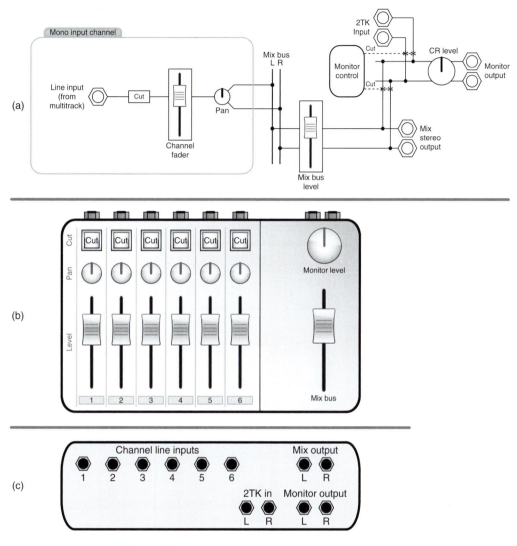

(a)

(b)

(c)

Figure 9.25 A basic console with monitor output facilities.

Additional controls

Figure 9.26 illustrates a master monitor section, very similar to that found on a large-format console. The various controls are:

- **Cut** – cuts the monitor output. Used in emergencies, like sudden feedback or noise bursts. Also used to protect the monitors from predictable clicks and thumps caused by computers being turned on and off, or patching.
- **Dim** – attenuates the monitor level by a fixed amount of dB, which is sometimes determined by a separate pot (around 18 dB is common). The monitor output will still be audible, but at a level that enables easier communication between people in the room or on the other end of a telephone.

- **Mono** – sums the stereo output into mono. Enables mono-coherence checks and can be used as an aid when dealing with masking.
- **Speaker selection** – a set of buttons to switch between the different sets of loud-speakers, i.e., the full-range, near-fields and mini-speakers. It is also possible to have an independent switch that toggles the sub-woofer on and off. A console with such a facility will have individual outputs for each set of speakers.
- **Cut left, cut right** – two switches, each cuts its respective output channel. Used during mix analysis and to identify suspected stereo problems. For example, a reverb missing from the left channel due to a faulty patch lead.
- **Swap left/right** – swaps the left and right monitor channels. Swapping the mix channels can be disturbing since we are so used to the original panning scheme. Also, in a room with acoustic stereo issues (due to asymmetry for example), pressing this button can result in altered tonal and level balances, which are even more disturbing.

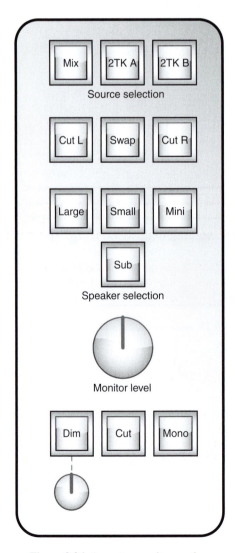

Figure 9.26 A master monitor section.

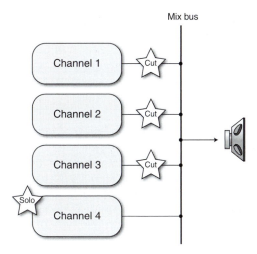

Figure 9.27 Destructive in-place solo. It is worth knowing that in practice the individual channels are cut by internal engagement of their cut switch.

This button is used to check stereo imbalance in the mix. If there is such an imbalance, pressing this button might make us feel like turning our head toward one of the speakers; if it is unclear toward which speaker the head should turn, the stereo image is likely to be balanced. Image shifting can also be the result of problems in the monitoring system, such as when the vocals are panned center but appear to come from an off-center position. If by pressing this button the vocals remain at the same off-center position, one speaker might be louder than the other or one of the speaker cables (or amplifier channels) might be attenuating the channel signal. If the vocal position is mirrored around the center, then it is clear that the shifting is part of the mix itself.

- **Source selection** – determines what feeds the monitor outputs. Sources might include the mix bus, external inputs or an aux bus. Sources might be divided into internal and external sources, and additional controls let us toggle between the two types.

Solos

Solo modes determine how the solo facility operates, that is, what exactly happens when a channel is soloed. There are two principal types of solo: destructive and nondestructive. Nondestructive solos can be either PFL, AFL or APL (the meanings of which will be covered shortly). The hierarchy of the different solo modes is as follows:

- Destructive in-place
- Nondestructive:
  - PFL
  - AFL or APL

Large consoles often support more than one type of solo mode, for example, a desk might support destructive solo, PFL and APL. A console can either support AFL or APL solo, but rarely both. Often manufacturers use the term AFL for a facility that is essentially APL. The active solo mode can be selected through a set of switches that reside on the master section. In certain circumstances, discussed soon, a desk will toggle momentarily to nondestructive mode.

Destructive in-place solo

Destructive solo might also be called *in-place, destructive in-place, mixdown solo* or *SIP* which stands for *solo in-place*. In destructive solo mode (Figure 9.27), whenever a channel is soloed all the other channels are cut. Therefore only the soloed channel (or channels) is summed to the mix bus (which is still monitored). It should be clear that both channel level and panning information is maintained with this type of solo.

Nondestructive solo

In nondestructive mode (Figure 9.28), no channel is being cut. Instead, the soloed channels are routed to a solo bus. As long as solo is engaged, the console automatically routes the solo bus to the monitors, cutting momentarily the existing monitor source (normally the mix bus). With nondestructive solo, the mix output remains intact as all the channels are still summed to the mix bus. PFL (pre-fade listen), AFL (after-fade listen) and APL (after-pan listen) signify the point along the channel signal path from which the signal is taken (see Figure 9.29). PFL takes a copy before the channel fader and pan pot; therefore both mix levels and panning are ignored. AFL takes a copy after the fader but before pan, so it maintains mix levels, but ignores panning. APL takes a copy after both fader and pan, so both level and panning are maintained. Only a desk that provides APL solo requires a stereo solo bus; for PFL and AFL solos, a mono bus will suffice.

A solo-bus level control is provided on the master section. It is recommended to set this control so that signals are neither boosted nor attenuated once soloed. Essentially, this involves leveling the solo bus with the mix-bus level. Any other setting and the mixing engineer might be tempted to approach the monitor level following soloing. Then, when solo is disengaged the monitored mix level will either drop or rise, which might again lead to alteration of the monitor level. The greatest danger of this is constantly rising monitor levels.

Solo safe

Solo safe (or *solo isolate*) provides a remedy to problems that arose as part of the destructive solo mechanism. More specifically, it prevents channels flagged as solo safe from being cut when another channel is soloed. A good example of channels that should be solo safe would be those hosting an audio group. If soloing a kick subsequently cuts the drum group to which it is routed, the monitors would output null (Figure 9.30a). To prevent them from being cut, the channels hosting the drum group are flagged as solo safe. The same null can occur if the audio group channels are soloed since this will cut the source channels, including the kick (Figure 9.30b). Unfortunately, a console cannot determine the individual source channels of an audio group, so in order to prevent them

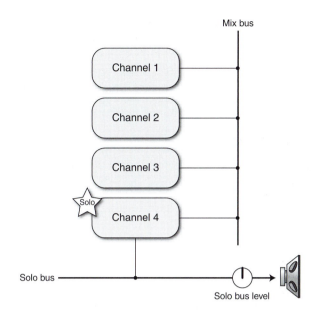

Figure 9.28 Nondestructive solo.

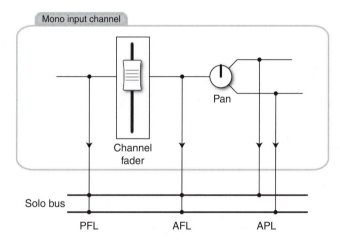

Figure 9.29 Depending on the console and the global solo mode, soloed signals can be sourced from one of three different points along the channel path.

from being cut, the console would automatically engage into nondestructive solo mode when solo-safe channels are soloed. Along with audio groups, effect returns are also commonly flagged as solo safe.

Which solo?

Destructive solo is considered unsafe during recording. Recording during mixdown only happens in two situations: when we bounce and when we commit our mix to a 2-track recorder. In these cases, any destructive solo will be imprinted on the recording, which

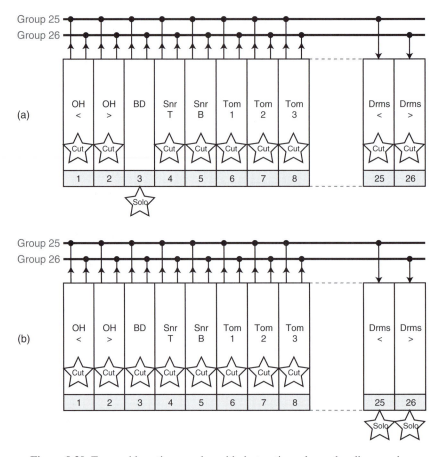

Figure 9.30 Two problematic scenarios with destructive solos and audio grouping. (a) Soloing a source channel will cut the group it is routed to. (b) Soloing the group hosting channels will cut the source channels.

is seldom our intention. With the exception of this risk, destructive solo offers many advantages over nondestructive solo. First, if the console involves a mono solo bus, all soloed signals are monitored in mono. This is highly undesirable, especially for stereo tracks and effect returns (e.g., overheads or stereo reverb). With destructive solo we listen to the stereo mix bus, so panning information is maintained. Another issue with nondestructive solo has to do with unwanted signals on effect returns. Consider a mix where both the snare and the organ are sent to the same reverb unit, and in order to work on the snare and its reverb we solo both. In this scenario, the reverb return still carries the organ reverb, since nothing prevents the organ from feeding its aux send. By soloing in destructive mode, the organ channel will be cut and so will its aux feed to the reverb. Destructive solo also ensures that the soloed signal level remains exactly as it was before soloing. In nondestructive solo, soloed signals might drop or rise in level depending on the level of the solo bus.

Destructive solo is favored for mixdown.

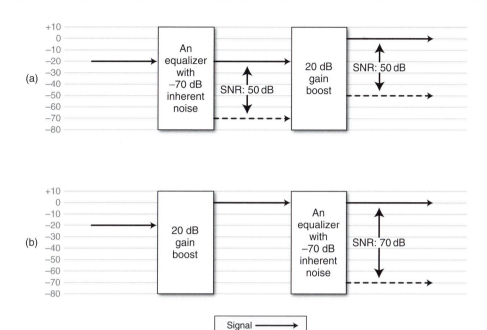

Figure 9.31 A demonstration of gain application and its effect on SNR. (a) The gain is applied after the noise is added, resulting in output signal with 50 dB SNR and noise at −50 dB. (b) The gain is applied before the noise is added, resulting in higher SNR and lower noise floor. The noise level of both the original signal and that of the gain stage are omitted from this illustration to keep it simple. We can assume that these are too low to be displayed on the given scale.

The only time PFL solo comes in handy during mixdown is when we want to audition the multitrack material with faders down. But as PFL ignores mix levels, it is not suitable when soloing more than one instrument.

Correct gain structure

Correct gain structure is a tactic that helps us to reduce unwanted noise, thus improving the overall signal-to-noise ratio of our mix. Analog components are noisy – microphones, preamps, compressors, equalizers, pan pots and analog tapes are just a few examples of components that add their inherent noise to the signal. It would be fair to say that from the point where an acoustic performance is converted into an electrical signal, to the point where the same signal flows on the mix bus as part of the final mix, hundreds of analog components have added their noise. High-quality gear, like large-format consoles, is built to low-noise specifications, but some vintage gear and cheap equipment can add an alarming amount of noise. Whatever quality of equipment is involved, simple rules can help us to minimize added noise.

The principle of correct gain structure is simple: never boost noise. Say for example that an equalizer has inherent noise at −70 dB, and we feed into it a signal with peak at −20 dB. The resultant output SNR (signal-to-noise ratio) will be 50 dB. If we then boost the equalizer output by 20 dB, we boost both the noise and the signal, ending up with the signal level at 0 dB but noise at −50 dB. The SNR is still 50 dB (Figure 9.31a). Now let us consider what happens if we boost the signal *before* the equalizer. The peak signal of −20 dB will become 0 dB as before, but the noise added by the equalizer would still be −70 dB. The output SNR will be 70 dB (Figure 9.31b).

So when boosting the level of something we also boost the noise of any preceding components. We can do a lot worse by attenuating something and then boosting it again. Figure 9.32 illustrates a signal at 0 dB going through two gain stages; one attenuates by 50 dB, the other boosts by 50 dB. Both stages add noise at −70 dB. The second stage will boost the noise of the first one. The input signal will leave this system with its level unaltered but with 50 dB extra noise.

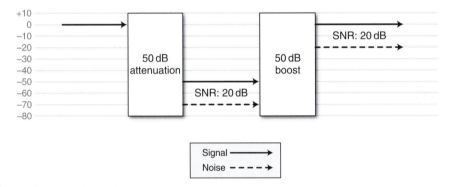

Figure 9.32 The penalty in SNR when boosting after attenuating. An incoming signal will leave the system at the same level, but the noise from the first gain stage will be boosted by the second gain stage, resulting in output noise level of −20 dB and SNR of 20 dB.

To protect the signal from unnecessary boosts, the law of correct gain structure states:

> *Set the signal to optimum level as early as possible in the signal chain, and keep it there.*

To keep the signal at optimum level, we ideally want all the components it passes through to neither boost nor attenuate it; in other words – they should all be set to unity gain. Let us look at some practical examples:

- If the channel signal is too low, is it better to boost it using the input-gain control or the channel fader?
- If the input to a reverb unit is too low, should the master aux send boost it or the input gain on the reverb unit?

- If the overall mix level is too low, should we bring the channel faders up or the mix level?

The answers to all these questions are the first-given option in each case, being the earlier stage in the signal chain. If we take the first question for example, boosting the signal using the channel fader will also boost the noise added by the compressors and equalizers preceding it. Setting the optimum level at the input-gain stage means that the compressor, the equalizer and the fader are fed with a signal at optimum level. If the channel fader is then used to attenuate the signal, it will also attenuate any noise that was added before it. If correct gain structure is exercised, the input gain of a reverb unit should be set to unity gain (0 dB), and the level to the reverb unit should be set from the master aux send.

Correct gain structure does not mean that signals should not be boosted or attenuated – it simply helps us to decide which control to use when there is more than one option. There are of course many cases where boosting or attenuating is the appropriate thing to do. For example, the makeup gain on a compressor, which essentially brings back the signal to optimum level after the level loss caused by the compression process.

The digital console

Digital consoles might resemble the look of their analog counterparts, but they work in a completely different way. Analog signals are converted to digital as they enter the desk and converted back to analog before leaving it. A computer – either a normal PC or purpose-built – handles all the audio in the digital domain, including processing and routing. The individual controls on the console surface are not part of any physical signal path – they only interact with the internal computer, reporting their state. The computer might instruct the controls to update their state when required (e.g., when a mix is recalled or when automation is enabled). Essentially, a digital desk is a marriage between a control surface and a computer.

The various controls can be assigned to many different facilities, for example, a rotary pot can control the signal level, pan position, aux send level, equalizer gain or even switch a compressor in and out. Most manufacturers build their consoles with relatively few controls and let the user assign them to different facilities at will. In addition, the user can select the channel to which various global controls are assigned. So instead of having a set of EQ controls per channel strip, there will be only one global set, which affects the currently selected channel (see Figure 9.33). While faders are motorized and buttons have an on/off indicator, pots are normally endless rotary controls and their position is displayed on a screen – from a small dot-display to a large, color, touch screen. The screen also provides an interface for additional functions like metering; scene store and recall; effect library management; and various configurations.

In this realm of assignable controls, each strip can control any channel. Essentially, one strip would suffice to control all the channels a desk offers, but this would be less

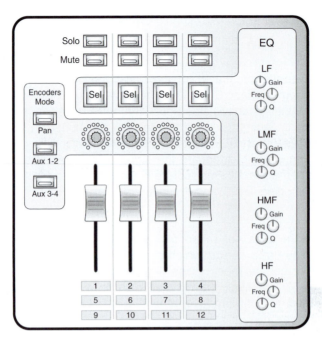

Figure 9.33 The layout of an imaginary mini-digital desk. Each strip has a single encoder, whose function is determined by the mode selection on the left. There is only one global set of EQ controls, which affects the currently selected channel.

convenient. Manufacturers provide a set of channel strips that control threefold, four-fold or other multiples of channels. For example, a desk with 16-channel strips might have 64 channels divided into four chunks (often called *layers* or *banks*). The user can select which layer is controlled by the channel strips at any given time (Figure 9.34). Effectively, all modern digital consoles are split consoles where many strips are orga-nized in a few layers. The strips can also be assigned to control master aux sends and group buses. (There is however, a dedicated fader for the mix bus.) Just like soft-ware mixers, the differences between a channel, an auxiliary bus and a group bus are blurred. With very few exceptions, the facilities available per channel are also being avail-able for the auxiliaries and groups – we can, for example, process a group with the same EQ or compressor available for a channel. Show me an analog console with four fully parametric EQ bands per group and I will show you a manufacturer about to go bankrupt.

Assignable controls enable digital consoles to be substantially smaller than an analog console but with the same facilities. However, their operation is far less straightforward than the what-you-see-is-what-you-get analog concept. Thus, not all digital consoles are indeed smaller. Leading manufacturers like NEVE intentionally build consoles as big and with as many controls as their analog equivalents in order to provide the feel and speed of an analog experience. Still, these consoles provide the digital advantage of letting users assign different controls to different facilities. Users can also configure each session to have a specific number of channels, auxiliaries or groups. The configuration of these large desks – which might involve quite some work – can be done at home, on a laptop and then loaded onto the desk.

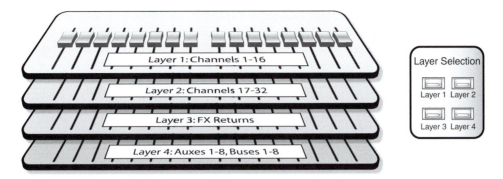

Figure 9.34 Layers on a digital desk. The 64 channels of this desk are subdivided into 4 layers, each represent 16 different channels. The layer selection determines which layer is displayed on the control surface.

Figure 9.35 The Yamaha O2R96 studio at SAE Institute, London. This desk provides 4 layers, each consisting of 24 channels (or buses). Had this desk's layers been unfolded, it would be fourfold in size. The monitors in this photo are the Mackie HR824s.

For the mixing engineer, digital desks aim to provide a complete mixing solution. Dynamic range processors and multiband equalizers are available per channel, very much like on large-format consoles. Most digital desks also offer built-in effect engines, reducing the need for external effect units (although these can be connected when needed in the standard way via aux sends). They let us store and recall our mixes (often called *scenes*) along with other configurations. A great advantage of digital consoles is that their automation engine allows us to automate virtually every aspect of the mix – from faders to the threshold of a gate. Perhaps the biggest design difference between digital and analog consoles has to do with insert points. Breaking the digital signal path, converting the insert send signal to analog, the insert return back to digital and making sure everything is in sync requires far more complex design and often more converters. Many desks offer insert points that, instead of being located on the digital signal path, are placed before the A/D conversion. It is still possible to use external processors that way during mixdown but in a slightly different way than on an analog desk.

10 software mixers

Computers have changed our lives, the music industry, audio engineering and mixing. DAWs are gaining popularity, and as additional new and improved plugins are released each year, more and more professional mixing engineers are embracing software. While it would be fair to discuss all the audio sequencers currently available in the market, it would be impractical in the context of this book. So, with no disrespect to other similar products, the applications presented in this book are Steinberg's Cubase, MOTU's Digital Performer, Apple's Logic and Digidesign's Pro Tools.

Audio sequencers let us mix *inside-the-box*. That is, they provide everything that is needed to complete a mix without external hardware (with the obvious exception of speakers). They can also be fully integrated with outboard gear and consoles when required. The software mixer provides the same core functionality as the mixing console: summing, processing and routing. For routing, each software mixer offers a generous amount of internal buses, used mainly as group and auxiliary buses. In audio-sequencers jargon these are simply called buses. All audio sequencers ship with processors and effects, either integrated into the software mixer or in the form of plugins that can be loaded dynamically. Third-party plugins extend the selection and can offer additional features and quality. All of the processing is done in the digital domain, with calculation carried out on the host CPU. Both summing and routing are relatively lightweight tasks; processors and effects are the main consumers of processing power, and the CPU speed essentially determines how many plugins can be used simultaneously in the mix. DSP expansions – either internal cards or external units – offer their own plugins that use dedicated hardware processors instead of the computer CPU. Digidesign's TDM platform, Universal Audio's *UAD*, t.c. electronic's *Powercore* and Focusrite's *Liquid Mix* are just a few examples.

The physical inputs and outputs provided by the audio interface are available within the software mixer. If mixing is done wholly inside-the-box, a mixing station would only require a stereo output. Since we often bounce the final mix (rather than record it via the analog outputs), this stereo output might only be used for monitoring.

Tracks and mixer strips

Unlike a console-based setup, with audio sequencers there is no separation between the multitrack and the mixer – these are combined in one application. The multitrack is represented as a sequence window (or *arrangement, edit, project* window), where we see the various tracks and audio regions. Whenever we create a new track, it appears in the sequence window, and a new **mixer strip** is created in the mixer window. The change in terminology is important – we lose the term channel, and talk about tracks and their associated *mixer strips* (Figure 10.1).

Tracks

Audio sequencers offer a few types of tracks. Those most likely to be used in mixing are:

- **Audio** – in the sequence window these contain the raw tracks and their audio regions (references to audio files on the hard disk).
- **Aux** – used mainly for audio grouping and to accommodate effects as part of an aux send setup. In the sequence window, these tracks only display automation data.
- **Master** – most commonly represent the main stereo bus. In the sequence window, these tracks only display automation data.

Two more types of tracks, though less common, may still be used in mixing:

- **MIDI** – used to record, edit and output MIDI information. Sometimes used during mixing to automate external digital units, or store and recall their state and presets. Might also be used as part of drum triggering.
- **Instrument** – tracks that contain MIDI data, which is converted to audio by a virtual instrument. On the mixer, these look and behave just like audio tracks. Might be used during drum triggering with a sampler as the virtual instrument.

Variations do exist. Cubase, for example, classifies its auxes as *FX Channel and Group Channel.*

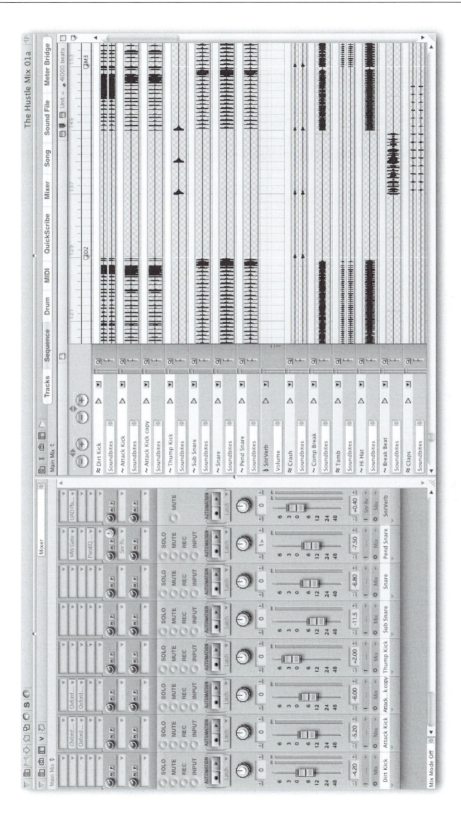

Figure 10.1 Digital Performer's mixer (left) and sequence window (right).

Mixer strips

Figure 10.2 shows the audio mixer strips of various applications, and the collection of different sections and controls each one provides. Audio, aux and instrument tracks look the same. Master tracks offer less facilities, while MIDI tracks are obviously different. We can see that despite user-interface variations, all the applications provide the same core facilities. The mixer strips are very similar to the channel strips found on a console - they have an input, the signal flows through a specific path and is finally routed to an output. Figure 10.3 shows the simplified signal flow of a typical mixer strip. Since tracks can be either mono or stereo, mixer strips can be either mono or stereo throughout. However, a mixer strip might have mono input which changes to stereo along the signal flow (usually due to the insertion of a mono-to-stereo plugin).

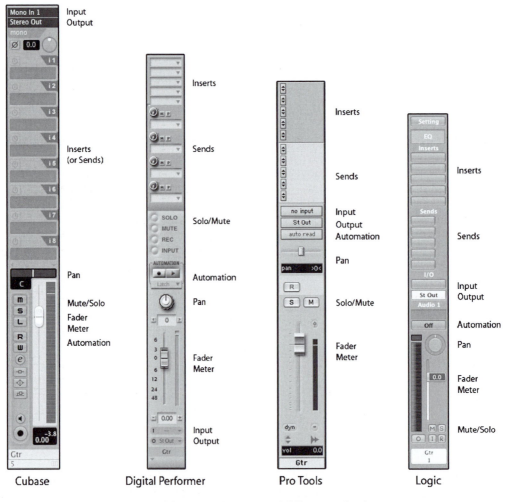

Figure 10.2 Audio mixer strips of different applications.

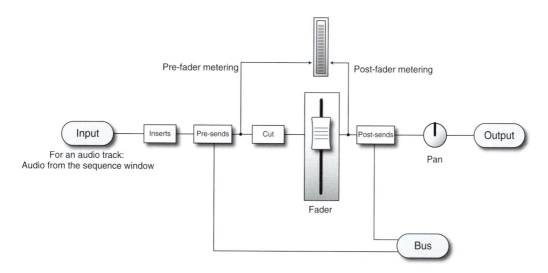

Figure 10.3 A signal flow diagram for a typical software mixer strip. Note that the single line can denote either mono or stereo signal path throughout - depending on the track.

The fader, mute button and pan control need no introduction. The solo function will be explored shortly, while both automation and meters have dedicated chapters in this book. Other sections are:

- **Input selection** – determines what signal feeds the mixer strip. Options being either buses or the physical inputs of the audio interface. However, only aux strips truly work that way. Audio strips are always fed from the audio regions in the sequence window. When an audio track is recorded, the input selection determines the source of the recorded material; but the audio is first stored onto files, and only then fed to the mixer strip (therefore any loaded processors are not captured onto disk).
- **Output selection** – routes the output of the mixer strip to either physical output or a bus. Some applications enable multiple selections so the same track can be routed to any number of destinations. This can be handy in very specific routing schemes (side-chain feed, for example). Commonly, mixer strips are routed to the main physical stereo output.
- **Insert slots** – let us load plugins in series to the signal path, to facilitate the addition of processors. However, insert slots are also used to add effects, as we shall soon see. The order of processing is always top to bottom.
- **Send slots** – similar to the traditional aux sends, they feed either a pre- or post-fader copy of the signal to a bus, and let us control the send level or cut it altogether. Each send slot can be routed to a different bus, and the selection is made locally per mixer strip (i.e., two mixer strips can have their first slot routed to a different bus).

Solos

Audio sequencers normally offer destructive in-place solo. Applications like Cubase and Pro Tools TDM also provide nondestructive solos. While some applications let us solo safe a track, others have a solo mechanism that automatically determines which channels should not be muted when a specific track is soloed. For example, soloing a track in Logic will not mute the bus track it is sent to.

Control grouping

Control grouping was mentioned in Chapter 9 as the basic ability to link a set of motorized faders on an analog desk. Audio sequencers provide much more evolved functionality than that. Each track can be assigned to one or more groups, and a group dialog lets us select the track properties we wish to link. These might include mute, level, pan, send levels, automation-related settings and so on. Figure 10.4 shows the control grouping facility in Logic.

Figure 10.4 Logic's control grouping. The various backing-vocal tracks all belong to control group 1 – titled BV. The linked properties are ticked on the group settings window.

Control grouping, or just *grouping* as it is often termed, can be very useful with editing. However, when it comes to mixing it is something of a love-hate relationship. On one hand, it removes the need for us to create an additional channel for an audio group. For example, if we have bass-DI and bass-mic, we often want the level of the two linked, and creating an audio group for just that purpose might be something of a hassle. On the other hand, control grouping does not allow collective processing over the grouped channels. There is very little we can do with control grouping that we cannot with audio grouping. Perhaps the biggest drawback of control grouping is its actual essence – despite grouping a few tracks, on occasion we may wish to alter some of them individually. Some applications provide more convenient means to momentarily disable a group or a specific control of a grouped track.

Routing

Audio grouping

Audio grouping provides the full freedom to both individually and collectively process. As opposed to consoles, software mixers do not provide routing matrices. To audio-group tracks, we simply set the output of each included track to a bus, then feed the bus into an auxiliary track. We can process each track individually using its inserts, and process the audio group using the inserts on the aux track. Figure 10.5 illustrates how this is done.

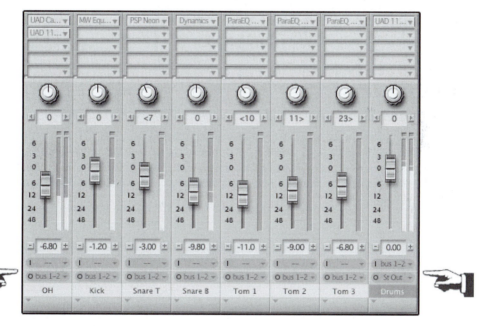

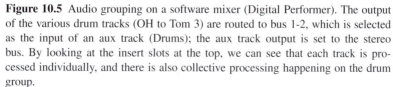

Figure 10.5 Audio grouping on a software mixer (Digital Performer). The output of the various drum tracks (OH to Tom 3) are routed to bus 1-2, which is selected as the input of an aux track (Drums); the aux track output is set to the stereo bus. By looking at the insert slots at the top, we can see that each track is processed individually, and there is also collective processing happening on the drum group.

Sends and effects

As already mentioned, the addition of effects is commonly done using the send facility. Individual tracks are sent to a bus of choice, and similar to audio grouping the bus is fed to an aux track. The effect plugin is loaded as an insert on the aux track, and we must make sure that its output is set to wet signal only. The aux track has its output set to the stereo mix. Figure 10.6 illustrates this.

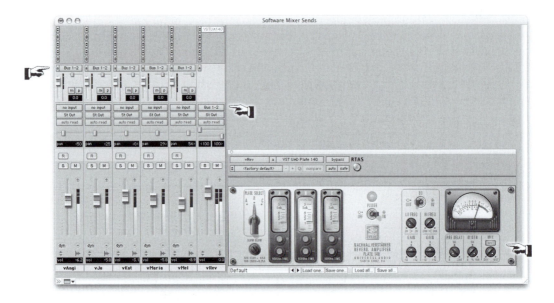

Figure 10.6 Sends on a software mixer (Pro Tools). Each vocal track is sent to bus 1-2. The aux track input is set to bus 1-2, and the output to the stereo bus. Also, the plugin wet/dry control is set to fully wet. In this setup, the audio tracks provide the dry signal (since each track has its output set to the stereo bus) and the wet signal is all the aux adds.

Other routing

The input/output routing facility can be useful for more than grouping and sends. For example, a compressor loaded on the insert slots is pre-fader, but what if we wish to gain-ride the vocals and only then apply compression? We can route the output of the vocal track to an aux and insert the compressor on that aux. Figure 10.7a shows how this is done in Logic. A small trick provides a better alternative to this situation: we can insert a gain plugin before the compressor, and gain-ride the former (Figure 10.7b). However, this might not be the ideal solution if we want to use a control surface to gain-ride the fader. Cubase provides an elegant solution to affairs of this kind – each mixer strip offers both pre- and post-fader inserts.

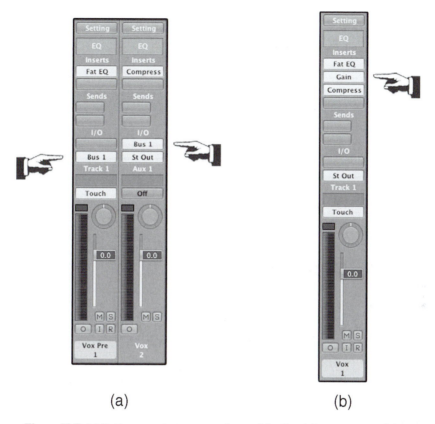

(a)   (b)

Figure 10.7 (a) Setting a track output to a bus and feeding it into an aux track lets us perform gain-riding before compression. The signal will travel through the EQ, the audio track fader (Vox Pre) which we gain-ride, the aux compressor and the aux fader (Vox). (b) The same task can be achieved if we place a gain plugin before the compressor and automate the gain plugin instead of the fader. The signal will travel through the EQ, the gain plugin which we automate, the compressor and the audio track fader.

Naming buses (housekeeping tip)

All audio sequencers let us label the physical inputs, the outputs and the buses. If we fully mix inside-the-box, we do not use physical inputs and we only use one stereo output, so the naming of these is less critical. But a complex mix might involve many buses, and very easily we can forget the purpose of each bus. Scrolling along the mixer trying to trace the function of the different buses is time-wasting. This time can be saved if we label each bus as soon as it becomes a part of our mix. Figure 10.8 shows the same setup as in Figure 10.6 but with the reverb bus labeled.

Naming buses makes mixer navigation easier.

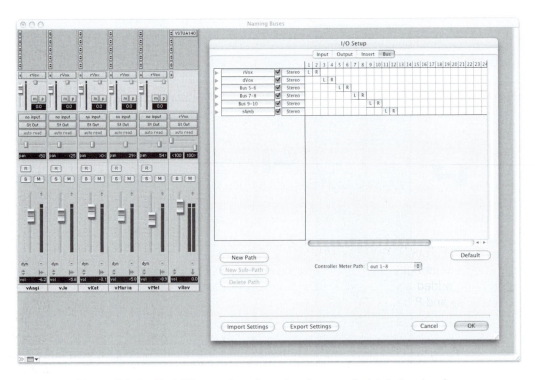

Figure 10.8 Pro Tools I/O setup (one place where buses can be labeled) and a mixer with a labeled bus.

The internal architecture

Integer notation

Digital waveforms are represented by a series of numbers. The sample rate defines the amount of samples per second, and each sample holds a number that represents the amplitude of the waveform at a specific fraction of time. The possible values used to represent the amplitude are determined by the amount of bits each sample consists of, along with the notation used to represent the numbers. If each sample is noted as a 16-bit integer, the possible values are 0 to 65 535 (negative integer notation is ignored throughout this section for simplicity). The highest amplitude a 16-bit sample can accommodate is represented by the number 65 535, which on a peak meter would light the full scale. Such a full-scale amplitude equals 0 dBFS, which is the highest possible level any digital system can accommodate. Mixing digital signals is done by simple summation of sample values. Summing two samples with the value of 60 000 should result in 120 000. But since 16 bits cannot hold such a large number, a digital system will trim the result down to 65 535. Such trimming results in clipping distortion – normally, an unwanted addition. The same thing happens when we try to boost a signal beyond the highest possible value.

Boosting by approximately 6 dB is done by doubling the sample value. Boosting a value of 40 000 by 6 dB should result in 80 000, but a 16-bit system will trim it down to 65 535.

Floating-point notation

Audio files are most commonly either 16- or 24-bit integer, and a D/A converter expects numbers in such form. However, audio sequencers handle digital audio with a different notation called *floating-point,* which is slightly more complex than the integer notation. With floating point, some of the bits (the *mantissa*) represent a whole number, while other bits (the *exponent*) dictate how this number is multiplied or divided. It might be easier to understand how the floating-point notation works if we define a simplified system where in a 4-digit number the three rightmost digits represent a whole number (the mantissa), and the leftmost digit defines how many zeros should be added to the right of the mantissa. For example, with the value of 3256, the whole number is 256 and 3 zeros are added to its right, resulting in 256 000. On the same basis, the value 0178 is equal to 178 (no added zeros). The most common floating- point notation, which has 24 bits of mantissa and 8 bits of exponent, is able to represent an enormous range of numbers, whether extremely small or extremely large. A 16-bit floating-point system supports much smaller and much larger values than a 16-bit integer system. As opposed to its integer counterpart, on a 16-bit floating-point system 60000 + 60000 does result in 120 000.

While the range of numbers supported by modern floating-point systems extends beyond any practical calculations mankind requires, there are some precision limitations worth discussing. In the 4-digit simplified system above, we could represent 256 000 and 178, but there is no way to represent the sum of both: 256 178. Floating-point can support very small or large numbers, but no number can extend beyond the precision of the mantissa. A deeper exploration into the floating-point notation reveals that each mantissa always starts with binary 1, so this 'implied 1' is omitted and replaced with one more meaningful binary digit. Thus, the precision of the mantissa is always one bit larger than the amount of bits it consists of. For instance, a 24-bit mantissa has an effective precision of 25 bits.

The precision of the mantissa defines the dynamic range of a digital system, where each bit contributes around 6 dB (more closely 6.02 or precisely 20log2). Many people wrongly conclude that the famous 32-bit floating-point notation has 193 dB of dynamic range, where in practice the 25-bit mantissa only gives us around 151 dB. Two samples on such a system can represent a very high amplitude or an extremely low amplitude, which can be around 1638 dB apart. However, when the two samples are mixed, the loud signals 'win' the calculation, and any signal 151 dB below it is removed. Just like in our 4-digit simplified system the result of 256000+178 would be 256 000.

What is in it for us?

The internal architecture of audio sequencers is based on the 32- bit floating-point notation (Pro Tools TDM also uses fixed-point notation, which is very similar). The audio files that constitute the raw tracks are converted from integers to 32-bit float on-the-fly during playback. The audio data is kept as floating-point throughout the mixer, and is only

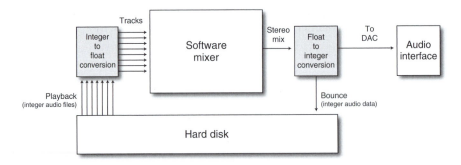

Figure 10.9 Macro audio flow in an audio sequencer and integer/float conversions.

converted back to integers prior to the D/A conversion or when bouncing to an integer based audio file (Figure 10.9). From the huge range of values 32-bit float numbers can accommodate, sample values within the application are in the range of −1.0 to 1.0 (the decimal point denotes a float number). This range, which can also be considered as −100 to 100%, was selected as it allows the uniform handling of different integer bit depths – a value of 255 in an 8-bit integer file, and a value of 65 535 in a 16-bit file are both full scale; thus both will be represented by 1.0 (100%).

We know already that a 16-bit integer system clips when two full- scale (65 535) samples are summed. Audio sequencers, with their 32-bit float implementation, have no problem summing two sample values of 1.0 (integer full-scale) – software mixers live peacefully with sample values like 2.0, 4.0, 808.0 or much larger numbers. With such an ability to overshoot the standard value range we can theoretically sum a million tracks at integer full-scale, or boost signals by around 764 dB, and still not clip on a 32-bit floating-point system. Practically speaking, even the most demanding mix would not cause clipping in the 32-bit floating-point domain.

Since 1.0 denotes integer full-scale, audio sequencers are designed to meter 0 dB when such a level is reached. We say that 1.0 is the reference level of audio sequencers, and it denotes 0 dBr. However, this is not 0 dBFS. Floating-point systems can accommodate values around 764 dB higher than 1.0; so essentially, 0 dBr denotes approximately −764 dBFS.

However, we are all aware that audio sequencers do clip, and that the resultant distortion can be quite obvious and nasty. The reason behind this is that at some point the float data is converted back to integer, and 1.0 is converted back to integer full-scale (65 535 for 16 bits). During this conversion, a value above 1.0 is trimmed back to 1.0, which causes clipping distortion just like when integers are trimmed. The key point to remember here, is that trimming – and its subsequent clipping distortion – can only happen during the float-to-integer conversion, nowhere else within the software mixer. This conversion is applied as the mix leaves the master track, therefore only overshooting signals at this stage will cause clipping distortion. While all mixer strips provide a clipping indicator with a threshold set to any value larger than 1.0, none of these clipping indicators denotes real clipping distortion – apart from the one on the master track.

> *Only clipping on the master track indicates clipping distortion. All other individual tracks cannot clip.*

As unreasonable as this fact might seem, it can be easily demonstrated by a simple experiment which is captured in Figure 10.10. We can set up a session with one audio and one master track, set both faders to unity gain (0 dB), then insert a signal generator on the audio track and configure it to generate a sine wave at 0 dB. If we then bring up the audio

Figure 10.10 This Pro Tools screenshot shows a signal first being boosted by 12 dB by the audio track fader, then attenuated by the same amount using the master fader. Despite the lit clipping indicator on the audio track, this setup will not generate any clipping distortion.

fader by, say, 12 dB, both clip indicators will light up (assuming post-fader metering), and clipping distortion will be heard. However, if we then bring down the master fader by 12 dB, the clipping distortion disappears, although the clipping indicator on the audio track is still lit up. By boosting the sine wave by 12 dB it overshoots and thus the clipping indicator lights up. But as the master fader attenuates it back to a valid level of 0 dB, it does not overshoot the master track following which real clipping distortion occurs.

Track 10.1: Kick Source
The source kick track peaks at −0.4 dB.

Track 10.2: Kick Track and Master Clipping
By boosting the kick by 12 dB it clips on the audio track. It also clips on the master track when its fader is set to 0 dB. The result is an evident distortion.

Track 10.3: Kick Track only Clipping
With an arrangement similar to that shown in Figure 10.10, the kick fader remains at +12 dB while the master fader is brought down to −12 dB. Although the post-fader clipping indicator on the audio track is lit, there is no distortion on the resulting bounce.

One thing worth knowing is that clipping-distortion will always occur when a clipping mix is bounced, but not always when it is monitored. This is dangerous since a supposedly clean mix might distort once bounced. Both Pro Tools and Cubase are immune to this risk as the monitored and bounced mixes are identical. However, when a mix clips within Logic or Digital Performer, clipping distortion might not be audible due to headroom provided by certain audio interfaces. Still, the mix will distort once bounced. Monitoring the master track clipping indicator in these applications is vital.

If individual tracks cannot clip, what is the point of having clipping indicators? A few possible issues make it in our interest to keep the audio below the clipping threshold. First, some processors assume that the highest possible sample value is 1.0 – a gate for example will not act on any signals above this value, which represents the gate's highest possible threshold (0 dB). Second, the more signals crossing the clipping threshold, the more likely we are to cause clipping on the master track. While we will soon see that this issue can be easily resolved, why cure what you can prevent? Last, some plugins – for no apparent reason – trim signals above 1.0, which generates clipping distortion within the software mixer. To conclude, flashing clipping indicators on anything but the master track are unlikely to denote distortion, yet it is a good practice keeping all signals below the clipping threshold.

| *It is a good practice keeping all signals below the clipping threshold.* |

Clipping indicators are also integrated into some plugins. It is possible, for example, that a boost on an EQ will cause clipping on the input of a succeeding compressor. It is also possible that the compressor will bring down the level of the clipping signal to below the clipping threshold, thus the clipping indicator on the mixer strip will not light up. The same rule applies to clipping within the plugin chain – they rarely denote distortion, but it is wise to keep the signals below the clipping threshold.

Bouncing revisited

It is worth explaining what happens when we bounce – whether it be intermediate bounces or the final mix bounce. If we bounce onto a 16-bit file, we lose 9 bits of dynamic range worth 54 dB. These 54 dB are lost forever, even if the bounced file is converted back to float during playback. Moreover, such bit reduction introduces either distortion or dither noise. While not as crucial for final mix bounces, using 16-bit for intermediate bounces simply impairs the audio quality. It would therefore be wise to bounce onto 24-bit files, especially when bouncing intermediately. Another issue is that converting from float to integer (but not the other way around) impairs the audio quality to a marginal extent. Some applications let us bounce onto 32-bit float files, which means that the bounced file is not subject to any degradation. However, this results in bigger files, which might not be supported by some audio applications, including a few used by mastering engineers.

> *Bouncing to 16-bit files: bad idea, unless necessary. 24-bit: good idea. 32-bit float: excellent idea for intermediate bounces when disk space is not an issue.*

Dither

Within an audio system, digital distortion can be the outcome of either processing or bit reduction. Digital processing is done using calculations involving sample values. Since memory variables within our computers have limited precision, sometimes a digital system cannot store the precise result of a specific calculation. For example, a certain fader position might yield a division of a sample value of 1.0 by 3, resulting in infinity of 3s to the right of the decimal point. In such cases the system might round off the result, say, to 0.333334. Since the rounded audio is not exactly what it should be, distortion is produced. Essentially, under these circumstances the system becomes nonlinear just like any analog system. Luckily, the distortion produced by calculations within our audio sequencer is very quiet and in most cases cannot be heard. The problems start when such distortion accumulates (a distortion of distortion of distortion etc.). Some plugin developers employ double-precision processing. Essentially calculations within the plugin are done using 64-bit float, rather than 32-bit float. Any generated distortion in such systems is introduced at far lower levels (around −300 dB). But in order to bring the audio back to the standard 32-bit float, which the application uses, the plugin has to perform bit reduction. Unless dither is applied during this bit reduction, a new distortion would be produced.

The reason bit reduction produces distortion is that whenever we convert from a higher bit depth to a lower one, the system truncates the least meaningful bits. This introduces errors into the stream of sample values. To give a simplified explanation why, if truncating would remove all the digits to the right of the decimal point in the sequence [0.1, 0.8, 0.9, 0.7], the result would be a sequence of zeros. In practice, all but the first number is closer to 1 than it is to 0. Bit reduction using rounding would not help either, since this would produce rounding errors – a distortion sequence of [0.1, 0.2, 0.1, 0.3].

Dither is low-level random noise. It is not a simple noise, but one that is generated using probability theories. Dither makes any rounding or truncating errors completely random. By definition, distortion is correlated to the signal; by randomizing the errors, we de-correlate them from the signal and thus eliminate the distortion. This makes the system linear, but not without the penalty of added noise. Nonetheless, this noise is very low in level – unless accumulating.

These samples were produced using Cubase, which uses 32-bit float internally. A tom was sent to an algorith-mic reverb. Algorithmic reverbs decay in level until they drop below the noise floor of a digital system. Within an audio sequencer, such reverb can decay from its original level by approximately 900 dB. The tom and its reverb were bounced onto a 16-bit file, once with dither and once without it.

Track 10.4: Rev Decay 16bit
The bounced tom and its reverb, no dithering. The truncating distortion is not apparent as it is very low in level.

Track 10.5: Rev Tail 16bit
This is a boosted version of the few final seconds of the previous track (the 16-bit bounced version). These few seconds are respective to the original reverb decay from around −60 to −96 dB. The distortion caused by truncating can be clearly heard.

Track 10.6: Rev Decay 16bit Dithered
The bounced tom and its reverb, dithered version. Since the dither is at an approximate level of −84 dB, it cannot be discerned.

Track 10.7: Rev Tail 16bit Dithered
The same few final seconds, only of the dithered version. There is no distortion on this track, but dither noise instead.

Track 10.8: Rev Decay 16bit Limited
To demonstrate how the truncating distortion develops as the reverb decays in level, track 10.4 is heavily boosted (around 60 dB) and then limited. As a result, low-level material gets progressively louder.

Track 10.9: Rev Decay 24bit Limited
This is a demonstration that truncating distortion also happens when we bounce onto 24-bit files. The tom and its reverb are bounced onto a 24-bit file. The resultant file is then amplified by 130 dB and limited. The distortion error becomes apparent, although after a longer period compared to the 16-bit version.

Tom: Toontrack *dfh Superior.*
Reverb: Universal Audio *DreamVerb*
Dither: t.c. electronic *BrickWall Limiter*

But the true power of dither is yet to be revealed: dither lets us keep the information stored in the truncated bits. The random noise stirs the very low-level details into the bits that are not truncated, thus the information is still audible even after bit reduction. The full explanation of how exactly this happens is down to the little-known fact that digital systems have unlimited dynamic range, but they do have a defined noise floor. This noise floor is determined, among other things, by the bit depth. In practice, a 16-bit system provides unlimited dynamic range, but it has its noise floor at around −96 dB. Human beings can discern information below that noise floor, but only to a point. If dither is not applied, the information in the truncated bits is lost, and distortion is produced. With dither, there is no distortion and the low-level details are maintained. However, at some point they will be masked in our perception by the noise.

> These samples were produced using Cubase. The original 8 bars were attenuated by 70 dB, then bounced onto 16-bit files, once with dither and once without it. Then the bounced files were amplified by 70 dB.
>
> **Track 10.10: 8Bars Original**
> The original 8 bars of drums and electric piano.
>
> **Track 10.11: 8Bars No Dither**
> The amplified no-dither bounce. Note how distorted everything is, and the short periods of silence caused by no musical information existing above the truncated bits.
>
> **Track 10.12: 8Bars Dither**
> The amplified dithered bounce. Note how material that was lost in the previous track can now be heard, for example, the full sustain of the electric piano. The hi-hat is completely masked by the dither noise.
>
> *Drums:* Toontrack *EZdrummer*
> *Dither:* t.c. electronic *BrickWall Limiter*

Theoretically, any processing done within a digital system can result in digital distortion. This includes compressors, equalizers, reverbs and even faders and pan pots. Due to its huge range of numbers, any floating point system is far less prone to digital distortion compared to an integer system. Within an audio sequencer, most of this distortion is produced around the level of -140 dB, and for the most part would be inaudible. Manufacturers of large-format digital consoles apply dither frequently along the signal path. This is something audio sequencers do not do, but many plugins do (as part of double-precision processing). Dither has to be applied in the appropriate time and place. There is no point in us randomly adding dither, since doing so would not rectify the distortion – it would just add noise on top of it. The only instance where we should dither is as part of bit reduction. Such bit reduction takes place every time we bounce onto 16-bit files. We have just seen that bit reduction also causes distortion when bouncing onto 24-bit files, so there is a rationale for dithering even when bouncing onto such files. Only if we bounce onto 32-bit float files (Cubase, for example, enables this) do we not need to dither. The rule of dither is therefore very simple:

| *Apply dither once, as the very last process in the mix.* |

The generation of dither noise is not a simple affair. Different manufacturers have developed different algorithms; most of them involve a method called *noise-shaping* that lets us minimize the perceived noise level. Since our ears are less sensitive to noise at specific frequencies, shifting some of the dither energy to those frequencies makes the noise less apparent. There are different types of noise-shaping, and the choice is usually done by experimentation.

Most audio sequencers ship with dither capabilities. It might be an individual plugin, a global option or part of the bounce dialog. In addition, more than a few plugins, notably limiters and loudness-maximizers, provide dither as well. Essentially, what the dither process does is add dither noise, then remove any information below the destination bit depth. Figure 10.11 shows a dither plugin. The plugin is loaded on the master track in Pro Tools to establish itself as the very last process (even after the master fader). We can see the typical dither parameters: destination bit depth and noise-shaping type. In this case, the plugin is preparing for a bounce onto a 16-bit file.

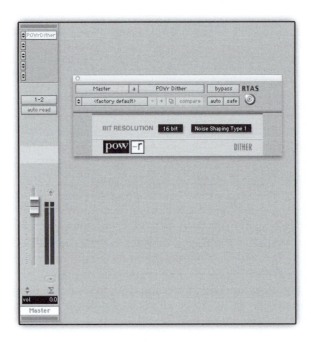

Figure 10.11 (a) Since the limiter on the master track is post-fader, any fader movements will result in varying limiting effect. (b) The individual audio tracks are first routed to an auxiliary, where limiting is applied, and only then sent to the main output, where dithering takes place post-fader. Note that in this setting, both the aux and the master faders succeed the limiter, so both can be automated. Automating the aux is preferred, so the master fader remains as an automation-independent scaling control.

Normalization

Figure 10.12 shows two audio regions. It is worth mentioning how the scale in Pro Tools works (or any other audio sequencer for that matter). The middle mark is complete silence, otherwise roughly −151 dB; the top mark is 0 dB. Then every halving of the scale is worth −6 dB. The majority of the display shows the high-level part of the signal – there is as much space to show level variations between −12 and −18 dB, as there is to −18 to

−151 dB. Consequently, it might seem that the right region is substantially lower in level than the left region; but in fact, the left region is only 18 dB lower in level.

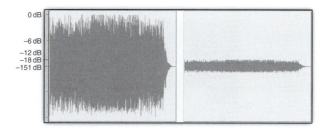

Figure 10.12 Waveform display scales. The left and right regions are identical, except that the right region is 18 dB quieter than the left one. The scales in editor windows are divided so each halving worth −6 dB of level. The majority of what we see is the loud part of the waveform.

We often want our signals to be at optimum levels. An offline process called normalization lets us do this – it boosts the signal level to maximum without causing any clipping. For example, if the highest peak is at −18 dB, the signal will be boosted by −18 dB. As an offline process, normalization creates a new file that contains the maximized audio. Normalization, however, has its shortcomings. During the process, rounding errors occur, resulting in the same distortion just discussed. This is especially an issue when 16-bit files are involved.

Normally, loud recordings are not placed next to quiet recordings like in Figure 10.12 – we often simply have a whole track low in level. A more quality-aware approach to normalization involves boosting the audio in realtime within the 32-bit float domain. This can be easily done by loading a gain plugin on the first insert slot of the low-level track. This is very similar to the analog practice of raising the gain early in the signal-flow path when the signal on tape was recorded too low.

The master fader

Due to the nature of float systems, the master fader in audio sequencers is, in essence, a *scaling fader*. It is so called since all it does is scale the mix output into a desired range of values. If the mix clips on the master track, the master fader can scale it down to an appropriate level. It is worth remembering that a master track might clip even if none of the individual channels overshoots the clipping threshold. In these situations, there is no need to bring down all the channel faders – simply using the master fader is the right thing to do. In the opposite situation where the mix level is too low, the master fader should be used to scale it up.

This is also worth keeping in mind when bouncing. The master fader (or more wisely a gain plugin on the master track) should be used to set the bounced levels to optimum. Determining the optimum level is easy if we have a numeral peak display – we have to reset the master track's peak display and play the bounce range once. The new reading on the peak display will be equal to the amount of boost or attenuation required to bring the signal to 0 dB (or to the safety −3 dB) (Figure 10.13).

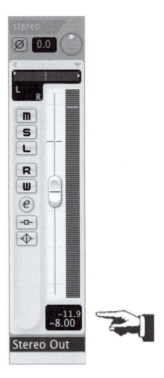

Figure 10.13 The master track in Cubase after the bounced range has been played. The top figure shows the peak measurement (−11.9 dB) and tells us that the master fader can be raised by nearly 11.9 dB before clipping will occur.

The master track in Pro Tools is worth mentioning. As opposed to all the other track types, a master track in Pro Tools has its inserts post-fader. This is in order to enable dithering at the very last stage in the digital signal chain. The consequence of this is that any dynamic processing applied on the master track, such as limiting, will be affected by the master fader. If we automate, or fade out the master fader, the amount of limiting and its sonic effect will vary (Figure 10.14a). Although seldom desired, such an effect can be heard on a few commercial mixes. To fix this issue, individual tracks are routed to a bus that feeds an auxiliary, and the limiting is applied on the auxiliary. The dithering is applied on the master track (Figure 10.14b). Here again, having both pre- and post-inserts per mixer strip, like in Cubase, would be very handy.

The playback buffer

Audio sequencers handle audio in iterations. During each iteration the application reads samples from the various audio files, processes the audio using the mixer facilities and plugins, sums all the different tracks and delivers the mix to the audio interface for

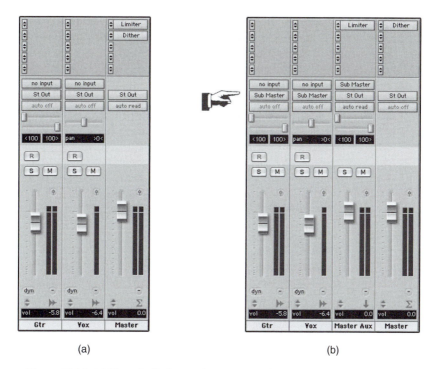

(a)   (b)

Figure 10.14 (a) Since the limiter on the master track is post-fader, any fader move-ments will result in varying limiting effect. (b) The individual audio tracks are first routed to an auxiliary, where limiting is applied, and only then sent to the main out-put, where dithering takes place post-fader. Note that in this setting, both the aux and the master faders succeed the limiter, so both can be automated. Automating the aux is preferred, so the master fader remains as an automation-independent scaling control.

playback. Each iteration involves a multitude of buffers; each represents a mono track or a mono bus. The playback buffer size determines how many audio samples each of these buffers holds. For example, if the playback buffer size is 480 samples, during each iteration 480 samples from each file are loaded into the track buffers; each plugin only processes 480 samples, and the different tracks are summed into the mix-bus buffer, which also consists of 480 samples. Once the mix buffer has been calculated, it is sent to the audio interface. If the buffer size is 480 samples and the playback sample rate is set to 48 000, it will take the audio interface exactly 10 ms to play each mix buffer. While the audio interface is playing the buffer, another iteration is already taking place. Each iteration must be completed, and a new mix-buffer delivered to the audio interface, before the previous buffer finished playing (less than 10 ms in our case). If an iteration takes longer than that, the audio interface will have nothing to play, the continuous playback will stop and a message might pop with a CPU overload warning. A computer must be powerful enough to process audio in realtime faster than it takes to play it back.

The playback buffer size determines the monitoring latency of input signals, and the smaller the buffer size the lower the input latency. This is important during recording, but not as much during mixdown where input signals are rarely involved. Smaller settings, however, also mean more iterations – each with its own processing overhead – so less plugins can be included in the mix. With a very small buffer size (say, 32 samples) some applications might become unpredictable, start producing clicks or skipped playback. However, depending on the application, large buffer size can affect the accuracy of some mix aspects, like automation. When choosing the buffer size, it should be large enough to allow sufficient amounts of plugins and predictable performance; anything larger than that might not be beneficial. A sensible buffer size to start experimenting with is 1024 samples.

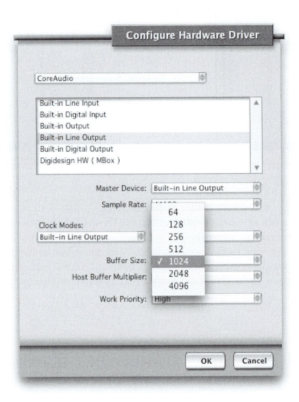

Figure 10.15 Digital Performer's configuration window where the playback buffer size is set.

Plugin delay compensation

We have just seen that plugins process a set amount of samples per iteration. Ideally, a plugin should be given an audio buffer, process it and output the processed buffer straight away. However, some plugins can only output the processed buffer a while after, which means their output is delayed. There are two reasons for such plugin delay. The first involves plugins that run on a DSP expansion like the UAD card. Plugins of this kind send the audio to the dedicated hardware, which returns the audio to the plugin after it has been processed. It takes time to send, process and return the audio; therefore delay is introduced. The second reason involves algorithms that require more samples than those provided by each playback buffer. For example, a linear-phase EQ might require around 12 000 samples in order to apply its algorithm, which falls short of the common 1024 samples the plugin gets per iteration. To overcome this, the plugin accumulates incoming buffers until these constitute a sufficient amount of samples. Only then, after some time has passed, the plugin starts processing the audio and outputs it back to the application.

Plugin delay, in its least harmful form, causes timing mismatches between various instruments, as if the performers are out of time with one another. A mix like that, especially one with subtle timing differences, can feel very wrong although it can be hard to pin-point timing as the cause. In its most harmful form, plugin delay causes combfiltering, tonal deficiencies and timbre coloration. For example, if plugin delay is introduced on a kick track, it might cause combfiltering when mixed with the overheads.

Luckily, most audio sequencers provide automatic plugin delay compensation. In simple terms, this is achieved by making sure that all audio buffers are delayed by the same amount of samples. When delay compensation is not available, we compensate for the delays manually – a long and annoying practice that can be impractical in complex mixes. Manual delay compensation is beyond the scope of this book. It is however, documented in some white papers available over the Internet.

11 Phase

What is phase?

Phase, in sound theory at least, describes the time relationship between two or more waveforms. Phase is measured in degrees, and can be easily demonstrated using sine waves. A sine wave starts when its cycle starts. This is denoted by 0°. One complete cycle signifies 360° (which is also 0°), half a cycle 180°, quarter of a cycle 90° and so forth. If one sine wave reached quarter of its cycle when a cycle begins on an identical waveform, the two are said to be 90° out of phase (Figure 11.1).

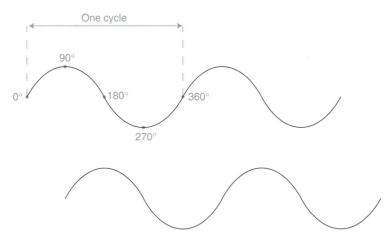

Figure 11.1 Two identical waveforms 90° out of phase with each other.

The degree of phase shift is dependent on the time gap between the two sine waves, but also on the frequency under question. Different frequencies have different phase shifts for the same time gap. Complex waveforms, like all recorded and synthesized waveforms (apart from sine waves), contain many frequencies. If two complex waveforms are delayed with respect to one another, each frequency will have its own degree of phase shift.

As stated in the last sentence, we only consider phase in relation to **similar** waveforms. We have to define *similar* first:

- **Identical waveforms** – as the name suggests, these are similar in every way and usually the outcome of duplication. For example, by duplicating a vocal track in an audio sequencer we get two identical waveforms. By sending a snare to two different groups we also create two duplicate signals.
- **Waveforms of the same event** – two microphones capturing the same musical event. For example, two microphones fronting an acoustic guitar; or a kick-mic and the overheads – although the kick-mic captures mostly the kick, and the overheads capture the whole drum kit, both recordings will have the same kick performance in them.

It is also important to define what is dissimilar: two takes of any instrument, no matter how similar in performance, do not count as similar waveforms. Based on this, double-tracking is not subject to phase consideration.

There is one exception to the above. A kick's sound is largely determined by its first cycle, which can be similar enough between two kicks – even if these were recorded using a different drum and microphone. In practice, phase mostly becomes a factor when we layer two kick samples.

There are three types of phase relationships between similar waveforms:

- **In phase** or **phase-coherent** – when the waveforms start at exactly the same time.
- **Out of phase** or **phase-shifted** – when the waveforms start at different times. For example, if we delay or nudge the duplicate of a vocal track, but not the original. Also, if two microphones are placed at different distances from an instrument, they would capture the same waveform, but the farthest microphone will do so slightly later.
- **Phase inverted** – when both waveforms start at exactly the same time, but their amplitude is inverted (i.e., mirrored across the zero line). Many people confuse phase-inversion with 180° phase shift, but the two are not the same. When two waveforms are 180° out of phase, they do not start at the same time and rarely their amplitude appears inverted (see Figure 11.2).

One thing that the Haas effect tells us is that an audible delay will only be perceived if the time gap between two similar waveforms is bigger than 30 to 40 ms. In this text, 35 ms will be used, and we will regard the *Haas window* as any time gap between 1 and 35 ms. We normally associate phase with time gaps that fall into the Haas window. In mixing terms, a time gap longer than 35 ms is an audible delay.

Problems

Phase problems occur when we sum similar waveforms that are either phase-shifted or phase-inverted. If the time gap involved is smaller than 35 ms, one should expect **comb-filtering** and its subsequent tonal alterations and timbre coloration. If two waveforms are

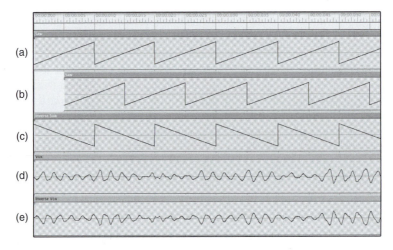

Figure 11.2 Five mono tracks in Digital Performer. (a) The original rising sawtooth. (b) The same sawtooth only 180° out of phase with the original. (c) The original sawtooth phase-inverted. (d) A vocal track. (e) The same vocal track phase-inverted.

phase-inverted with one another, they will **cancel** each other and cause level attenuation; if both are of equal amplitude, they will cancel each other completely to result in complete silence.

It should be said that all this holds true when the two waveforms are summed *internally* – either electronically or mathematically within the mixer. If each of the waveforms is played through a different speaker, the result would be completely different, as we shall soon see. However, if the stereo output is folded to mono, the same phase problems will arise.

It is worth knowing that as part of mono coherence checks, sometimes phase-inverting one channel of a stereo track improves its sound when the mix is folded to mono.

Phase problems on recorded material

Raw tracks, to our great misfortune, often involve some phase issues – especially those of recorded productions. One example is a guitar amplifier recording made with a microphone a few feet from the grill. With both the amplifier and the microphone close to a reflective floor, the direct and the later reflected sound are likely to produce noticeable combfiltering when they hit the microphone. In cases like this, combfiltering is built into a single track, and usually there is little the mixing engineer can do. However, when two or more tracks represent the *same take* of the *same instrument*, some enhancements can be made. Here are a few usual suspects for tracks that might benefit from phase correction:

- **Top/bottom** or **front/back tracks** – snares are often recorded with two microphones in a top and bottom arrangement, while kicks and guitar cabinets might be recorded with two, front and back microphones. Since the two microphones are positioned on opposite sides of the sound source, they are likely to capture opposite sound pressures (contrary to belief, this is not always the case, especially if the microphones are at different distances from the instrument). Recording engineers are aware of this and usually inverse the phase of one microphone before it is recorded. This way, the two tracks arrive ready for mixing in phase.

- **Close-mic** and **overheads** – it is possible to have the kick, snare or any other close-mic either phase-inverted or phase-shifted with the overheads. Again, a good recording engineer will take the overheads as a reference and make sure all drum tracks are phase-coherent with the overheads.
- **Mic** and **direct** – in addition to the conventional microphone recording, guitars (bass, in particular) might also be recorded using a DI box, or direct output from the amplifier. Being kept in the electronic domain, direct signals travel at almost the speed of light. Sound takes some extra time to travel through the air between the amplifier and the microphone, resulting in a slightly delayed instance of the mic track compared to the direct track. This is one example of phase-shifting that once corrected can result in dramatic timbre improvements.

These tracks demonstrate the effect of phase shift on the timbre of a bass guitar. Such phase shifts can be the outcome of mixing direct and mic tracks, or two microphones that were positioned at different distances from the source:

Track 11.1: Bass In Phase
The two bass tracks perfectly in phase.

Track 11.2: Bass 1 ms Shift
1 ms time gap between the two tracks results in loss of high frequencies, definition, and attack. (It is worth mentioning that this is the most primitive form of a digital low-pass filter–by summing a sound with a very short delayed version of itself, high frequencies are softened while low frequencies are hardly affected.)

Track 11.3: Bass 2 ms Shift
With 2 ms time gap, the loss of low frequencies becomes evident. The track sounds hollow and powerless.

Track 11.4: Bass 5 ms Shift
5 ms time gap produces a severe coloration of timbre, and a distant, weak sound.

- - - - - - - - - - - - - -

These two tracks show the effect of phase inverting one of two snare tracks:

Track 11.5: Snare Top and Bot
Both the snare top and bottom are mixed as recorded.

Track 11.6: Snare Top with Bottom Inverted
With the snare-bottom track inverted, some snare body is lost. The effect is not always similar to the one heard here, or as profound.

Phase problems are not always audibly obvious. In the case of phase-inversions, we go through the usual suspects and see what happens when we engage the phase-invert control. Sometimes the difference is noticeable, whether for good or bad, sometimes it is not. We can also zoom into associated tracks to look for phase-inversions. Zooming can also be useful when hunting for phase-shifts. To correct phase-shifts we nudge the late audio region backward in time until it is aligned with its associated track. We can also delay the early track until it is aligned with the late track, but this less-straightforward solution is mostly used when nudging is not available, like with a tape multitrack. When

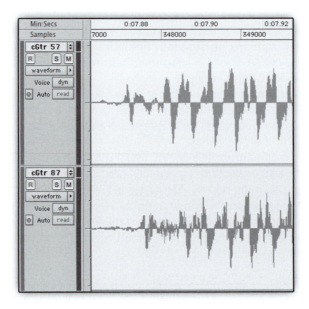

Figure 11.3 Phase shift on guitar recordings. This screenshot shows two guitar recordings that were captured simultaneously using microphones at different distances from the amplifier. The time gap, approximately 5 ms, is evident here.

aligning two phase-shifted tracks, it is not always possible to determine the exact nudging amount by eye, and different values result in different timbre. Fine phase corrections are mostly done by ear.

> *Phase checks and corrections are better carried out before mixing onset. Associated tracks are checked for phase-inversions and phase-shifts.*

Phase problems during mixdown

Rarely mixing engineers initiate phase problems, and even then it is often done by mistake. Here are a few examples how phase problems can be introduced during mixdown:

- **Lack of plugin delay compensation** – uncompensated plugin delay is an open invitation for phase problems. For example, if a compressor plugin delays the snare by 500 samples, chances are that when mixed with the overheads combfiltering will occur.
- **Digital to analog conversions on outboard gear** – digital units convert the audio from analog to digital and back, which takes some time. This introduces a short delay, but one long enough to cause combfiltering if the signal is then summed with identical track.
- **Very short delays** – short delays that fall into the Haas window might cause combfiltering if the original and the delayed sounds are summed internally.

- **Equalizers** – as part of their design, equalizers cause group delay for a specific range of frequencies. Phase problems are especially noticeable with large gain boosts.

Tricks

It's been said already that when two identical waveforms are not in-phase, but each is played through a different speaker, the result would be quite different from combfiltering. Two mixing tricks are based on such a stereo setup. With both, two identical mono signals are sent to a different extreme, and one of the signals is either delayed or phase inverted. To distinguish the two we will call the unaltered signal the *original signal*, and the copy, which is either delayed or phase-inverted, the *ghost copy*.

The Haas trick

The Haas trick was not invented by Helmut Haas, but it is, essentially, a demonstration of the Haas effect. Haas was interested in what happens when an initial sound is quickly succeeded by similar sounds from various directions. His findings teach us that the directivity of the sound is determined solely by the initial sound providing that (a) the successive sounds arrive within 1–35 ms from the initial sound. (b) The successive sounds are less than 10 dB louder than the initial sound. Although the successive sounds do not give any directivity cues, they still play a spatial role. The Haas trick simply involves an original signal panned to one extreme, and a ghost copy, which is delayed by 1-35 ms, sent to the other (see Figure 11.4).

The Haas trick is usually achieved in one of two ways. The first involves panning a mono track hard to one channel, duplicating it, panning the duplicate hard to the opposite channel

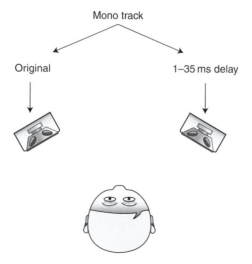

Figure 11.4 The Haas trick. A mono track is sent to both left and right, but one channel is delayed by 1–35 ms.

and nudging the duplicate by a few milliseconds (Figure 11.5). The second involves loading a stereo delay on a mono track, setting one channel to have no delay and the other to have a short delay between 1 and 35 ms.

The Haas trick results in a wide, open, spacious sound; although the sound can be localized to the non-delayed channel, there is some sense of unfocused stereo image. It can be applied during mixdown for three main purposes:

- **To fatten sounds panned to the extremes** – using the Haas trick on instruments already panned to the extremes can make them sound bigger, especially in a crossed arrangement. For example, it is common to double-track distorted guitars and pan each mono track to an opposite channel. Applying the Haas trick on each guitar track (sending the delayed duplicate to the opposite channel like in Figure 11.6) results in a fatter, more powerful effect.
- **As a panning alternative** – sometimes when panning a mono track, all the panning options seem less than ideal. For example, in a sparse arrangement of 3 mono tracks – vocal, bass and guitar – chances are that both the bass and the vocal will be panned center. Panning the guitar to the center, apart from resulting in a monophonic mix, will place it in the busiest, high-masking area of the mix. Panning the mono guitar to one side or another will cause stereo imbalance. By applying the Haas trick, we can open up the monophonic guitar sound, achieve some stereo width and place the guitar in a low-masking area.
- **More realistic panning** – the ear uses amplitude, time and frequency differences for the localization of sounds. A pan pot works by only altering amplitude; therefore the results are less natural. The Haas trick adds to the standard panning method time differences, and with a filter on the delayed channel we can also tuck on some frequency differences. However, this application is limited to instruments already panned to one extreme – we cannot pan the original signal anywhere else.

> **Track 11.7: dGuitars Stereo**
> Two distorted guitars, each panned hard to a different extreme, before applying the Haas trick.
>
> **Track 11.8: dGuitars Haas Crossed**
> The same two guitars after applying the Haas trick in crossed arrangement as shown in Figure 11.6. The delay times are set around 28 ms.
>
> *Plugin:* Digidesign *DigiRack Mod Delay II*

One of the settings we can control with the Haas trick is the **amount of delay** applied on the ghost copy. Different delay times give slightly different effects, so it is up to the ear to decide the best effect. One thing to consider is what happens when the mix is folded to mono. To be sure, the resultant combfiltering will have its impact (often loss of high frequencies), but some delay times, mostly longer ones, sum to mono more gracefully than others, so while checking in mono we look for the least-destructive delay time.

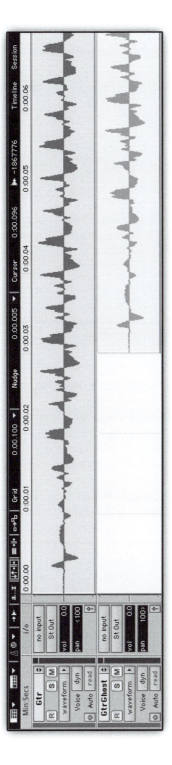

Figure 11.5 The Haas trick using duplicates. One way to achieve the Haas trick is by duplicating a mono track, nudging the duplicate a few milliseconds in time (30 ms in the screenshot above), then panning each track to an opposite extreme.

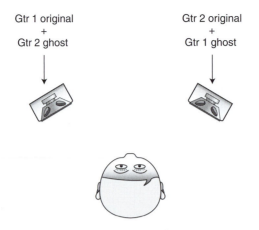

Figure 11.6 The Haas trick in crossed setup. Guitar 1 is sent to the left and its delayed copy to the right. Guitar 2 is sent to the right and its delayed copy to the left.

Track 11.9: dGtr Original
The original distortion guitar in mono.

Track 11.10: dGtr Haas 5 ms
The Haas trick applied on this track with the right channel delayed by 5 ms.

Track 11.11: dGtr Haas 30 ms
The Haas trick applied on this track with the right channel delayed by 30 ms. Compared to the previous track, the effect here appears wider and fuller.

Track 11.12: dGtr Haas 5 ms Mono
Summing track 11.10 (5 ms delay) to mono results in very evident combfiltering effect.

Track 11.13: dGtr Haas 30 ms Mono
There is still coloration of the sound when track 11.11 (30 ms delay) is summed to mono, but this coloration is neither as evident nor as obstructive as the effect with 5 ms delay.

Plugin: Digidesign *DigiRack Mod Delay II*

Another setting we can control is the **level** of the ghost copy. The Haas effect only applies if the ghost copy is less than 10 dB louder than the original signal. Depending on the delay time, it might not apply with even smaller figures than 10 dB. Assuming that both the original and the ghost copy are initially at the same level, boosting the level of the ghost copy will appear at first to balance the stereo image. With slightly more boost, a rather confusing, unnatural, yet interesting effect is achieved – it can be described as an audible delay where it is impossible to say which extreme is the delayed one.

Attenuating the ghost copy below the level of the original signal makes the effect less noticeable and the stereo image of the original signal somewhat less vague. Altogether, it is worth experimenting with different levels, and making the final decision based on taste and feel.

The following tracks demonstrate how the relative level of the ghost copy affects our perception of the Haas trick. When applied, the delay time of the ghost copy was set to 23 ms.

Track 11.14: Guitar Src
The source track used in the following samples. This is the mono recording with no delay involved.

Track 11.15: Guitar Panned
In the following tracks, the source guitar is panned hard left and the ghost copy hard right. This track, which is given for reference purposes, only involves the source track panned hard left, without the ghost copy.

Track 11.16: Guitar Haas 9 dB Down
The Haas effect is applied in this track with the ghost copy 9 dB below the original. The ghost copy on the right channel is not easily discerned. But comparing this track to the previous one would reveal the effect: In the previous track, the guitar is distinctly isolated on the left channel. In this track, the image still clearly appears to come from the left channel, but in a slightly more natural way. This is an example of how the Haas trick can yield more realistic panning and subtly richer sound. Yet, even in a sparse mix, the ghost copy on the right channel could be masked to an extent that would conceal the effect, therefore it might need to be louder.

Track 11.17: Guitar Haas 6 dB Down
In this track, the ghost copy is 6 dB below the original. It is only slightly more noticeable. The guitar image still clearly appears to come from the left channel.

Track 11.18: Guitar Haas 3 dB Down
With the ghost copy 3 dB below the original, it becomes more noticeable, and the resultant effect is richer. The guitar image still appears to come from the left channel.

Track 11.19: Guitar Haas 0
Both the original and the ghost copy are at the same level in this track. As Haas suggested, the sound image still appears to come from the left channel.

Track 11.20: Guitar Haas 3 dB Up
The ghost copy is 3 dB above the original. The effect is becoming confusing, although the image still appears to come from the left channel. It is interesting that the image should still come from the left channel if you move your head toward the right speaker.

Track 11.21: Guitar Haas 6 dB Up
With the ghost copy 6 dB above the original, the effect becoming very confusing indeed. Most listeners would still perceive the image as if coming from the left, but there is some sense of balance between the left and right speakers. One way or another, the overall effect is wide and it creates an unfocused image (which can be appealing if used in the right context).

Plugin: Digidesign *DigiRack Mod Delay II*

The Haas trick can be enhanced in two ways. First, applying a **filter** on the ghost channel – usually a low-pass filter – can bring about more natural results, and might even reduce combfiltering when the mix is folded to mono. This is especially useful when the Haas trick is used to achieve more realistic panning. Another enhancement can be used when the Haas effect is applied to fatten sounds. While using a delay on the ghost channel, we **modulate** the delay time with low rate and low depth (Figure 11.7). This gives a richer impact and bigger size to the effect. While the results are not always suitable, it is worth experimenting.

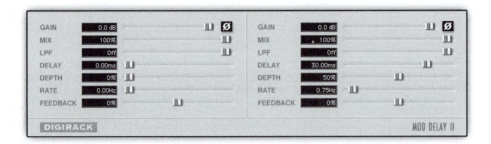

Figure 11.7 The modulated Haas trick using a delay plugin (Digidesign DigiRack Mod Delay II). This mono-to-stereo instance of the plugin involves a left channel that is neutralized with 0 ms delay time. The right channel is delayed by 30 ms and modulated using 50% depth and 0.75 Hz rate.

Track 11.22: dGtr Haas 30 ms Modulated
The same setup as in track 11.11, only that in this track the delay time is modulated using 50% depth and 0.75 Hz rate. The exact setup can be seen in Figure 11.7. Compared to track 11.11, the effect here is richer and provides a hint of movement.

Plugin: Digidesign *DigiRack Mod Delay II*

The Haas trick is often only expedient if used on one, maybe two instruments in the mix. When used on more, it can clutter the extremes and result in an overall unfocused stereo image – doing more harm than good to the mix. Like with many other mixing tricks, overdoing is inadvisable.

The out-of-speakers trick

With the Haas trick we delay the ghost copy. With the out-of-speakers trick we keep it time-aligned to the original signal, but we invert its phase (Figure 11.8). This means that the sound arriving at one ear is phase-inverted with the sound arriving at the other. In nature, this happens (for specific frequencies only) when the sound source is located to one side of the head. The two phase-inverted signals emitted from each speaker first travel to the near ear, but shortly after both will arrive to the far ear (Figure 11.9). The outcome of this is sound that seems to arrive at both sides of the head at the same time. Since different frequencies are localized slightly more forward or backward, the final effect is more of sound coming from around you, rather than simply from your left and right.

The out-of-speakers trick can make some people overwhelmingly excited when heard for the first time – not only that sound seems to arrive from outside the speakers, it also seems to surround you in a hallucinatory fashion. To add to the excitement, an instrument on which the out-of-speakers trick is applied will disappear completely when the mix is folded to mono (provided that the original and the ghost copy are at exactly the same level, like they usually are with this trick. While there is no problem setting exactly the

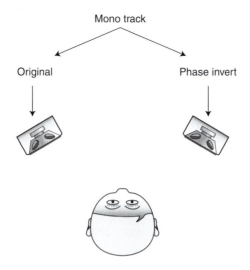

Figure 11.8 The out-of-speakers trick. A mono track is sent to both left and right, but one channel is phase-inverted.

same levels on a digital system, analog heads will have to put slightly more effort in here – usually when listening in mono).

Track 11.23: 8Bars both Mono
Both the drums and electric piano are monophonic and panned center.

Track 11.24: 8Bars ePiano OOS
The out-of-speakers trick is only applied on the electric piano. When listening in mono the piano should disappear.

Track 11.25: 8Bars Drums OOS
The out-of-speaker trick is only applied on the drums. When listening in mono the drums should disappear.

Track 11.26: 8Bars both OOS
The out-of-speaker trick is applied on both instruments. Listening to this track in mono should yield silence.

Plugin: Logic *Gain*
Drums: Toontrack *EZdrummer*

The trick, however, is only in full effect when the listener's head is on the central plane between the two speakers. Also, our ears only localize sounds coming from the sides based on low-frequency content. The wavelength of high frequencies is too short compared to the distance between our ears. High frequencies can easily change their phase a few times while traveling the distance between our ears. Scientifically speaking, 1 kHz is roughly the frequency below which side localization is experienced. Consequently, it is mostly frequencies below 1 kHz that give the impression of surrounding the head. The out-of-speakers trick is therefore mostly effective with low-frequency sounds.

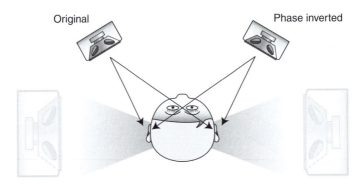

Original   Phase inverted

Figure 11.9 With the out-of-speakers trick, the original and inverted signals will reach first the ear closer to the speaker from which each signal is emitted, and shortly after, the other ear. This creates the impression that the sound is arriving from wide side angles.

The out-of-speakers trick causes low-frequency material to appear as if coming from all around us. Instruments to which the trick is applied disappear in mono.

These tracks demonstrate the varying effect of the out-of-speakers trick with relation to frequencies. An arpeggiated synth line involves a band-pass filter that is swept up, then down. The kick and hi-hats provide a center-image reference, on which the trick is not applied.

Track 11.27: Arp Mono
The arpeggiated line in mono, panned center.

Track 11.28: Arp Out of Speakers
The arpeggiated line with the out-of-speakers trick applied. It is possible to hear that the higher the band-pass filter is swept, the narrower the synth image becomes.

Track 11.29: Arp Toggle
The out-of-speakers effect is toggled every two beats. The greater effect at lower frequencies should be easier to discern here.

Plugin: Logic *Gain*

A sound that appears to surround your head in a hallucinatory fashion is clearly the definition of anti-focus. We can apply the out-of-speakers trick on an instrument, but we should not expect the instrument to relate well to the sound stage we are building. What's more, the instrument will disappear in mono, so it would not be wise to apply this trick on important instruments, say, the beat of a hip-hop track. The out-of-speakers trick is used as a special effect, a sonic gimmick, and is usually applied on the least-important instruments, or ones that appear for a very short time in the mix.

The out-of-speaker trick is used as a special effect, occasionally, and mostly on the least-important tracks.

To conclude this section, it should be noted that the out-of-speakers trick is not strictly vinyl proof. In fact, the worst-case scenario for vinyl cutting is phase-inverted low-frequency content on the left and right channels. For this reason, mastering engineers sum the low-end of the mix to mono, which cancels out, to some degree, any instruments to which this trick has been applied. If the mix is to be cut on vinyl, mixing engineers are advised to submit a vinyl edit in which this trick is spared.

12 Faders

Even before the invention of the multitrack, various microphone sources were balanced using faders and summed before being recorded. The summing devices used in early recording studios had faders all right, but these were rather bulky rotary knobs, often 3.5″ in diameter. When trying to balance more than two microphones, hands had to jump between one knob and another. One engineer to foresee the advantage of employing linear faders was the legendary Tom Dowd. By building a control section with small linear faders he could control each fader using a finger, and so balance more than two microphones at a time. Technically speaking, a fader is a device that can fade sound, whether it be rotary or linear. Today, the term 'fader' is widely associated with linear controls, while pots denote rotary control. This book follows this convention.

Faders are the most straightforward tools in the mixing arsenal. They are the main tools for coarse level adjustments, but, as a matter of fact, both equalizers and compressors are used for far more sophisticated or fine level control. Faders are the first thing we approach when starting to mix, and often part of the final fine level adjustments. It might seem unnecessary to say anything about faders – you slide them up, the level goes up, slide them down, the level drops. But even faders involve some science worth knowing. They are not as straightforward as one may think.

Types

Sliding potentiometer

The simplest analog fader is based on a sliding potentiometer. The amplitude of analog signals is represented by voltage, and resistance is used to drop it. Inside a fader there is a resistive track on which a conductive wiper slides as the knob is moved. Different positions along the track provide different resistance and thus different degrees of level attenuation. One shortfall of the sliding potentiometer is that it cannot boost the level of audio signals passing through it (however, the potentiometer can behave as if boosting by placing a fixed-gain amplifier after it). While the actual circuitry involved is more complex, we can consider this type of fader as one into which audio signal enters and leaves, as shown schematically in Figure 12.1.

Figure 12.1 A schematic illustration of a sliding potentiometer.

VCA fader

A VCA fader is a marriage between a voltage-controlled amplifier (VCA) and a fader. A VCA is an active amplifier through which audio passes. The amount of boost or attenuation applied on the signal is determined by incoming DC voltage. The fader, through which no audio flows, only controls the DC voltage sent to the amplifier. Figure 12.2 demonstrates this.

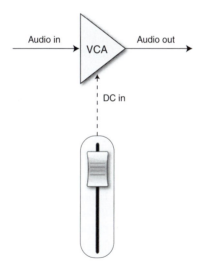

Figure 12.2 A VCA fader. The fader only controls the DC voltage sent to a VCA. The VCA is the component that boosts or attenuates the audio signal as it passes through it.

One advantage of the VCA concept is that many DC sources can be summed before feeding the VCA. A channel strip on an analog console can be designed so many level related functions are achieved using a single VCA. This shortens the signal path and reduces the amount of components in it, resulting in a better noise performance. SSL consoles, for example, are designed around this concept – DC to the VCA arrives from the VCA fader, VCA groups, cut switch and automation computer. (Figure 12.3 illustrates this.)

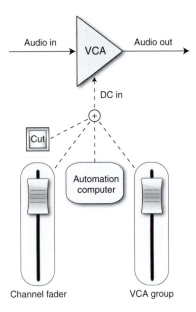

Figure 12.3 A single VCA can be used with many level-related functions.

Digital fader

A digital fader simply determines a coefficient value by which samples are multiplied. For example, doubling the sample value – a coefficient of 2 – results in approximately 6 dB boost. Halving the sample value – a coefficient of 0.5 – results in approximately 6 dB attenuation. Other coefficient values can be easily calculated based on a simple formula.

Scales

Figure 12.4 shows a typical fader scale. There are perhaps as many fader scales as manufacturers and software developers. While the actual numbers might be different, the principles discussed in this section are common to the majority of faders.

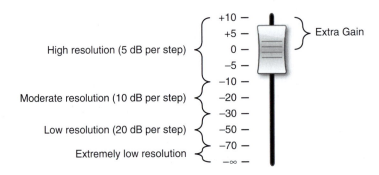

Figure 12.4 A typical fader scale.

The typical scale unit is dB, which bears a strong relationship to the way our ears perceive loudness. Most often the various steps are based on either 10 dB (which is subjectively a doubling or halving of perceived loudness) or 6 dB (which is approximately doubling or halving of voltage or sample value). The 0 dB point is also known as *unity gain* – a position at which a fader neither boosts nor attenuates the signal level. Any position above 0 dB denotes a level boost, while anything below denotes attenuation. Most faders can boost, therefore they provide an *extra-gain*. Common extra-gain figures are 6, 10 and 12 dB. We should regard this extra-gain as an emergency range – using it implies incorrect gain structure and possible degradations to the sound. Ideally, this extra-gain should not be used, with rare exceptions like when all the mix levels are set, and a specific track still calls for a bit of push above unity gain (using the line input gain is not always a sensible solution since it affects any dynamic processors on the channel). At the very bottom of the scale there is −∞ (minus infinity), a position at which the signal is inaudible.

One very important thing that becomes evident from Figure 12.4 is that although the different scale steps are evenly distributed, the dB gaps between them are not consistent. At the top of the scale (between −10 dB and +10 dB) each step is equal to 5 dB; at the bottom (−70 to −30 dB) each step is worth 20 dB. This means that sliding the fader at its lower end will cause more drastic level changes than sliding it on its higher end. Another way to look at this is that at the top end of the fader level changes are more precise – clearly something we want while mixing. This suggests that we want to have our faders in the high-resolution area of the scale, provided we do not use any extra-gain, the area between −20 and 0 dB is the most critical one. One useful function found on some audio sequencers is the ability to textually enter the required level or have more precise control over the fader position using modifier keys.

Another consequence of the uneven scale is that when we gain-ride a fader, the lower it goes the more drastic the attenuation becomes (and the opposite effect for raising). If we fade out a track, for example, we might have to slow down as we approach the bottom of the scale.

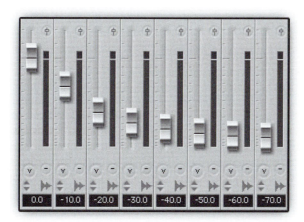

Figure 12.5 A demonstration of the uneven fader scale. These 8 Pro Tools faders are set at −10 dB intervals. It demonstrates how hard it can be to set any value within the −70 to −60 dB range, while quite easy within the −20 to −0 dB range. Modifier keys can accompany mouse movements for more precise results.

Working with faders

Level planning

A common question that probably every mixing engineer asks at least once is: so which fader goes up first and where exactly does it go? There are a few things we have to consider in order to answer this question.

Faders simply like to go up. It should not come as a surprise if throughout a mixing session one-by-one all the faders go up – a few times – ending up exactly at the same relative positions. This is the outcome of the louder-perceived-better axiom. Here is an example of how easily things can go wrong: you listen to your mix and find yourself unsure about the level of the snare. You then boost the level of the snare, it is likely to make it more defined so the move appears to be right. A few moments after, when working on the vocals, you realize that they do not stand out since the snare is nearly as loud, so you boost the vocals. The vocals should now stand up alright. Then, you find that the kick seems too weak since the snare is way louder. So you boost the kick. But now you are missing some drum ambiance so you boost the overhead, then the toms, then the cymbals, then the bass, then the guitars and before you know it you are back to square one – you are unsure about the snare again.

| *Faders like to go up.* |

There are a few ways to solve this cyclic-level syndrome. First, understand that faders are the least-sophisticated tools to make something stand up. Second, listening in mix-perspective minimizes the likelihood of individual level moves. Third, it takes some discipline to stick to the plan, especially when it comes to level boosts. After absorbing these three ideas, you might still find that faders like to go up, so the ultimate solution is this:

| *Leave some extra-gain available.* |

The extra-gain dead-end is shown in Figure 12.6. It happens when a fader is fully up, but the instrument can still use some gain. This scenario can be solved in various ways, but it would be much easier planning the levels right in the first place, so the extra-gain dead-end never happens. We said already that the fader's extra-gain is better kept for an emergency. So we normally do not want any faders above 0 dB at the early mixing stage. If we also take into account that faders like to go up, it might be wise to leave even more additional gain, so there is still some distance to go before we start using the extra-gain. For example, perhaps it would be wise to start the mix with the loudest instrument set to -6 dB (which gives us 6 dB of virtual extra-gain on top of the standard extra-gain).

Level planning requires setting the loudest instrument of the mix first, and then the rest of the faders in relation to it. An example of the opening mix moves for a production where the lead vocal is expected to be the loudest, would involve setting the lead vocal track to −6 dB, and then setting the overheads level with respect to the lead vocal. The vocal track can then be brought down, if desired. These things can be easily determined during the rough mix – if the rough mix ends with the highest fader at +4 dB, that fader

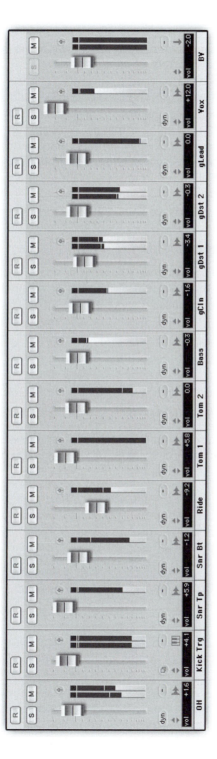

Figure 12.6 The extra-gain dead-end. The vocal track (Vox) is at the top position of +12 dB, which means that no additional level boost can be applied by the fader if needed.

should open the real mix at 0 dB (or slightly below it), then the rest of the mix is built in comparison to this level.

Set initial levels in relation to what you think will be the loudest instrument in the mix.

So that is the digital version of level planning. Analog mixing works slightly different, and we can borrow the analog wisdom for the benefit of digital. Unlike software sequencers, analog desks do not have a hard clipping threshold – levels can go above 0 dB. When planning levels on analog, engineers look at VU meters – a far better indication of loudness than the peak meters on a software mixer. One might mix a dance track in Logic, and start by setting the kick to −6 dB. Then when another instrument is introduced, it might need some extra-gain, and can even end up at the fader's dead-end. Employing a VU meter during the early stages of level planning can be beneficial in digital all the same.

The extremes-inward experiment

Sometimes level decisions are hard, and we find it difficult to ascertain how loud a specific instrument should be. The extremes-inward experiment can help in these situations. Accompanied by Figure 12.7, the process is as follows:

- Take the fader all the way down.
- Bring it up gradually until the level seems reasonable.
- Mark the fader position.
- Take the fader all the way up (or to a point where the instrument is clearly too loud).
- Bring it down gradually until the level seems reasonable.
- Mark the fader position.
- You should now have two marks that set the limits of a level window. Now set the instrument level within this window based on the importance of the instrument.

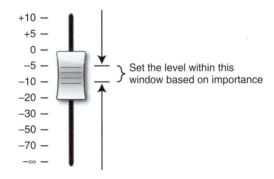

Figure 12.7 An illustration of the extremes-inward experiment.

If the result of the experiment is that the window is too wide, say more than 6 dB, it suggests that some compression or equalization might be beneficial.

13 Panning

How stereo works

On 14 December 1931, Alan Dower Blumlein, then a researcher and engineer at EMI, applied for patent number 394,325 called 'Improvements in and relating to Sound-transmission, Sound-recording and Sound-reproducing System'. This 22-page application outlined his ideas and vision to create a better sound reproduction system than the mono-phonic one used in those days. What Blumlein described as 'binaural sound' is what we refer to today as stereo. His original concept was so far ahead of its time that many people could neither understand it nor realize its potential. The first stereo record was published in 1958 – 16 years after Blumlein's mysterious death and 6 years after EMI's patent rights had expired.

The term 'binaural' denotes having or involving two ears. Our brain uses differences between the sound arriving to the left and the right ears to determine the localization of the sound source. The differences fit three criteria: amplitude, time (phase) and frequency. For example, if a sound is emitted from a trumpet placed to our right (Figure 13.1), the sound arriving to our right ear will be louder than the that arriving to our left ear. This is due to the fact that the head absorbs and reflects some of the sound energy traveling to the left ear. As sound takes time to travel (approximately one feet per millisecond), it will also reach the left ear slightly later. Finally, since high frequencies are not very good at

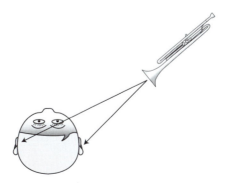

Figure 13.1 Sound produced from the right of the listener will have amplitude, time and frequency differences between the left and right ears.

diffracting, they will not bend around our head like low frequencies, and less will reach the left ear.

In order to simulate the binaural differences that occur in nature, we can use a dummy head with microphones at its ears or use a computer to do the calculations instead. (This is how 3D effects are implemented.) Either way, if we play such binaural material through headphones, we achieve a good sense of localization, with the ability to have sounds appearing as if coming from all around us, including the sides, behind and even up and down. Essentially, we fool our brain by providing two distinct sounds with the same differences that sounds in nature involve.

Track 13.1: Congas Circle
A demonstration of binaural processing. Using headphones, the congas should appear as if circling the head. Listening to this track through speakers would only make the congas shift between the right and left channels.

Plugin: Wave Arts *Panorama*
Percussion: Toontrack *EZdrummer*

Using two speakers instead of headphones does not achieve the same effect. In nature, the sound from a central source in front of the listener travels an equal path, thus reaching both ears at the same time, at the same amplitude and with the same frequency content. But a two-channel stereo setup does not have a center speaker, so in order to create a central image both speakers emit the same sound. Having no real center speaker, this central image is known as *phantom center*. If we could ensure that the sound from each speaker only arrived at the nearest ear, as in Figure 13.2a, the simulation would be perfect. But the sound from each speaker also arrives to the far ear and does so slightly later, while being quieter and having less high frequencies (Figure 13.2b). This late arrival confuses our brain and results in a slight smearing of the perceived sound image. This smearing effect can be demonstrated in every surround studio by comparing the sound emitted solely from the center speaker (real center) and that emitted at equal levels from both the left and right speakers (phantom center). How unfocused the latter might be can be quite surprising, especially in poorly tuned surround studios.

The best stereo perception is experienced when a listener sits on the central plane between a correctly configured pair of speakers. Sitting in such position is a requisite for

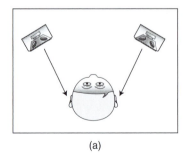
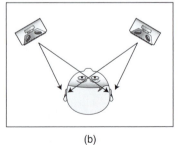

(a)   (b)

Figure 13.2 (a) A great stereo perception would be achieved if the sound from each channel reached its respective ear only. (b) In practice, sound from both speakers arrives to both ears, yielding a less accurate stereo image.

any mixing engineer, but most people listen to music in far less ideal locations. Although some of the stereo information is retained from the sweet spot, much of it is lost, especially when moving away from the central plane.

Pan controls

The pan pot

The first studio to have installed a stereo system was Abbey Road in London. Many other studios followed suit and soon enough all consoles had to accommodate this new feature. At first, consoles offered a three-state switch that could pan a mono signal either left, center or right. This type of switch is evident in many old mixes where drums are panned to one extreme, bass to the other and all sorts of other panning strategies that are today only employed as a creative effect.

While working on *Fantasia* in the late 1930s, Walt Disney's engineers wanted to create the effect of sound moving between the two sides of the screen. They derived from the previous work of Blumlein and Fletcher (the same Fletcher from the Fletcher-Munson curves, who also researched multichannel sound at Bell Labs) that as the level of one speaker continuously drops in relation to the other speaker, the image gradually shifts towards the louder speaker. Studio engineers could use this knowledge. They could feed the same signal into two channel strips, pan each to a different extreme, then attenuate one channel. Depending on the amount of attenuation, engineers could pan instruments across the whole stereo panorama (Figure 13.3). Problem was, that with

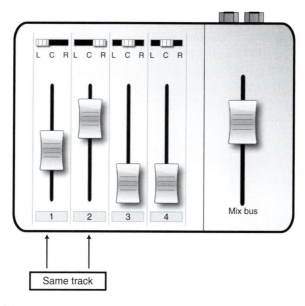

Figure 13.3 Manual panning. The same track is sent to two channels, each panned to a different mix extreme. When the signal that travels to one speaker is attenuated, the resultant image shifts to the other speaker.

the small consoles used back then, this was a huge waste of a channel. The *panoramic potentiometer* was invented, otherwise simply known as the *pan pot*.

A pan pot does something very similar to the setup in Figure 13.3. It splits a mono signal to left and right, and attenuates one of the channels by a certain amount. Effectively, this alters the relative level between the two speakers. Figure 13.4 illustrates what happens inside a pan pot. To pan something to one speaker only requires complete attenuation of the other channel. Center panning is achieved by sending the signal at equal level to both speakers.

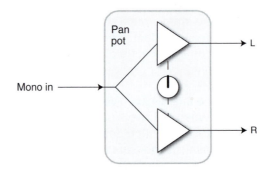

Figure 13.4 Inside a pan pot. The signal is split and each copy is sent to a gain control. The pot itself only controls the amount of attenuation applied on one channel or another.

The pan clock

Often hours are used to describe the position of a pan pot. Although not set in stone, the hours span from 7:00 to 17:00 with 7:00 being the left extreme (also hard left), 12:00 being center and 17:00 for the right extreme (hard right). Some use a 12-hour clock, representing 17:00 as 5:00, although for clarity the 24-hour notation will be used in this text. Also, some call the center dead-center or hard-center, in this text the term center signifies 12:00, and any panning position around it will be termed around the center (or slightly off-center).

Figure 13.5 The Pan Clock.

Different software sequencers use different numeric scales to represent pan positions. Pro Tools and Cubase use a 201-step scale, while Digital Performer and Logic use the MIDI-based 128-step scale. Where hours might appear in between the two speakers and their equivalent values on the numeric scales is shown in Figure 13.6.

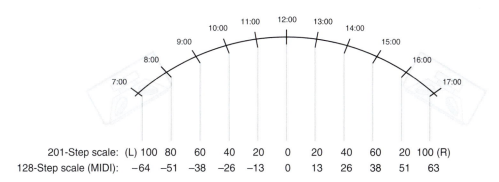

| 201-Step scale: | (L) 100 | 80 | 60 | 40 | 20 | 0 | 20 | 40 | 60 | 20 | 100 (R) |
| 128-Step scale (MIDI): | −64 | −51 | −38 | −26 | −13 | 0 | 13 | 26 | 38 | 51 | 63 |

Figure 13.6 The rough position of the hours on the stereo panorama, and the hour equivalent numeric values on the 201 and 128 scales.

Track 13.2: Pan Clock
This track involves 11 pink noise bursts, each panned to a different hour starting from hard left (7:00).

Panning laws

Not all pan pots behave the same. Depending on the pan law, the signal level might rise, remain consistent or drop as we twiddle the pot from the extremes toward the center. Just to complicate things, the panning behavior is different in stereo and mono. A console usually offers one pan law for all of its pan pots, although some inline desks have two different pan laws – one for the channel path and another for the monitor path. Some applications let the user choose the pan law used in the software mixer (Figure 13.7). In these applications the choice of pan law is done per project and not globally for the system (to keep backward compatibility). Changing the pan law halfway through a mix would alter the relative levels between instruments, so a choice must be made before mixing commences.

There are two main principles to be aware of in panning law. First, acoustics has it that if two speakers emit the same signal at the same level, a listener on the central plane will perceive a 3 dB boost of what each speaker produces. For example, if two speakers emit 80 dBSPL each, the perceived loudness will be 83 dBSPL. The second principle is concerned with how mono is achieved. When two channels are summed to mono, half the level of each is sent to both. (L = R = 0.5L + 0.5R.) For example, with a signal panned hard left, the left speaker might produce 80 dBSPL and the right speaker 0 dBSPL. Half

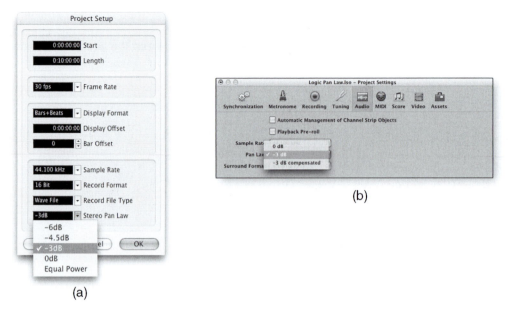

Figure 13.7 The pan law settings in Cubase (a) and Logic (b). At the time of writing, the pan law in both Pro Tools and Digital performer is fixed to −2.5 dB.

the level is approximately −6 dB; so when summed to mono each speaker will produce 74 dBSPL. Based on the first principle, the perceived loudness in this case will be 77 dBSPL. We can see from this example that if a signal panned to one extreme is played in mono, it drops by 3 dB. For centrally panned signals, mono summation makes no difference – half the level of each channel sent to both channels results in exactly the same level on both.

Thus, the level of centrally panned signals remains consistent whether played in stereo or mono. The fact that when summed to mono the center mix-image remains unaffected and the extremes drop by 3 dB, is not such a bad thing if we consider that most of the important instruments are panned center.

> *When summed to mono, the center image remains at the same level while the extremes drop by 3 dB.*

The four common pan laws −0, −3, −4.5 and −6 dB – are presented below for both stereo and mono. Before delving into each pan law, it is worth understanding, as mixing engineers, what it is we really care about. Depending on the pan law, mix levels might change as we pan. We generally do not want this to happen while we mix, as each pan move might call for subsequent fader adjustments. Since we mix in stereo, our only true concern is for levels to stay uniform as we pan across in stereo. Once the mix is bounced, the pan law becomes meaningless – the level balance printed on the master is determined and final.

How different laws behave in mono is of great importance in radio drama – signals panned from one extreme to another (for instance, a character crossing the stage) will be perceived differently on stereo and mono receivers.

Figure 13.8 illustrates how the perceived level varies for different pan positions with the **0 dB pan law**. Pan pots of this kind do not drop the level of centrally panned signals. A signal panned hard-left might cause an 80 dBSPL radiation from the left speaker. If the signal is then panned center, both speakers will produce 80 dBSPL, which results in a 3 dB increase in perceived loudness. An instrument will rise in level as we pan it from either extreme toward the center (phrased alternatively, the instrument level will drop as we pan from the center outward). While panning from left to right in stereo, an instrument will have a 3 dB center boost. In mono, which is not of much interest to us, the standard 3 dB extremes drop happens, and while panning from left to right an instrument will experience a 6 dB center boost.

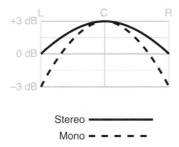

Figure 13.8 The perceived loudness of the 0 dB pan law.

The **−3 dB pan law** (Figure 13.9 compensates for the 3 dB center boost that occurs with the 0 dB law. Centrally panned signals on pots of this kind are sent to each speaker at −3 dB compared to the level of a signal panned to either extreme. An 80 dBSPL from one speaker for a signal panned hard-left will become 77 dBSPL from each speaker when the signal is panned center. This is a total perceived loudness of 80 dBSPL. When panning

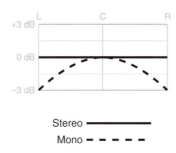

Figure 13.9 The perceived loudness of the −3 dB pan law.

from left to right in stereo, signals would remain at the same level. This is highly desirable while mixing, as pan adjustments rarely bring about subsequent fader adjustments.

With the −3 dB pan law, signals panned from left to right have a uniform level in stereo, but in mono the perceived level has a 3 dB center boost. For mono-critical applications, the −6 dB law in Figure 13.10 is used. It provides a uniform level in mono, but a 3 dB center dip in stereo. Similar to the 0 dB law, the −6 dB law is not recommended for stereo mixing, as fader adjustments might be needed after panning. Another pan law, the −4.5 dB, is a compromise between the −3 and −6 dB laws. In stereo, there is a 1.5 dB center dip, while in mono a 1.5 dB center boost (Figure 13.11). While neither in stereo nor in mono the signal level is consistent as it is panned across, the maximum error is 1.5 dB and not 3 dB like with the −6 and −3 dB laws.

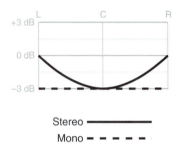

Figure 13.10 The perceived loudness of the −6 dB pan law.

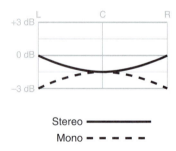

Figure 13.11 The perceived loudness of the −4.5 dB pan law.

Other variables might affect the choice of pan law. The first principle we discussed tells us that the power summation of two speakers produces a 3 dB increase on the central plane. In practice, the resultant loudness increase in less reverberant rooms might be 6 dB, but for low frequencies only. Having a low-frequency boost of 6 dB and a high-frequency boost of only 3 dB makes the −4.5 dB law a reasonable choice. Another issue is based on the assumption that most masking happens in the central area, which means that panning instruments to the sides can result in an apparent loudness increase. Another pan law, the −2.5 dB one, has a subtle +0.5 dB center boost to compensate for such a phenomenon. When a pan law choice is available, experiments can help us determine which pan law brings about uniform levels as sounds are panned across. The −3 dB law is very likely to be what we are after.

| *The −3 dB pan law is generally the best option when stereo mixing.* |

The following tracks demonstrate the different pan laws. Pink noise sweeps across the stereo panorama, and for each pan law two versions are produced (stereo and mono) and are then bounced. Perhaps the most important thing to observe is how the level varies on the stereo versions as the noise is swept between the center and the extremes.

Track 13.3: Pan Law 0 dB Stereo
Track 13.4: Pan Law 0 dB Mono
Track 13.5: Pan Law −3 dB Stereo
Track 13.6: Pan Law −3 dB Mono
Track 13.7: Pan Law −4.5 dB Stereo
Track 13.8: Pan Law −4.5 dB Mono
Track 13.9: Pan Law −6 dB Stereo
Track 13.10: Pan Law −6 dB Mono

The balance pot

As opposed to the pan pot, the input to a balance pot is stereo, as can be seen in Figure 13.12. The two input channels each pass through a separate gain stage before being routed to the stereo output. The pot position determines the amount of attenuation applied on each channel. It is important to note that a balance pot never cross-feeds the input signal from one channel to the output of another. This results in a serious mixing limitation – the image of the input stereo signal is tied to at least one extreme. This is illustrated in Figure 13.13. We can narrow the stereo width of the input signal, but we cannot set it free from ending at one extreme.

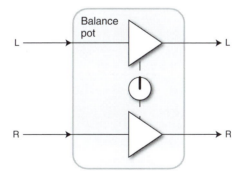

Figure 13.12 Inside a balance pot. Both the left and right inputs pass through a gain stage. But the input from one channel never blends into the output of the other.

In order to narrow the width of a stereo signal and place it freely across the stereo panorama, two pan pots are required (Figure 13.13c). Unfortunately, not all software mixers provide a dual-mono control for stereo tracks. Applications like Logic only provide a balance control, and users need to load a plugin in order to place stereo signals freely across the stereo panorama (Figure 13.14).

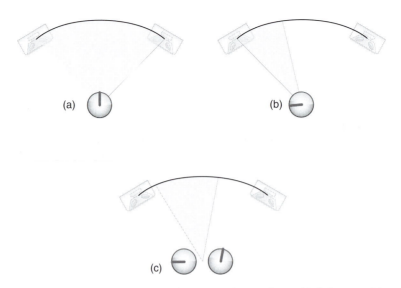

Figure 13.13 Both (a) and (b) show the resultant image of a specific balance position. As can be seen in (b), when the balance pot is turned left, the span of the resultant stereo image narrows but remains in the extreme of the stereo panorama. (c) If we need to pan anything, so the span of the stereo image does not remain in the extremes, we need two pan pots instead of a single balance pot.

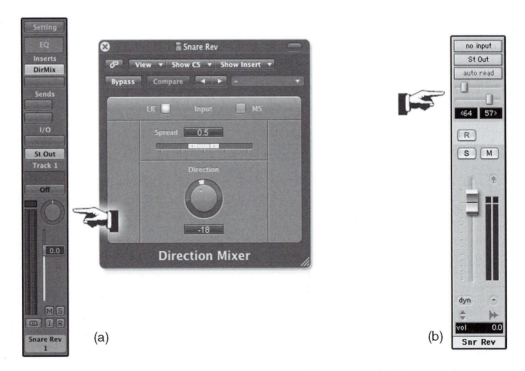

Figure 13.14 (a) Logic only offers balance control for stereo tracks. More control over stereo image can be achieved with the help of the *direction mixer* plugin. (b) Pro Tools offers dual-mono pan control for stereo tracks, which provides full functionality for any stereo imaging task.

Track 13.11: Balance Stereo Delay
This track involves a synth sent to a stereo delay.

Track 13.12: Balance Hard Left
The balance pot on the delay return is panned hard left. As a result the right channel of the delay is lost.

Track 13.13: Balance Hard Right
The balance pot on the delay return is panned hard right. As a result the left channel of the delay is lost.

Track 13.14: Balance No Can Do
Instead of using a balance pot on the delay return, a dual-pan control is employed. The left and right channels are panned to 10:00 and 14:00 respectively, narrowing the echoes towards the center. No balance control can achieve this, alas.

Plugin: PSP *84*

Types of tracks

Many instrument recordings are captured to more than one track. The raw tracks for each instrument can be classified like so:

- Mono
- Stereo:
  - Acoustic:
    - Coincident pair (XY)
    - Spaced pair (AB)
    - Near-coincident pair
  - Electronic
- Multiple mono

Mono tracks

A single instrument, a single take, a single track. A mono track can be panned anywhere across the stereo panorama. The main problem with a dry mono track is that it provides no spatial cue. Only in an anechoic chamber will signals arrive to our ears without some room response – without some space being noted. When a dry mono track is placed untreated, it can appear foreign to the depth and ambiance aspects of the mix, and can appear frontal. A reverb emulator, or other spatial effect, is commonly employed in order to blend mono tracks into the spatial field of the mix. Even a tiny amount of ambiance reverb can do the job. A real problem can arise when a mono track includes some room or artificial reverb. The monophonic reverb is unlikely to mix well with the stereo reverb of the mix ambiance. Sending the track, with its embedded reverb, to the ambiance reverb might not be wise – a reverb of a reverb seldom makes any spatial sense.

Mono tracks often benefit from the addition of some spatial effect,
which blends them into the ambiance of the mix.

Track 13.15: Drum Loop
The loop under discussion in the following two tracks.

Track 13.16: Drum Loop Dry
When the drum loop is mixed dry, it can appear foreign to the mix – it is focused and positioned at the very front of the sound stage.

Track 13.17: Drum Loop with Reverb
By adding some reverb, the drum loop blends into the mix better, and it is not as distinguishable as in the previous track.

Plugin: Audio Ease *Altiverb*

Stereo pairs

A stereo pair track denotes a pair of mono tracks, each represents one of two microphones arranged in an established stereo miking technique. The two microphones are positioned in a way that will cause amplitude and time differences between the signals captured by each microphone. When the microphones are panned to the opposite extremes, the differences between them result in spatial reproduction. Stereo miking techniques are divided into three families: coincident pair, spaced pair and near-coincident pair. While mostly a recording topic, each technique has an effect on the way we mix the given tracks.

The **coincident pair** technique, also known as XY, is based on two microphones angled apart with their diaphragms nearly touching. All but sounds arriving from the center will present different amplitude on each microphone. Due to the close proximity of the diaphragms, phase differences can be disregarded. Due to the phase similarity between the two microphones, this family is known to provide the **best mono-compatibility**. The two tracks can be summed internally without causing noticeable combfiltering, and thus, the two tracks can be panned inward from the extremes. This is often done in order to narrow the image width of an instrument, for example, drum overheads.

> *Coincident pair gives the best mono-compatibility, which allows us to narrow the image of the stereo source during mixdown.*

The coincident pair family includes stereo miking techniques like Blumlein, MS (both invented by Alan Blumlein) and Power-thrust (hyper-cardioids at 110°). Of all these techniques, MS (mid-side) is most worthy of discussion, as it requires a special decoding matrix. The technique is based on a directional microphone facing the sound source (M), and a bi-directional polar pattern picking up the sides (S). Only two tracks are recorded, but on a desk these are decoded onto three channels. The M is routed to one channel panned center. The S is routed to two, oppositely panned channels, one of which is phase-inverted (Figure 13.15). The relative level of S determines the amount of reverberation and the stereo spread of the source in the mix.

More on MS in Chapter 26.

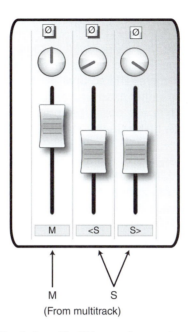

Figure 13.15 The MS technique. The M is routed to one channel panned center. The S is routed to two oppositely panned channels, one of which is phase-inverted.

The **spaced pair** technique is also known as AB (often incorrectly titled XY). The arrangement of a spaced pair involves two microphones spaced a few feet apart; due to this distance, phase differences occur. The microphones are usually omnidirectional so amplitude differences can be disregarded. Due to the phase differences between the two microphones, this technique is characterized by increased spaciousness and vague imaging. However, due to the same phase differences this technique is **not mono compatible**. Any inward panning of the two microphones might result in noticeable combfiltering, nearly forcing the two microphone tracks to be panned to the opposite extremes.

The coincident pair family is based on amplitude differences, while the spaced pair family on phase differences. The **near-coincident** family is a marriage of both – two microphones are angled and spaced, but only a few inches apart. Phase differences of a near-coincident recording do not present the same mono-compatibility as a coincident pair, but not as profound phase differences as a spaced pair. Panning the two microphones to anywhere but the extremes can result in some degree of combfiltering, which may or may not be noticeable.

In addition to the acoustic stereo pair, we can also get a stereo recording of a synthesizer or other electronic instrument. Whether or not the image of these can be narrowed can only be determined through experiment. But it is worth knowing that despite having a stereo output, some of these devices are monophonic inside. The stereo size these devices produce might be achieved by having the pure monophonic output on the left channel and a processed version of it on the right channel. This is known as *pseudo-stereo*. When the stereo effect fails to blend into the mix, we can always desert the processed channel and apply our own effect on the pure monophonic signal.

Multiple mono tracks

Multiple mono tracks are those that represent the same take of the same instrument, but do not involve an established stereo miking technique. Often each track captures a different timbre component of the sound source, so during mixdown we can shape the instrument sound using faders rather than processors. Some of these complementary tracks would be panned to different positions, most likely to the two extremes. An example would be two acoustic guitar tracks – sound-hole and neck.

Often still, these complementary tracks are panned to the same position, a practice that requires them to be in phase with one another. Examples include, snare-top and snare-bottom, bass-mic and bass-direct, or three microphones used to capture the full timbre of a double bass. However, identical panning positions are not obligatory. When appropriate, the tracks can be panned to different positions, which will widen and blur the stereo image of the instrument. For instance, we can achieve such a blurry-wide snare image by panning the snare-top and snare-bottom differently. The effect can be made subtler if one of the tracks is attenuated. This effect can also be achieved artificially by panning a duplicate of a mono track to a different position, while generously altering its tonal character.

Track 13.18: Snare Top Bot Same
In this track, the snare top and bottom microphones are panned to the same position. The snare image is not perfectly focused due to the overheads and room-mic, but its stereo position can easily be pointed out.

Track 13.19: Snare Top Bot Mirrored
The snare-top is panned to the same position as in the previous track, but the snare-bottom is panned to the opposite side. This results in a less focused snare image, where its exact position is harder to localize.

Drums: Toontrack *EZdrummer*

Just as not all the songs recorded for an album are actually used, not all multiple mono tracks must be used. Sometimes using both bass-mic and bass-direct call for treating each individually, and then both collectively – this might involve two equalizers, three compressors and a control group. Using one track only might not only be easier, but can also fall perfectly into the requirements of the mix. In the case of guitar-hole and guitar-neck tracks, sometimes excluding one track and adding reverb to the other would work better than mixing both tracks.

Combinations

With some recordings, like drums or orchestral, we get a blend of different types of track, for instance, stereo overheads, multiple mono kick and mono toms. In such cases, the stereo pair provides a reference image where each of the instruments can be localized. The respective mono tracks are often panned to the same position. For example, if the snare appears to come from 11:00 on the overheads, the snare track might be panned to that position. Again, this is not obligatory – different panning will result in wider, less focused images, which is sometimes sought after (Figure 13.16).

Figure 13.16 The stereo image of a snare, depending on how its mono track is panned compared to the overheads image. (a) The snare is panned to the same position as it appears in the overheads stereo image, resulting in an overall focused image for the snare. (b) A snare panned far away from its overhead position will have a wider, unfocused overall stereo image, which is sometimes sought after.

Panning techniques

An initial panning plan can be drawn just by looking at the track sheet. The nature of each track will usually hint at its rough panning position in the mix. The following is an example of an initial panning plan for a simple rock production:

- **Overheads** – 70% wide around the center. Audience view.
- **Kick** – center.
- **Snare** – to the same location as it appears on the overheads.
- **Snare reverb** – 40% wide, around the snare position.
- **Tom 1** – 14:00.
- **Tom 2** – 13:00.
- **Tom 3** – 10:00.
- **Bass** – center.
- **Vocals** – center or maybe slightly off-center to the left.
- **Power guitar I** – hard-left (+Haas trick).
- **Power guitar II** – hard-right (+Haas trick).
- **Flutes** – halfway toward the extremes.
- **Ambiance reverb** – 100% stereo width.

No plan is successful until committed. Clearly, the above will not be suitable for all productions that use such an arrangement, but it provides a possible starting point. A calculated approach, however, can backfire – different panning schemes can project very different feel, size and power. In complex arrangements, experiments are a necessity, and part of our role is to explore, discover and study the different panning possibilities.

Panning schemes can have a profound effect on the mix.

As the mix progresses and elements are added, we revisit the pans. Sometimes even small alterations result in big improvements – just shut your eyes and inch it until it latches into place. As already mentioned, panning alterations are often part of the very last refinements of a mix.

Panning and masking

We already know that internal summing is less forgiving compared to acoustic summing. When two instruments are panned center, their internal summing accents any masking interaction between them. However, if each is panned to a different extreme, they are summed acoustically, which makes masking interaction less harmful. As a result, panning instruments toward the extremes also increases their definition. On the very same basis, masking is more dominant when two fighting instruments are panned to the same side of the panorama; mirroring one of them (e.g., from 15:00 to 9:00) can improve the definition of both.

The sound stage

Essentially, the stereo panorama reflects the horizontal plane of the imaginary sound stage. One of the main roles of pan pots is to determine where instruments are positioned on the stage. Just like on a real stage, listeners expect to find the most important instrument in the center of the mix – be it the vocalist or the trumpet player. Also, if we imagine a five-piece rock band on a big festival stage, it is unlikely that the two guitarists will stand on the stage edges; more likely that each will be somewhere halfway to each side. Many more examples can be given, but contemporary music is not always mixed to the full realistic extent of a live performance – if the guitars sound better when panned each to a different extreme, one would be a fool not to pan them that way. The visual sound stage is a guide often more suitable for natural mixes.

Low-frequency instruments, notably the bass and kick, are usually panned center. One reason is that low-frequency reproduction demands much more power than the reproduction of high frequencies. Panning low frequencies off-center will cause uneven power consumption between the two channels, which can have many downgrading consequences on the combined stereo response. The way vinyls are cut is another reason. What's more, panning low-frequency instruments to the sides is somewhat pointless since low frequencies give very little directional cue (hence subwoofers can be positioned off the central plane). Yet, one can argue that both the kick and the bass have important content in their mids as well.

Having both the most important, and low-frequency instruments panned center, makes it inevitably the busiest area of most mixes, and where most masking happens. No panning police will arrest you for not panning the most important instruments hard center. Lead vocals, for example, can be panned slightly off-center; sometimes, even the bass and kick are panned that way. Most listeners do not listen on the central plane anyway, and even if they do, they are unlikely to recognize these subtle panning shifts.

The center is usually the busiest area in the mix. Instruments that would be panned center can be panned slightly off-center.

Track 13.20: Vocal Center
The vocal on this track is panned center (12:00).

Track 13.21: Vocal 1230
The slightly off-center (12:30) panning of the vocal in this track can easily go unnoticed, especially if the track is not compared to the previous one.

Track 13.22: Vocal 1300
The appreciation of the vocal panning in this track is dependent on the width of the monitoring setup. While mixing engineers with wide stereo setup could find this off-center panning disturbing, normal listeners could easily fail to notice the off-center vocals.

Track 13.23: Vocal 1400
With the vocal panned 14:00, it is hard to overlook the off-center image of the vocal. Both engineers using a narrow stereo setup and normal listeners could find this panning scheme disturbing.

Another important area of the mix is the extremes. We have said already that over-panning to the extremes can result in mixes that have busy extremes and center, but nothing in between. These W-mixes are the outcome of the tendency to pan every stereo signal and every pair of mono instruments to the extremes. Whilst it could be beneficial to pan overheads, stereo effects or a pair of mono guitars that way, most mixes will benefit from a **stereo spread balance**, where mix elements are panned to fill the sound stage.

Not all stereo elements and mono pairs should be panned to the extremes.

It is a common practice in surround film-mixing that the voice of an actor standing behind the camera is still panned to the front speakers. Had it come from the rear, viewers would notice the existence of the rear speakers, which might distract them from the visual experience. As a general rule, cinema sound should accompany the visual but never draw too much attention on its own – you should feel as part of the scene, but should not notice how this is done technically. The same principal applies with music mixing, the listener's realization that two distinct speakers are involved in the production of a sonic experience can distract from the spatial illusion. Highly creative panning techniques should be used sparingly and only when appropriate.

Another fact worth remembering is that the center of the mix is the least-focused area (due to the late arrival of sound from both speakers to the respective far ear), while the extremes provide the most-focused area (as only one speaker generates the sound in true mono). Consequently, as instruments are panned toward the extremes they become more focused. Combined with the curved nature of the mix front line, these instruments can also appear closer. This implies that the more an instrument is panned toward the extremes, the more it might benefit from the addition of a spatial effect – especially in the case of mono tracks.

The extremes = more focus, more definition, closer image.

When comparing the following tracks, the drum loop should appear closer, more distinguished and focused when panned to the right extreme:

Track 13.24: Drum Loop Panned Center
Track 13.25: Drum Loop Panned Hard Right

As per our discussion in Chapter 6, one of the main mixing objectives when it comes to the stereo domain is balance. Both **level and frequency** balance are our main concern. The off-center panning of individual instruments is guided by this objective, and rough symmetry is often sought. One guitar to this side, another guitar to the other; a shaker to this side, a tambourine to the other; first vocal harmony at 11:00, the second at 13:00. A stereo balance relies on both instrumentation and recordings, 'what instrument can we record that complements a shaker?' or 'shall we double-track the shaker?' are questions that should be asked before the mixdown stage. When an off-center track creates imbalance that no other track can repair, mixing engineers often utilize stereo effects. For example, by sending a delayed echo to the opposite of a dry instrument (while perhaps altering its tonality).

We can also associate stereo balance with **performance**. Say sixteenth notes are played on the hi-hats, and there are two guitars – one plays a sixteenth-note arpeggio, the other half notes. Panning the hi-hats and the arpeggio guitar to opposite sides will create rhythmical balance. But panning the hi-hats and the arpeggio guitar to the same side, and the half-notes guitar to the other will result in one extreme being rhythmically faster than the other. Subsequently, this will also affect the stereo balance between the two sides of the stereo panorama.

While balance is a general stereo objective, a perfectly stereo-balanced mix is essentially a mono mix. There is always a margin for some degree of disparity. If we take a drum kit for example, with its hi-hats and floor tom on opposite sides, imbalance is inevitable. But if a tambourine also forms part of the arrangement, it would make more sense to pan it opposite the hi-hats. As per our axiom, that percussives weigh less, sustained performance is more of a factor when it comes to stereo balance.

Effects panning

Stereo effects can be panned in various ways. The panning choice is mostly based on two aspects of the stereo panorama: width and center position. An ambiance reverb is the one type of a stereo effect that can benefit from full-width panning, especially when natural results and a wide sound stage are sought. But reverbs and delay lines used for creative purposes can benefit from a narrower image and off-center position. A classic example is the snare reverb, which is often different from the reverbs used for ambiance (or from the ambiance on the recording). When panned to the extremes, the wide snare reverb might compete with the ambiance and interfere with the spatial effect the latter provides. Combined, it might sound like the mix took place in a hall, while the snare was in a bathroom. Narrowing the reverb width will give a clear cue that it is not part of the general ambiance, but only an effect linked to the snare. This also promotes separation.

Matching the center of the snare and its reverb will emphasize this relationship. A mix with a snare panned to 11:00 might benefit from the snare reverb being panned around it, say, 10:00 to 12:00. Narrowing the width of a stereo effect can also be beneficial with stereo delays. The combination of centrally panned vocals with hard panned echoes makes the echoes noticeable, spatially random and distant from their parent. To make the effect more subtle and focused we can pan the echoes inward toward the location of the vocals.

Tracks 21.38–21.40 from Chapter 21 demonstrate various stereo delay-panning techniques. Tracks 23.86–23.93 from Chapter 23 demonstrate various reverb-panning techniques.

Mono effects can benefit from the opposite way of thinking. The closer the effect is to its parent instrument the less noticeable it will be, and we miss an opportunity to balance the stereo panorama. An example can be a mono delay applied on hi-hats. If the delay is panned to the same position as the hi-hats, the two are likely to collide. Panning the delay to the opposite side will increase the definition of both, while enriching the stereo panorama. One of the world's most famous delay lines is said to have been created by accident when Giorgio Moroder panned a bass line to one speaker and its delay to the other. This effect that originally appeared on Donna Summer's *I Feel Love*, has been popular ever since, and can be heard on countless dance mixes.

Track 13.26: Moroder Delay
As with *I Feel Love,* the bass line (8th-notes sequence) and the hats are panned hard left. The 16th-note delays of both the bass line and hats are panned hard right.

Track 13.27: Moroder No Delay
Same as the previous track but without the delay.

Track 13.28: Moroder Delay Only
In this track, the source bass line and the hats are muted, but not their delay. The delay is evidently a 16th-note behind the main beat.

Track 13.29: Moroder All Center
This is the result of panning all tracks, including the delay, to the same position. The effect is distinctively different than that on track 13.26.

Plugin: PSP *84*

Panning for Volvos

There is a bone of contention when it comes to panning that is worth mentioning just because it's unusual. People in the United States argue that the left side of the mix is more important than the right side, while the Brits claim the opposite. Reason being is that each nation has its driver's wheel on a different side. If this issue is to be taken into account then... yeah, one side of the mix might be more important than the other, and US releases should be mirrored compared to UK ones. Then comes the question: Isn't this a mastering issue?

Beyond pan pots

Autopanners

An autopanner pans audio back and forth between the left and right sides of the stereo panorama in a cyclic way. The operation of an autopanner might involve these parameters:

- **Rate** – when defined in Hz, determines the amount of cycles per second; each cycle consisting of the audio being panned from the center to one extreme, to the other, and back to the center. Alternatively, the duration of a cycle might be defined in a tempo-related unit, like a bar or half note. With slow rate (roughly below 1 Hz) the left and right movement is clearly recognizable. This recognition is lost with higher rates, where the effect resembles a Leslie cabinet.
- **Depth** – defines in percentage how far toward the extreme the signal will be panned. With 100% the varying panning position will reach the extremes, 50% halfway to the extremes and so forth. The higher the depth setting, the more apparent the effect.
- **Waveform** – defines the shape of the pan modulation. A sine wave will vary the position between left and right in a sinusoidal way. A square wave will make the audio jump between the two sides.
- **Center** – defines the center position of the modulation. With the center set halfway to the left extreme, it is possible for the audio to only drift back and forth between the center and the left extreme.

Track 13.30: Beat Source
The source track used for the following tracks.

Track 13.31: Beat Autopan Tri Depth 100
Triangle waveform modulation, 100% depth, rate is tempo-synced to one bar (center, as with all other samples here, is set to 50%, which is 12:00).

Track 13.32: Beat Autopan Sine Depth 100
Sine waveform, 100% depth, rate is tempo-synced to one bar.

Track 13.33: Beat Autopan Sine Depth 50
Sine waveform, 50% depth, rate is tempo-synced to one bar.

Track 13.34: Beat Autopan Sine 5 Hz
Sine waveform, 100% depth, rate is set to 5 Hz. The cyclic left and right movement becomes vague due to the fast rate.

Track 13.35: Beat Autopan Sine 10 Hz
Sine waveform, 100% depth, rate is set to 10 Hz. The fast rate creates a unique effect, but one that doesn't resemble autopanning. It is interesting to note the kick, which appears at different positions every time it hits.

Plugin: MOTU *Autopan*

The traditional autopanner effect – a slow, obvious left and right movement – has lost popularity over the years. Perhaps due to being gimmicky and not playing any vital mixing role. Today the tendency is to use the effect subtly, or occasionally in specific sections of the song (mostly for interest's sake). An interesting motion effect can be achieved when more than one track is autopanned simultaneously, and all the tracks complement one another to produce a relatively balanced stereo image. For example, having three guitars moving across the stereo panorama, with each starting at a different position and perhaps autopanned with a different rate.

Autopanners can be connected to other processors in order to create cyclic morphing between two effects. Instead of routing the autopanner outputs to the mix, each can be routed to a different bus, which feeds a different processor. For example, the left output might feed a reverb, while the right output feeds a delay. This will make one effect come when the other goes and vice versa. We can also use only one output of the autopanner

Figure 13.17 The MOTU *AutoPan plugin.*

for an effect that we simply want to come and go (however, using a slow tremolo might be easier to setup for this purpose).

Further reading

Alexander, Robert Charles (1999). *The Inventor of Stereo: The Life and Works of Alan Dower Blumlein*. Focal Press.

14 Equalizers

In the early days of telephony, the guys at Bell Labs faced a problem: high frequencies diminished over long cable runs, a fact that made the voice during long-distance calls dull and hard to understand. They set to designing an electronic circuit that would boost the high frequencies on the receiving end, making the sound on both ends of the line equal. The name given to this circuit was: equalizer.

The equalizers used in mixing today are not employed to make one sound equal to another, but to manipulate the frequency content of various mix elements. The frequency virtue of each individual instrument, and how it appears in the overall frequency spectrum of the mix, is paramount when mixing. Operating an equalizer is easy, but understanding frequencies and how to manipulate them is perhaps the greatest challenge mixing has to offer. This is worth repeating:

> *Understanding frequencies and how to manipulate them is perhaps the greatest challenge mixing has to offer.*

Equalization and filtering is one area where digital designs overpower their analog counterparts. Many unwanted artifacts in analog design are easily rectified in the digital domain. Digital equalizers provide more variable controls and greater flexibility. For example, it is very common to have variable slopes on a digital equalizer – a rare feature of analog designs. Digital filters can also have steeper or narrower responses. Theoretically, such advantages should not exist. However, cost and noise issues make digital-like analog designs impractical to build. This, by no means, implies that digital equalizers sound better than analog ones. But equalization is one area in mixing that has experienced a radical upgrade with the introduction of digital equalizers, and more specifically software plugins.

Applications

Although equalizers are not the only tools that alter frequencies, they are the most common. In simple terms, equalizers change the tonality of signals. This simple ability is the concept behind many, many applications:

Balanced frequency spectrum

This is a very important aspect of a mix. It was suggested in Chapter 6 that despite a vague definition of what tonal balance is, a deformed frequency response is unlikely to go unnoticed. Having overemphasized or lacking frequency ranges, whether wide or narrow, is one of the most critical faults equalizers are employed to correct. They also help us to narrow or widen the frequency size of instruments or shift them lower or higher on the frequency spectrum.

Shaping the presentation of instruments

Equalizers give us comprehensive control over the tonal presentation of each instrument. We can make sounds thin or fat, big or small, clean or dirty, elegant or rude, sharp or rounded and more. Perhaps the king of tonal presentations is the kick – there are just so many ways we can equalize it, and so many options to choose from. Whether the kick or any other instrument, tonal presentation is one of the more creative aspects of mixing.

Separation

Rarely do the frequencies of various mix elements not overlap. A sub-bass (low-frequency sine wave) being a rare exception. Most other instruments will end up in an aggressive masking war that can, in some mixes at least, get quite bloody. When two or more instruments are fighting for the same frequency range, we can find it hard to discern one instrument from the other. As long as instruments are mixed together they inevitably mask one another; we can ease the conflict by cutting from instruments any dispensable frequencies, and sometimes even the less-essential frequencies. Had we limited each instrument to a unique frequency range, there would be no mixing engineers at all – mixes would sound horrific. We only combat masking until we can separate one instrument from another, and all instruments are defined to our satisfaction. Depending on the nature and density of the arrangement, this may or may not involve serious equalization.

Definition

Definition is a subset of separation. No separation, low definition – simple. But we also associate definition with how recognizable instruments are (provided we want them recognizable) or how natural they sound (provided we want them to sound natural). For example, in a vocal and piano arrangement there might be great separation between the two; but if the piano sounds as if it is coming from the bottom of the ocean, it is poorly defined. On the same basis, we might say that the hi-hats are not well defined if they are missing essential highs.

Convey feelings and mood

Our brain associates different frequencies with different emotions. Bright sounds tend to convey a livelier, happy message, while dark sounds might be associated with mystery or sadness. As our voice deepens with adulthood, some believe that low-frequency attenuation of the human voice gives a more youthful impression. Equalizers can be used to make vocals sweeter, a snare more aggressive, a trumpet mellower, a viola softer, etc.

Portishead. *Dummy.* Go! Discs, 1994.

This album involves one of the most distinct equalization works, tailored to create an imaginary world infused with bittersweet moods, mystery and mellowness. A prime example of characteristic and creative equalization for emotional purposes.

Creative use

Equalizers are not employed solely for practical reasons. Creative use entails less natural or 'correct' equalization that can put instruments in the limelight, give a retro-feel or create many fascinating effects. This Portishead album demonstrates the use of creative equalizing wonderfully.

Interest

Automating EQs is one way to tuck some interest into the mix. A momentary telephone effect on vocals, the relaxation of lows during a break, the brightening of the snare during the verse are just a few examples.

Depth enhancements

Low frequencies bend around obstacles and are not readily absorbed. High frequencies are exactly the opposite. These are some of the reasons why we hear more low-frequency leakage coming from venues. Our brain decodes dull sounds as if coming from further away, a reason why our perception of depth is stretched when underwater. We use this darker-equals-further phenomenon to enhance, if not to perfect, the front/back impression of various instruments.

Track 14.1: EQ and Depth
This track is produced with a LPF swept down and then up. Note how the less high frequencies there are the further back the drums appear (this is also the outcome of the level reduction caused by the filter).

Plugin: Digidesign *DigiRack EQ 3*
Drums: Toontrack *EZdrummer*

More realistic effects

Based on the same idea that low frequencies are not readily absorbed, reverb tails contain less high frequencies over time. The frequency content of a reverb also hints at the nature of the room. Basic equalizers are often built into reverb emulators, yet we can apply them externally. We will see later that filters can reduce the size impression of a reverb.

Stereo enhancements

Frequency differences between the sound arriving to our left and right ears are used by our brain for imaging and localization. Having similar content on the left and right channels, but equalizing each differently would widen the perceived image, often creating the impression of a fuller and bigger sound. It was already mentioned in Chapter 13 that pan pots only utilize level differences between left and right. Stereo equalization can reinforce the pan pot function to enhance left/right localization (Figure 14.1).

Figure 14.1 Stereo equalization (the McDSP FilterBank P6). A rather unique feature of the FilterBank range is the dual-mono mode (engaged on the left) for stereo inputs. Note that each control has a pair of sliders, each for a different channel. Only band 3 is enabled, and the different settings for each channel are evident in the two curves on the frequency-response display. Such a dual-mono feature enables stereo widening, and more realistic localization.

Fine level adjustments

Faders boost or attenuate the full frequency spectrum of signals. Equalizers let us boost only parts of the spectrum. Many instruments can be made louder by a boost of around 3 kHz – our ears' most sensitive area. The advantage of doing this is that other parts of the spectrum remain intact, thus on these parts masking interaction with other instruments is not affected. In addition, altering the level of some instruments (e.g., bass) can have an unmistakable effect on the overall frequency balance. Only boosting a specific range can be far less explicit, yet effective. Needless to say, it works for attenuation all the same, but we have to remember that as opposed to faders, EQs can easily downgrade the quality of the treated sounds, so using EQs to adjust levels should be reserved for fine adjustments and done in the context of masking interaction.

Better control over dynamic processors

Dynamic range processors are more responsive to low frequencies than high ones. The reasons for this are explained later in Chapter 16 when we deal with compressors. Provided all notes on a bass guitar are played at an equally perceived level, the lower the note the harder it will be compressed. A kick bleed on a snare track can cause false triggering on a gate. These are just two examples of many problems that an equalizer can rectify when hooked to a dynamic range processor – a combination that also facilitates some advanced mixing techniques. This is expounded upon in the following chapters.

Remove unwanted content

Rumble, hiss, ground-loops, mains hum, air conditioning, the buzz of a light dimmer and many other extraneous noises can be imprinted on a recording. Spill is also a type of unwanted content, and a gate cannot always remove it. For example, gating is rarely an option when we want to remove the snare bleed from underneath the ride. Equalizers let us filter these types of unwanted sounds.

Compensate for a bad recording

The recording industry would probably have survived had a mysterious virus struck all equalizers and knocked them out for a year or so – we would simply have to spend more time miking and use better instruments in better studios. In the meantime, not all recordings are immaculate, so we are happy to have our equalizers unaffected by mysterious viruses.

The frequency spectrum

Humans are capable of hearing frequencies between approximately 20 Hz and 20 kHz (20 000 Hz). Not all people are capable of hearing frequencies up to 20 kHz and our ability to hear high frequencies decreases with age. There are more than a few audio engineers who cannot hear frequencies above 16 kHz, although this does not seem to affect the quality of their mixes.

One very important aspect of our frequency perception is that it is not linear. That is, 100-200 Hz is not equally perceived as 200-300 Hz. Our ears identify with the most fundamental pitch relationship – the octave. We recognize an octave change whenever a frequency is doubled or halved. For instance, 110 Hz is an octave below 220 Hz; 220 Hz is an octave below 440 Hz. While the frequency bandwidth between 110 and 220 Hz is an octave, an identical frequency bandwidth between 10 110 and 10 220 Hz is not even a semitone. Mixing-wise, attenuating 6 dB between 10 110 and 10 220 Hz has little effect compared to attenuating 6 dB between 110 and 220 Hz.

Our audible frequency range covers nearly ten octaves, having the first octave between 20 and 40 Hz and the last octave between 10 000 and 20 000 Hz. Octave-division, however, is commonly done by continuously halving 16 kHz. Many audio engineers develop the skill to identify the ten different octave ranges, which most of us are able to do with a bit of practice. Keen engineers practice until they can identify 1/3 octave ranges. Although ear-training is beneficial in any area of mixing, it is almost a must when it comes to frequencies. A few gifted people have perfect pitch – the ability to identify (without any reference) a semitone, and even finer frequency divisions.

One of the most routine debates in audio engineering is whether or not people are capable of hearing frequencies above 20 kHz. Science defines hearing as our ability to discern sound through our auditory system, i.e., through our ear canal, eardrum and an organ called the cochlea, which converts waves into neural discharges that our brain decodes. To date, there has been no scientific experiment that proved that we hear frequencies above 20 kHz. There have been many that have proved that we cannot. There are studies showing that transmitting high-level frequencies above 20 kHz via bone conduction (metal rods attached to the skull) can improve sound perception among the hearing impaired. But at the same time it has been proven that the high-frequency content itself is not discerned. Sonar welding systems utilizing frequencies above 20 kHz (like those used in eye surgery) can pierce a hole in the body. But this is not considered hearing. In the physiological sense of hearing, whatever testing methods researchers used, no subject could perceive frequencies above 20 kHz.

It is well understood, though, that frequencies above 20 kHz can contribute to content below 20 kHz and thus have an effect on what we hear. This can be explained by two phenomena called inter-modulation and beating; both can produce frequencies within the audible range as material above 20 kHz passes through a system. Our ears are no such system, but analog and digital systems are. To give an example, inter-modulation happens in any nonlinear device – essentially, any analog component. There is always at least one nonlinear stage in our signal chain – the speakers. Also, if any harmonics are produced within a digital system (due to clipping for example), harmonics above the Nyquist frequency (half the sampling rate) are mirrored. For this reason, digital clipping at sample rates of 44.1 kHz can sound harsher than at 88.2 kHz.

Regarding the low limit of our audible range, frequencies below 20 Hz are felt via organ resonance in our body. But these are not discerned via our auditory system (they would be equally felt by a deaf person).

To conclude: No research has shown that humans are capable of hearing frequencies below 20 Hz or above 20 kHz, but it is well understood that such frequencies can have an effect on what we hear or feel.

While engineers talk in frequencies, musicians talk in notes. Apart from the fact that we often need to communicate with musicians, it is beneficial to know the relationship between frequencies and notes. For example, it is useful to know that the lowest E on a bass guitar is 41 Hz, while the lowest E on a standard guitar is an octave above – 82 Hz. We sometimes set high-pass filters (HPFs) just below these frequencies. Also, frequencies of notes related to the key of the song might work better than other frequencies. We can also deduce the resonant frequency of an instrument by working out the note that excites this resonance. Perhaps the most important frequencies to remember are 262 Hz (middle C), 440 Hz (A above middle C) and 41 Hz (lowest E on a standard bass guitar). Every doubling or halving of these frequencies will deliver the same notes on a different octave. While these frequencies are all rounded, a full table showing the relationship between frequencies and notes can be found in Appendix A.

Spectral content

A sine wave is a simple waveform that involves a single frequency. Waveforms like a square-wave or a sawtooth generate a set of frequencies – the lowest frequency is called the *fundamental*, all other frequencies are called *harmonics*. The fundamental and harmonics have a mathematically defined relationship in both level and frequencies. Figure 14.2 shows the spectral content of a 100 Hz sawtooth.

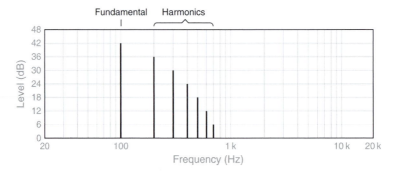

Figure 14.2 The harmonic content of a 100 Hz sawtooth.

Waveforms like sines, square waves and sawtooths constitute the core building blocks of synthesizers. However, the natural sounds we record have far more complex spectral content, which involves more than just a fundamental and its harmonics. The spectral content of all instruments consists of four components (or partials). Combined together, these four partials constitute a half of an instrument's timbre (the other half is the dynamic envelope). These are:

- **Fundamental** – the lowest frequency in a sound. The fundamental defines the pitch of the sound.
- **Harmonics** – frequencies that are an integer multiple of the fundamental. For example, for a fundamental of 100 Hz, the harmonics would be 200, 300, 400, 500 Hz and so forth. Harmonics are an extremely influential part of sounds – we say the fundamental gives sound its pitch, harmonics give the color.
- **Overtones** – frequencies that are not necessarily an integer multiple of the fundamental. A frequency of 150 Hz, for example, would be an overtone of a 100 Hz fundamental. Instruments like snares tend to have a very vague pitch, but produce lots of noise – a typical sound caused by dominant overtones.
- **Formants** – frequencies caused by physical resonance that do not alter with relation to the pitch being produced. Formants are major contributors to sonic stamps – for example, our ability to recognize each person's voice.

The spectral content of each instrument spans from the fundamental frequency, up to and beyond 20 kHz (even a kick has some energy above 10 kHz). It is worth mentioning that there might be some content below the fundamental due to body resonance or subharmonics caused by the nonlinearity of recording devices. One remarkable faculty of our brain is its ability to **reconstruct missing fundamentals** – whenever the fundamental or the low-order harmonics are removed our brain uses the remaining harmonics to conclude what is missing. This is the very reason why we still recognize the lower E on a bass guitar as the lowest note, even if our speakers do not produce its fundamental frequency of 41 Hz. This phenomenon has a very useful effect in mixing and will be discussed in greater detail soon.

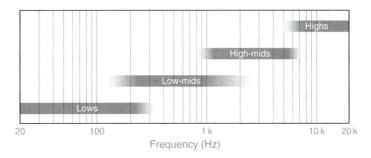

Figure 14.3 : The basic four-band division of the audible frequency spectrum.

Bands and associations

We already know from Chapter 6 that the audible frequency spectrum is divided into four main bands: lows, low-mids, high-mids and highs. For convenience, the illustration shown in Chapter 6 is reproduced in Figure 14.3. The frequency spectrum can also be subdivided into smaller ranges; each with its own general characteristics. The list below provides very rough guidelines to these different ranges. All the frequencies are approximate and expressed in Hz:

- **Sub sonic** (up to 20) – the only instruments that can produce any content in this range are huge pipe organs found in a few churches across the world. This range is not heard but it is felt. Although used in motion picture cinemas for explosions and thunders, it is absent from music masters.
- **Low bass** (20–60) – also known as very lows, this range is **felt** more than heard and is associated with power rather than pitch. The kick and bass usually have their fundamental in this range, which is also used to add sub-bass to a kick. A piano would also produce some frequencies in this range.
- **Mid bass** (60–120) – within this range we start to perceive tonality. Also associated with **power**, mostly that of the bass and kick.
- **Upper bass** (120–250) – most instruments have their **fundamentals** within this range. This is where we can alter the **natural tone** of instruments.
- **Low-mids** (250–2k) – mostly contain the very important **low-order harmonics** of various instruments, thus their **meat, color**, and a big part of their **timbre**.
- **High-mids** (2k–6k) – our ears are very sensitive to this range (as per the equal-loudness curves), which contains **complex harmonics**. Linked to **loudness, definition, presence** and **speech intelligibility**.
- **Highs** (6k–20k) – contain little energy for most instruments, but an important range all the same. Associated with **brilliance, sparkle** and **air**.

The subjective terms in the list above are just a few of many we associate with the various frequency ranges. We also have terms to describe deficiency and excess of various ranges. We use these terms in verbal communication, but we might also use them in our heads – first we decide that we want to add spark and then translate it to a specific frequency range. These terms are not standardized, and different people might have different ideas about particular terms. One thing is certain – the frequency ranges we associate these terms with are very rough. The body of a bass guitar is very different from the body of a flute. Figure 14.4 lists these terms.

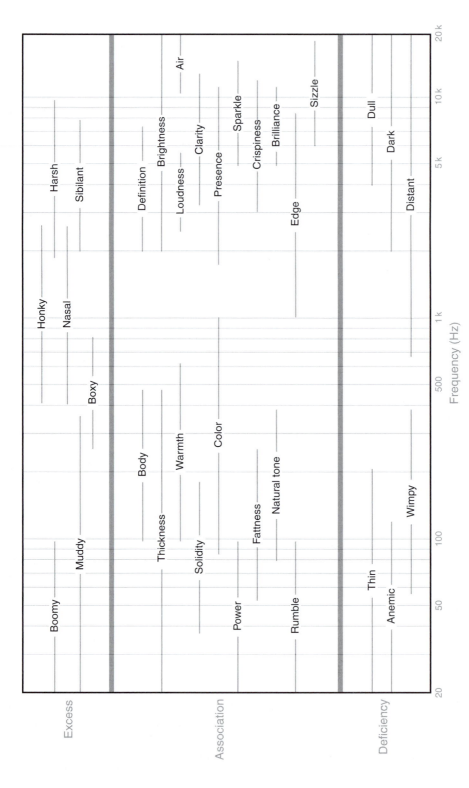

Figure 14.4 Subjective terms we associate various frequency ranges with, and excess or deficiencies in these ranges. The terms are not standardized, and the frequency ranges are rough.

Sibilance

Sibilance is the whistling sound of s, sh, ch and sometimes t. 'Locks in the castle' has got two potentially sibilant consonants. Languages like German and Spanish have more sibilant constants than English. The same sentence in German – 'Schlösser im schloss' – has got four potentially sibilant consonants. Sibilance spans between the 2 and 8 kHz range and can be emphasized by certain microphones, tube equipment or just standard mixing tools like equalizers and compressors. Emphasized sibilance pierces through the mix in an unpleasant and noticeable way. It would also distort on radio transmission and when cut to vinyl. It is important to note that not all speakers produce the same degree of sibilance. One must be familiar with his/her monitors to know when sibilance has to be treated and to what extent.

Track 14.2: Sibilance
Since the original vocal recording was not sibilant, an EQ is employed to draw some sibilance from the vocal. The sibilance is mainly noticeable on 'circles', 'just', and 'pass'. Some untreated vocal recordings can have a much more profound sibilance than the one heard on this track.

Types and controls

Filters, equalizers and bands

Faders attenuate or boost the whole signal. Put another way, the whole frequency spectrum is made softer or louder. Equalizers attenuate or boost only a specific range of frequencies, and by that they alter the tone of the signal. Early equalizers could only attenuate (filter) frequencies; later designs could boost as well. Regardless, all equalizers are based on a concept known as filtering, thus the terms 'equalizer' and 'filter' are used interchangeably. In this text, we will use the term filter to mean a circuit that acts with reference to a single frequency, and equalizer as a device that might consist of a few filters.

A filter might be in charge of a specific range of frequencies known as *band*. A typical equalizer on a large-format console has four bands: LF, LMF, HMF and HF (low frequencies, low-mid frequencies, high-mid frequencies and high frequencies). There would normally also be a high-pass filter (HPF) and a low-pass filter (LPF). Such an equalizer is shown in Figure 14.5. Many software plugins and digital equalizers provide more than four bands, with some plugins offering ten or more. Such equalizers are known as *paragraphic* EQs, and are a hybrid between parametric and graphic equalizers (both explained soon). In many cases, a graph on the plugin window shows the combined frequency response of all bands. A paragraphic EQ plugin is shown in Figure 14.6.

It's probably the right moment to include a conceptual discussion about the difference between the two types of equalizers shown in Figures 14.5 and 14.6. It can be argued that we rarely need more than what is offered by the analog equalizer shown in Figure 14.5 – it provides two pass filters to remove content, two shelving filters to change the general tonality of sounds and two parametric equalizers for more focused treatment. Essentially,

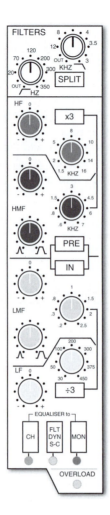

Figure 14.5 A diagram of the SSL 4000G+ equalizer. The equalizer section consists of four bands: the LF and HF bands are shelving filters; the LMF and HMF are parametric equalizers. At the top are high- and low-pass filters.

most productions before the DAW age were mixed with equalizers that are no more lavish than the one above. Software plugins with a multitude of bands, say ten, can fool people into thinking that standard equalization requires that many bands. It is true that sometimes many bands are useful, but this is more often the case with problematic recordings. For general tone shaping, it is often seasoned engineers that have sufficient auditory skill to take advantage of such a magnitude of bands. Equalization is not the easiest aspect of mixing, and introducing more bands can make it harder. It is better to hit the right spot with one band than miss it with three.

Another potential problem with paragraphic plugins is the response plot they present. These can easily divert the attention from the actual sonic effect of equalization and lead people to equalize by sight alone. For example, one might over-boost on an equalizer simply because a 6 dB boost on its frequency plot looks graphically small. If you ever

Figure 14.6 A paragraphic equalizer plugin (Digidesign's *DigiRack 7-band EQ 3*). This plugin provides seven bands. Two are high and low filters; and the other five are parametric filters, although the LF and HF bands can be switched into shelving mode.

boosted on a paragraphic plugin then looked at the gain value to see why the effect is so extreme or little, you've probably been equalizing by sight. Professional engineers working on analog desks hardly ever use their eyes while equalizing – you will notice that the gain scales in Figure 14.5 are not even labeled. McDSP lets the user choose whether or not to present the response plot, promoting hearing-based equalization for those who fancy it. It might be wise for other manufacturers to allow us to switch off the response plot for the same reason.

Frequency-response graphs

We use frequency-response graphs to show how a device alters the different frequencies of the signal passing through it (see Figure 14.7). On these gain-vs.-frequency graphs, the frequency axis covers the audible frequency range between 20 Hz and 20 kHz. The fact that our perception of pitch is not linear is evident on the frequency scale – 100-200 Hz has the same width as 1-2 kHz. The spacing between the grid lines corresponds to their range (10 Hz steps for tens of hertz, 100 Hz for hundreds of hertz and so forth).

The filter in Figure 14.7 is known as a *brick-wall filter*. It has a step-like response that divides the frequency spectrum into two distinct bands, where all frequencies to one side are removed and all frequencies to the other are kept. A filter with such a vertical slope is hypothetical – it cannot be built – and even if it could be, it would bring about many unwanted side effects. Yet, the term brick-wall filter is sometimes used to describe a

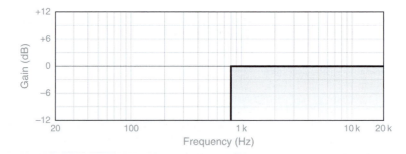

Figure 14.7 A brick-wall filter on a frequency-response graph. The filter in this graph removes all frequencies below 800 Hz, but lets through all the frequencies above. This type of brick-wall filter only exists in theory.

digital filter with a very steep slope, although never vertical. Equalizers obey scientific law and achieve their effect gradually.

There are so many terms and circuit designs involved in filters that it can be very hard to keep track of them all. To generalize, in mixing we use the following types: **pass, shelving** and **parametric**. Each type has a recognizable shape when shown on a frequency-response graph. Regardless of the filter we use, there is always a single reference frequency (e.g., 800 Hz in Figure 14.7), and in most cases we have control over it. This frequency has a different name for each type of filter.

Pass filters

The circuitry of a pass filter can be as simple as a mere combination between a capacitor and a resistor. The reference frequency of pass filters is called the *cut-off frequency*. Pass filters allow frequencies to one side of the cut-off frequency to pass, while continuously attenuating frequencies to the other side. A *high-pass filter* (HPF) allows frequencies higher than the cut-off frequency to pass, but filters frequencies below it. A *low-pass filter* (LPF) does the opposite – it lets what is below the cut-off frequency through, while filtering what is above it. Figure 14.8 illustrates this.

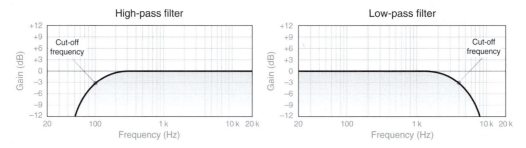

Figure 14.8 A high-pass and a low-pass filter.

A HPF is also referred to as a *low-cut filter*, and a LPF as a *high-cut filter*. There is no known difference between the contrasting terms, although it is easier to talk about pass filters (rather than cut filters), and more common to use the HPF and LPF abbreviations. The abbreviations LCF and HCF might leave some baffled.

It is easy to notice that the cut-off points in Figure 14.8 are not where the curve starts to bend (the transition frequency). Indeed, the cut-off frequency on pass filters is where 3 dB of attenuation occurs. For example, it is 100 Hz for the HPF in Figure 14.8. As a consequence, we can see that a range of frequencies has been affected despite being higher than the cut-off frequency (or lower in the case of LPF). Why −3 dB? It is a significant point in the science of filters, and any other reference frequency would make things more complex to design. For us, it is not such a big deal anyway – the effect above −3 dB is subtle compared to the effect below −3 dB.

Some filters have a fixed cut-off frequency and only provide an in/out switch. In many cases, the fixed frequency of a HPF would be 80 Hz, which is right below the lowest note of a regular guitar (E, 82 Hz) and the second harmonic of the lowest note on a bass guitar (E, 41 Hz). However, the majority of pass filters we use in mixing let us sweep the cut-off frequency.

The following tracks demonstrate the effect of a LPF and a HPF on drums. The cut-off frequency used in each sample is denoted in the track name. All slopes are 24 dB/oct:

Track 14.3: Drums Source
The source, unprocessed track used in the following samples:

Track 14.4: HPF 50 Hz (Drums)
Track 14.5: HPF 100 Hz (Drums)
Track 14.6: HPF 250 Hz (Drums)
Track 14.7: HPF 2 kHz (Drums)
Track 14.8: HPF 6 kHz (Drums)

Track 14.9: LPF 12 kHz (Drums)
Track 14.10: LPF 6 kHz (Drums)
Track 14.11: LPF 2 kHz (Drums)
Track 14.12: LPF 250 Hz (Drums)
Track 14.13: LPF 100 Hz (Drums)

And the same set of samples with vocal:

Track 14.14: Vocal Source

Track 14.15: HPF 50 Hz (Vocal)
Track 14.16: HPF 100 Hz (Vocal)
Track 14.17: HPF 250 Hz (Vocal)
Track 14.18: HPF 2 kHz (Vocal)
Track 14.19: HPF 6 kHz (Vocal)

Track 14.20: LPF 12 kHz (Vocal)
Track 14.21: LPF 6 kHz (Vocal)
Track 14.22: LPF 2 kHz (Vocal)
Track 14.23: LPF 250 Hz (Vocal)
Track 14.24: LPF 100 Hz (Vocal)

Plugin: McDSP *FilterBank F2*
Drums: Toontrack *EZdrummer*

Another characteristic of a pass filter is called slope. It determines the steepness of the filter curve. The slope is expressed in dB per octave (dB/oct), with common values being a multiple of six. A gentle slope of 6 dB/oct means that below the cut-off frequency each consecutive octave experiences an additional 6 dB of gain loss. With an aggressive slope of 30 dB/oct, it will only take two octaves before frequencies are attenuated by more than 60 dB (a point that can be perceived as effective muting). Generally, the steeper the slope the more unwanted effects the filter produces. Figure 14.9 shows different slopes.

The 6 dB multipliers are set in stone in analog EQs, and rarely can we alter the slope of a filter. An analog pass filter often has a fixed slope of either 12, 6 or 18 dB/oct (in rough order of popularity). A slope of 36 dB/oct is considered extremely steep. It is easier in digital designs to offer a variety of slopes, and they are not always bound to 6 dB/oct steps. It should be mentioned that if we do not have slope control, we could achieve our desired slope by combining two filters. For example, we can achieve a 12 dB/oct response by combining two 6 dB/oct filters with the same cut-off frequency. As the signal travels through the first filter, the first octave experiences maximum attenuation of 6 dB; as it travels through the second filter, the same octave experiences an additional 6 dB of attenuation, resulting in a summed response of 12 dB/oct. In the analog domain, this involves connecting two pass filters sequentially. With a paragraphic plugin, this involves having two bands set to the same pass response. In all cases, the cut-off frequencies of the two filters should be identical.

The following tracks demonstrate the effect of different filter slopes on drums and vocals. The cut-off frequency and slope used in each sample are denoted in each track name (slopes are in dB/oct):

Track 14.25: HPF 250 Hz Slope 6 (Drums)
Track 14.26: HPF 250 Hz Slope 12 (Drums)
Track 14.27: HPF 250 Hz Slope 18 (Drums)
Track 14.28: HPF 250 Hz Slope 24 (Drums)

Track 14.29: LPF 6 kHz Slope 6 (Drums)
Track 14.30: LPF 6 kHz Slope 24 (Drums)

Track 14.31: HPF 250 Hz Slope 6 (Vocal)
Track 14.32: HPF 250 Hz Slope 12 (Vocal)
Track 14.33: HPF 250 Hz Slope 18 (Vocal)
Track 14.34: HPF 250 Hz Slope 24 (Vocal)

Track 14.35: LPF 6 kHz Slope 6 (Vocal)
Track 14.36: LPF 6 kHz Slope 24 (Vocal)

Plugin: McDSP *FilterBank F2*
Drums: Toontrack *EZdrummer*

Although this might seem self-explanatory, a few people wrongly assume that once a signal is passed through a filter, a second pass through the same filter will have no effect. Filters always attenuate; never remove completely. If we take a pass filter, for example, during the first pass the cut-off frequency will be attenuated by 3 dB, then during the second pass it will be attenuated by an additional 3 dB.

Many of us are familiar with the pass filters on synthesizers that have both cut-off and resonance control. Resonance provides a boost around the cut-off frequency, and gives

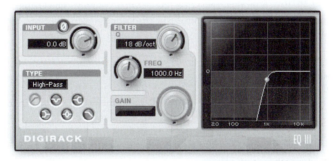

Figure 14.9 Different slopes on a HPF. These four instances of the Digidesign DigiRack EQ 3 show a different slope each. From top to bottom: 6, 12, 18 and 24 dB/oct. The cut-off frequency is set to 1 kHz and is clearly indicated with a white circle.

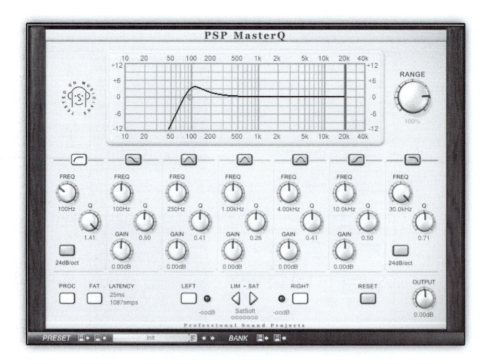

Figure 14.10 Pass filter resonance (the PSP MasterQ plugin). The Q control on the HPF section acts as a resonance control. The resonance is seen as a bump around the cut-off frequency.

Figure 14.11 The Universal Audio NEVE 1073 EQ plugin. This plugin, which emulates the sound of the legendary analog NEVE 1073, has a frequency response that deviates from the perfect theoretical shape of filters, a fact that contributes to its appealing sound. The high-pass filter (rightmost control) involves changing resonance for each of the four selectable cut-off frequencies.

some added edge to the transition range. Resonance is highly noticeable when the cut-off frequency is swept, and most DJ mixers incorporate resonant filters. Resonance is not an extremely common feature in mixing equalizers, although some, like the PSP MasterQ in Figure 14.10, do provide it. However, it might still be 'secretly' incorporated into some designs, often those that offer very few controls and do not reveal their frequency-response graph (typical analog or the digital plugins that emulate them). One such example is the Universal Audio NEVE 1073 EQ shown in Figure 14.11. Generally speaking, equalizers of this kind have a frequency response that is far from textbook perfect (like Figure 14.8), a fact that often contributes to their appealing sound. Even if

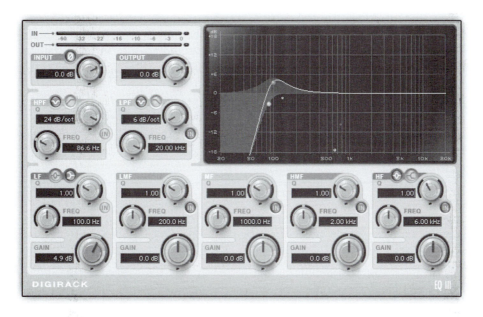

Figure 14.12 Combining two bands to create a resonant filter (the Digidesign DigiRack 7-band EQ 3). A parametric filter (LF band) is used to create a bump around the cut-off frequency of a HPF, resulting in a response typical to a resonant filter.

an equalizer does not offer a resonance control, we can achieve this characteristic by adding a parametric filter around the cut-off frequency. Figure 14.12 illustrates how this is done. The only limitation with such a setup is that both reference frequencies have to be adjusted if we want to sweep the resultant response.

Track 14.37: HPF Sweep No Resonance
A HPF set to 250 Hz with no resonance starts sweeping up in frequency after the second bar.

Track 14.38: HPF Sweep Resonance
Similar arrangement as in the previous track, only with resonance.

Plugin: PSP *MasterQ*
Drums: Toontrack *EZdrummer*

Shelving filters

Most people have used shelving filters. These are the bass and treble controls found in our domestic hi-fi systems, also known as tone controls. Shelving filters, as we know them today, were conceived by Peter Baxandall in the late 1940s. They are so called as their response curve can, in inspiring moments at least, remind us of shelves.

As opposed to pass filters, which only cut frequencies, shelving filters can also boost. The reference frequency of shelving filters divides the spectrum into two bands. To one side, frequencies are undisturbed; to the other, frequencies are either attenuated or boosted by a constant amount. A gain control determines that amount. As per our discussion about vertical response slopes and that they aren't possible, there is always a transition band between the unaffected frequencies and those affected by the set gain. Figure 14.13 shows the four possible versions of shelving filters.

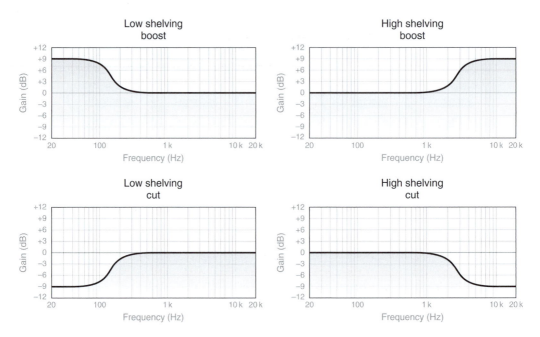

Figure 14.13 The four versions of shelving filters. For boost, +9 dB of gain is applied; for attenuation, −9 dB.

When it comes to defining what is the reference frequency of shelving filters, we encounter a lot of ambiguity. Designers might choose one of the three main possibilities (illustrated in Figure 14.14) to define it. Some define it in the traditional engineering sense as the point at which 3 dB of gain is reached – the familiar *cut-off frequency*. However, this tells us little about the real effect of the filter, which roughly happens when the set amount of gain is reached. Since this is often what we are after, some designers use this point as a reference, and it is called the *corner frequency*. To add to the confusion, it is also possible for the reference frequency to be halfway on the transition range – a frequency we can regard as the *center frequency*. Of all three options, it can be argued that the corner frequency is the most intuitive one to work with.

All shelving filters offer control over the gain amount, often ranging between -12 and +12dB. Most filters in mixing also offer control over the shelving frequency. Some filters also offer slope control, which determines the steepness of the slope in the transition band. Just like with pass filters, the actual response of shelving filters might deviate from the curves shown in Figure 14.13. In fact, they are likely to deviate. One very common response involves a contrast resonance around the transition frequency, i.e., a section of the curve that bends in the opposite direction to the normal response. Such a response can be seen in Figure 14.15.

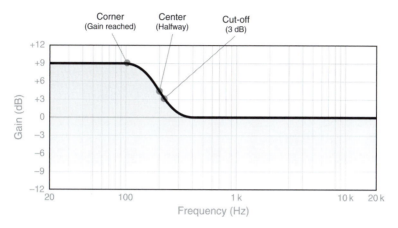

Figure 14.14 Three possible options for the shelving frequency. The corner frequency is where the set gain is reached. The center frequency is halfway through the transition band. The cut-off frequency is the traditional 3 dB point.

The following tracks demonstrate the effect of different boost and attenuation amounts on low- and high-shelving filters (LSF and HSF). The cut-off frequency and amount of gain used in each sample are denoted in each track name:

Track 14.39: LSF 250 Hz 3 dB Down (Drums)
Track 14.40: LSF 250 Hz 6 dB Down (Drums)
Track 14.41: LSF 250 Hz 12 dB Down (Drums)
Track 14.42: LSF 250 Hz 20 dB Down (Drums)

Track 14.43: LSF 250 Hz 3 dB Up (Drums)
Track 14.44: LSF 250 Hz 9 dB Up (Drums)

Track 14.45: HSF 6 kHz 3 dB Down (Drums)
Track 14.46: HSF 6 kHz 9 dB Down (Drums)
Track 14.47: HSF 6 kHz 20 dB Down (Drums)

Track 14.48: HSF 6 kHz 3 dB Up (Drums)
Track 14.49: HSF 6 kHz 9 dB Up (Drums)

Track 14.50: LSF 250 Hz 3 dB Down (Vocal)
Track 14.51: LSF 250 Hz 6 dB Down (Vocal)
Track 14.52: LSF 250 Hz 12 dB Down (Vocal)
Track 14.53: LSF 250 Hz 20 dB Down (Vocal)

Track 14.54: LSF 250 Hz 3 dB Up (Vocal)
Track 14.55: LSF 250 Hz 9 dB Up (Vocal)

Track 14.56: HSF 6 kHz 3 dB Down (Vocal)
Track 14.57: HSF 6 kHz 9 dB Down (Vocal)
Track 14.58: HSF 6 kHz 20 dB Down (Vocal)

Track 14.59: HSF 6 kHz 3 dB Up (Vocal)
Track 14.60: HSF 6 kHz 9 dB Up (Vocal)

Plugin: Sonnox *Oxford EQ+Filters*
Drums: Toontrack *EZdrummer*

Figure 14.15 A typical contrast resonance on a shelving filter (the Universal Audio *Cambridge EQ*). Both the low- and high-shelving filters are engaged in this screenshot, and it might be easier to discern them if we imagine a cross-line at 700 Hz. The low-shelving filter, set to type A, has a single contrast bend around the transition frequency. The high-shelving filter, set to type C, has two of these bends – one around the transition frequency, the other around the corner frequency.

Parametric filters

In 1972 at the AES, George Massenburg unveiled the parametric equalizer – a revolutionary circuit that he designed with help from fellow engineers. Although the concept of band-pass and band-reject filters (primitive types of parametric filters) was already well established, parametric equalizers became, and still are, *de facto* in mixing.

Like shelving filters, parametric filters can also cut or boost. Their response curve is reminiscent of the shape of a bell, as can be seen in Figure 14.16. The reference frequency is called the *center frequency* and we can sweep it higher or lower. The gain determines the maximum amount of boost or cut reached at the center frequency. The two cut-off points are 3 dB away from the center frequency (3 dB below for boost, 3 dB above for cut). The bandwidth is measured between these two cut-off points, and we express it in octaves.

Had we expressed the bandwidth in hertz (for example 600 Hz for the graphs in Figure 14.16), the effect of the filter would alter as the center frequency is swept, where the higher the frequency the less the effect. Consequently, the bell shape would narrow as the center frequency is swept higher. The reason for this has to do with our nonlinear pitch perception. To demonstrate this again, 600 Hz between 200 and 800 Hz equals two octaves (24 semitones); the same 600 Hz between 10 000 and 10 600 Hz is only a semitone. There is no comparison between affecting two octaves and a semitone.

Although the bandwidth on some equalizers is expressed in octaves, it is far more common to use a parameter called Q (*Quality Factor*). Q can be calculated by the mathematical expression Fc/(Fh–Fl), where Fc is the center frequency, Fh and Fl represent the high and low cut-off frequencies, respectively. The higher the Q the narrower the shape of the bell. Roughly speaking Q values range from 0.1 (very wide) to 16 (very narrow). Three different Q settings can be seen in Figure 14.17.

In this book, the term wide Q denotes a wide bell response that is achieved using low Q settings (like 0.1). The term narrow Q denotes a narrow response that is the outcome of high Q setting (like 16).

Track 14.61: Pink Noise Automated Q
This sample, which unmistakably resembles the sound of a sea wave, is the outcome of an equalized pink noise. The initial settings include a 9 dB boost at 1 kHz with the narrowest Q of 10. Due to the narrow Q and the large boost, it is possible to hear a 1 kHz whistle at the beginning of this track. In the first 8 seconds, the Q widens to 0.1, a period at which the whistle diminishes and both low and high frequencies progressively become louder. For the next 8 seconds, the Q narrows back to 10, a period when the augmentation of the 1 kHz whistle might become clearer.

Plugin: Digidesign *DigiRack EQ 3*

The shape of the bell gives the filter much of its characteristics, and it is not surprising that many variants exist. One important aspect is whether or not there is a dependency between the gain and the *Q*. With some designs, the bell narrows with gain (a behavior described as *proportional-Q*). As a consequence, changing the gain might also require

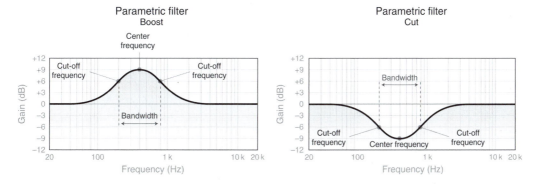

Figure 14.16 A parametric filter. Both response graphs involve 9 dB of gain (boost or cut) and a center frequency at 400 Hz; 3 dB below the center frequency are the two cut-off frequencies at 200 and 800 Hz. The 2-octave bandwidth (200-800 Hz) is measured between the two cut-off points.

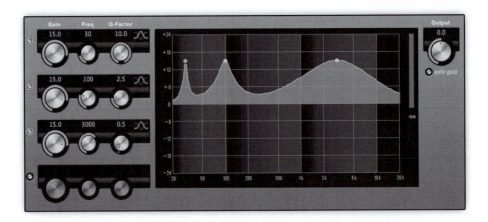

Figure 14.17 Different Q settings (the Cubase StudioEQ). Three bands are engaged in this screenshot, all with a gain boost of 15 dB. The lowest band (leftmost) shows a response with a narrow Q (10). The middle band response is achieved with a moderate Q (2.5). The widest Q (0.5) is applied on the highest band. The different bandwidths can be visualized by looking at the +12 dB grid line between the cut-off points.

adjustment to the *Q*. Equalizers of this type tend to sound more forceful as they become sharper with higher gain settings. A design known as *constant-Q* provides an alternative where the bandwidth is (nearly) constant for all gain settings. This produces a softer effect that often brings about more musical results.

In the following tracks, a parametric filter is applied on drums. The center frequency is set to 1 kHz, the Q to 2.8 and the gain is automated from 0 up to 16 dB and back down to 0. Notice how in the first track, which involves proportional Q, the operation of the filter seems more obstructive and selective, whereas in the second track, which involves constant Q, it seems more subtle and natural.

Track 14.62: Proportional Q
Track 14.63: Constant Q

Plugin: Sonnox *Oxford EQ+ Filters*
Drums: Toontrack EZdrummer

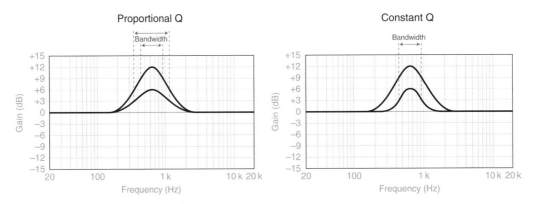

Figure 14.18 Proportional vs. constant Q. With proportional Q, the bandwidth varies with relation to the gain; with constant Q, it remains the same.

Most of the sounds we are mixing have rich and complex frequency content that mostly spans from the fundamental frequency to 20 kHz and beyond. The timbre components of various instruments do not exist on a single frequency only, but stretch across a range of frequencies. One of the accepted ideas in equalization is that sounds that focus on a very narrow frequency range are often gremlins that we would like to remove. To accommodate this, designs provide an attribute called *boost/cut asymmetry*, which grew in popularity in recent years. In essence, an asymmetrical filter of this type will use wider bell response for boosts, but a narrower one for cuts. We, the users, see no change in the Q value. This type of asymmetrical response is illustrated in Figure 14.19.

Figure 14.19 Cut/boost asymmetry (the Sonnox Oxford Equalizer and Filters plugin). Both the LMF and HMF are set with extreme gain of -20 and +20 dB, respectively. The Q on both bands is identical (2.83). The asymmetry is evident as the cut response is far narrower than the boost response. This characteristic is attributed to the EQ-style selection seen as Type-2 in the center below the plot. Other EQ styles on this equalizer are symmetrical.

Sounds that focus on a very narrow frequency range are more likely to be flaws that we will want to eliminate.

A variation of a parametric filter is known as notch filter. We use this term to describe a very narrow cut response, like the cut in Figure 14.19. This type of response is often used to remove unwanted frequencies, like mains hum, or strong resonance.

A summary of pass, shelving and parametric filters

It would be worthwhile at this stage to summarize the various filters we have looked at so far. The McDSP plugin in Figure 14.20 will help us recap.

Figure 14.20 The McDSP FilterBank E6.

A **pass filter** continuously removes frequencies to one side of the cut-off frequency. Normally we have control over the **cut-off frequency**, and occasionally we can also control the slope. The HPF and LPF are bands 1 and 6, respectively, in Figure 14.20.

A **shelving filter** boosts or attenuates frequencies to one side of the corner frequency. Normally we can control the **corner frequency** (or any other reference frequency if different) and **gain**. The shelving bands in Figure 14.20 are 2 and 5. The *FilterBank* E6 also provides control over the peak and dip resonance and the slope.

A **parametric filter** normally provides **gain, frequency** and **Q**. The set gain is reached at the center frequency, the Q (or octave bandwidth) determines the width of the bell between the two cut-off points. Bands 3 and 4 in Figure 14.20 are parametric filters. As a note on terminology, a parametric filter that offers variable gain, frequency and Q is known as a (*fully*) parametric equalizer. A parametric filter that only offers variable gain and frequency is known as a *semi-parametric* or *sweep* EQ.

The response characteristics of various filter designs give each equalizer its unique sound. Deviations from the textbook-perfect shapes, among other factors, gave many famous analog units their distinctive and beloved sound. To our delight, it is not uncommon nowadays to come across professional plugins that provide different types of characteristics to choose from and more control over the equalizer response.

Graphic equalizers

A graphic equalizer (Figure 14.21) consists of many adjacent mini-faders. Each fader controls the gain of a bell-response filter with fixed frequency, acting on a very narrow band. The Q of each band is fixed on most graphic equalizers, yet some provide variable Q. The frequencies are commonly spaced either an octave apart (so 10 faders are used to cover the audible frequency range) or 1/3 of an octave apart (so 27–31 faders are used). Graphic equalizers are so called because the group of faders gives a rough indication of the applied frequency response.

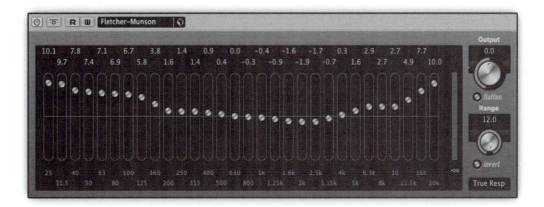

Figure 14.21 A 30-band graphic equalizer plugin (the Cubase *GEQ-30*). The fader settings shown here are the outcome of the Fletcher-Munson preset, which is based on the equal-loudness counters.

Graphic equalizers are very common in live sound, where they are used to tune the venue and prevent feedback. However, they are uncommon in mixing due to their inherent limitations compared to parametric equalizers. The multitude of filters involved (up to 31) and the less quality demanded for live applications mean that many hardware units compromise on quality for the sake of cost. Arguably, software plugins can easily offer a graphic EQ of better quality than most analog hardware units on the market. But there are not many situations where such a plugin would be favored over a parametric equalizer.

Notwithstanding, graphic equalizers are the standard tools in **frequency training**, where the fixed and limited amount of frequencies is actually an advantage. As pink noise is played through the equalizer, one person boosts a random band and another person tries to guess what band has been boosted. The easiest challenge involves pink noise, focusing on a limited amount of bands (say eight) and generous boost such as 12 dB. Things get harder with lower gain boosts, cutting instead of boosting, adding more bands and playing real recordings rather than noise. Highly trained engineers can identify gain changes as small as 3 dB in 1/3 octave spacing. Trained ears make equalization easier as we can immediately recognize which frequencies need treating. Any masking issues are readily addressed, and we have a much better chance of crafting a rich and balanced frequency spectrum.

We can train our ears even when alone. For example, playing a kick through a graphic equalizer, going octave by octave and attentively listening to how a boost or cut affects the kick's timbre is a beneficial exercise. It goes without saying that this can also be done with a parametric equalizer, but graphical equalizers are tailored for the task.

| *Graphic equalizers are of great benefit in frequency training.* |

Dynamic equalizers

Dynamic equalizers are not currently widespread and are often associated with mastering applications. Yet, the plugin revolution means that we should expect to see more of them in the future, and they can be just as beneficial in mixing as they are in mastering.

As opposed to the standard (static) equalizer, where the amount of cut or boost on each band is constant, the same amount on a dynamic equalizer is determined by the gain intensity on each band. Put another way, the louder or softer a specific band becomes, the less or more cut or boost is applied. Dynamic equalizers are something of a marriage between a multiband equalizer and a multiband compressor. For each band we often get the familiar compressor controls of threshold, ratio, attack and release. Unlike multiband compressors, these do not control the gain applied on the frequency band, but the amount of boost or cut on the equalizer of that band. Figure 14.22 is a diagram of a one-band dynamic equalizer, while Figure 14.23 shows a multiband dynamic EQ.

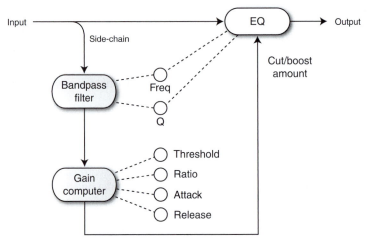

Figure 14.22 Basic diagram of single-band dynamic EQ. The frequency and Q settings determine both the equalizer settings and the pass-band frequencies that the gain computer is fed with. Basic compressor controls linked to the gain computer dictate the amount of cut or boost applied on the EQ.

Dynamic equalizers are very useful when we want to correct varying frequency imbalance, such as those caused by the proximity effect. We can set the EQ to cut more lows when these increase in level. We can also reduce finger squeaks on a guitar only when these happen. There are many more fascinating applications, but these are explored in

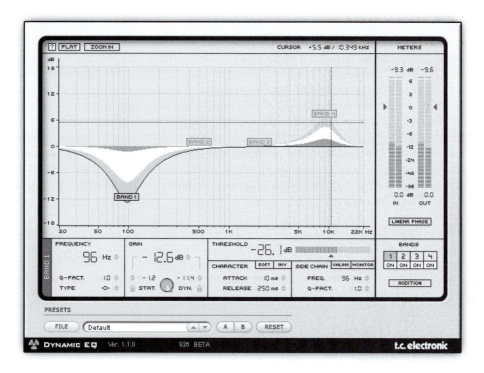

Figure 14.23 Dynamic EQ plugin (the t.c. electronic *Dynamic EQ*). Both bands 1 and 4 are active in this screenshot. Looking at each band, the darkest area (most inward) shows static equalization, the light area shows dynamic equalization and the gray area (most outward) shows the band-pass filter response curve.

Chapter 16, which is about compressors, as they are more common at present than dynamic equalizers.

In practice

Equalization and solos

In its majority, equalization is done in mix-context. Solving masking or tuning an instrument into the frequency spectrum is done with relation to other instruments. Equalizers and solo buttons are not exactly best friends. Instruments that sound magnificent when soloed can sound very bad in the mix. The opposite is even more likely – an instrument that sounds bad when soloed can sound great in the mix. If we apply high-pass filtering to soloed vocals, we are likely to set a lower cut-off point than we would had we listened to the whole mix – in isolation we only hear what the filter removes from the sound, but we cannot hear how this removal can give the vocals more presence and the mix more clarity. Yet, there are several situations where we equalize while soloing. Mostly this involves initial equalization and instances where the mix clouds the details of our equalization. But it is worth remembering that:

| *Equalizing a soloed instrument can be counter-effective.* |

Upward vs. sideways

An old engineers' saying is that if a project was recorded properly, with the mix in mind, there would be little or no equalization to apply. To some recording engineers, this is a flagship challenge. We say once sounds are recorded that they exist in their purest form and any equalization is interference with this purity. In subtle amounts, equalization is like make-up. But in radical amounts it is like a pretty drastic plastic surgery – it can bring dreadful results. I like to compare equalizers to funfair mirrors – gently curved mirrors can make you slimmer, broader, taller or shorter, which sometimes make you look better. But the heavily curved mirrors just make you look funny and disproportional. Having said that, as part of the sonic illusion we provide in some mixes, equalizers are used generously for extreme manipulations of sounds. The natural vs. artificial debate is very relevant when it comes to equalization.

There is no doubt that some equalizers lend themselves better to our artistic intentions than others, but they all share common issues. The more drastically we employ them the more we stand to lose in return. In that sense, a perfect EQ is one that has a flat frequency response, and thus no effect at all. In this ideal state, we are more likely to impair the sound with:

- More gain
- Narrower Q settings
- Steeper slopes
- More angular transition bends

Still, equalizers are not that injurious after all – they are widely used in mixes and yield magnificent results. The point is that sometimes we can achieve better results with a slight change of tactic – one that simply involves less drastic settings.

Say for example we want to add attack to a kick. Sid read in some book that the kick's attack is at 6 kHz. So he dials a parametric EQ to that frequency, and the more attack he wants the more he boosts the gain. Sid equalizes *upward*. He ends up with narrow Q and +12 dB of gain. Nancy knows that very few specific sounds focus on a narrow frequency range, and these are often gremlins that we want to remove, not emphasize. So she dials a frequency of 6 kHz as well, but with wider Q and lower gain. Nancy works *sideways*. Nancy boosted less, but a wider range of the kick's attack, so the overall impact is roughly the same, only with less artifacts. This example is illustrated in Figure 14.24a. We can give the same example with other types of filters. We can set a high-shelving filter at one frequency with more gain or set it to a lower frequency with lower gain (Figure 14.24b). Similarly, we can set a HPF with a steep slope at one frequency or we can set a gentler slope at a higher frequency (Figure 14.24c). The performance of pass filter is often evaluated by how well they handle the transition area, especially with steeper slopes. Cheap designs can produce very disturbing side effects around the cut-off frequency. The three examples in Figure 14.24 can be summarized as:

> *Q (for parametric filter) and frequency (for pass or shelving) can be traded for gain or slope, resulting in less obstructive equalization.*

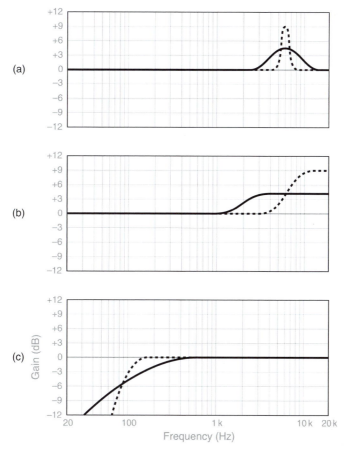

Figure 14.24 Equalization alternatives. In all these graphs, the dashed curves involve more drastic settings than the solid curves. The dashed and solid curves could bring about similar results, although the dashed curves are likely to yield more artifacts. This is demonstrated on a parametric filter (a), shelving filter (b) and pass filter (c).

Track 14.64: Kick Q No EQ
The source drums with no EQ on the kick.

In the next two tracks, the aim is to accent the kick's attack. Clearly, the two tracks do not sound the same, but they both achieve the same aim:

Track 14.65: Kick High Gain Narrow Q
The settings on the EQ are 3.3 kHz, +15 dB of gain and a Q of 9. The resonant click caused by these settings might be considered as unflattering.

Track 14.66: Kick Lower Gain Wider Q
The settings are 3.3 kHz, +9 dB of gain and a Q of 1.3. The attack is accented here, yet in a more natural way than in the previous track.

Plugin: Sonnox *Oxford EQ*
Drums: Toontrack *EZdrummer*

We must not forget that like with many other mixing tools, sometimes we are more interested in hearing the edge – subtlety and transparency is not always what we are after. For example, in genres like death-metal, equalizers are often used in what is considered a radical way, with very generous boosts. The equalizer's artifacts are used to produce harshness, which works well in the context of that specific music. Some equalizers have a very characteristic sound that only sharpens with more extreme settings.

Equalizers and phase

The operation of an equalizer involves some delay mechanism. The delays are extremely short, well below 1 ms. *Group delay* is a term often used, suggesting that only specific frequencies are delayed – while not strictly precise, it is faithful to the fact that some frequencies are affected more than others. Regardless, the delay mechanism in equalizers results in unwanted phase interaction. Just like two identical signals that are out of phase result in combfiltering, we can simplify and say that an equalizer puts some frequencies out of phase with other frequencies, resulting in phasing artifacts. We always see the frequency-response graph of equalizers, but rarely see the frequency-vs.-phase graph, which is an inseparable part of the operation of an equalizer. Figure 14.25 shows such a graph for a parametric equalizer.

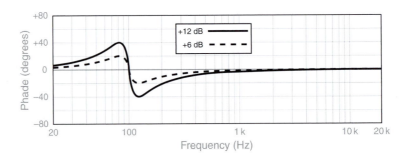

Figure 14.25 : The phase response of a boost on a parametric equalizer. The equalizer center frequency is 100 Hz. The solid line shows a boost of 12 dB, the dashed line shows a gain boost of 6 dB.

There are two important things we can learn from Figure 14.25. First, the fact that the higher the gain the stronger the phase shift. Second, we can discern that frequencies near the center frequency experience the strongest phase shifts, which diminish with further frequencies. This behavior, although demonstrated on a parametric EQ, can be generalized to all other types of EQ – the more gain there is the more severe phase artifacts become, with most effect around the set frequency.

> *The more gain the more severe phase artifacts become.*

The phase artifacts produced by an equalizer are sometimes perceived by a trained ear as a subtle impairment of sound rather than as something very obvious. We do, however, often associate these artifacts with the term resonance.

Track 14.67: Snare Source
The source snare used for the following samples.

The following tracks all involve varying degrees of gain boost at 500 Hz. The EQ artifacts, which become more severe with gain, can be heard as a resonance of around 500 Hz:

Track 14.68: Snare 500 Hz 10 dB Up
Track 14.69: Snare 500 Hz 15 dB Up
Track 14.70: Snare 500 Hz 20 dB Up

Similar artifacts can be demonstrated using varying slopes of a HPF (represented in the following track names as dB/oct). Note that the applied filter is not of a resonant nature, yet a resonance around 500 Hz can be discerned:

Track 14.71: Snare HPF 500 Hz Slope 12
Track 14.72: Snare HPF 500 Hz Slope 24
Track 14.73: Snare HPF 500 Hz Slope 48

Plugin: Logic *Channel EQ*
Snare: Toontrack *EZdrummer*

One interesting question is this: what happens with phase when we cut rather than boost? Figure 14.26 shows the same settings as in Figure 14.25, only this time for cut rather than boost. We can see that the only difference is that the two response graphs are mirrored, but the phase extent remains the same. It is a myth that equalizers cause more phase shift when we boost – there is nothing in the science of either analog or digital filters to support such a claim. However, it is correct that we notice the phase shift more when we boost, for the simple reason that we also make the phase artifacts louder. It is therefore not surprising that many mixing engineers prefer to attenuate rather than boost when equalizing, and that many sources give such a recommendation. Also, when boosting we risk clipping.

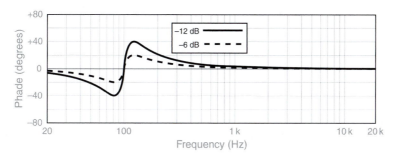

Figure 14.26 The phase response of a cut on a parametric equalizer. This is the same equalizer as in Figure 14.25, only this time with gain cut instead of boost.

When possible, it is better to attenuate rather than boost.

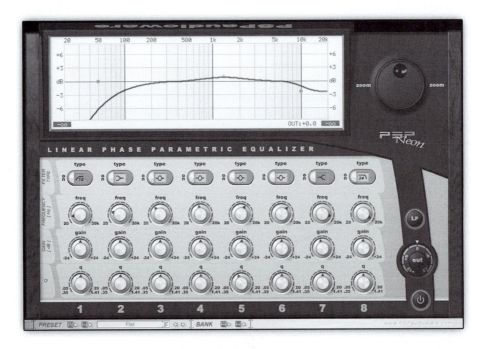

Figure 14.27 A linear phase equalizer (the PSP *Neon*). This plugin offers eight bands with selectable filter type per band. It is worth noting the LP (Linear Phase) button, which enables toggling between linear phase and standard mode.

There is one type of *digital* equalizer that has a flat phase response – the **linear phase equalizer** (Figure 14.27). Digital filters are based on mathematical formulae. These formulae have stages, and the audio travels through the different stages. By making the formula of a filter symmetrical, as audio travels through one side of the formula its phase is shifted, but once going through the mirrored side, it shifts back in phase to its original in-phase position. One issue with linear phase equalizers is that they require an extensive processing power and a large buffer of audio (bigger than the typical 1024 samples often provided by the audio sequencer). Thus, they are CPU consuming and introduce plugin delay. Designers have to compromise between phase-accuracy, processing power and the delay the plugin introduces.

There can be something very misleading about the concept of linear phase equalizers, since one might conclude that a linear phase equalizer is a perfect equalizer. In practice, both standard and linear-phase equalizers have many other unwanted byproducts (ringing, lobes and ripples to name a few). A linear phase design only rectifies one artifact of equalization, not all of them.

Linear phase equalizers tend to sound more 'expensive'. They excel at retaining detail, depth and focus. They are generally less harsh and likely to be more transparent. But many of the unwanted artifacts we get from standard equalizers (including those we associate with phase) are also a product of linear phase equalizers. Transients, for example, might not be well handled by the linear phase process. In fact, there are situations where standard equalizers clearly produce better results. Thus, linear phase equalizers provide an alternative, not a superior replacement.

> *Linear phase equalizers are better at retaining detail, depth and focus, but at the cost of processing. Like standard equalizers, they can also produce artifacts, and thus provide an alternative but not a superior choice.*

Track 14.74: Snare 500 Hz 20 dB Up

This track involves a linear phase equalizer with the same settings as in Track 14.70. A very similar resonance to that in Track 14.70 now has a longer sustain. In fact, its length is stretched to both sides of the hit. Many would consider the artifact on this track as worse than the standard equalizer version on Track 14.70. Indeed, due to their design, when linear-phase equalizers produce artifacts, these tend to sound longer and more noticeable. This is a characteristic of all linear-phase equalizers, not just the one used in this sample.

Plugin: Logic *Linear Phase Equalizer*
Snare: Toontrack *EZdrummer*

The following tracks involve a comparison between linear phase and standard EQ processing, and both are the result of boosting 12 dB on a high-shelving filter with its center frequency set to 2 kHz. Note how the highs on the standard version contain some dirt, while these appear cleaner and more defined on the linear-phase version.

Track 14.75: Guitar Standard EQ
Track 14.76: Guitar Linear Phase EQ

Plugin: PSP *Neon HR*

The frequency yin-yang

Figure 14.28 shows what I call the frequency yin-yang. I challenge the reader: what type of filter is this? Is it a +12 dB high-shelving filter brought down by 6 dB? Or is it a −12 dB low-shelving filter brought up by 6 dB? The answer is it can be both, and following our discussion about phase shift not varying between cut and boost, the two options should sound identical.

Regardless in which of the two ways the frequency yin-yang is achieved, it teaches us an extremely important thing about our frequency perception: provided the final signal is at the same level, boosting the highs or reducing the lows has the same effect. To make something brighter we can either boost the highs or attenuate the lows. To make something warmer we can boost the low-mids or attenuate from the high-mids up. While this concept is easily demonstrated with shelving filters, it works with other filters all the same. For example, to brighten vocals we often apply HPF, and despite removing a predominant low-frequency energy, the vocals can easily stand out more. Adding both highs and lows can be achieved by attenuating the mids. Surely, in many cases we have

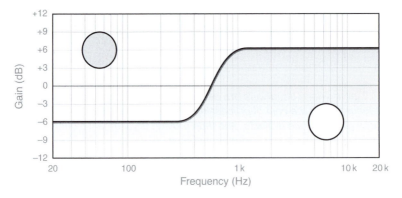

Figure 14.28 The frequency yin-yang.

a very specific range of frequencies we want to treat, but it is worth remembering that sometimes there is more than one route to the same destination.

The following tracks demonstrate the frequency yin-yang as shown in Figure 14.28; the two tracks are perceptually identical:

Track 14.77: Yin
This is the outcome of a high-shelving filter with a center frequency at 600 Hz, 12 dB boost and −6 dB output level.

Track 14.78: Yang
This is the outcome of a low-shelving filter with a center frequency at 600 Hz, 12 dB attenuation and +6 dB output level.

- - - - - - - - - - - - - - -

Track 14.79: Vocal Brighter
This is an equalized version of Track 14.14. What can be perceived as brightening is the outcome of a HPF (−6 dB point at 200 Hz, 12 dB/oct slope) and +1.8 output gain.

Track 14.80: Vocal Warmer
This equalized version of the previous track can be perceived as warmer. In practice, −4 dB was pulled around 3.5 kHz (and +1.3 of output gain).

Plugin: PSP *Neon*

One more example is essential here. If we stop thinking about individual treatment, we can apply the frequency yin-yang at mix level as well. Say we want to add attack to a kick. Instead of boosting the attack frequency on the kick, we can attenuate the same frequency on masking instruments. In fact, a common technique based on this concept is called *mirrored equalization*. It involves a boost response on one instrument and mirrored response on another. This helps reduce frequency overlapping, and the same effect can be achieved with less gain on each filter. Figure 14.29 illustrates this.

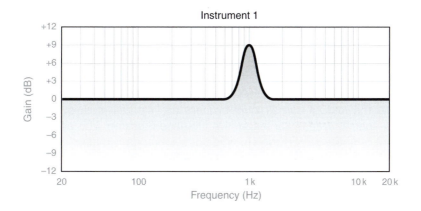

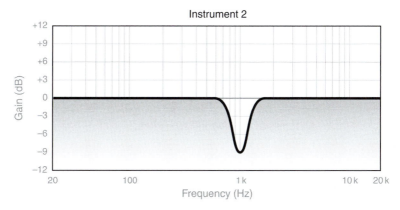

Figure 14.29 Mirrored equalization. A boost on one instrument is reinforced by mirrored response on a masking instrument.

Equalization and levels

By altering the level of a specific frequency range we also alter the overall level of the signal. As per our axiom, louder-perceived-better, A/B comparisons can be misleading – we might think it is better when boosting on the equalizer and worse when attenuating. In order to allow fair A/B comparisons, some equalizers provide an output level control. After equalizing, we match the level of the processed signal to the unprocessed signal, so our critical judgment is purely based on tonality changes, not level changes. While it is understood why output level control is not found on the small EQ sections of even the largest consoles, it is unclear why many plugin developers overlook this important feature.

> *Equalization alters the overall level of the signal and can deceive us, so we think that boosts sound better.*

A/B comparison aside, the louder-perceived-better principle can lead to immediate mis-judgments as we equalize, before compensating for the gain change. The risk is the same – we might think boosting improves the sound purely due to the overall level increase. The frequency yin-yang can minimize that risk; by taking the attenuation route we are less likely to base our evaluation on the overall level factor. It was just mentioned that attenuating the lows of vocals could increase their presence. By attenuating a specific frequency range we reduce masking on that specific range. Reducing the lows of the vocals, for example, would increase the low-end definition of other instruments. If by any chance the loss of overall level bothers us, we can always compensate for it. By boosting a specific frequency range, we increase masking. The equalized instrument becomes more defined, but on the specific boosted range it masks more. This is why it is recommended that we consider attenuating before boosting.

The psychoacoustic effect of taking away

In the first instance, our ears tend to dislike the attenuation or removal of frequencies. By attenuating or removing we take away something and make instruments smaller. Our brain might need some time to get used to such changes. By way of analogy it is like a drastic haircut – it might look a bit weird on the first day, but we get used to it after a while. Both listening in mix-perspective and giving the equalization effect some time to settle in can be beneficial when we take away some frequency content.

Attenuation or filtering frequencies might be right for the mix, but might not appear so at first.

One specific technique that can be used to combat this psychoacoustic effect is making an instrument intentionally smaller than appropriate, forgetting about the equalization for a while, then going back to the equalizer and making the instrument bigger. The ear tends to benefit from this approach, which allows us fairer judgment when we are deciding exactly how to set the equalizer.

Track 14.81: dGtr No EQ
The source track for the following samples, with no EQ applied.

Track 14.82: dGtr First EQ
The EQ settings in this track involve a HPF (144 Hz, 12 dB/oct) and a high-shelving filter (9 kHz, −8 dB). When played in straight after the previous track, this guitar might appear smaller but now playing the previous track again would reveal that the EQ in this track removed both rumble and high-frequency noise that were imprinted on the previous track.

Track 14.83: dGtr Second EQ
This is an equalized version of the previous track, with the same EQ settings as in the previous track (an equalizer with the same setting was inserted serially). Again, this track appears smaller than the previous one. Also, comparing this track to the previous would reveal that some high-frequency noise still existed in the previous track.

Track 14.84: dGtr Third EQ
This is an equalized version of the previous track, this time involving a band-pass filter (between 100 Hz and 2 kHz). Now listening to Track 14.82 would make it sound bigger compared to this one. Comparing this and the unequalized track would make the latter sound huge.

Plugin: Digidesign DigiRack EQ 3

Applications of the various shapes

Having a choice between pass, shelving and parametric filters, we employ each for a specific set of applications. The most distinct difference between the various shapes puts pass and shelving against the parametric filter. Both pass and shelving filters affect the extremes. Parametric filters affect a limited, often a relatively narrow bandwidth, and rarely we find them around the extremes. In a more specific context, the basic wisdom is this:

- **Pass filters** – used when we want to **remove** content from the extremes. For example, low-frequency rumble.
- **Shelving filters** – used when we want to alter the overall **tonality** of the signal (partly like we do on a hi-fi system) or to **emphasize** or **soften the extremes**. For example, softening exaggerated low-frequency thud.
- **Parametric filters** – used when what we have in mind is a **specific frequency range** or a **specific spectral component**. For example, the body of a snare.

For many, common sense will dictate the order in which these filters should be introduced into a mix: use pass filters first to remove unwanted content, then use shelving for general tonality alterations and finally use parametric filters for more specific treatment. The following sections detail the usage of each filter type.

HPF

HPFs are very common in mixes of recorded music. For one, any low-frequency gremlins that are not part of the instrument's timbre, like rumble or mains hum, are removed. Then, recorded tracks tend to contain a greater degree of lows than needed in most mixes (partly an outcome of the proximity effect). When the various instruments are mixed, the accumulating mass of low-end energy results in muddiness, lack of clarity and ill definition. HPFs tidy up the low-end by removing any dispensable lows or low-mids. Doing so clears some space for the bass and kick, but more importantly it can add clarity and definition to the treated instrument. This is worth stressing:

> *Despite removing spectral content, HPFs increase clarity and definition and can make the treated instrument stand out more in the mix.*

A HPF might be applied on every single instrument. Vocals, guitars and cymbals are common targets. Vocals, nearly by convention, are high-passed at 80 Hz – a frequency below which no contributing vocal content exists. Higher cut-off frequencies are used to remove byproducts of the proximity effect or some body that might not have a place in the mix. Many guitars, especially acoustic, occupy unnecessary space on the lows and low-mids, which many other instruments can use instead. When acoustic guitars play a rhythmic role, they are often filtered quite liberally with most or all of their body removed. Cymbal recordings often involve some drum bleed that cannot be gated (e.g., removing the snare from underneath the ride), so the filter also acts as a spill eliminator. Pianos, keyboards, snares or any other instrument can benefit from low-end filtering all the same. We sometimes even filter the kick and bass in order to mark their lowest frequency boundary, and in turn that of the overall mix.

The frequencies involved in this tiding up process are not strictly limited to the lows. A HPF on cymbals, for example, might be well within the low-mids. A possible approach is to simply sweep up the cut-off frequency until the timbre of the treated instrument breaks, then back it off a little. Usually the busier the mix the higher frequency HPFs reach. Over-filtering can create a hole in the frequency spectrum or reduce warmth. To rectify this, we can pull back the cut-off frequency on some instruments.

One interesting characteristic of HPFs is that they can be pushed quite high before we notice serious impairment to the treated instrument. The reason for this has to do with our brain's ability to reconstruct missing fundamentals. What a HPF removes, the brain reconstructs – we clear space in the mix, but do not lose valuable information.

Track 14.85: No HPF (aGtr)
The source track for the following samples:

In the following tracks, a 12 dB/oct HPF is applied with different cut-off frequencies (denoted in the track name). Virtually each of the following degrees of filtering might be appropriate in a mix. Note how despite filtering the fundamentals of the notes, our ears have no problem recognizing the chords:

Track 14.86: HPF 150 Hz (aGtr)
Track 14.87: HPF 250 Hz (aGtr)
Track 14.88: HPF 500 Hz (aGtr)
Track 14.89: HPF 1 kHz (aGtr)
Track 14.90: HPF 3 kHz (aGtr)

Plugin: McDSP *FilterBank F2*

> *Due to the brain's ability to reconstruct missing fundamentals, HPFs can be used quite generously.*

HPFs can manipulate the characteristics of reverbs. Generally speaking, the size, depth and impression of a reverb all focus on the lower half of the spectrum (lows and low-mids). The highs mostly contribute some definition and spark. By filtering the lows we can reduce the size of a reverb and its resultant depth. The higher the cut-off frequency is set the smaller the size becomes. Although we have full control over the size and depth of a reverb when using a reverb emulator, these factors are imprinted into a reverb captured on recordings. We will soon see how useful high-pass filtering can be for overheads.

In the following tracks, percussion is sent to a reverb, which is filtered by a HPF with various cut-off frequencies (denoted in the track name). Note how the dimension of the space shrinks the higher the cut-off frequency is (this is also the consequence of the overall level attenuation caused by the filter):

Track 14.91: No HPF (Reverb)
Track 14.92: HPF 400 Hz (Reverb)
Track 14.93: HPF 1 kHz (Reverb)
Track 14.94: HPF 2 kHz (Reverb)
Track 14.95: HPF 4 kHz (Reverb)

Track 14.96: Percussion Dry
This is the dry percussion used in the previous samples. Compare this track to the previous one and note the appealing enhancement the filtered reverb provides.

Plugin: PSP *MasterQ*, Audio Ease *Altiverb*
Percussion: Toontrack *EZdrummer*

A HPF can reduce the size and depth of reverbs.

HPFs are also used to remove pops – low-frequency air bursts caused by Ps, Bs or any other plosive sounds like an emphasized 'who'. The higher the cut-off frequency is the more the pop is reduced, but also the more likely the timbre is to suffer. In the days of tapes, removing these pops required filtering the whole vocal track and compromises had to be made. Audio sequencers offer a much more elegant solution – filtering the pop only. This is achieved using off-line processing, where we simply select each pop, apply filtering, then crossfade to prevent clicks. Figure 14.30 illustrates this.

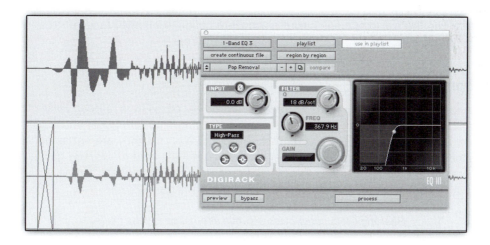

Figure 14.30 Pop removal using off-line processing. The top track, before processing, shows an evident pop. An off-line HPF is set with a cut-off frequency of 368 Hz. Had the filter been applied in real time throughout the track, much of the vocals' thickness would have been lost. The off-line process was only applied on a limited selection that included the pop and some margins for the crossfades. The crossfades were then applied to prevent clicks. The resultant bottom track shows apparent energy loss of the pop.

LPF

As opposed to their twins, LPFs are somewhat neglected when it comes to mixing. Unfortunately, our ears do not reconstruct the higher harmonics once lost, so we find the removal of high frequencies more disturbing. Some instruments get away with it. For example, filtering above 10 kHz from a kick could be unnoticeable, despite the fact that some energy does exist there. Instruments with richer harmonic or overtone content are more sensitive to low-pass filtering.

LPFs are used for two principal tasks. The first is the removal of **hiss** or **high-frequency noise**. The second is to mark the high-frequency boundary of a specific instrument. A HPF and LPF are occasionally combined to form a flexible band-pass filter (Figure 14.31). The general idea is to restrict an instrument to a limited band, which can increase separation. One example of where this can be beneficial involves distorted guitars. These often have excess of both low-end rumble and high-end noise. Being forceful maskers as they are, band-limiting them also reduces their masking interaction to a limited range of frequencies. Other instruments can also benefit from such an approach.

Shelving filters

While both pass and shelving filters affect the extremes, pass filters remove while shelving filters soften or emphasize. Needless to say, a shelving filter can boost and a pass filter cannot. One important characteristic of shelving filters is that they are not as aggressive as pass filters – while with shelving filters we have ultimate control over the gain amount, with pass filters we get the less flexible slope control. Shelving filters can be set to have very little effect and their response curves are generally gentler.

| *Shelving filters can provide a subtle alternative to pass filters.* |

The various applications of pass filters can easily become a shelving job if we can exchange the word 'remove' with 'soften'. For example, sometimes we only want to soften the body of vocals, not get rid of it altogether. Sometimes we want to reduce the size of a reverb, but still keep some of its warmth. The high-frequency noise on distorted guitars might only call for moderating. Sure enough, we can always use a combination between pass and shelving – pass to remove unwanted content (e.g., rumble), shelving to soften wanted content (e.g., body).

Figure 14.31 A HPF and LPF combined to form a band-pass response (the MOTU *Masterworks EQ*). The two filters are set to remove frequencies below 200 Hz (HPF) and above 5 kHz (LPF). Band-limiting instruments in a similar way can increase separation and reduce masking.

Shelving filters are often associated with the terms thick and thin (for the lows), bright and dark (for the highs). We are all familiar with the shelving effect from our hi-fis. Having the same ability to shape the tonality of each individual instrument is a great opportunity, which is often overlooked. Two key challenges that the frequency domain involves are balanced spectrum and separation between various instruments. Referring back to Figure 6.2, it was suggested that we take an abstract approach and imagine where each instrument is located on the frequency spectrum, how wide it is and whether there are any lacking frequency ranges in our mix. In the voyage to balanced spectrum and separation, we might want to nudge various instruments up and down on the frequency spectrum. We might also want to narrow or widen their frequency range. This practice is often referred to as **tuning** an instrument into the frequency spectrum. Keeping the same abstract approach, we could really use the ability to nudge the lower or higher boundary of an instrument.

Shelving filters are great tools for doing this, and Figure 14.32 illustrates how. Essentially, if you want an instrument to take more of the highs, boost using high shelving; if you want it to take less, attenuate. Using the very same principle, if you want an instrument to take more of the lows, boost on a low shelving; for less of the lows, attenuate. The only trick is to hit the right shelving frequency. The advantage that shelving filters have over pass or parametric is that they can be set for a very subtle effect (unlike pass) and affect a wide frequency range with a flat response (while parametric are more oriented toward a specific range of frequencies).

Shelving filters let us tune instruments into the frequency spectrum, help create separation and balance the overall frequency spectrum.

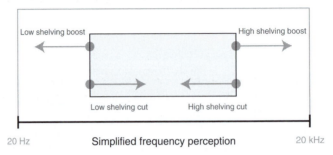

Figure 14.32 Tuning an instrument frequency range using shelving filters. Low shelving affects the low boundary: boost lowers it, cut makes it higher. High shelving affects the high boundary: boost makes it higher, cut lowers it.

Track 14.100: Drums No EQ
The source track for the following samples, in which only the snare is equalized:

Track 14.101: Drums HSF Up
A 3 dB boost at 1.8 kHz on a high-shelving filter creates the impression that the snare's high boundary was extended up in the frequency spectrum.

Track 14.102: Drums HSF Down
A −4 dB at 1.8 kHz on a high-shelving filter creates the impression that the snare's high boundary moved down in the frequency spectrum.

Track 14.103: Drums LSF Up
A −3 dB at 170 Hz on a low-shelving filter creates the impression that the snare's low-frequency boundary moved higher.

Track 14.104: Drums LSF Down
A 4 dB boost at 200 Hz creates the impression that the snare's low boundary moved lower.

Plugin: Digidesign *DigiRack EQ 3*
Drums: Toontrack *EZdrummer*

Parametric filters

Parametric filters are used when we wish to treat a specific range of frequencies. Often these specific ranges are associated with some spectral component of the treated instrument (e.g., body) or with a general sonic quality (e.g., nasal). Although not limited to the midrange, often this is where parametric filters work best.

Recognizing and comprehending the essence of various frequency ranges takes a life-long experience. A popular trick can help us locating the right **frequency** when using parametric filters. Say we are trying to find the attack of a kick. The technique involves a generous boost with a narrow *Q*, then sweeping the center frequency across the spectrum. The frequency at which the kick's attack is most profound is the frequency we are after. Then, we adjust the gain and *Q*. While many mixing resources mention this technique, it has some strong opposition as well. For one, it hinders critical evaluation skills, which are based on the process of listening, analyzing, then taking action. It can be compared to the less experienced guitarist playing all sorts of possible chords in the search for the right one. Then, there is the same risk of liking the boost too much, ending up with a higher gain than appropriate. Having said that, this technique can still be helpful for the less skilled engineers and with extremely tricky material.

Although **gain** settings are often straightforward, even here there is more than meets the eyes. We have already discussed the upward-vs.-sideways principle, where we demonstrated that gain can sometimes be traded for Q. How much effect an equalizer has is a function of both the gain and Q settings. Perhaps surprisingly, exactly the same settings and response curve on two different equalizers might not produce identical results. While the gain on an equalizer can have a noticeable effect with 0.5 dB of boost, certain equalizers offer up to 18 dB of gain. A 15 dB boost could be harrowing, but it can also be musical. For these reasons it is not easy to generalize what typical gain settings are – it depends on many factors such as the frequency, Q, the source material and the equalizer being used. Some people regard 1 dB as gentle, 3 dB as conservative, 6 dB as moderate, 9 dB as serious and, 12 dB as extreme.

One useful equalization technique involves distributing the equalization load across different frequencies. More specifically, instead of equalizing around one frequency only, we also equalize around its harmonic (an octave above). Figure 14.33 demonstrates this. This lets us reduce the gain on the first frequency and shift that gain to the harmonic. Compared to using one band only, each of the two bands will have a less drastic response and less phase artifacts. Also, having the equalization take place at two points often has the advantage of overcoming masking.

| *Harmonics equalization can reduce artifacts and can be less prone to masking.* |

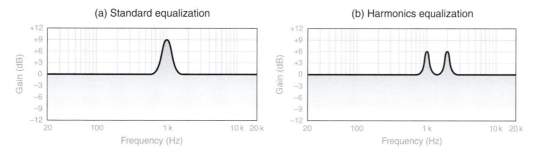

Figure 14.33 Harmonics equalization. (a) Standard equalization where a single range of frequencies is equalized. (b) Harmonics equalization where the same range is equalized with a softer response, but its harmonic is also equalized.

The source track for the following samples is Track 14.64. The aim is to accent the kick's attack:

Track 14.105: Kick Standard EQ
This track involves a single band parametric filter set to 15 dB of boost around 2 kHz, with *Q* set to 6.7. This results in an evident click, but also a noticeable resonance around 2 kHz.

Track 14.106: Kick Harmonic EQ
Instead of equalizing a single frequency, three parametric filters are employed with their center frequency set to 2, 4 and 8 kHz. Each band is set to around 8 dB of boost, and all *Q* settings are 6.7. The result is still an accented click, but one that is cleaner and less obstructive than in the previous track.

Plugin: Sonnox *Oxford EQ*
Drums: Toontrack *EZdrummer*

Q settings are also a vital part of equalization. A *Q* too wide might take on-board unrelated frequencies, a narrow *Q* might not be effective enough. Narrow *Q* is often used when we want to treat a very narrow bandwidth, often an unwanted noise or a resonant frequency. Wide *Q* tends to sound more natural and is mostly used to emphasize wanted parts of the frequency spectrum and for fine tonal alterations. Accordingly, wide *Q* is more common with boosts and narrow *Q* with cuts.

Equalizing various instruments

Spectral components

We say that we can divide the timbre of each instrument into four parts: lows, low-mids, high-mids and highs. Learning how each of these bands affects the tonality of an instrument is one of the first things mixing engineers should grasp. It only takes a few minutes to experiment with how each band affects the tonality of an instrument – the experience is invaluable. However, the equalization of each instrument is not as simple as that – subtle alteration of very specific frequencies can have dramatic effect. By way of analogy, the equalization of each instrument is like a safe waiting to be cracked. It would be fair to make no sub-divisions at all – it is simply 20 Hz–20 kHz and what we are after can be anywhere. Once familiar with the four different bands, we progress and learn how smaller ranges affect the timbre of various instruments. Each instrument has a few influential spectral components – frequency ranges highly vital for its timbre. Learning these takes a bit more time as different instruments have different spectral components at different ranges.

The rest of this chapter covers common instruments, their spectral components and other frequency ranges that might be of interest to us.

The information presented in this chapter hereafter, including any frequencies and advice, are guidelines only. Each recorded instrument produces different sound, each microphone and its position capture different frequency content. The spectral complexity of the sounds we treat is immense. The true essence of equalization cannot be realized by mere guidelines – we must develop critical evaluation skill, which in the case of equalization takes time. I urge the reader to experiment with the information presented below, but in no way take it for granted.

Vocals

It can be argued that the term vocals alone cannot encompass the great variety of voices people have. For one, the voices of males and females are very different. Then, each person has so many unique qualities in his/her voice that every singer can be considered a completely different instrument. It should be fair to assume that vocal equalization starts by eliminating the wrongs – the likes of muddiness, honky sound and sibilance if apparent (which de-essers are likely to rectify better). Being the key element of many productions, vocals must be most prominent. It is vital for vocals to overcome any instrument that masks them. This might entail equalization not only on the vocals, but also on the masking instrument. Different vocal parts have different roles in mixes, so at times we make them sweet, at times warm and at times aggressive. Generally speaking, the more powerful the vocal part is, the more equalization will be used.

Figure 14.34 shows the possible frequency range of the human voice. The long gradient bar shows the possible frequency content of both fundamentals and harmonics. The bounded rectangle denotes the possible fundamentals range, which spans between 82 and 1398 Hz (E2-F6). No person can produce this full range of notes. Also, the extremes of this range can only be produced by skilled opera singers. To give an example, Mariah Carey's vocal range spans approximately 123-830 Hz (B2-G#5). Typical pop singers cover a far less impressive range than that. Above and below the bar are frequency ranges relevant to vocals. It is worth noting that equalizing a specific range can yield either positive or negative results. For example, by boosting the high-mids we might add presence, but at the same time add unwanted sibilance. By attenuating the low-mids we might add clarity but lose warmth and make the sound thin.

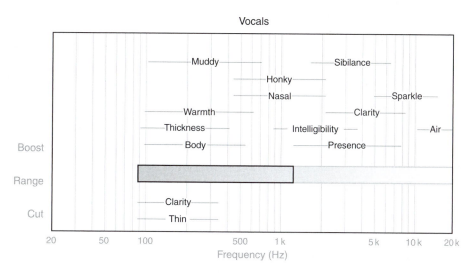

Figure 14.34 The frequency range of the human voice and relevant frequency ranges. The gradient bar denotes the range of possible fundamentals and harmonics. The possible fundamental range is confined in a rectangle. The associations above the bar are likely to be emphasized by a boost in their frequency ranges, the associations below by a cut.

Track 14.107: Vocal No EQ
The unequalized track.

Track 14.108: Vocal HPF 280 Hz
A 12 dB/oct HPF at 280 Hz removes some mud but also some body. The vocal here is clearer compared to the previous track.

Track 14.109: Vocal HSF 9 dB Up
This is an equalized version of the previous track. A 9 dB boost at 10 kHz adds some air and sparkle to the vocal.

Track 14.110: Vocal 1 kHz Boost
This is an equalized version of Track 14.108. An 11 dB boost at 1 kHz adds some nasal character to the voice.

Track 14.111: Vocal 1 kHz Boost
This is an equalized version of Track 14.108. A 6 dB boost at 3 kHz adds some presence and clarity.

Track 14.112: Vocal 300 Hz Dip
This is an equalized version of Track 14.108. A 9 dB dip at 300 Hz also adds some clarity.

Track 14.113: Vocal 130 Hz Boost
This is an equalized version of Track 14.108. An 8 dB boost at 130 Hz adds some warmth.

Plugin: Sonnox *Oxford EQ*

Overheads

The overheads glue all the individual drums together. When the overheads are too quiet, the drums tend to sound very artificial and can come across as lifeless. There are three general approaches to the involvement of overheads in mixes. What differs from one approach to the other is the amount of low-end filtering:

- **Main stereo pair** – the overheads are responsible for the majority of the drum sound, with possible minor reinforcement from close-mics, mostly kick and snare. This is typical in conventional jazz mixes and more natural-sounding productions. No or extremely subtle low-end filtering is applied.
- **First among equals** – an equal share between the overhead and the close-mics. This is common in contemporary productions where kicks and snares are very prominent. A HPF might be positioned in the frequency range between the kick and the snare, mostly to clear the low-end for the kick's close-mic and reduce spatial cues that send the various drums backward.
- **Cymbals only** – the overheads are responsible for the cymbals only, close-mics for all other drums. This least-natural approach entails extreme high-pass filtering (even into the high-mids), intended to remove as much drum content (and room) as possible. This technique can be a lifeboat when the overheads are seriously flawed – nothing new in low budget and home recordings.

In addition to the drum kit itself, overhead recordings also include the room response (the reverb). This creates an illusion of space in the mix, which is more or less similar to that of the recording space. The problem is that sometimes there is too much room, sometimes the space is just not appealing – most domestic rooms and even some studio spaces can produce a weary mix ambiance. HPFs let us correct, to some extent, such recordings.

As previously mentioned, the size, depth and impression of reverbs focus on their lows. One particular problem is that low-frequency drums, notably the kick, tend to excite most rooms more than, say, the cymbals. This is normal as low frequencies take longer to absorb. Yet, it is a definite attribute of overheads recordings due to the small live rooms often used. As an outcome, the reverb from the overheads imposes some depth on the kick, toms and the snare. Such depth might be unwanted, and no matter how the close-mics are mixed, it can be hard to get the drums to the in-your-face position. Filtering the lows from the overheads reduces spatial information and helps bring the individual drums forward in the depth field. The more we filter the smaller we make the room and the closer the image becomes. In order to eliminate more of the individual drums but retain more of the cymbals' timbre, a steep filter is often needed. Combined with the fact that overheads are a broadband signal that involves transients, a high-quality HPF is a requisite for the task. For more gentle results, a shelving filter can be used instead of a pass filter – it allows us to keep more of the warmth and spatial impression that the overheads contribute, but still reduces their size and depth contribution.

By taking the cymbals-only approach, we can filter a flawed room from the overheads. If this dries out the mix ambiance, we can always use a technique called **ambiance reconstruction** – sending the individual drums into a reverb emulator to recreate the missing ambiance. We have full control over the reverb we choose, but we have to make sure that it matches the high-frequency reverb still present on the overheads.

Unless recorded in an exceptionally good room, the low end of the overhead might muddy the mix. Both the lows and low-mids can benefit from some attenuation that will clear some space for the kick (close-mic), bass and the fundamentals of other instruments. However, a caution must be taken as the same lows might contribute to general warmth and the sense of ambiance in the mix.

The highs of the overheads contribute some shimmer. If no cymbal mics are mixed, the overheads might be the ones to govern the high-end of the mix. Thus, how bright or dull we make the overheads can also mean how bright or dull the mix is, and in turn how balanced the overall mix is.

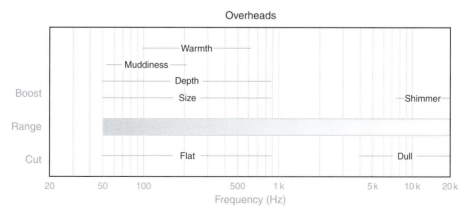

Figure 14.35 The frequency range of overheads and relevant frequency ranges.

Track 14.114: Overheads No EQ
The unequalized track, used for all the following samples:

Track 14.115: Overheads HPF 100 Hz
An 18 dB/oct with −9 dB point at 100 Hz. The filter removes much of the ambiance and attenuates the kick.

Track 14.116: Overheads HPF 420 Hz
A steep 36 dB/oct HPF is set to 420 Hz. This type of processing could be used for the cymbals-only approach as it filters much of the kick and the snare. Depending on the mix, even cut-off frequencies higher than 420 Hz could be appropriate.

Track 14.117: Overheads HSF Up
A 2.4 dB boost at 7.8 kHz adds shimmer.

Plugin: Universal Audio *Cambridge Equalizer and Filters*
Drums: Toontrack *EZdrummer*

Kick

Early in this chapter the kick was lauded as the king of tonal presentation. If you listen to a collection of productions, you will find kicks in many shapes and forms – the thunderous, the rounded, the 909, the 808, the woody, the typewriter style, the basketball, the pillow-like, to name a mere few. How we shape the sound of a kick is one of the most creative decisions we have to make; we control aspects like how solid, how punchy, how thick and how snappy. But the sound of the kick is also a practical affair. Being the admiral of the rhythm, the kick has immense weight in the ability of the mix to move people. A weak kick on a hip-hop mix means a weak hip-hop tune; ditto for dance; and mostly so for rock, pop and metal. When it comes to mixing kicks, equalizers play a huge and powerful role.

Kicks have two main spectral components – the impact and the attack. These two components correspond closely with the lows and high-mids, respectively. Within these two ranges, different frequencies have a different effect. For example, 60 Hz might produce more oomph, while 90 Hz more thud. Dance and hip-hop might benefit from robust

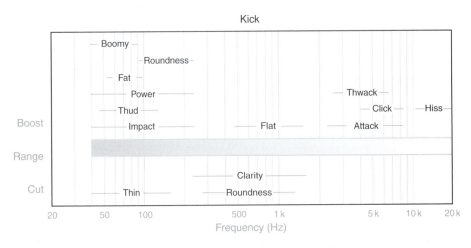

Figure 14.36 The frequency range of a kick and relevant frequency ranges.

oomph, while rock and pop might benefit from a solid thud. On the high-mids, the higher the frequency the more the click. A healthy boost around 8 kHz can produce the typewriter click that many associate with heavy metal. The low-mids of the kick have little to offer, and attenuating them can clear resourceful space for the low-order harmonics of other instruments. On the highs, kicks have very little valuable energy, which is sometimes entangled with hiss. Often rolling-off the highs from a kick would go unnoticed.

One important aspect of equalizing a kick is how it interacts with the bass. As the two instruments are competing for the lows, sometimes we have to sacrifice the impact of one for the benefit of the other. Another important issue with mixing kicks is that there is always a risk of mishandling the low-end due to limited monitors or poor acoustic environments. Under such conditions, it pays to spend some time going through the stabilizing options described in Chapter 4.

Track 14.118: Kick No EQ
The unequalized track, used for all the following samples.

Track 14.119: Kick LSF 160 Hz Up
A 6 dB boost at 160 Hz on a low-shelving filter adds some thud.

Track 14.120: Kick 344 Hz Dip
This is the result of −10 dB at 344 Hz.

Track 14.121: Kick 1600 kHz Boost
The added thwack on this track was achieved by a 9 dB boost at 1.6 kHz.

Track 14.122: Kick 5 kHz Boost
A 6 dB boost at 5 kHz adds some attack.

Track 14.123: Kick 7 kHz Boost
A 6 dB boost at 7 kHz also adds some attack, but with more click.

Track 14.124: Kick Two Bands
This track is the outcome of −8 dB at 270 Hz on a high-shelving filter, together with −10 dB at 261 Hz.

Plugin: Universal Audio *EX-1*
Kick: Toontrack *EZdrummer*

Snare

Perhaps the prince of rhythm, the tonality of the snare is also an important aspect of many mixes. Snare recordings are often less than ideal. It is often only high-budget projects that can afford snare comparisons, fresh skins, accurate tuning, appropriate muffling, a suitable microphone and correct placement. In all other projects, snares are prone to a few issues that might need correction – excess rattle and resonant ring are two examples.

Although the spectral components of a snare are not as distinct as those of a kick, we often talk about body and presence, while also considering aspects like snap and crispiness. One of the playful parts of snare mixing entails tuning them into the frequency spectrum, i.e., deciding how high or low they sit, also with relation to other instruments. A dark snare will be more distant and laid-back. A bright snare will be more upfront and active. Automating snare sounds (between verse and chorus for example) has rationale in more than a few mixes.

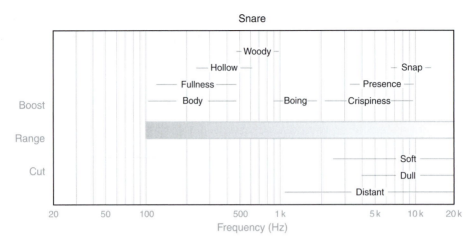

Figure 14.37 The frequency range of a snare and relevant frequency ranges.

Track 14.125: Snare No EQ
The unequalized track, used for all the following samples.

Track 14.126: Snare 150 Hz Boost
A 6 dB boost at 150 Hz adds body.

Track 14.127: Snare 400 Hz Dip
This is the result of a −6 dB at 400 Hz.

Track 14.128: Snare 910 Hz Boost
A 6 dB boost at 910 Hz.

Track 14.129: Snare 4700 Hz Dip
A −6 dB at 4.7 kHz.

Track 14.130: Snare HSF 10 kHz Boost
A 4 dB boost at 10 kHz on a high-shelving filter.

Plugin: PSP *Neon*
Snare: Toontrack *EZdrummer*

Toms

While often only playing occasionally, toms might just be blended into the overall drum sound, but can also be mixed to have a distinct powerful impact. Toms, notably floor toms, can get a bit out of control unless we contain their lows. Also, it is important to observe timbre differences between the various toms – poorly tuned toms that are recorded with different microphone brands can sound as if each belongs to a different kit.

Toms have some similarity to kicks in the way we mix them, in that they have more defined pitch and far longer decay. Needless to say, a high tom will have its spectral components higher on the spectrum than a floor tom. At the very lows there is usually rumble that might need filtering. The fullness and thud are around the higher-lows, 200

Hz or so. The attack, like with kicks, is on the high-mids. It is also worth seeing whether cutting the highs reduces cymbal spill without affecting the sound of the toms.

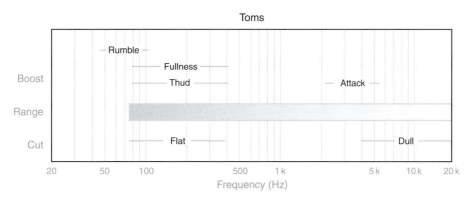

Figure 14.38 The frequency range of toms and relevant frequency ranges.

Track 14.131: Tom No EQ
The unequalized track, used for all the following samples.

Track 14.132: Tom 100 Hz Boost
A 9 dB boost at 100 Hz adds some thud and tone.

Track 14.133: Tom 400 Hz Dip
A −7 dB at 400 Hz.

Track 14.134: Tom 3 kHz Dip
A −9 dB at 3 kHz reduces the attack.

Plugin: McDSP *G Equalizer*
Tom: Toontrack *EZdrummer*

Hi-hats and cymbals

The hi-hats are often the brightest instrument in the mix. As such, they play a big part in our perception of how bright the mix is altogether. One common problem with hi-hats is that they sound detached from the rest of the mix. To elaborate, they are perceived in the spectrum as the brightest instrument, but there is some empty frequency range below them. This is often the outcome of overemphasized highs. One characteristic of many pleasing mixes is that the hats are not glaring bright – they just shimmer. It is worth remembering that it might be the overheads contributing to this aspect in the mix and not close-mics.

Cymbals have very similar characteristics to hi-hats, but as they only play occasionally they play a less crucial part in the overall frequency balance. The larger the cymbal the lower its fundamental would be. Hi-hats, for example, can have some content below 500

Hz; rides go lower than that. These lower ranges of cymbals, which often involve bleed, can sometimes be pulled without affecting too much of the timbre.

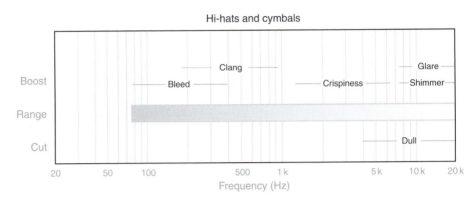

Figure 14.39 The frequency range of hi-hats and cymbals, and relevant frequency ranges.

Track 14.135: Cymbals No EQ
The unequalized track, used for all the following samples.

Track 14.136: Cymbals HPF 400 Hz
An 18 dB/oct HPF at 400 Hz. Such a treatment would doubtfully be noticeable with the rest of the mix playing along.

Track 14.137: Cymbals 400 Hz Boost
A 6 dB boost at 400 Hz adds some clang, mainly to the ride.

Track 14.138: Cymbals 1 kHz Boost
A 4 dB boost at 1 kHz adds some crisp.

Track 14.139: Cymbals HSF 6 kHz Up
A 4 dB boost at 6 kHz on a high-shelving filter adds some glare.

Plugin: Digidesign *DigiRack EQ 3*
Cymbals: Toontrack *EZdrummer*

Bass

Usually one of the trickiest instruments to mix is the bass – whether a guitar or a synth. For one, recordings tend to be very different from one another. But the real problem with bass sounds is that their tone often changes with relation to the note being played. For example, the lower E (41 Hz) produces substantially more low frequencies than the E two octaves above it (164 Hz). Solid tone can be achieved with equalizers, although both dynamic equalizers and compressors (especially multiband) can do a better job. There are also different playing techniques and different bass sounds in different genres. The bass in funk, for example, is miles away from the bass in metal.

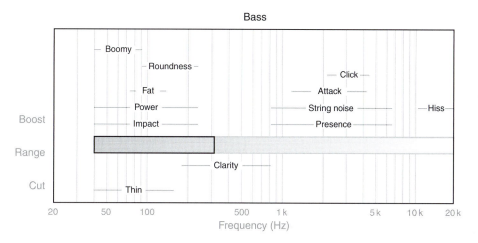

Figure 14.40 The frequency range of a bass guitar and relevant frequency ranges.

There are two principal aims when mixing a bass: solid lows power, and definition. Neither of these is easy to achieve. Being of sustained nature in most productions, the bass is often the instrument that fills the low-end. Too much lows, and the mix is boomy; too little, and the mix is thin (and as mentioned this characteristic can change with relation to the note being played). Then, the bass and the kick are often in competion for the lows. Most bass sounds (whether recorded or synthesized) have relatively little energy on the high-mids and highs, and some have no energy at all above, say, 5 kHz (specific recordings can have almost nothing even below that). On the low-mids there are usually enough instruments fighting for space. So defining the bass might require some experimentation. Any mixing tool that generates harmonics (for example: distortions and enhancers) can be extremely useful in providing some definition for bass sounds.

Track 14.140: Bass No EQ
The unequalized track, used for all the following samples.

Track 14.141: Bass LSF 230 Hz Boost
A 7 dB at 230 Hz on a low-shelving filter adds some power and fattens the bass.

Track 14.142: Bass 500 Hz Boost
A 7 dB boost at 500 Hz.

Track 14.143: Bass 1 kHz Boost
A 6 dB boost at 1 kHz adds some presence.

Track 14.144: Bass HSF 1 kHz Down
A −10 dB at 1 kHz on a high-shelving filter removes some attack and presence.

Plugin: Cubase *StudioEQ*

Acoustic guitar

Acoustic guitars can play various roles in the mix. On some mixes, often sparse ones, the guitar is one of the main instruments, usually along with vocals. In such circumstances, we often want the guitar sound to be rich and full-bodied. On other mixes, the acoustic guitars only provide a reinforcement of the harmony and rhythm, which means that their body can be less important. Very commonly acoustic guitars are treated with HPFs. The cut-off frequency is set depending on how much space there is, often starting from around 80 Hz – just below the lower E (subject to standard tuning). In many commercial mixes the body of the guitar is fully removed, leaving it playing more of a rhythmical role than a harmonic one. When appropriate, the cut-off frequency is set so high that all that is left is a very thin, somewhat metallic sound – the played chords can hardly be discerned. Another aspect of equalizing an acoustic guitar involves treating its body resonance and could result in unwanted accent of a specific spectral area or an increased level and sustain of specific notes. Finger squeaks are another thing we might want to reduce using equalization. We can also control the timbre of the guitar by accenting or easing second- and third-order harmonics, which are found on the low-mids. The very highs of an acoustic guitar can be very sensitive to boosts, and even a small push from 10 kHz and above can make the sound appear thin and cheap. For this reason, if an acoustic guitar needs to be brightened, it could benefit from a wide-Q bell, rather than a shelving EQ.

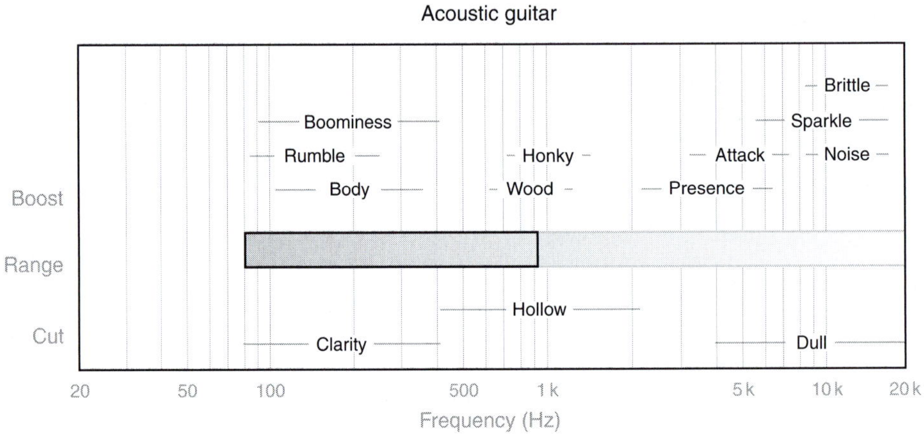

Figure 14.41 The frequency range of acoustic guitar and relevant frequency ranges.

Track 14.145: aGtr HPF 220 Hz
The acoustic guitar in Track 14.85, with a 12 dB/oct HPF at 220 Hz. This is the source track to be equalized in the following samples:

Track 14.146: aGtr 300 Hz Boost
A 4 dB at 300 Hz adds some body, but also a hint of boominess.

Track 14.147: aGtr 1 kHz Boost
The honkiness on this track is the result of a 7 dB boost at 1 kHz.

Track 14.148: aGtr 4 kHz Boost
A 2 dB with wide Q at 4 kHz is sufficient to add noticeable presence.

Track 14.149: aGtr HSF 7 kHz Up
This is the result of a 3 dB boost at 7 kHz on a high-shelving filter.

Plugin: PSP *MasterQ*

Clean electric guitar

Electric guitars are an inseparable part of rock music, and on many productions, provide the harmonic backbone. Most of them arrive to mixing already equalized – whether by the tone controls on the guitar itself or those on the amplifier. Often our aim with clean electric guitars is making them well defined, yet without masking other instruments. The way the guitar is played is a factor in the way it is mixed. Some involve strumming chords while others are played in a phrasal fashion; some are played more rhythmically while others more melodically. The latter require more attention for spectral balance between different notes. Often we get more than one guitar, and unless different on the recordings, we can distinguish them using equalization. Although to a lesser degree than with acoustic guitars, electric guitars can also benefit from low-end filtering, which is dependent on the density of the arrangement.

Track 14.150: cGtr No EQ
The source track to be equalized in the following samples.

Track 14.151: cGtr HPF 230 Hz
A 6 dB/oct HPF at 230 Hz removes some of the guitar's body, but in mix context this type of treatment could yield clarity.

Track 14.152: cGtr 350 Hz Dip
This track is the result of a 3 dB attenuation at 300 Hz.

Track 14.153: cGtr 1 kHz Boost
Again, the honkiness on this track is the result of a 7 dB boost at 1 kHz.

Track 14.154: cGtr 4 kHz Boost
A 4 dB boost at 4 kHz adds some presence and attack.

Track 14.155: cGtr HSF 5 kHz Up
A 5 dB boost at 5 kHz on a high-shelving filter adds some spark.

Plugin: McDSP *FilterBank E6*

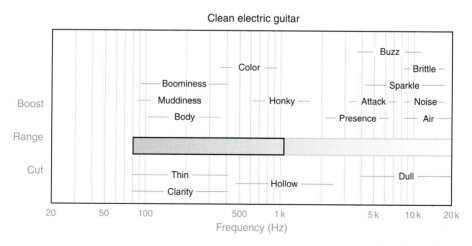

Figure 14.42 The frequency range of clean electric guitar and relevant frequency ranges.

Distorted guitar

Perhaps the frequency-richest instrument is the distorted guitar. This is both a disadvantage and an advantage. It is a disadvantage since distorted guitars are masking animals – most of them have dominant energy from the very lows to the very highs, and they can easily cloud everything else. It is an advantage since we can shape their sound in many fascinating ways. Distorted guitars are super-EQ-sensitive. A boost of 3 dB on the high-mids can easily make an EQ whistle. They are responsive all the same to small level cuts. A quick listen to commercial tracks would reveal that many distorted guitars are elegantly powerful, but not pushy powerful. Put another way, they are not just thrown into the mix at a loud level, but crafted more wisely to blend with other instruments. In many cases distorted guitars also involve some stereo effect or another.

Low-end rumble is often removed or attenuated. How much and up to what frequency the filter goes is done by ear. It is worth remembering that we can first clean the mix (going a bit higher with the filter) and later add a few more lows if we still feel these are missing. Much of the sound of distorted guitars has to do with the actual distortion being used, and there are as many flavors of guitar distortions as there are notes on a guitar. Often the high-end contains grainy noise that the mix can live without. The mids can also benefit from softening, which generally clears some space for all other instruments. But taking too much of the mids creates a very distant sound that is mostly associated with some metal sub-genres. Another consequence of softening the mids is that the guitars become louder as the mix is played louder. This is the nature of any instrument with pulled mids, but it makes more sense with distortion guitars since they are more controlled in quiet levels, but emphasized at louder levels (where things tend to get messier anyway).

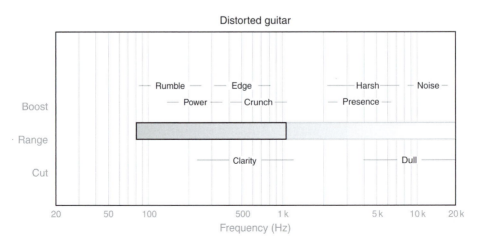

Figure 14.43 The frequency range of distorted guitar and relevant frequency ranges.

Track 14.156: dGtr Source
This track is a combination of a HPF (115 Hz, 24 dB/oct), a LPF (10.7 kHz, 12 dB/oct) and a high-shelving filter (11.5 kHz, −11 dB), all applied on Track 14.81. This is the source track to be equalized in the following samples.

Track 14.157: dGtr 200 Hz Dip
A 6 dB attenuation at 200 Hz. This attenuates even further some low-frequency rumble.

Track 14.158: dGtr 500 Hz Dip
This track is the result of a 6 dB attenuation at 500 Hz.

Track 14.159: dGtr 1500 Hz Boost
A 4 dB boost at 1.5 kHz.

Track 14.160: dGtr 4 kHz Dip
A 4 dB attenuation at 4 kHz.

Plugin: MOTU *MasterWroks EQ*

15 Introduction to dynamic range processors

Chapters 16-20 deal with compressors, limiters, gates, expanders and duckers. All fall into a group of processors called *dynamic range processors*. Dynamic range processors have common ground worth discussing as a background to the succeeding chapters.

Dynamic range

Dynamic range is defined as the difference between the softest and loudest sounds a system can accommodate. Our auditory system, a digital system (like an audio sequencer or a CD) and an analog system (like a loudspeaker or a tape) are all associated with dynamic range. Absolute level measurements are expressed in various dB notations – dBSPL, dBu, dBFS, etc. Dynamic range, like any *difference* between two level measurements, is expressed in plain dB. The table below shows the dynamic range of various systems.

System	Approximate dynamic range (dB)
Software mixer (25 bits)	150
Best A/D converters available	122
Human hearing	120
Dynaudio Air 25 Active Monitors	113
Neumann U87 Condenser Microphone	110
CD, or 16-bit integer audio file	96
Tape with noise reduction	75
FM radio	65
Vinyl	60

Say, the quietest moment of an orchestral piece involves a gentle finger tap on the timpani, which generates 10 dBSPL. Then at the loudest moment the orchestra generates 100 dBSPL. That is a musical piece with 90 dB of dynamic range. A U87 microphone placed in the venue would theoretically be able to capture the full dynamic range of the performance, including both the quietest and the loudest moments (although during the quiet moment the microphone noise might become apparent and during the climax the microphone might distort). All but the cheapest A/D converters can accommodate 90 dB of dynamic range, so when the microphone signal is converted to digital and recorded onto a computer the dynamic range of the performance is retained. The full dynamic range will also be retained if the performance is saved onto a 16-bit integer file, burnt onto a CD or played through good studio monitors.

However, we will encounter a problem cutting this performance on vinyl – we have to fit 90 dB of dynamic range into media that can only accommodate 60 dB. There are a few things we can do. We can just trim the top 30 dB, so every time the orchestra becomes loud it is muted – an idea so senseless that no processor is actually utilized to operate this way. We can just trim the bottom 30 dB of the performance, so every time the orchestra gets really quiet there will be no sound. It might also not be the wisest idea, although we do something similar every time we bounce from our audio sequencer (25 bits) into a 16-bit file – we trim the bottom 9 bits (reducing from 144 to 96 dB).

How much is 60 dB? Readers are welcome to experiment: use pink noise from a signal generator, and bring it down by 60 dB using a fader. Chances are that at −60 dB you would not hear it. Now boost your monitor level slightly in order to hear the quiet noise, then slowly bring the fader back up to 0 dB. Chances are you will find the noise fairly loud, and you might not even fancy bringing the fader all the way up; 60 dB is regarded in acoustics as the difference between something clearly audible to something barely audible. We will discuss this further in Chapter 23 on reverbs. The 60 dB experiment might be disappointing for some – you might wonder where did half of our 120 dB hearing ability go. The auditory 120 dB was devised from what was thought to be the quietest level we can hear – 0 dBSPL – and the threshold of pain which is around 120 dBSPL. In practice, most domestic rooms have ambiance noise at roughly 30 dBSPL; 120 dBSPL is the level of an extremely loud rock concert or a dance club when you stand a meter away from the speaker. Under such devoted circumstances you might not be able to hear a shouting person next to you. A shouting voice is roughly 80 dBSPL. We perceived 90 dBSPL as twice as loud, which is already considered very loud. If there is such a thing as the dynamic range of everyday life, it is somewhere between 30 and 70 dBSPL – 40 dB of dynamic range. Still in a studio control room, it is practical to talk about levels between 20 dBSPL and 110 dBSPL.

Having said that, we might still want our vinyl to retain both the loudest and softest moments of our orchestral performance, and trimming would not let us do so. What we can do is **compress** the dynamic range. We can make loud levels quieter – known as **downward compression**; or we can make soft levels louder – **upward compression**. By way of analogy, it is like squeezing a foam ear-plug before fitting it into our auditory canal – all the material is still there, but compressed into smaller dimensions.

Say, we have the opposite problem: we have an old vinyl that we want to transfer onto a CD. We might want to convert 60 dB of dynamic range into 96 dB. A process called **expansion** enables this. We can make quiet sounds even quieter – **downward expansion**; or we can make loud sounds even louder – **upward expansion**. It is similar to what happens when we take the ear-plug out – it returns to its original size.

By and large, downward compression is the dominating type of compression in mixing, so by convention 'downward' is omitted – a compressor denotes a downward compressor. It is the same with expansion – downward expansion is more common, thus 'downward' is omitted. This book follows these conventions.

Now how does this relate to mixing? Well, very little but at the same time quite a lot. It is the mastering engineer's responsibility to fit the dynamic range of a mix onto different kinds of media (and they do compress vinyl cuts). Nowadays, with the decreasing popularity of tape machines, we hardly change between one system and another – during mixdown, we care very little about the overall dynamic range of our mix. However, very often while we mix we want to make loud or quiet sounds louder or quieter. It is rarely the overall dynamic range we have in mind – what we really want to control is **dynamics**.

Dynamics

The term dynamics in mixing is equivalent to the musical term – variations in level. Flat dynamics mean very little or next to no variations in level; vibrant dynamics mean active and live variations; wild dynamics mean excessive fluctuations that are somewhat out of control. We distinguish between **macrodynamics** and **microdynamics**.

In this book, **macrodynamics** are regarded as variations in level for events longer than a single note. We talk about macrodynamics in relation to the changing level between the verse and the chorus, the level variations between snare hits, bass notes or vocal phrases.

Microdynamics are related to level variations that happen within each note being played due to the nature of an instrument, for example, the attack and decay of a snare hit. We associate microdynamics with the **dynamic envelope** of sounds, which entails level variations that happen within each note (or hit) being played. The dynamic envelope constitutes the second half of an instrument's timbre (first half being the spectral content). We can associate different envelopes with different instruments, as Figure 15.1 demonstrates. It is worth knowing that most instruments have some initial transient or level burst, as part of their dynamic envelope. Very often we employ dynamic range processors to control the microdynamics in these envelopes or reshape them to alter the instrument's timbre.

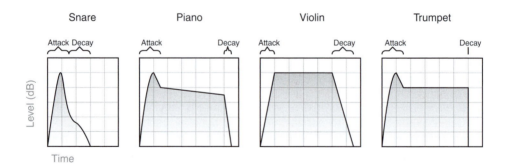

Figure 15.1 The typical dynamic envelope of various instruments. A snare has a quick attack caused by the stick hit; the majority of impact happens during the attack stage, and there is some decay due to resonance. A piano has also got an attack bump due to the hammer hitting the string; as long as the key is pressed the sustained sound drops slowly in level; after the key is released, there is still a short decay period before the damping mechanism reaches full effect. The violin dynamic envelope in this illustration involves legato playing with gradual bowing force during the initial attack stage; the level is then kept at a consistent sustain level, and as the bow is lifted, string resonance results in slow decay. In the case of a trumpet note, the initial attack is caused by the tongue movement; consistent air flow results in a consistent sustain stage, and as the air flow stops, the sound drops abruptly.

It is worth noting the terminology used in Figure 15.1 and throughout this book. The attack is the initial level build-up and includes the fall to sustain level. Decay is used to describe the closing level descent. This is different from the terminology used for a synthesizer's ADSR (attack, decay, sustain, release) envelope.

Dynamic range processors in a nutshell

Transfer characteristics

Say, we regard an electrical wire as a processor, and its ends to be the input and output. Theoretically, signals entering the wire pass through unaffected. For any given input level we get the same output level. We can say that the level ratio between the input and output is 1:1. Such a ratio is known as *unity gain*. We can draw a graph that shows the relationship between the input and output levels. Such a graph is called a *transfer characteristics graph*, or *input-output graph*, and is said to show the *transfer function* of a processor. It shows the input levels on the horizontal axis and the output level on the vertical axis. The scale is given in decibels; 0 dB denotes the reference level of the system – the highest possible level of a digital system (which cannot be exceeded) or the standard operating level of an analog system (which can be exceeded). Figure 15.2 shows the transfer function of an electrical wire.

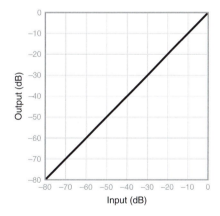

Figure 15.2 The transfer function of an electrical wire. There is no change in level between the input and the output – in other words, unity gain throughout.

As opposed to wires, dynamic range processors do alter the level of signals passing through them – input and output levels might be different. We already know that a compressor reduces dynamic range, so we can expect input levels to be brought down. Figure 15.3 shows the transfer function of a compressor. We can see that the output's dynamic range is half that of the input's.

Had all dynamic range processors behaved like the compressor in Figure 15.3, they would all be rather limited (in fact, the earliest dynamic range processors were limited very much in this way). Treating all levels uniformly is rarely what we want, even in cases where it is the

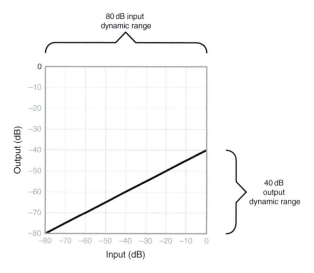

Figure 15.3 The transfer function of a compressor. The output dynamic range is half the input dynamic range.

whole mix we are treating. Most of the time, we only want to treat a specific level range – commonly either the loud or quiet signals. Moreover, dynamic range processors provide additional treatment control with functions like attack and release that benefit from more focused level treatment. To enable us to treat selectively, all dynamic range processors allow us to limit treated levels and untreated levels using a parameter called **threshold**.

The threshold divides the full input level range into two sections. Depending on the processor, the treated level range might be above or below the threshold. Figure 15.4

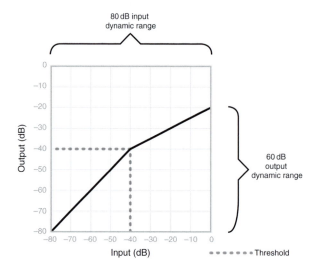

Figure 15.4 A compressor with threshold. The threshold is set to -40 dB. Levels below the threshold are not treated, and the input to output ratio is 1:1. Above the threshold, input signals are brought down in level. In this specific illustration the input to output ratio above the threshold is 2:1.

shows a compressor with the threshold set at −40 dB. Being a downward compressor, loud levels are made softer, but only above the threshold – levels below the threshold are unaffected. Note that the output dynamic range has been reduced. It is also worth noting that the input to output ratio above the threshold is 2:1, resulting in the top 40 dB of input range (−40 to 0 dB) turning 20 dB at the output (−40 to −20 dB). We will discuss ratios in greater detail in the following chapters.

The function of different processors

Figure 15.5 shows the transfer function of various dynamic range processors, and how levels are altered between the input and output. Here is a quick summary of what each of these processors does:

- **Compressor** – reduces the level of signals above the threshold, making loud sounds quieter.
- **Limiter** – ensures that no signal exceeds the threshold by reducing any signals above the threshold down to the threshold level.
- **Upward compressor** – boosts the level of signals below the threshold, making quiet sounds louder.
- **Expander** – reduces the level of signals below the threshold, making quiet sounds quieter.
- **Upward expander** – boosts the level of signals above the threshold, making loud sounds even louder.

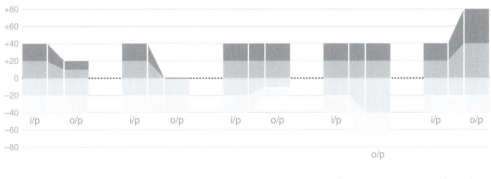

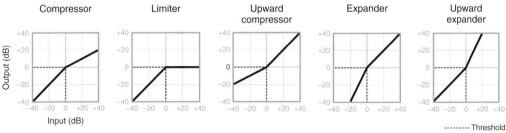

Figure 15.5 Sample transfer functions of compressors, limiters and expanders, and their effect on input signals. The bottom graphs show the transfer function of each processor, all with threshold set to 0 dB. The top row illustrates the relationship between the input and output levels.

All the processors above apply some ratio while transforming levels, making the level change dependent on the input level. For example, the compressor above has a 2:1 ratio above the threshold; +40 dB at the input is reduced to +20 dB (20 dB level change), while +20 dB is reduced to +10 dB (10 dB level change). The ratio of reduction is identical, but the level change is different.

Gates work on a slightly different principle – all signals below the threshold are reduced by a fixed amount known as range. With large range settings (say −80 dB), signals below the threshold become inaudible and are said to be muted. A ducker also reduces the signal level by a set range, but only does so for signals above the threshold. Figure 15.6 shows the function of a gate and a ducker, which can be summarized as follows:

- **Gate** – attenuates all signals below the threshold by a fixed amount known as range. Drastic attenuation makes everything below the threshold inaudible.
- **Ducker** – attenuates all signals above the threshold by a fixed amount known as range.

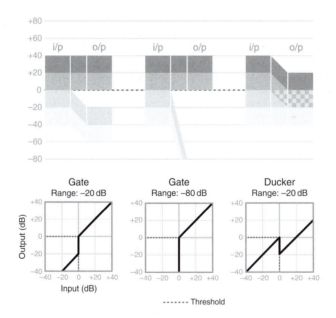

Figure 15.6 The transfer function of a gate with different range settings and a ducker. With small range settings, the gate simply attenuates everything below the threshold. With large range settings, the gate is said to mute signals below the threshold. A ducker attenuates signals above the threshold by a set range.

Pumping and breathing

Dynamic range processors alter dynamically the level of input signals and as a result can produce two artifacts known as pumping and breathing. **Pumping** is caused by quick noticeable variations of level. We usually associate pumping with loud level variations, such as those that can be the outcome of heavy compression or limiting. **Breathing** is the audible effect caused by varying noise (or hiss) levels. Often it is the quiet level variations that produce such artifacts, mostly due to the operation of gates or expanders.

Tracks 15.2 and 15.3 demonstrate pumping. The threshold on the compressor is set to −45 dB, attack set to slow and ratio to 60:1. In all of these tracks, the kick triggers the heaviest compression. It is worth noting the click on the very first kick, which is caused by the initial, drastic gain reduction.

Track 15.1: Original (No Pumping)

Track 15.2: Pumping (Fast Release)
The quick gain reduction and recovery is evident here. For example, the level of the hats fluctuates severely.

Track 15.3: Pumping (Medium Release)
The medium release causes slower gain recovery, which means levels fluctuate less. Although not as wild as in the previous track, the level changes are still highly noticeable.

Track 15.4: Pumping (Slow Release)
The slow release means that the gain hardly recovers before the next kick hits. The resultant effect cannot really be considered as pumping, since the gain fluctuations are too slow. However, it is hard to overlook how such slow recovery adds some dynamic movement to the loop.

Track 15.5: Breathing
The fluctuating noise level is associated with the term 'breathing'. The vocal and the underlaid noise pass through a gate/compressor – both respond to the vocal. The gate causes the noise to come and go, while the varying amount of gain reduction on the compressor results in fluctuating noise level.

Plugin: Universal Audio *GateComp.*

Exit Music (For a Film)
Radiohead. *OK Computer.* **Parlophone, 1997.**
Mixed by Nigel Godrich.

Pumping is usually an unwanted artifact. They say once rules are learned, they can be broken. The drums on this track are pumping, and the effect grows with the building intensity of the song (with most severe pumping between 3:22 and 3:37). In this specific track, the pumping effect creates a chaos-like feel that complements the music in its emotional climax.

16 Compressors

Perhaps the most misused and overused tool in mixing is the compressor – an especially worrying thing considering how predominant compressors are in contemporary mixes. Compressors, to a large extent, define much of the sound of contemporary mixes. It is not a secret that compressors can make sounds louder, bigger, punchier, richer, more powerful, and if they were kitchenware they would probably also make food tastier. Neither is it a secret that since compressors were introduced per channel on analog consoles, the amount of compression in mixes has grown significantly – from transparent compression to evident compression to heavy compression to hypercompression to ultracompression, so that by the start of the new millennium compression had been applied in so many mixes to such massive degrees that some people think it absurd:

- Are these VU meters broken?
- No.
- So why aren't they moving?
- Ah! It's probably my bus compression. Sounds like the radio, doesn't it?

If used incorrectly or superfluously, compressors suppress dynamics. Dynamics are a crucial aspect of a musical piece and a key messenger of musical expression. How impossible would our life be if we could not alter the loudness of our voice? How boring would our singing be? A drummer would sound like a drum machine, and every brass would resemble a car horn. If anything is to blame for the lifeless dynamics in contemporary music, it is the ultracompression trend.

Luckily, in recent years more and more mixing and mastering engineers are leaving the ultracompression club and reverting back to more musical compression. There is nothing here to suggest that compressors should not be used, or used subtly. Many mixes still benefit from a hard-working compressor on nearly every track. The real secret is to retain musical dynamics. We also have to distinguish the tightly controlled dynamics of pop music and the loss of dynamics caused by ultracompression. More than a few mixes present tightly controlled dynamics, yet instruments sound alive. Ultracompression is not a requisite for commercial success; in fact, it is a threat.

One specific problem novice engineers have is that they do not know how the dynamics of a mix should sound. All they have for comparison is the dynamics of a mastered mix or, worst still, the dynamics of a mastered mix played through the radio. The latter can be squashed dead compared to a pre-mastered mix. By trying to imitate the sound of a mastered album or the radio, the novice can easily downgrade the final master. Reviving a lifeless mix during mastering is hard; but if a mix has vibrant dynamics, mastering engineers can treat them as much as they want or are asked to, and would probably be able to do so much more skillfully than anyone else.

| *One must **pay attention to musical dynamics** when compressing.* |

People often underrate how hard compressors are to master. Making a beat punchy is not as easy as simply compressing it. There are more than a few controls we can tweak and most of them are correlated. In order to obtain excellent results from a compressor, we must first understand how it works, the function of each control and how these affect the dynamic behavior of each treated instrument. A compressor is such a powerful and versatile tool that within the many applications it has, some have opposing natures (e.g., softening transients or emphasizing transients). Perhaps after reading this chapter, those who used compressors in the past will feel that they have been flying a spaceship when they thought they were driving a car. But the knowledge provided here can make the difference between the flourish techniques of the seasoned pro and the random trials of the novice. The consequences on the mix are profound.

The course of history

One problem in early radio transmissions was that if program levels exceeded a certain threshold, the transmitter would overload and blow – never a good thing, particularly during the climax of a thrilling sport broadcast. Radio engineers had to observe levels meticulously and ride the gain of live broadcasts so that levels did not exceed the permitted threshold (Figure 16.1). While these engineers did their best, only a fortune-teller could predict sudden level changes, and it was safe to assume that no fortune-teller would

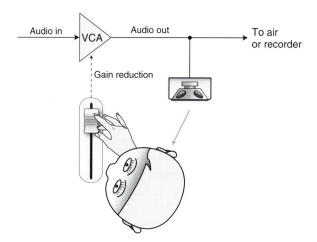

Figure 16.1 Manual gain-riding. The engineer gain-rides levels with relation to the speaker output. In this illustration, a VCA fader is employed.

apply for the job. Even if they did, there is a limit to just how quickly one can respond to sounds or physically move a fader – sudden level changes would still be an issue. While **peak control** was aimed at protecting the transmitter, gain-riding was also performed in order to **balance** the level of the program, making the different songs and the presenter's voice all consistent in level.

By the 1920s, people like James F. Lawrence Jr (who later designed the famous LA-2A) already had ideas about building a device that would automate the gain-riding process. The concept was to feed a copy of the input signal to a side-chain, which, based on the incoming level, would determine the required amount of gain reduction. The side-chain was connected to a gain stage, say a VCA, which applied the actual gain reduction on the signal (Figure 16.2). The name given to these early devices was a *leveling amplifier* (suggesting more of a balancing function) or a *limiter* (suggesting more of a peak control function). However, early models had very slow response to sudden level changes, thus they did not really limit the level of the signal – they behaved very much like today's compressors. With advances in technology, true limiting was eventually possible, and the distinction between compressors and limiters had been made.

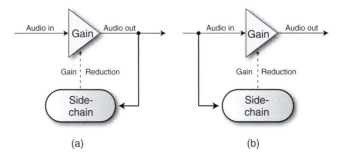

(a)   (b)

Figure 16.2 (a) A feedback-type compressor. The input into the side-chain is taken past the gain stage, which resembles the way manual gain-riding works. Early compressors benefited from this design as the side-chain could rectify possible inaccuracies of the gain stage (for example, when it did not apply the required amount of gain reduction). However, this design has more than a few limitations; for instance, it did not allow a look-ahead function. (b) A feed-forward type compressor has the input into the side-chain taken before the gain stage. As most modern compressors are based on this design, it will be discussed in the rest of this chapter.

Studio engineers quickly borrowed compressors for their recording studios as, just like radio engineers, they had to gain-ride live performances. At the time, it was necessary to contain peaks so that when music was cut on-the-fly to disc, it would not distort. Later, it was utilized to protect from tape saturation, and even later to protect digital systems from clipping. Compressors were also employed to even out the dynamics of a performance, for example, when vocals change from soft to loud. By the 1960s, units like the Urei 1176 LN, Fairchild 670 or the LA-2A were already a common sight in control rooms.

Compressors alter both the dynamic envelope of the source material and, like any other nonlinear device, deposit some distortion – together we get the distinctive effect of compression. The original intention of compressors was to alter the dynamic range while leaving as little audible effect as possible. However, it soon became apparent that the effect of compression, including that caused by an overloaded tape, can be quite appealing. One sonic pioneer who understood this was Joe Meek, who instead of concealing the effect of compressors used it as part of his distinctive sound. Among his many highly acclaimed recordings was the first British single ever to top the US billboard chart – *Telstar* by The Tornados (1962). Nowadays, the effect of compression can be heard in nearly every mix.

> *Sometimes we want the compressor to be transparent. Sometimes we use it for its distinctive effect.*

When it comes to the use of a compressor as a level balancer, the end to this brief history is quite interesting. The compression effect becomes more evident with heavier compression. If a vocal performance changes from crooning to shouting, the effect will be more noticeable on the shouting passage, which would be compressed more. While after compression the performance might appear to be consistent in level, the shouting passage will have a distinct compression effect that will be missing from the crooning passage. To combat this, manual gain-riding often takes place before compression. This takes us back to the start.

The sound of compressors

No two compressors sound alike, certainly no two analog compressors. We all know that some can work better for drums, some can work better on vocals; some provide added warmth, some an extra punch; some can be transparent while others produce a

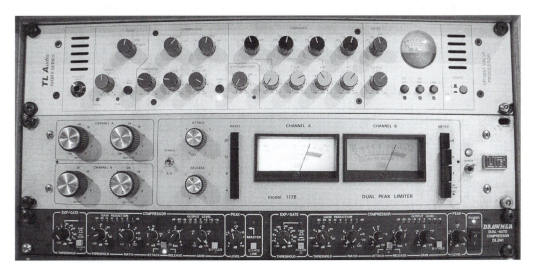

Figure 16.3 Top: the TL Audio *VP-5051*. Middle: the Urei *1178* (the stereo version of the 1176). Bottom: Drawmer *DL241*. The three have distinctively different sounds (courtesy of SAE Institute, London).

very obvious effect. Each compressor has a character, and in order to have a character something must be different from its counterparts.

We can generalize what we want from a *precise* compressor, here are just a few points:

- We want compression to start at a consistent point with relation to the threshold.
- We want the gain stage to act uniformly on all signal levels and with consistent response times.
- We want predictable performance from the ratio function.
- We want to have the ability to dial attack and release times of our choice, even if these are very short.
- We want the attack and release envelopes to be consistent and accurate.

In practice, there is no problem in building a digital compressor that will be precise. The algorithm for such a compressor is available freely over the Internet, so every software developer can code such a compressor quite easily. But none of these compressors will have character – they would all sound the same. Analog designs have a character due to their lack of precision, and each compressor is inaccurate in its own unique way. How the ratio behaves above the threshold or the nature of the attack and release functions defines much of the compressor sound. It is mostly these aspects that earned vintage models much of their glory. In order to introduce some character into digital compressors, designers have to choose where and how to introduce deviations from the precise design.

Principle of operation and core controls

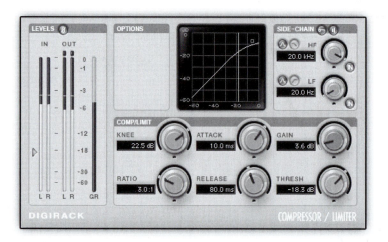

Figure 16.4 A compressor plugin. The Digidesign *DigiRack Compressor/Limiter Dyn 3.*

The explanation of how compressors work is often over-simplified and inaccurate. To really understand how this popular tool works, it would be beneficial to look at its internal

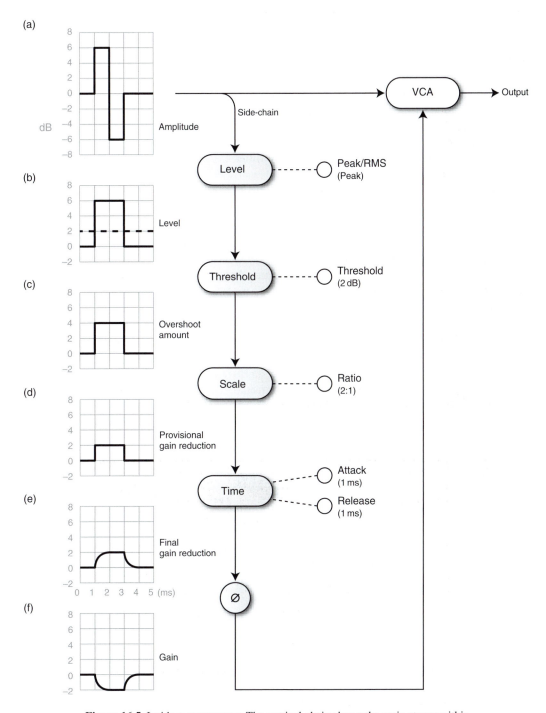

Figure 16.5 Inside a compressor. The vertical chain shows the main stages within the side-chain, and the controls link to each stage. The nature of the signal between one stage and another is shown to the left of the chain, along with a sample graph that is based on the settings shown under each control name.

building blocks – which essentially consists of a few simple stages. Figure 16.5 focuses on the main stages within a compressor and how the various controls we tweak are linked to each of the internal stages. While this illustration is based on the modern VCA analog design, other compressors (including digital ones) employ the very same concept.

Gain

The gain stage is responsible for attenuating (or in some cases also boosting) the input signal by a set amount of dB (which is determined by the side-chain). The gain stage in Figure 16.5 is based on a VCA, but there are other types of gain stages a compressor might be built around. It pays to learn the differences between them, if only because many of today's digital compressors try to imitate the sound of old analog designs:

- **Vari-mu** – the earliest compressor designs were based on variable-mu tubes (valves). Mu, in simple terms at least, is a form of amplification factor that can be used to make a vari-mu tube into a variable gain amplifier. Vari-mu designs have no ratio control; they provide incremental amounts of gain reduction with level (a behavior similar to soft knee). But this only happens up to a point, when the compressor returns to its characteristic linearity. This characteristic works well for percussive instruments as loud transients are not clamped down. Vari-mu designs have faster attack and release times than optical designs, but they are not as fast as VCA designs or FET.
- **FET** – as small transistors started replacing the large tubes, later compressor designs were based on field effect transistors (FETs). They offered considerably faster attack and release times than vari-mu and incorporated a new feature – ratio. Just like vari-mu designs, the compression ratio tends to return to linearity with a very loud input signal.
- **Opto** – the side-chain of an optical compressor controls the brightness of a bulb or LED. On the gain stage there's a photo-resistive material, which affects the amount of applied gain. Despite light being involved, the actual components are slow compared to musical dynamics, thus optical compressors exhibit the slowest response times of all compressors. Moreover, their attack and release curves are less than precise (especially with older designs), giving these compressors a very unique character. Optical designs are known to produce a very noticeable effect, which can be quite appealing.
- **VCA** – of all their analog equivalents, solid-state voltage-controlled amplifiers (VCAs) provide the most precise and controllable gain manipulation. Their native accuracy broadened the possibilities compressor designers had, and made VCAs favorite in most modern designs.
- **Digital** – digital compressors work using a set of mathematical operations. When it comes to precision, a digital compressor can be as precise as it gets. Their response time can be instant, which means no constrains on attack and release times. Digital compressors can also offer perfectly precise ratio, attack and release curves.

Level detection and peak vs. RMS

As the signal enters the side-chain, it first encounters the level stage. This is where its bipolar amplitude (Figure 16.5a) is converted into a unipolar representation of level (Figure 16.5b). At this point the level of the signal is determined by its peak value. Instead of peak-sensing, it might be in our interest to have RMS-sensing, so the compressor responds to the loudness of incoming signals rather than to their peaks (vocals are often better compressed that way). An RMS function (or other averaging function) might take effect at this stage. By way of analogy, it is as if we replaced a peak meter within the level stage with a VU meter.

A compressor might support peak-sensing only, RMS-sensing only or have the ability to toggle between the two. Some compressors also allow us to dial a setting between the two. If no selection is given, the compressor is likely to be RMS.

Threshold

The threshold defines the level above which gain reduction starts. Any signal exceeding the threshold is known to be an *overshooting signal* and would normally be reduced in level. The more a signal overshoots the more it is reduced in level. Signals below the threshold are generally unaffected (we will look at two exceptions to this later). The threshold is most often calibrated in dB.

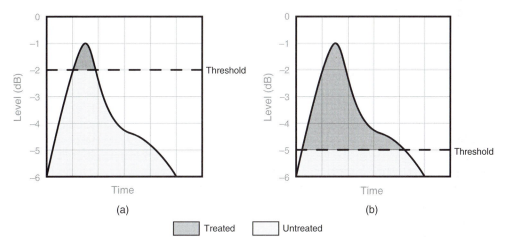

Figure 16.6 Threshold setting. (a) Higher threshold setting means that a smaller portion of the signal is treated. (b) Lower threshold results in a larger portion of the signal being treated.

Track 16.1: Dropping Threshold
This track involves a continuously dropping threshold, from 0 dB down to −60 dB at a rate of −7.5 dB per bar. The compressor is configured to bring down signals above the threshold as quickly as possible (fastest attack −10 µs) and as much as possible (highest ratio – 100:1). Therefore, the lower the threshold the larger the portion of the signal being brought down in level, resulting in a gradual decrease of overall level.

Track 16.2: Dropping Threshold Compensated
This track is similar to the previous one, except after compression has been applied the track level is gradually boosted to compensate for the loss of level caused by the compressor. This reveals a better increasing compression effect over time.

Plugin: Digidesign *DigiRack Compressor/Limiter Dyn 3*
Drums: Toontrack *EZdrumeer*

A threshold function on a compressor comes in two forms: a **variable threshold** or a **fixed threshold**. A compressor with variable threshold provides a dedicated control with which the threshold level is set. A compressor with fixed threshold (Figure 16.7) has an input gain control instead – the more we boost the input signal, the more it will overshoot the fixed threshold. This is similar to the way compression is achieved when overloading tapes. To compensate for the input gain boost, an output gain control is also provided. One advantage of the fixed threshold design (for analog units anyway) is that the noise introduced by the gain stage is often attenuated by the output gain. On variable threshold designs the same noise is usually boosted by the output (or make-up) gain in order to compensate for the applied gain reduction.

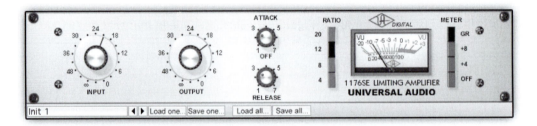

Figure 16.7 The Universal Audio *1176SE*. This plugin for the UAD platform emulates the legendary Urei 1176 LN. It has fixed threshold and the degree of compression is partly determined by the input gain amount.

As can be seen in Figure 16.5c, the threshold stage is fed with the level of the side-chain signal. The output of this stage is the *overshoot amount*, which indicates by how much an overshooting signal is above the threshold. For example, if the threshold is set to 2 dB and the signal level is 6 dB, the overshoot amount is 4 dB. If the signal level is below the threshold, the overshoot amount is 0 dB.

Ratio

Ratio can be compared to gravity. Gravity affects the extent to which objects are pulled towards the ground. There is less gravity on the moon, so astronauts find it easier to jump higher. Ratio determines the extent to which overshooting signals are reduced toward the threshold. The lower the ratio the easier it is for signals to jump above the threshold. Figure 16.8 shows the effect of different ratios on an input signal.

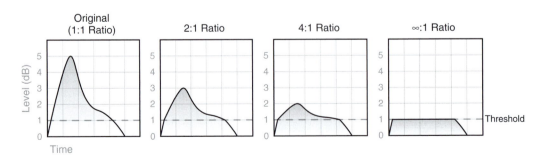

Figure 16.8 The effect of different ratios on the level envelope of a waveform.

Physics aside, once the signal overshoots the threshold, the ratio control determines the ratio between input level changes and output level changes (as the input:output notation suggests). For example, with a 2:1 ratio, an increase of 2 dB above the threshold for input signals will result in an increase of 1 dB above the threshold for output signals. The ratio determines how overshooting signals are scaled down. A 1:1 ratio (unity gain) signifies that no scaling takes place – a signal that overshoots by 6 dB will leave the compressor 6 dB above the threshold. A 2:1 ratio means that a signal overshooting by 6 dB is scaled down to half of its overshoot amount and leaves the compressor at 3 dB above the threshold. A 6:1 ratio means that a signal overshooting by 6 dB is scaled down to a sixth and leaves the compressor at 1 dB above the threshold. The highest possible ratio is ∞:1 (infinity to one), which results in any overshooting signal being trimmed to the threshold level. Often, a ratio of ∞:1 is achieved with a compressor using an extremely high ratio like 1000:1. Figure 16.9 illustrates these four scenarios.

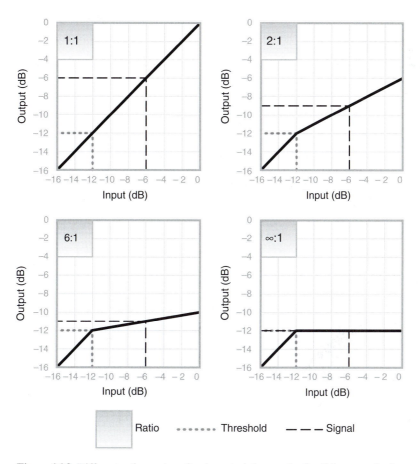

Figure 16.9 Different ratios on transfer characteristics graphs. In all these graphs the threshold is set to -12 dB and the input signal is -6 dB (6 dB above the threshold). With a 1:1 ratio the output signal leaves the compressor at the same level of -6 dB (6 dB above the threshold); with a 2:1 ratio, the signal leaves at -9 dB (3 dB above the threshold); with a 6:1 ratio the signal leaves at -11 dB (1 dB above the threshold); and with ∞:1 the signal leaves at the same level as the threshold.

Track 16.3: Raising Ratio
This track involves a threshold at −40 dB and a continuously rising ratio from 1:1 up to 64:1. The ratio doubles per bar, starting with 1:1 at the beginning of the first bar, 2:1 at the beginning of the second bar, 4:1 at the beginning of the third bar and so on. The higher the ratio the harder overshooting signals are reduced toward the threshold, resulting in a gradual decrease in the overall level.

Track 16.4: Raising Ratio Compensated
This track is identical to the previous one, only that the loss of level caused by the compressor is compensated using a gradual gain boost. Here again, the increased compression effect over time is easily discerned.

Plugin: Digidesign *DigiRack Compressor/Limiter Dyn 3*
Drums: Toontrack *EZdrumeer*

One characteristic of vintage compressors, notably vari-mu and FET designs, is that the ratio curve maintains its intended shape, except for loud signals when it deviates toward unity gain. This type of ratio curve tends to complement transients, as the very loud peaks are not tamed – compression is applied, but less dynamics are lost. Drum heaven. Many digital compressors imitate such a characteristic behavior using similar ratio curves. The McDSP *CB1* in Figure 16.10 is one of them.

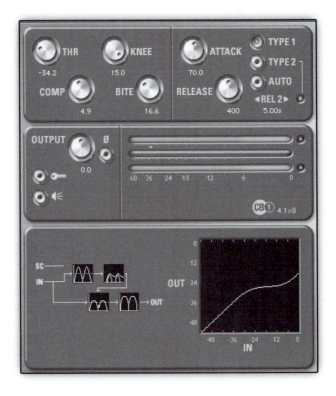

Figure 16.10 The McDSP CB1 plugin. The input-output plot shows that compression starts around the threshold and then develops into heavy limiting. However, around −12 dB the ratio curve changes course and starts reverting to unity gain. Ratio slopes of this nature, which deviate from the textbook-perfect shape, can provide a characteristic sound that is usually a trait of vintage analog designs.

The ratio is applied within a compressor by the scale stage (Figure 16.5). The scale stage is the operational heart of the compression process – this is where the overshoot amount is converted into the provisional gain reduction. The output of the scale stage is nearly fit to feed the gain stage. The ratio essentially determines the percentage by which the overshoot amount is reduced to 0 dB. A 1:1 ratio means that the overshoot amount is fully reduced to 0 dB, which means that no gain reduction takes place. With a 2:1 ratio, the overshoot amount is reduced by half. For example, 4 dB of overshoot becomes 2 dB.

Attack and release

Modern compressors can respond to sudden level changes instantly. However, quick response is not always sought. For example, in order to retain some of the instrument's natural attack we often want to let some of the initial level burst pass through the compressor unaffected (or lightly affected). In order to do so, we need to be able to slow down the compressor response times. Similarly, if a healthy amount of gain reduction drops too fast, the gain recovers too quickly to produce pumping. To prevent this, we need a way of controlling the rate at which gain reduction drops.

The attack and release are also known as *time constants* or *response times*. The attack determines how quickly gain reduction can rise, while release determines how quickly gain reduction can fall. Essentially, a longer setting on either will simply slow down the rate at which gain reduction increases (attack) or decreases (release). For example, in Figure 16.5d the gain reduction rises instantly from 0 to 2 dB and then falls back instantly to 0 dB. Having 1 ms of both attack and release means, in our case at least, that it takes 1 ms for the gain reduction to rise and fall (Figure 16.5e).

Both the attack and release times are typically set in milliseconds. Attack times usually range between 0.010 ms (10 μs) and 250 ms. Release times are often within the 5-3000 ms (3 seconds) range. It is important to understand that both times determine how quickly the gain reduction can change, and not the time it takes it to change. In practice, both define how long it takes the gain reduction to change by a set amount of dB. For example, 1 second of release time might denote that it takes the gain reduction 1 second to drop by 10 dB. Accordingly, it would take it half a second to drop by 5 dB. This behavior can be compared to the ballistics of a VU meter. A quick peak drop would appear gradually on a VU meter. When peaks drop slowly, the VU reading will be very similar to that of a peak meter. The attack and release have a very similar effect on gain reduction.

Figure 16.11 shows the effect of different attack and release settings on a waveform. On all graphs the ratio is 2:1, and the threshold is set to 6 dB. The original input signal (a) rises instantly from 6 to 12 dB, then drops instantly back to 6 dB. In all cases the overshoot amount is 6 dB, so with the 2:1 ratio the full gain reduction amount is 3 dB. We can see in (b) that if there is no attack and no release the overshooting signal is constantly reduced by 3 dB. When there is some attack and release (c-e), it takes some time before full gain reduction is reached and then some time before gain reduction ceases. It is worth noting what happens when the release is set and the original signal drops from 12 dB back to 6 dB. Initially, there is still 3 dB of gain reduction, so the original signal is still attenuated to 3 dB below its original level. Slowly, the gain reduction diminishes and only after the release period has passed does the original signal return to 6 dB. This is the first exception for signals below the threshold being reduced in level. The gain reduction graph below (d) helps us to understand why this happens.

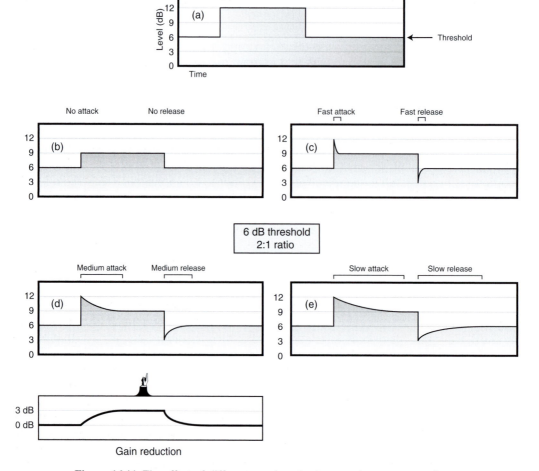

Figure 16.11 The effect of different attack and release settings on a waveform. (a) The original levels before compression. (b–e) The resultant levels after compression.

Track 16.5: Noise Burst Uncompressed
This track involves an uncompressed version of a noise burst. First, the noise rises from silence to −12 dB, then to −6 dB, then it falls back to −12 dB and back to silence.

The compressor in the following tracks was set with its threshold at −12 dB and a high ratio of 1000:1. Essentially, when full gain reduction is reached, the −6 dB step should be brought down to −12 dB.

Track 16.6: Noise Burst Very Fast TC
With the attack set to 1 ms and the release to 10 ms. we can hardly discern changes in level caused by the attack and release functions. Still, there is quick chattering when the level rises to −6 dB and when it falls back to −12 dB.

Track 16.7: Noise Burst Fast TC
This track involves an attack time set to 10 ms and release to 100 ms. The click caused by the attack can be heard here, and so can the quick gain recovery caused by the release.

> **Track 16.8: Noise Burst Medium TC**
> 25 ms of attack and 250 ms of release. A longer attack effect can be heard, and the gain recovery due to the release is slower.
>
> **Track 16.9: Noise Burst Slow TC**
> The most evident function of the time constants is achieved using slow attack and release – 50 and 500 ms, respectively. Both the drop in noise level due to the attack and the later rise due to the release can be clearly heard on this track.
>
> *Plugin*: Sonnox *Oxford Dynamics*

Some compressors offer a switchable **auto attack** or **auto release**. When either is engaged, the compressor determines the attack or release times automatically. Mostly this is achieved by the compressor observing the difference between the peak and RMS levels of the side-chain signal. It is worth knowing that in auto mode neither the attack nor the release is constant (like when we dial the settings manually). Instead, both change with relation to the momentary level of the input signal. For example, a snare hit might produce a faster release than a xylophone note. Thus, auto attack and release do not provide less control – they simply provide an alternative kind of control. Auto release, on respected compressors at least, has an excellent reputation.

Much of a compressor's character (or any other dynamic range processor for that matter) is determined by the attack and release functions, in particular, their **timing laws**. The timing laws determine the rate of change the attack and release apply on gain reduction, and in their most simple form these can be either exponential or linear. There is some similarity here to exponential and linear fades, only that fades are applied on the signal itself, while the attack and release are applied on gain reduction. Generally speaking, exponential timing laws tend to sound more natural and less obstructive. Linear timing laws tend to draw more color and effect, which is often associated with the sound of some favorite analog units. Very few compressors give us control over the compressor's timing laws; the compressor of the Sonnox *Oxford Dynamics* in Figure 16.13 is one of the few that does (the 'NORMAL' field above the plot denotes exponential law).

> The following tracks demonstrate the differences between exponential and linear timing laws. It should be obvious that in linear mode the compression effect is far more evident. It produces noticeable distortion and even some chaos. Although maybe not to such an extreme degree, this characteristic sound is sometimes what we are after.
>
> **Track 16.10: Metal Uncompressed**
> **Track 16.11: Metal Exponential Mode**
> **Track 16.12: Metal Linear Mode**
>
> *Plugin:* Sonnox *Oxford Dynamics*
> *Drums:* Toontrack *EZdrummer*

As seen in Figure 16.5, both the attack and release are controls linked to the time stage. The input to the time stage is the provisional amount of gain reduction. The time stage slows down sudden changes to that gain reduction. As long as the input to the time stage is higher than its output, the gain reduction will keep rising at the rate set by the attack. The moment the input is lower than the output, the gain reduction starts to drop at the rate set by the release. Figure 16.12 illustrates this.

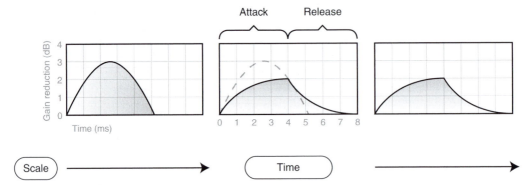

Figure 16.12 The input to the time stage arrives from the scale stage and denotes the provisional amount of gain reduction. The attack within the time stage slows down the gain reduction growth. Note that even when the input to the time function starts to fall (2.5 ms), the output still rises as the provisional amount of gain reduction is still higher than the applied amount. Only when the provisional amount falls below the applied amount (4 ms), does the latter start to drop. At this point, the attack phase ends and the release phase starts.

One crucial thing to understand is that attack and release are applied on gain reduction, and that the time stage, where these are applied, is unaware of the threshold setting. Many sources state incorrectly that the release only occurs when the signal level drops below the threshold. In reality, both the attack and release affect gain reduction (and in turn the signal level) even when the signal level changes above the threshold. Figure 16.13 demonstrates this.

> *Despite what many sources suggest, the release function is not related to the signal dropping below the threshold.*

The top waveform in Figure 16.13 is known as a noise burst, and is commonly used to demonstrate the time function of dynamic range processors. However, it could be just as easily demonstrated using a voice rising from levels already above the threshold and then falling back to levels above the threshold (something like ah-Ah-AH-Ah-ah, where only the ah is below the threshold). The fact that the attack and release affect level changes even when these happen above the threshold means that the dynamics are controlled to a far greater extent than they would be, if compressors only acted when signals overshoot or drop below the threshold. The attack and release would still be applied on the AH just like on the Ah. Ever so often, vocals do fluctuate in level after overshooting the threshold, and the release and attack still affect our ability to balance them. Also, attack and release will still act on signals fluctuating above the threshold even when it is set very low.

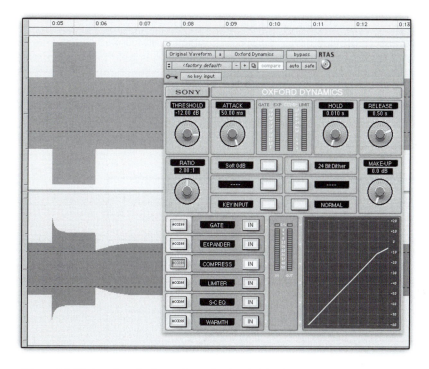

Figure 16.13 Attack and release above the threshold. The top waveform rises from −6 to 0 dB, then falls back to −6 dB. This waveform passes through an *Oxford Dynamics* plugin with the shown settings, and the post-compression result is the bottom waveform. The compressor threshold is set to −12 dB (dashed lines) so all the level changes in the top waveform happen above the threshold. On the bottom waveform we can see the action of both the attack and release. Analog compressors behave in the same way.

Track 16.13: Noise Burst Low Threshold
A compressed version of Track 16.5, with the threshold set to −40 dB and ratio to maximum. You should hear the first attack when the noise rises to −12 dB, and then again when it rises to −6 dB (which happens above the threshold). When the noise falls back to −12 dB (a level above the threshold), the gain recovery caused by the release function is evident.

- - - - - - -

The following two tracks involve a threshold at −50 dB and a ratio of 1.5:1. Apart from the very first 300 ms and the closing silence, the vocal is always above the threshold. The only difference between the two tracks is the attack and release settings. There is no doubt that the two are distinctly different, demonstrating that both the attack and release affect level variations above the threshold:

Track 16.14: Vocal Compression I
Track 16.15: Vocal Compression II

Plugin: Sonnox *Oxford Dynamics*

Hold

Some compressors provide a hold parameter, which is also linked to the time function. In simple terms, hold determines how long gain reduction is held before the release phase starts. Figure 16.14 illustrates this. In practical terms, the function of hold on a compressor is achieved by altering the release rate, so during its early stages gain reduction hardly changes. Although the results are only similar to those shown in Figure 16.14, the overall impression is still as if gain reduction is held.

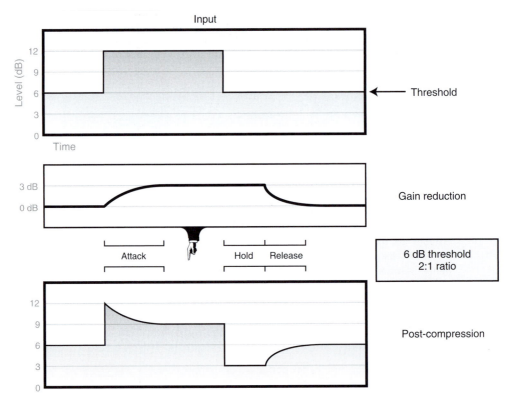

Figure 16.14 The hold function on a compressor. This simplified illustration shows the input levels at the top, the amount of applied gain reduction in the middle and the post-compression level at the bottom. We can see that once the original level drops, the compressor holds the gain reduction for a while before release.

Phase inverse

To be fair, the phase stage in Figure 16.5 might not be part of a real-world compressor. It was included to simplify the explanation. The output of the time stage is the final gain reduction, and it involves a positive magnitude (Figure 16.5e). What the gain stage expects is the amount of gain that should be applied on the signal, not the amount of *gain reduction*. So in order to reduce the input level by 3 dB, the gain stage should be fed with −3 dB. In order to convert the gain reduction to gain, the magnitude of the former is mirrored around the 0 dB line using a simple phase inversion (Figure 16.5f).

Additional controls

Now that we have established the basic principles behind the operation of compressors, let's take a look at a few more controls. Figure 16.15 shows the addition of the controls we will look at in the next sections.

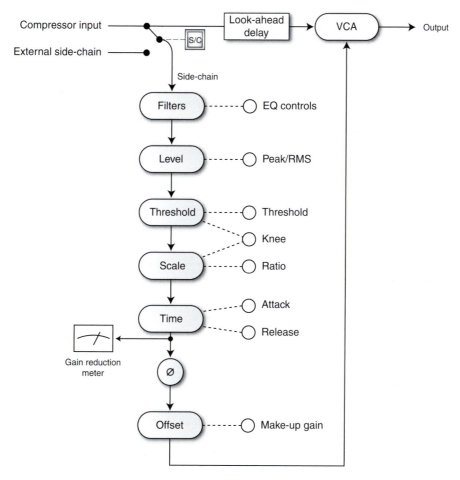

Figure 16.15 Detailed insight into a compressor.

Make-up gain

A compressor's principal function is to make the louder areas of sound softer. As a result, the perceived loudness of the compressed signal is likely to drop. To compensate for this, the make-up gain control (sometimes called *gain* or *output*) simply boosts the level of the output signal by a set amount of dB. The boost is applied uniformly to the signal, independently of any other control setting – both signals below and above the threshold are affected. Compressors produce this effect by either biasing the side-chain's gain amount before it is applied by the gain stage or simply by amplifying the signal after the gain stage.

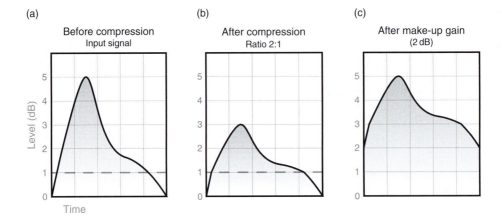

Figure 16.16 Make-up gain. (a) The input signal before compression. (b) The signal after compression with 2:1 ratio, but before make-up gain. (c) The output signal after 2 dB of make-up gain. Note that in this example the peak measurement of both the input (a) and the output (c) signals is identical. Yet (c) will be perceived as louder.

As per our louder-perceived-better axiom, when we do A/B comparison there is a likelihood that the compressed version will sound less impressive due to the loudness drop. Makeup gain is often set so whether the compressor is active or not, the perceived signal loudness remains constant. This way, any comparison made is fair and independent of loudness variations.

Some compressors have an **automatic make-up** gain. A make-up gain control might still be provided if the automatic make-up gain can be switched off. Compressors with auto make-up gain calculate the amount of gain required to level the input and output signals based on various settings like threshold, ratio and release. The auto make-up gain is independent of the input signal and only varies when the compressor controls are adjusted. Arguably, there is no way an automatic make-up gain will match flawlessly the dynamics of all instruments and the possible levels at which these have been recorded. In practice, auto make-up gain often produces a perceived loudness variation when the compressor is bypassed. When able to do so, people will often choose to turn this function off.

Hard and soft knees

On the ratio curve, the knee is the threshold-determined point where the ratio changes from unity gain to the set ratio. So called because it takes little imagination to see that on a transfer characteristics graph this curve is reminiscent of a sitting person's knee. The type of compressors we have discussed so far work on the **hard-knee** principle – the threshold sets a strict limit between no treatment and full treatment. The sharp transition between the two provides a more intrusive compression and a more distinctive effect. We can soften such compression by lengthening the attack and release, but the longer settings do not always complement the compressed material.

The **soft-knee** principle (also termed *over-easy* or *soft ratio*) enables smoother transition between no treatment and treatment – gain reduction starts somewhere below the threshold with diminutive ratio, and the full compression ratio is reached somewhere above the threshold. While a hard-knee compressor toggles between 1:1 and 4:1 as the signal overshoots, on a soft-knee compressor the ratio gradually grows from 1:1 to 4:1 in a transition region that spreads to both sides of the threshold. This second exception for signals below the threshold being reduced in level is illustrated in Figure 16.17.

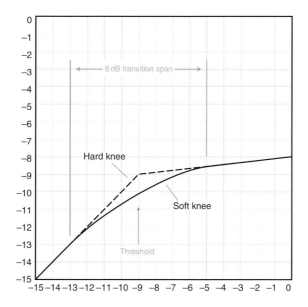

Figure 16.17 Hard and soft knees. With hard knee, the ratio of compression is attained instantly as the signal overshoots. With soft knee the full ratio is achieved gradually within a transition region that spreads to both sides of the threshold.

Soft knee is useful when we want more transparent compression (often with vocals). The smooth transition between no treatment and treatment minimizes the compression effect, which in turn lets us dial a higher ratio. Having soft-knee also frees the attack and release from the task of softening the compression effect and lets us dial shorter times on both (which can be useful for applications like loudening). When we are after the compression effect, a hard knee would be more suitable.

> *Soft knee for more transparent compression. Hard knee for more effect.*

Compressors might provide hard knee only, soft knee only or a switch to toggle between the two. Some compressors also provide different degrees of knee rates, which, via a control, let us determine the dB span of the transition region with relation to the input scale. For example, in Figure 16.17 the transition span is 8 dB, with 4 dB to each side of the threshold. The compression in this case will start with gentle ratio at –13 dB, and the full ratio will be achieved at –5 dB. A transition span of 0 dB denotes hard-knee function (such a setting can be seen on the *Oxford Dynamics* in Figure 16.13 below the attack control).

In order to produce soft-knee behavior, an analog compressor has to alter both the threshold and the scale functions. A precise soft knee is difficult to produce in the analog domain: the threshold might not fall within the center of the knee, and the ratio is likely to diverge outside the transition region. Digital compressors have no problem in exhibiting a perfect soft knee like the one in Figure 16.17.

Track 16.16: Vocal Uncompressed
This uncompressed vocal track involves noticeable level variations.

Track 16.17: Vocal Hard Knee
Although the level fluctuations were reduced, some still exist, mainly due to the operation of the compressor. The working compressor can be easily heard on this track.

Track 16.18: Vocal Soft Knee
Compression settings identical to the previous track, only with a soft knee (40 dB transition span). The level fluctuations in the previous track are smoothened by the soft knee, and the operation of the compressor is far less evident.

- - - - - - - -

One issue with soft knee is that the compression starts earlier (below the threshold), and so does the attack. Since the attack starts earlier, less of the natural attack is retained. Therefore, soft knee might not be appropriate when we try to retain the natural attack of sounds, and with it longer attack times might be needed. Notice the loss of dynamics, life and some punch in the soft knee version:

Track 16.19: Drums Hard Knee
Track 16.20: Drums Soft Knee

Plugin: Sonnox *Oxford Dynamics*
Drums: Toontrack *EZdrummer*

Look-ahead

Compression can be tricky with sharp level changes, like those of transients. In order to contain transients a compressor needs a very fast response. This is not always possible. For one, the gain stage of some compressors, optical ones for instance, is often not fast enough to catch these transients. Then, even if a compressor offers fast response times, the quick clamping down of signals might not produce musical results. It would be great if the side-chain could see the input signal slightly in advance so it would have more time to react to transients. The look-ahead function enables this.

One way to implement look-ahead on analog compressors is by delaying the signal before it gets processed. The delay, often around 4 ms long, is introduced after a copy is sent to the side-chain (Figure 16.18). This way, a transient entering the compressor will be seen immediately by the side-chain, but will only be processed shortly after. This enables longer (more musical) attack times since there is a few milliseconds gap between the compression onset and the actual processing of the signal that triggered it. By way of analogy, if we could have look-ahead in tennis, it would be like freezing the ball for a while as it crosses the net, so a player could perfectly reposition after seeing where the ball is heading.

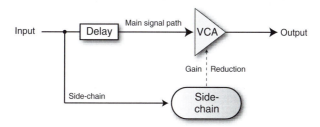

Figure 16.18 A look-ahead function on a hardware compressor. After a copy is sent to the side-chain, the signal is delayed on the main signal path, giving the control circuit more time to respond to the signal.

On a hardware compressor a look-ahead function will be switchable. When look-ahead is engaged, the output signal is delayed as well. These few milliseconds of delay rarely introduce musical timing outflow, but can lead to phase issues if the compressed signal is mixed with a similar track (snare top, snare bottom for example). Software compressors only introduce output delay when the buffer size is smaller than the look-ahead time. Although this is often the case, auto delay compensation will prevent any phase or timing issues.

Stereo linking

A stereo hardware compressor is essentially a unit with two mono compressors. Unless otherwise configured, these two mono compressors work independently in what is known to be a dual-mono mode. Let us consider what happens when stereo overheads are fed into a compressor in such a mode. A floor tom is likely to be louder on the microphone that covers one side of the drum kit – say the right side. When the tom hits, the right compressor will apply more gain reduction than its left counterpart. As a result, the left channel will be louder and the overheads image will shift to the left. Similarly, when the left crash hits, the overheads image will shift to the right. So in dual-mono mode every time a drum hits on either sides of the panorama, the stereo image might shift to the opposite side. Stereo linking interconnects both compressors, so an identical amount of gain reduction is applied to both channels. With stereo linking engaged, as the floor tom hits, both sides are compressed to an identical extent and no image shifting occurs.

In all but very few cases, stereo linking is engaged when a stereo signal is compressed.

> **Track 16.21: Drums Stereo Link On**
> The compression triggered by the tom causes the kick, snare and hi-hats to drop in level, but their position on the stereo image remains the same.
>
> **Track 16.22: Drums Stereo Link Off**
> With stereo linking off, image shifting occurs. With every tom hit, the kick, snare and hi-hats shift to the left. As the tom decays, these drums slowly return to their original stereo position.
>
> *Plugin:* PSP *VintageWarmer 2*
> *Drums:* Toontrack *EZdrummer*

There are various ways to achieve stereo linking; for example, the stereo input might be summed to mono before feeding both the left and right side-chains. The problem with this approach is that phase interaction between the two channels might produce a disrupted mono sum. To combat this, some compressors keep a stereo separation throughout the two side-chains and take the *strongest win* approach – the heaviest gain reduction product of either channel is fed to the gain stage of both. With the strongest-win approach, it still makes sense to have different settings on different channels – by setting lower ratio on the right channel we make the compressor less sensitive to right-side events (like a floor tom hit).

As opposed to their hardware counterparts, software compressors rarely provide a stereo linking switch. A software compressor knows whether the input signal is mono or stereo, and for stereo input, stereo linking is automatically engaged. Some compressors still provide a switch to enable dual-mono mode for stereo signals. Such a mode might be required when the stereo signal has no image characteristics, like in the case of a stereo delay.

External side-chain

By default, both the gain stage and the side-chain are fed with the same input signal. Essentially, the signal is compressed with relation to its own level. Most compressors let us feed an external signal into the side-chain, so the input signal is compressed with relation to a different source (Figure 16.19). For example, we can compress a piano with relation to the level of the snare. Hardware compressors have side-chain input sockets on

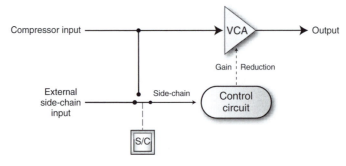

Figure 16.19 External side-chain. An external signal can feed the side-chain, so the compressor input is compressed with relation to an external source.

their rear panel, while software sequencers let us choose an external source (often a bus) via a selection box in the plugin window. A switch is often provided to toggle between the external side-chain and the native, internal one. There is also usually a switch that lets us listen to the side-chain signal via the main compressor output (overriding momentarily the compressed output).

Side-chain filters and inserts

More than a few compressors let us equalize the side-chain signal (the section to the right of the plot in Figure 16.4 is one example). As can be seen from Figure 16.5, such equalization occurs on the side-chain only (not the main signal path) and affects both the native and external side-chain sources. Some analog compressors also provide a side-chain insertion point, enabling processing by external devices. Soon we will see the many applications for side-chain equalization.

Compressor meters

Three metering options are commonly provided by compressors: input, output and gain reduction. When it comes to mixing, people often say: 'Listen, don't look'. True – we should be able to set all the compressor controls based on what we hear but recognizing the subtleties in compressor action takes experience, and the various meters can often help, especially when it comes to initial rough settings. Fine and final adjustments are done by ear.

Out of the three meters, the most useful is the **gain reduction meter**, which shows the applied amount of gain reduction. The gain reduction reading combines the effect of all the compressor controls like threshold, ratio, attack, release, peak or RMS, etc. The main roles of this meter are:

- To teach us **how much** gain reduction is applied (which hints at the most suitable amount of make-up gain).
- To teach us **when** gain reduction takes place. Mostly we are interested in:
  - When it starts.
  - When it stops.
- To provide a visual indication of the **attack** and **release** activity (which can help in adjusting them).

To give an example, when our aim is to balance the level of vocals, we want the gain reduction meter to move with respect to the level fluctuation that we hear on the unprocessed voice (or see on the input meter). If the attack or release is too long, the meter will appear lazy compared to the dynamics of the signal, and so might the compression.

The terms *compression* and *gain reduction* are often used interchangeably, where people think of the amount of compression as the amount of gain reduction. When people say they compress something by 8 dB, they mean that the gain reduction meter's highest reading is 8 dB. Associating the gain reduction reading with the amount of compression is not a good practice though. If the gain reduction meter reads steady 8 dB (say due to a hypothetical release time of 10 hours), no compression is taking place – there is simply constant gain attenuation of 8 dB. For the most part, such a compressor behaves like a fader – apart from when the gain reduction climbs to 8 dB, the output signal keeps all its input level variations, whether these happened above or below the threshold. In order for compression to occur, the amount of gain reduction must vary over time, the gain reduction meter must move, and the faster it moves the more compression takes place. There is a similarity here to sound itself – sound is the outcome of changes in air pressure; steady pressure, whether high or low, does not generate sound. Likewise, the amount of compression is determined by changes in gain reduction. We might want to consider the amount of compression as an average of the absolute difference between the peak and RMS gain reduction. But that's a story for a different book.

In addition, we should also consider the range within which this meter moves. If vocals overshoot the threshold between phrases, but within each phrase the gain reduction meter only moves between 4 and 6 dB, then the effective compression is only 2 dB. At the beginning and end of each phrase the signal is still compressed on its way to and back from 4 dB, but this compression is marginal compared to the compression taking place between 4 and 6 dB. We will call the range where gain reduction varies *dynamic compression* (4–6 dB in our case), and use the term *static compression* for the static range (0–4 dB in our case), where gain reduction only occurs on the way to or back from the dynamic compression. Figure 16.20 illustrates the differences between the two. Static compression can happen either due to slow release or a threshold set too low. If the latter is the case, it might be wise to bring up the threshold, so the gain reduction meter constantly moves between 0 dB and the highest reading. This would minimize static compression. However, there are cases where static compression makes perfect sense, like when compression is employed to shape the dynamic envelope of sounds, and the threshold is set intentionally low to enable more apparent attack.

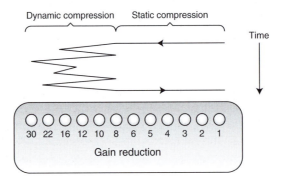

Figure 16.20 Dynamic and static compression. The dynamic compression happens in the range where the gain reduction meter varies. The static compression only happens as gain reduction rises to or falls from the dynamic compression range.

The subjective observation of light, moderate or heavy compression is based on the gain reduction meter. We just saw that the amount of compression is far more complex than the basic reading this meter offers. It is also unsafe to translate any observations to numbers, as what one considers heavy the other considers moderate. We can generalize that 8 dB of compression on the stereo mix is quite heavy. For instruments, people usually think of light compression as somewhere below 4 dB, heavy compression as somewhere above 8 dB and moderate compression as somewhere in between these two figures (Figure 16.21). We are talking about dynamic compression here – having 20 dB of constant gain reduction is like having no compression at all.

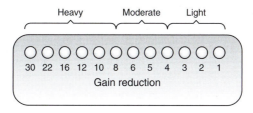

Figure 16.21 A gain reduction meter. The descriptions above the meter are an extremely rough guideline of what some people might associate different amounts of gain reduction with.

You might have noticed that the numbers in Figure 16.21 are positive and laid from right to left. This is the opposite direction to the input and output meter and suggests that gain reduction brings the signal level down, not up. Also, on compressors with no dedicated gain reduction meter, gain reduction is displayed on the same meter as the input and output levels (with a switch to determine which of the three is shown). Since on most meters the majority of the scale is below 0 dB, this is also where gain reduction is shown. Very often a gain reduction of 6 dB will be shown as −6 dB.

To be technically correct, a decrease of a negative value is an increase, so a gain reduction of −6 dB essentially denotes a gain boost of 6 dB. Many designers overlook that, and despite having a dedicated meter labeled 'gain reduction', the scale incorrectly shows negative values.

The **input meter** can be useful when we set the threshold. For example, if no signal exceeds −6 dB except for a peak we want to contain, this is where the threshold might go. We can also determine the action range of input levels by looking at this meter (more on action range below). **The output meter** gives a visual indication of what we hear and ensures that the output level does not exceed a specific limit, like 0 dBFS. The output meter can also help us in determining rough make-up gain without having to toggle the compressor in and out.

Controls in practice

Threshold

Figure 16.22 shows possible levels of a non-percussive performance – a vocal phrase perhaps. We can see that the levels fluctuate between the 0 and −10 dB range. We will call this range the *action range* and it can help us determine the threshold setting for

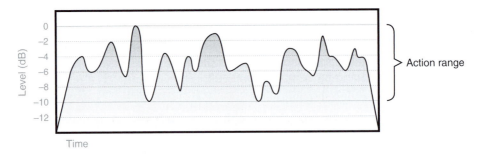

Figure 16.22 The action range of a non-percussive performance. Apart from the initial rise and final drop, the signal level is only changing within the 0 to −10 dB range.

specific applications. It should be made clear that setting the threshold anywhere above 0 dB has little point – unless the knee is soft, no compression will take place. Setting the threshold within this range will yield selective compression – only the signal portions above the threshold will be treated. Sometimes this is what we want, like in cases we only want to treat loud parts of the signal. However, the danger in selective compression is that the transition between treatment and no treatment happens many times, and the compression becomes more obvious. Setting the threshold to the bottom of the action range (−10 dB in Figure 16.22) will result in uniform treatment for all the level changes within it. While this might not be suitable with every compressor application, when a performance involves an evident action range, its bottom might provide a good starting point for threshold settings.

One key point to consider is what happens when we set the threshold below the action range. Say we first set the threshold to the bottom of the action range at −10 dB, and set the ratio to 2:1. The highest peak in Figure 16.22 hits 0 dB, so by overshooting 10 dB it will be reduced by 5 dB. Also, if we assume that the compressor timing rates are based on 10 dB and the release is set to 10 ms, as the same peak dives to −10 dB, it will take the 5 dB of gain reduction 5 ms to recover. Now if we reduce the threshold to −20 dB, the same peak overshoots by 20 dB and will be reduced by 10 dB – twice as before. Also, as the gain reduction is now 10 dB, it will take it 10 ms to recover. Effectively, by lowering the threshold we increase the amount of gain reduction, which results in further depression of dynamics. In addition, we slow down the release (and attack) response. Also, having the threshold at −20 dB will result in 5 dB of static compression as everything between −20 and −10 dB would be constantly compressed (apart from during the leading rise and closing fall).

Let us discuss three common compression applications and their threshold settings. These are containing peaks, level balancing and loudening (condensing levels). Figure 16.23 shows the level of a hypothetical performance, say vocal. Each column represents a different application, and the threshold setting for each is shown on the top row. The top row also shows the input signal, which is identical in all three cases. In this specific performance there is one peak and one level dip.

The left column (a) is concerned with containing levels. The high threshold is set around the peak base, above all other levels. The compression reduces the peak level, but no

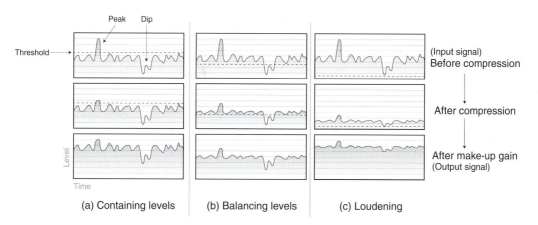

Figure 16.23 Threshold settings for common compression applications. (a) Containing levels. (b) Balancing levels. (c) Loudening. The columns from top to bottom show the input signal and the threshold setting, then the compressed signal before make-up gain and the output signal after make-up gain.

other portion of the signal. Reducing the peak lets us boost the overall level of the signal, and by matching the original peak with the compressed one, moderate- and low-level signals are made louder. Note that in this case the loss of dynamics is marginal.

> The term containing levels is being used here as a more general term for containing peaks. Containing peaks is concerned with preventing levels from exceeding a predefined limit – a job for a limiter really. We might want to contain the louder downbeats of a strumming guitar, although these might not peak.

The center column (b) demonstrates possible compression for level balancing. The threshold is set to a moderate level right above the dip, but below all other portions of the signal. The idea is to pull everything above the dip down toward it, which is exactly what happened after compression. Note, however, that compared to the compression shown on the left column, the signal has lost more of its dynamics. The make-up gain offsets the signal, so moderate levels are back to their original area (around the middle line of the input graph). If we look at the output graph, we can see that the peak got pulled down toward this middle line, while the dip got a gentle push up from the make-up gain. Again, one possible issue with this technique is the uneven compression effect it deposits due to selective compression – the dip gets louder, but as opposed to the rest of the signal its dynamics remain unaffected. If the applied compression results in audible effect, it will not be evident on the dip. To overcome this, we can lower the threshold so even the dip gets a taste of the compressor effect.

The right column (c) exhibits a compression technique used when we wish to make instruments louder or when we want to condense their dynamics. The low threshold is set at the base of the signal, so all but the very quiet levels get compressed. Two things are evident from the post-compression graph: First, the levels are balanced to a larger extent as the dip was treated as well. After compression, the peak, dip and moderate levels are highly condensed. Thus, this technique can also be used for more aggressive level balancing. Second, we can see that the signal has dropped in level substantially. To compensate for that, make-up gain is applied to match the input and output peaks. In this case, the output signal will be perceived as louder than the input signal, since its average level rose by a fair amount. The main issue with this technique is that of all three applications, it led to the greatest loss of dynamics.

There is more to threshold settings than these general applications. Sometimes we know which part of an instrument's dynamic envelope we should treat. On a snare, for example, it might be the attack, so the threshold is set to capture the attack only. Figure 16.24 illustrates this. One important thing to note is that the lower the threshold the sooner we catch the attack build-up. In turn, this lets us lengthen the attack time, which as we shall soon see is often beneficial.

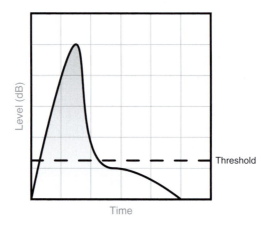

Figure 16.24 Catching the attack of a snare. The threshold is set so the attack is above it, but the decay is below.

Ratio

A compressor with ratios above 10:1 behaves much like a limiter – with 10:1 ratio a signal overshooting by a significant 40 dB is reduced to a mere 4 dB above the threshold. Yet, even a compressor with ∞:1 ratio is not really a limiter (signals can still overshoot due to long attack or the RMS function), so ratios higher than 10:1 have a place in mixes. Mastering ratios (on a compressor before the limiter) are often gentle and kept around 2:1. Mixing ratios can be anything – 1.1:1 or 60:1. Yet, the ratio 4:1 is often quoted as a good starting point in most scenarios.

Logic has it that the higher the ratio the more the compression applied and the more obvious its effect. One important thing to understand is that the degree of compression diminishes as we increase the ratio. As can be seen in Figure 16.25, turning the ratio from 1:1 to 2:1 for an 8 dB overshoot results in gain reduction of 4 dB. Turning the ratio from 8:1 to 16:1 for the same overshoot only results in an additional gain reduction of 0.5 dB. A more practical way to look at this is that changes of lower ratios are much more significant than the same changes for higher ratios – a ratio change from 1.4:1 to 1.6:1

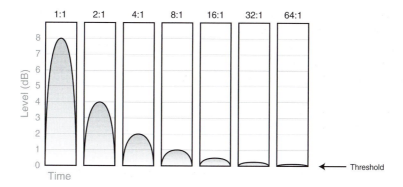

Figure 16.25 The diminishing effect of increasing the ratio. The number above each graph shows the ratio. The threshold is set to 0 dB.

will yield more compression than a ratio change from 8:1 to 16:1. Most ratio controls take this into account by having an exponential scale (for example, 1:1 to 2:1 might be 50% of the scale, 2:1 to 4:1 will be an additional 25% and so forth).

A demonstration of the diminishing effect of increasing the ratio can be heard in Track 1.3 (Rising Ratio). The majority of level lost, and the increasing effect of compression is mostly evident in the first bar, where the ratio rises from 1:1 to 2:1, and in the second bar, where the ratio further rises to 4:1. The change in the following bars is marginal compared to the change in these two first bars.

We can make some assumptions as for the possible ratios in the three applications discussed in Figure 16.23. In the first case – containing levels – a high ratio, say 10:1, could be suitable. With the high threshold set to capture the peak only, the ratio simply determines the extent by which the peak is brought down, with a higher ratio being more forceful. In the second case – balancing levels – the moderate threshold is set around the average level. The function of the ratio here is to determine the balancing extent. Since most of the signal is above the threshold, a high ratio would result in a noticeable effect and possible oppression of dynamics. However, a low ratio might not give sufficient balancing, so a moderate threshold, say 3:1, could be suitable here. In the third case – loudening – the low threshold already sets the scene for some aggressive compression. Even a moderate ratio can make the compression so aggressive that it will suppress dynamics. Therefore, a low ratio, say 1.4:1, might be appropriate.

The relationship between threshold and ratio

In the three examples above the lower the threshold the lower the ratio, which brings us to an important link between the two controls. Lowering the threshold means more compression, while lowering the ratio has the opposite effect. Often these two controls are fine-tuned simultaneously, where lowering one is followed by the lowering of the other. Ditto for rising. The idea is that once the rough amount of required compression is achieved, lowering the threshold (more compression) and then lowering the ratio (less compression) will result in roughly the same amount of compression, but one with a slightly different character.

Although lower threshold or higher ratio result in more compression, the effect of each is different. Lower threshold results in more compression as larger portions of the signal are compressed. High ratio results in more compression on what has already been compressed. Put another way, lower threshold means more is affected, while higher ratio means more effect on the affected. If the threshold is set too high on a vocal take, even a ratio of 1000:1 will not balance out the performance. If the ratio is set too low, the performance will not balance, no matter how low the threshold is. Finding the right balance between threshold and ratio is one of the key nuts to crack when compressing.

> *Threshold defines the extent of compression. Ratio defines the degree. The two are often fine-tuned simultaneously.*

This might be the right point to talk about how the two are utilized with relation to how evident the compression is. We already established that the threshold is a form of discrimination between what is being treated and what is not. In this respect, the higher the threshold the more the discrimination. We can consider for example the compression of vocals, when our task is to make it as transparent as possible. Chances are that with moderate threshold and moderate ratio, the effect of compression would be more evident than if we set the threshold very low, but applied a very gentle ratio, 1.2:1 for example (Figure 16.26). Since we already established that the attack and release are also applied on level changes above the threshold, we know that the timing characteristics are applied even if the threshold is set very low. All of this is said just to stress

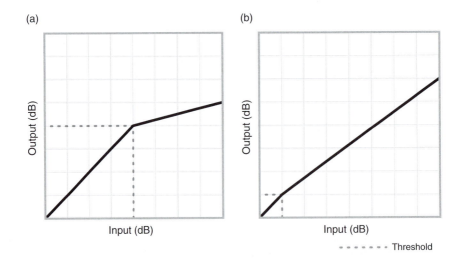

Figure 16.26 The effect of compression with relation to the threshold and ratio settings. The scale in these graphs represents the full dynamic range of the input signal. (a) Moderate threshold and ratio. One should expect more evident compression here as the threshold sets a discrimination point between untreated signals and moderately treated signals. (b) Low threshold and ratio. One should expect less evident compression here as all but the very quiet signal will be treated gently.

that lower threshold does not necessarily mean more evident effect – this is determined by both the threshold and ratio settings, as well as other settings described below.

Attack

One reason for longer attack times was given earlier – to prevent the suppression of an instrument's natural attack. But we don't always want to keep the natural attack. Sometimes we want to soften a drum hit, so kicks, snares and the likes are less dominant in the mix. In such cases, shorter attack is dialed.

> *The longer the attack time the more we retain the instrument's natural attack. However, this is not always wanted.*

The way longer attack helps retaining the natural attack is easily demonstrated on percussive instruments. Figure 16.27 illustrates the effect of different attack times on snare. We can see that with no attack, the natural attack is nearly lost as the compressor squashes it down quite drastically (clearly, this is also dependent on the threshold and ratio setting). Then the longer the attack the more of the natural attack is retained. Long attack hardly affects the natural attack, or in this specific snare case – hardly affects anything at all.

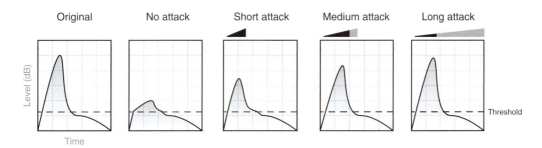

Figure 16.27 Different attack times on a snare. The attack build-up is shown below each caption. The longer the compressor attack time the more of the instrument's natural attack is retained.

We might instinctively think: if we want to retain the natural attack of an instrument, why not just set the attack time as high as possible? The long attack in Figure 16.27 demonstrates why this is unwise. We can see that by the time the signal has dropped below the threshold the attack has barely built up to cause any significant gain reduction – it nearly canceled out the compression effect, making the whole process rather pointless. Based on Figure 16.12, we know that shortly after the snare signal starts to drop, the applied gain reduction also starts to drop, and at this point the attack phase stops and the compressor enters the release phase. This point on Figure 16.27 is where the attack build-up turns gray, and we can see that the potential full effect is not achieved for either the medium or the long attack settings.

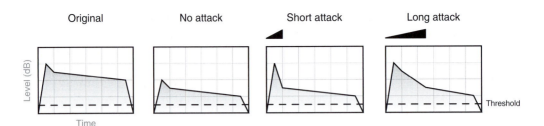

Figure 16.28 Different attack times on a piano key hit. The long attack time caused alteration of the level envelope, which would distort the timbre of the instrument.

Track 16.23: Snare Uncompressed
The original snare track used for the following samples. For the purpose of demonstration, the compressor threshold in the following tracks was set to −15 dB and the ratio to ∞:1.

Track 16.24: Snare Attack 10 microseconds
The extremely short attack on the compressor immediately reduces the natural attack of the snare, simply resulting in overall level reduction.

Track 16.25: Snare Attack 1 ms
With 1 ms of attack, the result is not much different than in the previous track. Yet, a very small part of the natural attack manages to remain, so there is a notch more attack on this track.

Track 16.26: Snare Attack 10 ms
10 ms is enough for a noticeable part of the natural attack to pass through, resulting in compression that attenuates the late attack and the early decay envelopes.

Track 16.27: Snare Attack 50 ms
An even larger portion of the natural attack passes through here, resulting in gain reduction that mainly affects the decay.

Track 16.28: Snare Attack 100 ms
100 ms of attack lets the full natural attack through. Essentially, little gain reduction is applied and it affects the early decay envelope. Compared to the uncompressed track, a bit of the early decay was reduced in level here.

Track 16.29: Snare Attack 1000 ms
The majority of sound in each of these snare hits is no longer than 500 ms. 1 second is too long for the compressor to respond, so no gain reduction has been applied at all, resulting in a track that is similar to the uncompressed one.

Plugin: PSP *MasterComp*
Drums: Toontrack *EZdrummer*

Even when the signal remains above the threshold after the attack phase, a longer attack time then required is rarely beneficial. We can demonstrate this on the piano key hit in Figure 16.28. The short attack brings the gain reduction to full impact slightly after the natural attack has declined, and the resultant dynamic envelope resembles the input signal envelope. This is not the case with the long attack, which clearly altered the dynamic envelope and resulted in unwanted timbre alteration. This example teaches us what is very likely to be a musical attack time for instruments of percussive nature: one that only affects the natural attack without altering the dynamic envelope succeeding it.

Another reason why a longer attack would be beneficial involves the compression of overheads. If the kick is the main instrument to trigger the compression and the attack is set too short, not only the kick's natural attack, but also any other drum played underneath it (like the hi-hats) would be suppressed. If this happens on the downbeat, downbeat hats would be softer than other hats. This might be suitable with more exotic rhythms, but rarely for the contemporary acid-house track or punk-rock or gangsta rap or most of everything else. A longer attack will let the downbeat hats sneak in before the compressor acts upon it.

Track 16.30: Drums Uncompressed
The uncompressed version of the drums used in the following samples.

Track 16.31: Drums Attack 5 microseconds
Notice how the hi-hats' level is affected by the compression applied on the kick and snare. Also, the short attack suppresses the natural attack of both the kick and the snare.

Track 16.32: Drums Attack 16 ms
16 ms of attack is enough to prevent the kick and snare compression from attenuating the hats. In addition, the natural attack of both these drums is better retained.

Plugin: McDSP *Compressor Bank CB1*
Drums: Toontrack *EZdrummer*

One of the things we have to consider when it comes to attack times is how much gain reduction actually takes place. The more the gain reduction is, the more noticeable the attack will be – there is no comparison between an attack on maximum gain reduction of 1 dB and that of 8 dB. Generally speaking, heavy gain reduction could benefit from longer attack times, so large upswings take place gradually rather than promptly, making level changes less drastic. In addition, fast attack and serious gain reduction can easily produce an **audible click**. Although such attack clicks are sometimes used to add definition to kicks, they are mostly unwanted.

Nevertheless, there is a catch-22 here since a longer attack lets some of the level through before bringing it down (and with heavy compression it is quite a long way down). So with longer attack times level changes might be less drastic but more noticeable. This is often a problem when heavy compression is applied on less percussive sounds like vocals. If the attack is too short, the vocals dynamics flatten, but if the attack is made longer, there might be a **level pop** on the leading edge of each phrase. It is sometimes possible to find a compromise in these situations, and when suitable, raising the threshold or lowering the ratio can also help. Yet, a few compressor tricks described later in this chapter can give a much more elegant solution and far better results.

Track 16.33: Vocal Attack 1 ms
This very short attack time yields instant gain reduction, which makes the compression highly noticeable. Essentially, each time the voice shoots up, a quick level drop can be discerned.

Track 16.34: Vocal Attack 5 ms
5 ms of attack means that the gain reduction is applied more gradually, producing far less level fluctuations due to the operation of the compressor. This attack time is also short enough to reasonably tame the vocal upswings.

Track 16.35: Vocal Attack 7 ms
There are still some quick gain reduction traces in the previous track, and 7 ms of attack reduce these further. However, the longer attack means that the voice in this track manages to overshoot higher.

Track 16.36: Vocal Attack 30 ms
30 ms are too long, and the compressor misses the leading edge of the vocal upswings. In addition, the compressor does not track the level variations of the vocal, making level fluctuation caused by the compressor more noticeable.

Track 16.37: Vocal Attack Pop
This track is also produced using 30 ms of attack, but with lower threshold and higher ratio so as to draw heavier gain reduction. Profound level pops can be heard on 'who', 'running' and 'pass'.

Plugin: Sonnox *Oxford Dynamics*

Another consequence of fast attack is **low-frequencies distortion**. The reason being that the period of low frequencies is long enough for the compressor to act within each cycle rather than on the overall dynamic envelope of the signal. Figure 16.29 shows the outcome of this, and we can see the attack affecting every half a cycle. The compressor has reshaped the input sine wave into something quite different, and as a result distortion is added. The character of such distortion varies from one compressor to another, and it can be argued that analog compressors tend to generate more appealing distortion. In large amounts, this type of distortion adds rasp to the mix. But in smaller, more sensible amounts it can add warmth and definition to low-frequency instruments, notably bass instruments.

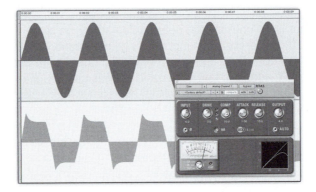

Figure 16.29 Low-frequency distortion due to fast attack settings. The top track is a 50 Hz sine wave passing through the McDSP *Analog Channel 1* shown on the bottom right. The compressor setting involves 1 ms of attack and 10 ms of release. The rendered result is shown on the bottom track, where we can see the attack acting within each half cycle.

> **Track 16.38: Bass Uncompressed**
> The source bass track used in the following tracks.
>
> **Track 16.39: Bass LF Distortion Fast TC**
> The low-frequency distortion was caused due to the fast attack (10 μs) and fast release (5 ms).
>
> **Track 16.40: Bass LF Distortion Fast Attack**
> Although the release is lengthened to 100 ms, the 10 μs attack still produces some distortion. When a bass is mixed with other instruments, a degree of distortion as such can add definition and edge.
>
> **Track 16.41: Bass LF Distortion Slow TC**
> With the release at 100 ms and the attack lengthened to 2 ms, there is only a faint hint of distortion.
>
> *Plugin:* Digidesign *DigiRack Compressor/Limiter Dyn 3*

The main purpose of the hold control is to rectify this type of distortion. By holding back the gain reduction for a short period, the attack and release are prevented from tracking the waveform. The cycle of a 50 Hz sine wave is 20 ms; often the hold time is set around this value.

The attack settings can affect the tonality of the compressed signal, where longer attack times tend to soften high frequencies. Alternatively, this can be seen as low-frequency emphasis and added warmth. The reason for this is demonstrated in Figure 16.30. High frequencies are the result of quick level changes, low frequencies of slow ones. A long release can slow the rate at which the signal level rises, which softens the quick level rise of high frequencies.

Release

Release and attack have a few things in common. A very short release can also **distort low-frequencies** for the same reason short attack can. Here again, the hold control can reduce this type of distortion. Long release tends to soften high frequencies on the same principle that long attack does. A very short release can cause **audible clicks**, although as opposed to attack clicks, it is hard to think of any practical use for these release-clicks. Release settings also become somewhat more important with heavier compression, as the release is applied on larger amounts of gain reduction.

> *Short release times can cause low-frequency distortion and audible clicks.*

> **Track 16.42: Bass LF Distortion Fast Release**
> The distortion on this track is the result of a fast release of 5 ms and a relatively long attack of 2 ms.
>
> *Plugin:* Digidesign *DigiRack Compressor/Limiter Dyn 3*

One thing we have to consider is that the gain recovery during the release phase some-times happens during silent sections. For example, if you try to sing aloud 'when we eat' and then 'when we go', you are likely to have cut the 'eat' while fading out the 'go'. Quick vocal drops like 'eat' can be shorter than 50 ms, and a release longer than that period

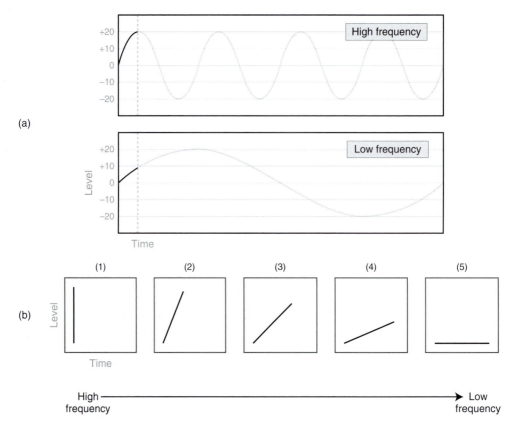

Figure 16.30 Level change rate and frequencies. (a) We can see that the slope of the
high-frequency waveform is steeper. By the time the high-frequency sine wave rose
by 20 dB, the low-frequency sine wave only rose by 10 dB. (b) Faster level changes
are associated with high frequencies. A long attack on a compressor can turn a level
rise like (2) into (3), which softens high frequencies. We will see later that a gate can
result in an abrupt level jump like (1), which generates high frequencies.

might end up affecting silence. Although the release setting is important to control vocal
drops, we have to acknowledge that some of the gain recovery can happen on inaudible
sections. This suggests that the release can be more important for instruments that rarely
dip into silence, like overheads and rhythm guitars. These types of instruments are also
more likely to experience **pumping** which is caused by short release times.

| *Short release can cause pumping.* |

As per our discussion on the difference between gain reduction and compression, we
know that a very long release time (hypothetically, ten hours) will cause steady gain
reduction, which is exactly the opposite of compression. Both long attack and release
slow down changes in gain reduction, which makes a compressor more of a fader.
Subsequently, the effect of compression becomes less noticeable. By way of analogy,
if during manual gain-riding the engineer's fader movements are extremely slow, the

input level variations will hardly alter. The true effect of compression only happens when the gain reduction tracks change in input levels. Short release times result in faster gain recovery, which in turn increases the average output level and the perceived loudness.

> *Short release times result in more compression, more evident effect and loudness increase.*

Two compressed versions of Track 16.30. Note that the compression is more evident in the short release version, which might also be perceived as louder. With 2000 ms of release, most of the gain reduction is static, and the result is very similar to the uncompressed version.

Track 16.43: Drums Release 50 ms
Track 16.44: Drums Release 2000 ms

Plugin: Focusrite *Liquid Mix (Trany 1a emulation)*
Drums: Toontrack *EZdrummer*

The previous attack section started with a snare demonstration, so let us start by using a snare to demonstrate the function of release. We established that the longer the attack is the more the natural attack is retained. The effect of release can easily be misunderstood as it is exactly opposite in nature – the longer is the release the less of the natural decay is retained. The release control determines the rate at which the gain reduction can drop, or in other words, how quickly the gain recovers. Once the input signal level drops, a longer release will slow down the recovery rate, so more gain reduction will be applied on signals. This effect was demonstrated in Figure 16.11 on a noise burst; Figure 16.31 shows this happening on a snare. It is interesting to note the effect of medium release, which quite notably reshapes the decay. This extra bump will alter the timbre of a snare

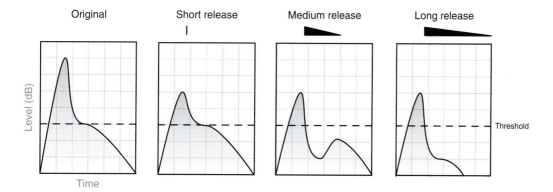

Figure 16.31 Release times on a snare. With short release the decay is unaltered compared to the original signal. With medium release, it takes some time for the gain to recover, and only then the natural decay returns to its original levels. With long release, the decay is attenuated and very little of it is retained.

in an unflattering way. So often we seek to either leave the decay as is or attenuate it altogether.

> The following tracks demonstrate the effect of different release times. In all of them a 50 ms of attack allows most of the natural attack to pass through, leaving the gain reduction to mainly affect the decay of the snare. The longer the release, the slower the gain recovers and consequently the more the decay is attenuated:
> **Track 16.45: Snare Release 100 ms**
> **Track 16.46: Snare Release 500 ms**
> **Track 16.47: Snare Release 1000 ms**
> **Track 16.48: Snare Release 2000 ms**
> **Track 16.49: Snare Release 3000 ms**
>
> Note that in Track 16.46 it is possible to hear the decay recovering after being reduced in level straight after the attack. This is a demonstration of decay reshaping similar to that caused by the medium release shown in Figure 16.31.
>
> *Plugin:* PSP *MasterComp*
> *Drums:* Toontrack *EZdrummer*

In the case of vocals, we might want fast release so that the already dropping signal is not attenuated further by the compressor. But both attack and release are also concerned with level changes above the threshold, so in the case of vocals their function is often more concerned with tracking level fluctuations. As already suggested, the two are set to be fast enough to track these fluctuations.

> *Longer release on a compressor means less of the instrument's natural decay and can cause timbre alterations.*

No discussion about the release control would be complete without mentioning its strong link to the rhythmical nature of the song. When compression is applied on instruments with percussive nature, the release dictates the length of each hit and how it decays. Different release settings can make a world's difference to how an instrument locks to the natural rhythm of the music. For example, a snare that hits every half-note might feel right in the mix if its perceived length is set to a quarter-note. There are no hard and fast rules – we have to experiment with the release settings until it feels right.

> These two tracks demonstrate the effect release can have on the groove of rhythmical material. The effect might be easier to comprehend if one tries to conduct or move the head in relation to the beat.
>
> **Track 16.50: Drums Release Groove I**
> The long release on this track results in gain reduction that does not track the groove of the drums. The relatively slow gain recovery interferes with the natural rhythm.
>
> **Track 16.51: Drums Release Groove II**
> A shorter release setting yields compression which is much more faithful to the rhythm.
>
> *Plugin:* Digidesign *Smack!*
> *Drums:* Toontrack *EZdrummer*

After a snare hit, the gain reduction meter shows when (and if) gain reduction falls back to 0 dB. Usually, we want this to happen before the next snare comes along, or the next

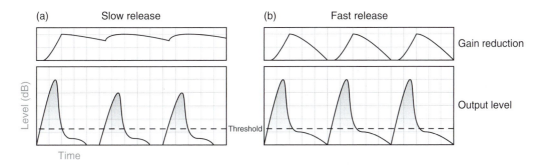

Figure 16.32 Release and snare spacing. (a) With slow release, the gain reduction does not recover before the next snare hit, causing attenuation to the consecutive hits and softening their attack. (b) With fast release, the gain reduction is back to 0 dB before the next hit, so the compressor treats all hits uniformly.

hit will be attenuated by the ongoing gain reduction (Figure 16.32). One would naturally ask: but what happens if the drummer plays half-notes for the most part, and just before each chorus plays eight-notes? A release setting to fit the length of a half-note will be too long for eight-notes and will de-emphasize them. Chances are that these eight-notes are there to squeeze some excitement in before the chorus, so such a de-emphasis is contrary to what we want. However, setting the release to fit eight-notes might not be suitable for the half-notes, which are played for most of the song. As always, there are two main options here: either to compromise on a setting that will be neither perfect nor horrible on both note lengths or simply automate the release so all the hits feel right.

> *In the case of percussive performance, we want gain reduction to*
> *fully drop before the next note arrives.*

Track 16.52: Snare II Uncompressed
In this uncompressed version, the velocity of each snare hit is identical.

Track 16.53: Snare II Release 1000 ms
With 1 s of release, the gain never fully recovers between the various snare hits. Notice how the first hit is louder than all other hits, and how the last hit, which is the closest to the hit preceding it, is the quietest of all hits.

Plugin: Sonnox *Oxford Dynamics*
Drums: Toontrack *EZdrummer*

Peak vs. RMS

Compared to peak sensing, an RMS compressor responds more slowly to changes in the input signal levels. In a way, the RMS averaging is a form of attack and release that is applied before the threshold stage. Having a strong link to the way we perceive loudness, RMS is very useful for instruments with a less percussive nature, like vocals.

However, their slow response makes them somewhat less suitable for percussive instruments, which change in level very quickly. If *catch me if you can* is the name of the game, a compressor will do a better job in peak mode.

| *Peak can work better on a percussive performance.* |

Since the RMS function slows quick level changes, the whole compression happens in a more relaxed manner that makes the compressor less evident. Peak level detection leads to more obtrusive operation and often more aggressive results. If aggression or power is what we are after, peak mode can be the right choice. If transparency or smoothness is our aim, RMS comes in handy. Notwithstanding, some RMS compressors can be very aggressive, powerful and punchy, while a compressor in peak mode can also do a decent job on vocals.

| *RMS for transparency, peak for a firmer control and more effect.* |

Peak-sensing allows the compressor to respond much quicker to transients. This might lead to extra loss in dynamics, for example, when compressing drums. Notice the loss of dynamics, attack and punch in the peak version:

Track 16.54: Drums RMS
Track 16.55: Drums Peak

In the tracks above, no settings have changed apart from the Peak/RMS parameter. This makes the comparison somewhat unfair since the peak version would natively present a harder, more evident compression. The next track involves the same settings as in Track 16.55, only with a longer attack; the result is very similar to the RMS version.

Track 16.56: Drums Peak Longer Attack

Plugin: Cubase *Compressor*
Drums: Toontrack *EZdrummer*

Side-chain control

The interaction between the threshold, ratio, attack and release gives us extensive control over the compressed signal. The ability to alter the side-chain signal adds further control and a few more compression applications. The side-chain signal can be altered via two facilities: either the built-in side-chain filters or any processors we apply on the external side-chain. Applying processors of our choice on the external side-chain gives us ultimate control. Yet, in some cases the built-in filters would suffice.

The nature of the signal feeding the side-chain can be one of two: the original input, which we can equalize, or a different signal altogether. The latter is used for specific applications or as part of a unique effect. More on this soon. For now, let us see the benefits of modifying the original input.

One of the main characteristics of compressors is that they are more responsive to low frequencies. Put another way, low frequencies tend to excite compressors more than high frequencies. One reason for this is that, due to their long period, low frequency signals spend more time above the threshold. Another reason is the fact that low frequencies have to be louder in order to be perceived as loud as higher-frequencies (as per the equal loudness curves). This characteristic of low frequencies is the very reason why compressors have a reputation for dulling sounds – as low frequencies are being compressed, the level of high frequencies is also reduced.

> *Low frequencies tend to trigger compression more than high frequencies, a fact that can yield dulling of compressed instruments.*

To give one practical example, when overheads are compressed the kick is most often the first drum to trigger the compression. So whenever the kick hits, the cymbals are also reduced in level. In addition, if the threshold is set too low (or the ratio too high), we could defeat the kick. However, settings that are appropriate for the kick might not trigger any compression when the cymbals are played. Effectively, not all of the drum kit is being compressed uniformly. A typical consequence of this is crashes that drop in level in a step-like fashion every time the kick or the snare hits (this is often something we check when compressing overhead, and both the attack and release can affect this behavior). The same issue exists with any type of broadband sound – low and high frequencies are not compressed to the same extent. The way to solve this is applying some low-frequency filtering on the side-chain. By applying, for example, a HPF, we make the compressor response more balanced to broadband signals like overheads.

> *A high-pass filter on the side-chain can bring about more uniform compression of broadband material.*

Track 16.57: Drums II Uncompressed
The source track for the following samples.

Track 16.58: Drums II No SC Filter
The compressor is set to the fastest attack and release in order to draw the most obvious effect. Both the kick and snare trigger heavy and fast changes in gain reduction, which yields fluctuation in the crashes' level.

Track 16.59: Drums II SC Filter
A 24 dB/oct HPF at 2 kHz is set to filter the side-chain. This essentially means that the kick does not trigger gain reduction anymore. Notice how the crashes' level is smoother as a result. Natively, filtering the low frequencies from the side-chain yields less gain reduction, so in this track the threshold was brought down and the ratio up so the gain reduction meter closely hits the same 10 dB as in the previous track.

Track 16.60: Drums II SC Only
This is the filtered SC signal that triggered the compression in the previous track.

Plugin: McDSP *Channel G*
Drums: Toontrack *EZdrummer*

Then there is also an issue with a performance that varies in pitch. We can give the example of vocals. The low-mids of a vocal performance vary noticeably with relation to the pitch being sung. As the melody travels up and down, the lower notes trigger more compression. The higher-mids of a vocal performance tend to provide more solid energy, which is less dependent on the sung pitch. Applying a HPF on the vocals' side-chain can yield far more consistent compression. Perhaps the best demonstration of this problem has to do with bass guitars. The low E on a bass guitar (41 Hz) produces radically more low energy than the E two octaves above (165 Hz), therefore the low E triggers more compression. Notwithstanding, the higher E will be more defined as both its fundamental and harmonics start higher on the frequency spectrum. Having the low E less defined yet more compressed suggests how imbalanced the bass might sound. We can give the same example with the violin, trumpet or virtually any instrument – the lower the note the more the compression. Equalizing the side-chain can improve the compressor response to different notes. The actual filter being applied is often a high-pass filter (that can be of any slope), but low shelving equalizers can also be suitable for this purpose. Sometimes, even a parametric EQ would work.

> *Side-chain equalization, more specifically low-frequency filtering, can yield uniform compression for different notes.*

There is another way to look at the above – if all frequencies below 500 Hz are filtered from the side-chain, the same frequencies are likely to be louder as they trigger less compression. If we boost 4 kHz on the side-chain, 4 kHz will be attenuated more (this is de-essing in essence). This leads to an interesting phenomenon – equalizing the side-chain affects the tonality of the compressed signal. Generally speaking, one should expect the *opposite effect* to the settings dialed on the side-chain EQ. But the applied tonal changes are far more sophisticated than simply the opposite of the side-chain EQ – it is compression-dependent equalization, otherwise a form of **dynamic equalization**. Apart from having to consider how the side-chain equalization affects the tonality of the compressed signal, we can use it to fine-tune the frequency spectrum of instruments. It takes experience to master this technique, but it is worth the effort.

> *Side-chain equalization alters the tonality of the compressed signal and enables dynamic equalization that is more sophisticated than the standard static equalization.*

Applications

The following is a summary of compressor applications:

- Making sounds bigger, denser, fatter
- Accentuating the inner details of sounds
- Adding warmth
- Containing levels
- Balancing levels
- Loudening
- Softening or emphasizing dynamics
- Reshaping dynamic envelopes:
  - Adding punch
  - Accenting the natural attack
  - Accenting transients
  - Reviving transients
  - Reconstructing lost dynamics
  - Emphasizing or de-emphasizing decays

- Bringing instruments forward or backward in the mix
- Making programmed music sound more natural
- Applying dynamic movement
- Ducking
- De-essing
- Applying dynamic equalization

Along with new applications listed here, some of the already discussed applications are covered in greatest depth in the coming sections.

Accentuating the inner details of sounds

Compression invariably condenses the levels of signals. One reason we find the effect so appealing is that it tends to emphasize the inner details of sounds. The small nuances of any performance – the lip movements of the vocalist or the air flow from a trumpet – become clearer, which is largely satisfying. Like anything in mixing, overdoing can have a counter effect. It is like luminance: in the right amount it can enhance an image and make its details clearer, but too much of it results in blinding white.

Balancing levels

Figure 16.33 shows the possible levels of three mix elements before and after compression. We can see that before compression the level fluctuations of the different instruments are quite problematic: at points the snare is the loudest, at points the bass, at points the vocals; the snare level varies, sometimes the bass is louder than the vocals, sometimes vice versa. It's a complete mess. No matter how much we try to balance such an arrangement using faders, the relative levels will always change over time and will do so quite radically. However, after we compress each instrument we can balance

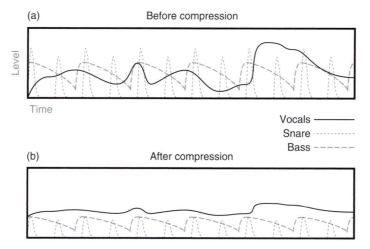

Figure 16.33 Relative levels before and after compression. (a) Before compression the fluctuations in level make these three instruments impossible to balance. (b) After compression the balancing task is easier.

them so their relative levels are consistent. Moreover, we can make them closer in level to one another, causing the impression of a louder, more solid mix.

Balancing of this sort usually happens very early in the mixing process, so as to make the tracks more manageable and promote some sense in their relative levels. Usually we are not very fussy at this stage about perfecting the compression – rough settings would suffice to give reasonable balancing. As we progress from coarse to fine, we give more attention to the individual compression applied on each instrument.

Out of the three instruments shown in Figure 16.33, balancing the vocals would be of prime importance as they exhibit the wildest level variations. Perhaps when working on the snare, balancing level would be the second objective after adding punch. The level balancing approach of each instrument is likely to be different – we might want the vocal compression to be transparent while drawing more effect from the snare compression. If we have to generalize what could be sensible settings for level balancing, the following guidelines could provide a possible starting point:

- **Threshold** – set at the bottom of the action range, so all level fluctuations are captured.
- **Ratio** – dependent on the extent of overshoots and to what degree we want to retain dynamic sense. High ratio can squash an instrument, while low ratio might not be sufficient to balance the levels. In the figure above, chances are that the highest ratio will be applied on the vocals as they vary in levels quite noticeably. Also, vocals have more settled dynamics, so softening these will be less noticeable.
- **Attack and release** – for less percussive instruments set to fast, so more compression is applied and level changes are quickly tracked. At the same time, we must observe low-frequency distortion and how the natural attack and decay are affected. For percussive instruments we have to set longer attack and release to keep the dynamics.
- **Hold** – if available, can help reducing low- frequency distortion while letting us shorten the attack and release times.
- **Knee** – soft, as mostly when balancing levels we are not after the effect of compression.
- **Level detection** – depending on the instrument. Probably RMS for the vocals and peak for the snare.
- **Side-chain filter** – likely. Attenuating or filtering the side-chain lows of all three instruments can bring about more accurate results.
- **Make-up gain** – set so A/B comparison is fair (i.e., the perceived level will not change as we bypass the compressor).

Track 16.61: Snare HH Unbalanced
The uncompressed source track, with loud snare hits followed by softer ones.

Track 16.62: Snare HH Balanced
Only the snare is compressed on this track. It involves a threshold at −30 dB, 4.4:1 ratio and soft knee. The attack, hold and release are adjusted in order to try and maintain the timbre of the loud hits. The attack ended up at 0.7 ms, hold at 300 ms and release at 290 ms. Also, a side-chain shelving filter is applied to attenuate the lows, along with a parametric 4 dB boost around 3 kHz which controls the degree by which the louder hits are reduced in level. The gain on the parametric side-chain filter was used for the final balancing of the hits.

Plugin: Sonnox *Oxford Dynamics*
Drums: Toontrack *EZdrummer*

One decision we often have to make is to what extent balancing should be done. If all the snare and kick hits are at the same level, they could sound like the product of a drum machine. Fortunately, the timbre of a drum changes with relation to the hit velocity, so some natural stamp is retained even if the individual hits are evened in level. Yet, a more natural production could benefit from a relaxed balancing approach. In more powerful productions or more powerful sections of productions, many mixing engineers tend to even out the drum hits in a mechanical fashion. We have to remember that when it comes to drums there is a fine line between natural and sloppy. A good drummer will exhibit some consistency, some variations and some consistency of variations. But if each hit is random in level, the performance comes across as amateur. Since the less sloppy the more even, a perfectly even drum velocity can give quite a powerful impression, although not necessarily a natural one. In addition, identical hits level might translate in our mind to a drummer that hits the drums as hard as possible.

Drum hits even in velocity can give the impression of a powerful performance, although might not be suitable for natural genres.

Loudening

Loudening is achieved by increasing average signal level, while maintaining its peak. During mixdown, this is done less in the mastering sense of maximizing the overall loudness of a stereo mix, but more with the objective of condensing the dynamics of individual instruments, which often creates a more powerful impact. If a specific instrument fails to compete with other instruments in the mix, some compression can make it cut through. Loudening is also a means of balancing levels:

- **Knee** – soft knee is obligatory here as the compression effect caused by a hard knee will limit our ability to dial the extreme settings on other controls. The lower the knee span the more evident the compression, but very high knee settings will limit our ability to push the levels higher. We generally try to have a knee span as wide as the action range of the input signal; 30 dB span can be a good starting point.
- **Ratio** – set to maximum. The soft knee will provide gradual limiting, and the threshold control is used to determine the degree of true limiting.
- **Threshold** – when the threshold is set to half the knee span below 0 dB, the full knee effect is reached at 0 dB, and only signals above this level will be limited. Any additional lowering of the threshold will result in an identical extent of limiting range for loud input signals. For example, with a 40 dB soft-knee span and the threshold set to −20 dB, limiting will affect input signals at 0 dB and above (Figure 16.34a). Then bringing the threshold 10 dB down to −30 dB will yield limiting to the top 10 dB of the input signal (Figure 16.34b).
- **Make-up gain** – identical to the threshold, but inverted in magnitude. For example, for a threshold at −12 dB, the make-up gain would be 12 dB. Since the ratio is set to maximum, this matched setting brings the maximum output level to 0 dB.
- **Attack and release** – just like with balancing levels, short for less percussive and longer attack for more percussive. Low-frequency distortion must be observed.
- **Hold** – like with balancing levels, can reduce low-frequency distortion.

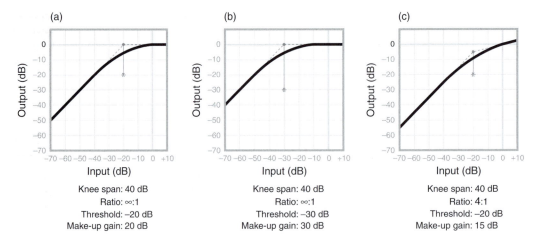

Figure 16.34 Three approaches to loudening. (a) The knee is reached at 0 dB. Any input signals above 0 dB will be limited. (b) A more aggressive compression is achieved by bringing the threshold down and the make-up gain up respectively. The full knee is reach at −10 dB, so limiting starts for any input signals above −10 dB. (c) A softer effect can be achieved by lower ratio settings. This approach retains the dynamics of the input signal better than the other two.

- **Level detection** – as we really need as little compression effect as possible, RMS is favored here.
- **Side-chain filter** – likely. Attenuating or filtering the side-chain lows can bring about more accurate results and reduce pumping.

This specific approach involves an increasing compression on louder signals. Although it can result in extra power, it does not excel at retaining dynamics. We can tweak the different settings to achieve less aggressive compression that would retain dynamics better. For example, we can reduce the ratio to 6:1; the make-up gain will have to be brought down, respectively, to compensate for the extra output level (Figure 16.34c).

Track 16.63: Bass Loudening I
This is a compressed version of Track 16.38. The threshold was set to −20 dB, ratio to the maximum of 1000:1 and the soft knee span to 40 dB. The attack, hold and release are all kept as short as possible – longer attack would mean higher signal peaks (which would limit the make-up gain) and longer release would mean less compression and less loudness. Notice the loss of bass attack compared to the source track. The make-up gain is set to +15.4 dB in order to match the peak reading of the source and this track.

Track 16.64: Bass Loudening II
Same settings as in the previous track, only with the threshold down to −30 dB, ratio to 2.5:1 and make-up gain to +15 dB. This track is slightly quieter than the previous one, but involves a bit more attack.

Plugin: Sonnox *Oxford Dynamics*

Reshaping dynamic envelopes

The level changes a compressor applies reshape the dynamic envelope of sounds. In most simple terms, the threshold defines which part of the envelope is reshaped and the ratio defines to what extent it is reshaped. Attack and release give us far more control over envelope reshaping – we can alter the natural attack and decay in different ways.

The first application we will discuss is **adding punch**. Before showing how a compressor can achieve that, we should give one view on what punch is. The definition of punch in dictionaries includes words like thrust, blow, forcefulness and power. A characteristic all these words share is that – they don't give the impression of slow motion. They have to be abrupt to have an impact. Many of us know how restful sounds become as we open the release on a synthesizer. Staccato for frisky, legato for relaxed. One of the basic principles of music is that fast equals energy. It is hard to write a mellow love song at 160 BPM, like it is hard to write a dance tune at 72 BPM. Consequently, as the tempo gets faster, the notes get shorter.

How does this relate to compression? We can shorten sounds using a compressor by attenuating their decay. It should be quite obvious by now how to do this – we set the attack to let the natural attack pass through and long release in order to attenuate the natural decay. Figure 16.35 illustrates this on a snare. We can enhance the effect by shortening the attack, so some compression catches the natural attack, condensing it to give more impact.

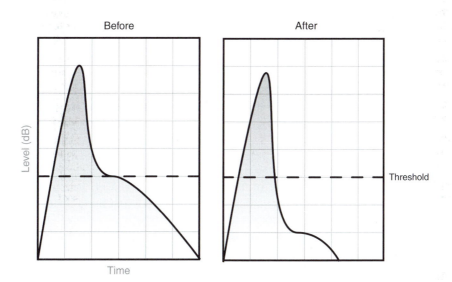

Figure 16.35 Adding punch to a snare using a compressor. The attack is set long enough to retain the natural attack, while the release is set long so gain reduction brings down the natural decay.

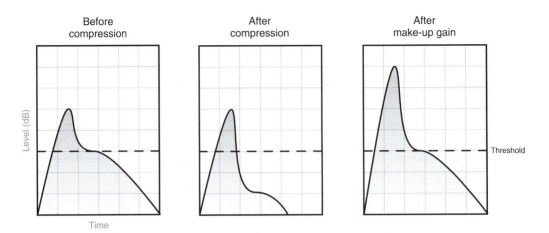

| Before compression | After compression | After make-up gain |

Figure 16.36 Accenting the attack of a snare using a compressor. The attack is set long enough to retain the natural attack, while the long release brings down the decay. After applying make-up gain, the attack is made louder, but the decay returns to its original level.

We can use exactly the same principle to **accent the natural attack, accent transients, revive transients or reconstruct lost dynamics** (due to over-compression for example). All we need to do is apply some make-up gain after bringing down the decay. Figure 16.36 illustrates this. We can see that after the addition of make-up gain the attack is emphasized compared to the input signal before compression.

There is more in the dynamics reshaping examples above than first meets the eye. Compression is a sophisticated way of **bringing instruments forward or backward** in the mix. Perhaps we want to bring the kick forward, but raising the fader detaches it from the mix and makes it too dominant. On the same basis, we might want to pull the kick back a bit, but even gentle lowering of the fader results in a faint kick. One of the problems with using faders is that they affect the full frequency spectrum of sounds. Consequently, this affects the frequency balance of the mix and the masking interaction between instruments. If we want the kick forward in the mix, why not just bring up its attack? If we want it further away, why not just soften its attack or perhaps its decay? Only affecting a part of the dynamic envelope means we do not have to consider issues like masking as much. Earlier we saw how equalizers can provide a surgical alternative to faders by only affecting limited frequency ranges in sounds. Compressors (and other dynamic range processors) can achieve something similar, except they affect limited time (or length) of sounds. In essence, compressors can be used for sophisticated level adjustments as some kind of surgical faders. This can also bring instruments closer together in levels.

> *Compression can be used as a surgical way to bring instruments forward and backward in the mix.*

> **Track 16.65: Reshaping Source**
> The source drums used in the following samples. The snare in this track is not compressed.
>
> **Track 16.66: Reshaping More Attack**
> For the purpose of demonstration, the attack of the snare is accented to an exaggerated degree. First, the ratio is set to the maximum of 100:1, both time constants to minimum and the threshold is brought down to −40 dB. Then, the attack is lengthened, and the longer it became the more of the natural snare attack passed through; it ended up at the maximum value of 300 ms. (This extremely long attack was required due to the extreme gain reduction of around 15 dB. There are less than 200 ms between the first and second hit, but remember that the attack time determines the rate – not the time – it takes gain reduction to rise.) The release was then lengthened to 1 second so as to track the natural snare decay. The gain reduction was set to +11 dB in order to bring the snare decay back to its original level.
>
> Notice how the snare has more punch now, and how it moved forward in the mix. Also, notice that as opposed to the previous track, the attack of each hit is not consistent in level. This is due to the release not recovering fast enough between the hits. The sooner each consecutive hit arrives the less attack it has. For example, the first hit has the loudest attack, and the one straight after it has the softest. There is little we can do about this since shorter release means that the decay of the snare would alter; either bringing the threshold or the ratio down in order to reduce the amount of gain reduction (and thus the quicker the gain would recover) would mean less natural attack compared to decay. We will soon see how this can be resolved using another compressor.
>
> **Track 16.67: Reshaping Less Attack**
> The settings involved in reducing the attack of the snare are completely different from the ones used in the previous track. This time, we only want the compressor to reduce the natural attack without affecting the decay. The threshold is set to −13 dB, a level above which the snare only overshoots by 5 dB. The ratio is set to 3.2:1. Both the attack and release are set to minimum −14 µs and 5 ms, respectively. No make-up gain is applied. The result of this compression is a snare with less punch, positioned further back in the depth field.
>
> **Track 16.68: Reshaping Reviving Attack**
> For the sake of demonstrating how a compressor can revive lost transients, the attack-lacking snare from the previous track is fed through another compressor with the same setting as in Track 16.66. How it regained some healthy attack is hard to miss.
>
> *Plugin:* Digidesign *DigiRack Compressor/Limiter Dyn 3*
> *Drum:* Toontrack *EZdrummer*

One case we have not touched upon yet is how compression can be used to **emphasize decay**. Although compression brings levels down, we already saw how we can accent the natural attack by attenuating the decay, then raising the make-up gain. To emphasize decay, we attenuate the attack but leave the decay as is, then the make-up gain brings the whole signal up. This can be extremely useful when we want reverbs to be more present. The compressor settings in this case involve low threshold, high ratio, fast attack and fast release so the gain reduction recovers as quickly as possible. Figure 16.37 illustrates this. The very same settings can be used to **lengthen the sustain** of instruments, notably guitars. It should be noted, however, that compressing natural reverbs is known to impair the spatial information within them – a faithful sense of space is dependent on the reverb's natural dynamics (notably its decay). If reverbs are to appear natural, extra care has to be taken while compressing them. Conversely, when reverbs are used in a creative context, compression can easily improve the effect.

While percussive instruments are a good source for demonstrating the attack and release controls, we should not forget that microdynamics control can also be useful for less percussive instruments. Taking an acoustic guitar as an example, sometimes its sound is too jumpy due to, say, the use of a small diaphragm condenser microphone. This is not in the sense that it fluctuates in levels, it might simply sound 'liver than life'. We can use a compressor to smooth spiky microdynamics and make the overall sound more rounded.

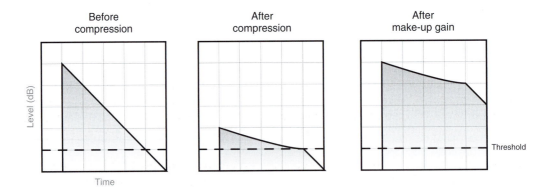

Figure 16.37 Emphasizing the decay of a reverb. The low threshold, high ratio and fast attack reduce the reverb quickly and drastically. As the reverb decays, the gain reduction recovers quickly due to quick release settings. After applying make-up gain, the reverb decay becomes evidently louder.

De-essing

One way to treat sibilance is attenuating the voice whenever sibilance occurs. This is done by feeding into the side-chain a signal that rises significantly in level whenever sibilance occurs. To achieve this, we boost the sibilance frequency on the side-chain signal, while often also filtering frequencies below it. Whenever the sibilance peaks, the side-chain signal overshoots the threshold and the compressor pulls down the overall voice. One issue with this manual approach is that it attenuates not only the sibilance, but the rest of the voice as well, including its body and higher harmonics. Another issue is that the sibilance on low-level passages might not be loud enough to overshoot the threshold, so only the sibilance on loud passages is treated. A well-designed dedicated de-esser takes into account these two issues, so it is most likely to do a better job.

Making programmed music sound more natural

The dynamic complexity of real performance is beyond comparison to what any computer can simulate. To give an example, synthesizers have attack and release controls, but none of them affect a chord already being played. On a real piano, the sound of every chord alters and is being altered by the chord succeeding it. Another example would be a crash hit – a resting crash being hit would sound different than a hit on a crash already swinging. Snares and many other instruments behave very much the same way. If a programmed musical event is a note, chord or a drum hit, the dynamics of a programmed sequence are applied on each event at a time, but there is no dynamic link between the different events. Compression works on all events as a whole, thus it can be used to link them in a real-life fashion.

A classic example would be the machine gun or buzzing sound of programmed snare rolls. Such mechanical arrangements cause each hit to stand out as an individual hit. We can blur the gaps between the hits and meld them together using a compressor. Generally, we want fast attack to make each hit stand out less, and fast release so the decay is retained.

Applying dynamic movement

We have discussed how useful compressors can be for balancing the varying levels of a performance. Unless used to reshape dynamic envelopes, compressors are often associated with the restriction of dynamics. But it is not a secret that a mix with static dynamics can be quite boring. Compressors can be used to get in some action.

To give an extreme example, we can consider an organ playing harmony. The keyboard player changes a chord each bar and all the chords are at reasonably the same level. We can feed this organ into a compressor and feed the kick into the side-chain. Set low threshold, moderate ratio, fast attack, and a release that will cause full gain recovery by the time the next kick hits. Every time the kick hits, the compressor will attenuate the organ, which will then rise in level until the next kick arrives. It is automatic automation. By attenuating the organ with every kick hit, we also clear some frequency space for the kick. We can also do something similar in order to create more sophisticated drum loops. We can go even further by adding synced delay to an external side-chain signal, so the dynamic movement is triggered by one event, then locks to the tempo.

A compressor as a ducker

A compressor used in the way described above – to attenuate a signal in relation to another signal – behaves like a ducker. A compressor is not really a ducker as the amount of attenuation a compressor applies is dependent on the overshoot amount, whereas a ducker applies a fixed amount. With a compressor, the attenuated signal fluctuates in level in relation to the side-chain signal, which is not the case with duckers. (Figure 20.5 illustrates these differences). Percussive instruments are generally less of an issue, since they do not tend to fluctuate as much above the threshold. Also, we can always lengthen the release to deprive gain-reduction changes, but this might not yield musical results once the side-chain signal drops below the threshold. Although for steady ducking it is best to use a ducker, compressors can be, and are, employed as well.

A full exploration of the fascinating ducking applications is presented in Chapter 20. For now, let's look at a small trick that can turn any compressor into a real ducker. What we are trying to prevent is the gain reduction from fluctuating with relation to the input signal. We could use long hold or release times, but we need these to control the gain recovery once the side-chain signal drops below the threshold.

We have to employ a moderate track. Say we want the vocal to duck the guitars. The moderate track is fed with pink noise, which is gated with relation to the vocal track. We set the gate to open every time the vocal crosses a specific threshold. The moderate track is not mixed to the main mix, but instead routed to the side-chain of the compressor on the guitar track. When the vocal overshoots, the gate opens and the pink noise will trigger the compression of the guitar. The pink noise can be regarded as steady in level, so the guitars will always be attenuated by the same amount. Figure 16.38 shows a snapshot of this setup.

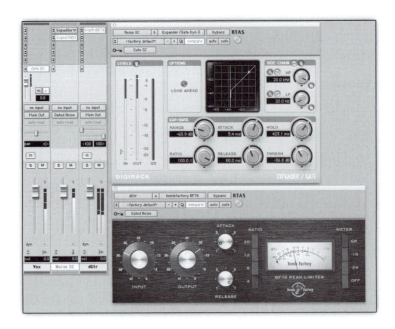

Figure 16.38 Making a compressor behave like a ducker. The vocal track is sent to a bus called Gate SC. The moderate track has a signal generator on its first insert slot (generating pink noise) followed by a gate that has its side-chain input set to the Gate SC bus. This moderate track is not mixed, but routed to a bus called Gated Noise instead. On the distorted guitar track a compressor is loaded, with its side-chain set to the Gated Noise bus.

A quick and dirty way to achieve a similar effect involves limiting the side-chain signal so it does not fluctuate in level once the limiter threshold is exceeded. If we dial a ratio of ∞:1 on the ducking compressor, the amount of gain reduction will be determined by how far below the limiter threshold the compressor threshold is set. The problem is that we have no control over changes in level between the two thresholds (limiter and compressor), so this method cannot be considered as pure ducking.

Tricks

Parallel compression

Parallel compression is one of the oldest and more common tricks in mixing. Parallel compression was already implemented as part of the Dolby A noise reduction system, which was introduced in 1965. In 1977, Mike Bevelle published an article in *Studio Sound* magazine, which gave this technique much of its publicity. Bob Katz coined the term parallel compression, with other names being *side-chain compression* (which is confusing since it is not the compressor side-chain we are compressing) and the *NY compression trick* (NY engineers were notorious for using this technique). It is usually applied on drums, bass and vocals, although it can be applied on virtually any instrument.

The compression we have discussed so far in this chapter is downward compression – loud signals are brought down in level. One fundamental problem with this practice is that our ears are more sensitive to the bringing down of loud signals than they are to soft signals being brought up. Also, the compression artifacts tend to be more noticeable as they are triggered by loud level signals.

The idea of parallel compression is quite simple – instead of bringing high levels down, we bring the quiet levels up. The ear finds this all the more natural, and the artifacts triggered by low-level signals make the compression effect less evident. In addition, parallel compression retains dynamics much better than downward compression as transients are not brought down – if anything their bottom is beefed up. Being more transparent, parallel compression lets us drive the compressor even harder when we are after a stronger effect. Also, we can adjust the degree of impact using a single fader rather than a multitude of compressor controls.

> *Parallel compression lets us make sounds even bigger while retaining their dynamics.*

Parallel compression involves a simple setup: Take a copy of a signal, compress it, and blend the compressed version with the uncompressed original (Figure 16.39). Generally speaking, the original remains consistent in level and the compressed version is layered below it at a level of our choice. Often we do this by bringing up the compressed version from the bottom fader position.

There are various ways to achieve that within an audio sequencer. If it is a single track in question, we can simply duplicate the track and compress the duplicate. However, if we change something on the original track (e.g., the EQ), we also have to change the duplicate. A more popular approach involves sending the track in question – or a group of tracks – to a bus; the bus is fed into two auxes, and a compressor is only loaded onto one aux (Figure 16.40). We then alter the level of the compressed aux track. We must make sure that there is no delay between the original and the compressed versions or combfiltering will occur – auto delay compensation grants this would not happen.

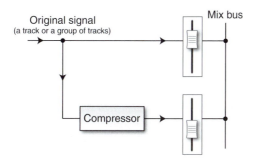

Figure 16.39 Parallel compression is a blend between a compressed and an uncompressed version of the same source. The level of the compressed version is what we alter for more or less effect.

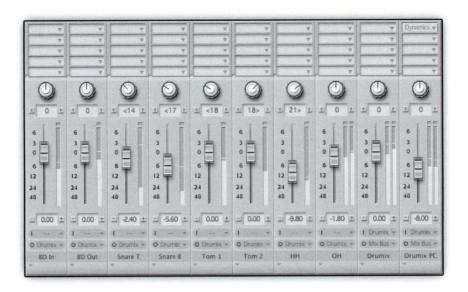

Figure 16.40 Parallel compression in Digital Performer. The output of various drum tracks is routed to a bus called 'drumix'. The bus feeds two auxiliaries – drumix and drumix PC. On the latter a compressor is loaded, and it is blended lower in level with the drumix auxiliary.

While additional routing is required when parallel compression is done on an analog desk, at least one aux track in an audio sequencer could be spared had compressor plugins given the ability to mix the compressed and the uncompressed input signals (something very similar to the standard wet/dry control on a reverb emulator). While this requires very little effort to implement, only a few compressors provide such control. The PSP *VintageWarmer 2* in Figure 16.41 is one of them.

Parallel compression is a form of upward compression – instead of making louder sounds quieter (downward compression), we make quiet sounds louder. But parallel compression is different from upward compression in that low-level signals are boosted in a linear fashion (1:1 ratio), the most drastic ratio happens right above the knee, and from there up the ratio curve slowly returns to 1:1. Figure 16.42 illustrates this. We can consider parallel compression as having the most effect on medium-level signals as they cross the threshold. In practice, however, the threshold is set quite low below the action range, so it is eventually the quiet signals that get most of the effect. In essence, the quieter the signal is the more it is brought up in level; high level signals are hardly boosted. This can be very useful for level balancing.

We can see from Figure 16.42a that the ratio transition around the knee can be quite drastic. We can achieve a softer transition in two ways: either bring down the ratio or bring down the level of the compressed version. As we bring down the compressed version in level, the resultant sum will morph toward the unity gain curve of the uncompressed version. With the compressed version all the way down, the resultant sum is identical to the uncompressed version.

Figure 16.41 The PSP VintageWarmer 2. The Mix control (to the right of the Brick Wall switch) lets us blend the unprocessed input with the compressed version, making parallel compression a very easy affair.

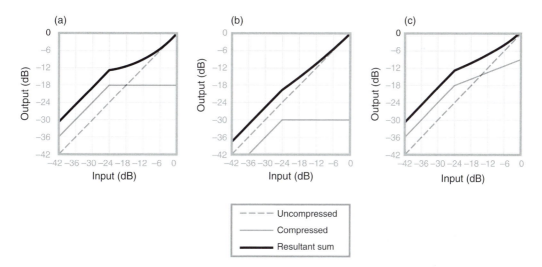

Figure 16.42 Parallel compression and resultant ratio curves. (a) The compressed version, with 1000:1 ratio, is 6 dB louder than the uncompressed version in order to demonstrate the shape of the resultant ratio curve. We can see that from the 1:1 ratio below the threshold point, the ratio slowly returns to 1:1 as the signal gets louder. (b) The same settings as in (a), only with a compressed version which is 6 dB below the uncompressed version. We can see a smoother ratio curve above the knee. (c) The same settings as in (a), only with 2:1 ratio. Again, we can see a smoother ratio curve above the knee.

The attack time of the compressor is set to be as fast as possible, so sharp transients are not overemphasized (unless this is what we want). If transparency is what we are after, the release time is set to minimize pumping and low-frequency distortion. However, a powerful effect can be achieved if the compressed version is pumping and distorting, and again, we control how much of this effect blends in using the level of the compressed version.

The parallel compression technique is almost matchless when applied to a group of instruments that has already been dynamically treated, like a drum mix. Applying a standard compression on a drum mix can easily impair the dynamics of each source track. For example, while setting the drum mix compression we have to find the right attack and release times that will retain the dynamic character of each drum. Parallel compression makes the whole affair of adding power less restricting.

Parallel compression is also one of the rare techniques where a compressor might be used in an aux send arrangement rather than the standard inserts. Taking a drum mix for example, we can create an alternative submix using sends and feed the send-bus to the parallel compressor. Then if we want a bit more weight on the kick, the kick send level is brought up.

Track 16.69: PC Drums Uncompressed Layer
The uncompressed version of the drums.

Track 16.70: PC Drums Compressed Layer
The compressed version of the drums in the previous track. The compressor is set for heavy effect with fastest time constants.

The following tracks are simply a blend of the two previous tracks. The uncompressed layer is always mixed at the same level, with the compressed layer always a few dB below it (denoted in each track name). Notice how much better the attack and dynamics are retained compared to standard compression:

Track 16.71: PC Drums 21 dB down
Track 16.72: PC Drums 15 dB down
Track 16.73: PC Drums 9 dB down
Track 16.74: PC Drums 3 B down

- - - - - - - - -

The same set of tracks, but this time involving a bass:

Track 16.75: PC Bass Uncompressed Layer
Track 16.76: PC Bass Compressed Layer
The intentional distortion on this track is the outcome of fast time constants.

Track 16.77: PC Bass 21 dB down
Track 16.78: PC Bass 15 dB down
Track 16.79: PC Bass 9 dB down
Track 16.80: PC Bass 3 B down

Plugin: Universal Audio *Fairchild*
Drums: Toontrack *EZdrummer*

Serial compression

Serial compression, or *multi-stage compression* as it is sometimes called, is also an old technique. It involves connecting two (or more) compressors in series. In an analog studio this calls for extra routing and the loss of either a channel strip or an external unit. There is no such overhead with audio sequencers – setting up serial compression simply involves loading two plugins in consecutive insert slots.

The idea behind serial compression is this: If we have two tasks to achieve, why not distribute each task to a different compressor and have the settings on each compressor optimal for its task? For example, if we want to add punch to a snare, but not all the snare hits are at the same level, we can use the first compressor to balance the level of the performance and the second to add punch. This mix-and-match approach is also beneficial since some compressors are better for specific tasks – one compressor might be better at balancing, while another at adding punch.

> **Track 16.81: Reshaping Balancing Attack**
> You probably remember the problem in Track 16.66, where the attack of the different hits varied in relation to hit spacing. To solve this, another compressor is inserted following the compressor that accented the attack. The balancing compressor has its threshold set at −18.3 dB, ratio of 2.8:1, soft knee, 15 ms attack, 5 ms release and 3.2 dB of make-up gain.
>
> *Plugin:* Digidesign *DigiRack Compressor/Limiter Dyn 3*
> *Drums:* Toontrack *EZdrummer*

More on compressors

Compressors, depth and focus

We have seen in the previous sections how manipulation of the dynamic envelope of instruments can send them backwards or forward in the depth field. Part of this phenomenon has to do with an increase or decrease in overall level of the treated instrument. There is another important aspect of compression that links it to depth and focus. It involves recordings that include ambiance or reverb.

The relationship between reverbs, depth and focus will be expounded upon in Chapter 23 on reverbs. For now, it is sufficient to say that the more reverb there is compared to the dry sound, the further away and the less focused sounds will appear. Compression can either increase or decrease the level of a reverb already printed on a recording. It is easy to demonstrate this with a drum recording that contains ambiance, like overheads or a drum-mix. If the compressor is set for short attack and release, the reverb would become louder, making the drums appear further back and their image more smeared. If the compressor is set with an attack long enough to let the natural attack through, and longer release so some gain reduction is still applied on the ambiance in between hits, the drums would appear closer and their image would sharpen.

Experienced or attentive readers have probably noticed the effect of compression on depth and focus in virtually any previous samples that included a drum kit. For example, in Track 16.2 the drums move further back as the threshold drops. Here are three more tracks demonstrating this:

Track 16.82: Drums III Uncompressed
The source track to be compressed in the following two tracks.

Track 16.83: Drums III Backwards
Short time constants and gain reduction peaking at around 7 dB are enough to noticeably increase the ratio of ambiance level. This sends the drums backward in the depth field and makes their image less focused.

Track 16.84: Drums III Forwards
The long attack and moderate release in this track are set to let the kick and snare pass through, but maintain some gain reduction after each hit, thus reducing the level of the ambiance. Both drums appear more frontal and focused.

Plugin: Sonnox *Oxford Dynamics*
Drums: Toontrack *EZdrummer*

Setting up a compressor

There are various ways to tackle the initial settings of a compressor. The method mentioned here provides a step-by-step approach that can be helpful in many situations, notably when compressing vocals or drums. People new to compression are most likely to find this technique useful. It can also be used as an *aural exercise* – it makes the function of each control appear rather obvious. This method is based on three main stages – the extent, degree, and timing, which in relation to controls are threshold; ratio; attack and release.

- **Initial settings** – the idea is to have the most obvious compression once the threshold is brought down in the next step. All controls but the threshold are set to produce the hardest possible compression.
  - Threshold all the way up (so no compression takes place initially).
  - Ratio all the way up.
  - Attack and release as short as possible.
  - If there is any option regarding peak/RMS or soft/hard knee, these are better set at this point based on the nature of the instrument. For the purpose of the aural exercise, peak and hard knee are preferred.
- **Determine extent** – **threshold** goes down. Due to all the other settings, as soon as signals overshoot, the compression becomes very obvious and on many compressors the fast attack and release will cause audible distortion. We can easily identify what portion of the signal is being compressed and adjust the threshold accordingly. It can be beneficial to set the threshold slightly below what seems to be the appropriate setting, as later it can lead to more uniform compression.
- **Determine degree** – **ratio** goes down from maximum. Once we know the extent of compression we can determine its degree.
- **Determine attack** – the attack is made longer from its initial short settings.
- **Determine release** – the release is also made longer.

There is a sense to the order in which controls are set with this method. Changes to the threshold and ratio will cause changes to the amount of gain reduction, which in turn affect the overall attack and release times (the more gain reduction the slower the attack and release). So the attack and release are set after the threshold and ratio. The attack is set before the release, since long release settings might interfere with succeeding attack phases. While this method can get very close to what we need, usually fine-tuning each control will follow.

Here is a demonstration of the steps involved in this method, using the uncompressed vocal in Track 16.16. The compressor is initialized as described above, and to make the effect more obvious the knee is set to hard.

Track 16.85: Vocal Steps Threshold
The threshold was brought down, ending at −22 dB. This level is slightly lower than what might seem needed to allow longer attack later on.

Track 16.86: Vocal Steps Ratio
The ratio is brought down from 1000:1 to 3:1, a ratio at which the vocal is reasonably balanced.

Track 16.87: Vocal Steps Attack
One problem with the previous track is that the function of the attack was too obvious, so it has been lengthened to 1.85 ms. The release is left untouched at the minimum value of 5 ms.

Plugin: Sonnox *Oxford Dynamics*

Multiband compression

The compressor we have discussed so far is known as a broadband compressor – it attenuates the whole input signal, with its full frequency spectrum. We already saw some issues this can bring about – it can cause the dulling of sounds due to high-frequency attenuation when low frequencies trigger the compression. Also, when we compress a bass guitar, low notes will result in more gain reduction, which also affect the mid and high frequencies that give the bass guitar much of its definition. When we de-ess vocals, we attenuate not only the sibilance, but also the body and the higher harmonics of the voice.

A multiband compressor, or a *split-band compressor*, splits the input signal into different bands, then lets us compress each band individually (Figure 16.43). For example, for de-essing we can define a band between 3 and 8 kHz, and only sibilant peaks at this frequency range will be compressed, without affecting the rest of the frequency spectrum. We can employ the same principle when trying to de-emphasize the resonance of an acoustic guitar.

The manufacturing of an analog multiband compressor is more costly as it involves more than one compressor and a quality crossover. While being a standard tool in mastering studios, multiband compressors were never high up in the mixing priority list. Software plugins cost nothing to manufacture, so these made multiband compressors more widespread than ever today (Figure 16.44).

Most multiband compressors have a predefined number of bands and they let us set the crossover frequencies between the different bands. A dedicated compressor per band should offer most related controls. Some controls might still be global and affect all bands.

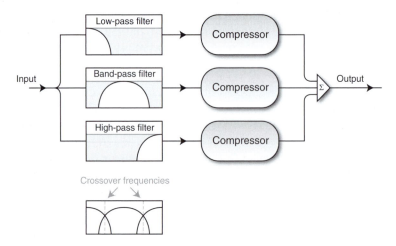

Figure 16.43 A block diagram of a multiband compressor. In this illustration, a 3-band compressor is shown. The input signal is split into three, and filters are used to retain the respective frequencies of each band. Then each band is treated with its own compressor, and the three bands are summed together to constitute the output signal. At the bottom, we can see the combined effect of the filters and the crossover frequencies between the different bands.

Figure 16.44 The Universal Audio *Precision Multiband*. This plugin offers a compressor, expander or a gate for each of the five bands. On the left are the controls for the compressor of the selected LF. The plot in the middle shows the frequency division between the five bands. The meters on the right show the gain reduction per band.

Multiband compressors provide great advantages when the signal in question is of a broadband nature, like a stereo mix. Compressing each band individually provides a greater degree of control over the signal. This is why multiband compressors are so popular in mastering. Nevertheless, every instrument being mixed can be divided into a few separate frequency bands that each play a different role, and we can benefit from compressing each band individually rather than the instrument as a whole, for example, the lows of a kick and the attack on its high-mids. The softer the drummer hits the kick the less are the lows compared to the click of the beater. We can compress the low band while compensating with some make-up gain specifically on that band. This will make the lows-to-attack ratio more consistent with different hit velocities. We can apply the same principle on a bass guitar in order to achieve more solid frequency balance for different notes. We can also reverse this idea in order to make a programmed kick more natural. Unless velocity layers are used, kick samples have the same ratio between lows and high-mids. We can set a low threshold and a low ratio on the high-mid band; so the higher the hit velocity is, the more the high-mid attack will be compressed. This will alter the ratio between the lows and attack like in the real world. Last but not least, multiband compression is an excellent way to rectify the proximity effect on vocal tracks – the low-end boost resulting from the vocalist moving closer to the microphone. Compressing the low bands would rectify these variations.

Another advantage of using multiband compressors is that while only affecting a specific frequency band, they tend to interfere less with the overall tonality of the processed signal. In addition, the compression on each band can be pushed harder since the compression artifacts are introduced on a limited range of frequencies. For example, fast release on the lower band will cause distortion, but we can dial a faster release on the higher bands, which will make these bands louder. By making each band louder separately, we can make signals even louder than we could with a broadband compressor. It is a divide and conquer approach applied to the frequency domain. Although they usually involve a bit more effort to set (and a little more auditory skill), multiband compressors give us more control over the dynamic treatment of sound.

When a multiband compressor is not available, we can set up a manual arrangement using an array of filters and broadband compressors. Say we need 2-bands, we can send a track to a bus and bring the bus back into two different tracks. Put a low-pass filter on one and a high-pass filter on the other, then set both to the same cut-off frequency. Next, we insert a compressor on each track, and send both at the same level to the mix bus. Had multiband compressors been as simple as explained so far, this manual setup would work flawlessly. However, multiband compressors involve some additional science in the way they sum the different bands back together. For example, they might align the phase of the different bands. Manual multiband compression does work, but a true multiband compressor should sound better.

> **Track 16.88: Bass Multiband**
> This track demonstrates how a multiband compressor can be utilized to reduce squeaks. In the original bass recording (Track 16.38), there are two noticeable nail squeaks on the A/G# notes, and then a few more toward the end. Squeaks as such might be masked when the guitar is mixed with other instruments, but they can become annoyingly noticeable if the bass is equalized, distorted or processed using an enhancer. The compressor is set with a single active band between 444 Hz and 3.3 kHz. Heavy gain reduction occurs on this band only when the squeaks happen. Compare this track to the original to see the difference.
>
> **Track 16.89: Bass Multiband SC**
> Most multiband compressors let us solo each band. This is the soloed version of the active band in the previous track, with its compressor disabled. Essentially, this signal feeds the side-chain of the compressor on the band used.
>
> *Plugin:* Universal Audio *Precision Multiband*

Before or after the EQ?

What seems like an eternal mixing question is: should a compressor come before or after an equalizer? Had there been a determined answer, this question would not be asked. The answer is: It depends. Let's see on what.

We already established that low frequencies excite compressors more than high frequencies. There are additional situations where we clearly do not want the compressor to respond to a specific frequency range of the input signal, in which case an equalizer has to be placed before the compressor. For example, we might want to filter the lows of vocals or overheads before compressing them. The problem with a pre-compressor equalization is that any changes to the equalizer affect the characteristics of the compression. So once the compressor is set, touching the equalizer might require a revisit to the compression settings.

So the wisdom is this: for *compression-related equalization*, we should put an equalizer before the compressor. Any other equalization should be done after the compressor. This includes equalization that alters the tonality of an instrument or one that compensates for the tonal alterations cause by the compressor.

> *All but compression-related equalization should be done after the compressor.*

Many compressors nowadays let us equalize the side-chain signal. When such a facility exists, equalizers are most likely to be placed after the compressor. There is a huge benefit of using side-chain equalization rather than pre-compressor equalizers. With pre-compressor equalizers, the tonality of the signal is affected by two equalizers – the one before and the one after the compressor. However, the pre-compressor equalizer is said to be locked for the reasons explained above. None of this happens with side-chain equalization – the tonality of the signal is solely controlled by a single equalizer placed after the compressor. Perhaps the only shortfall of using side-chain equalizers is that they often provide less bands, or control, than a standard equalizer. These might not be sufficient for our needs, so pre-compressor equalizers might still be employed.

> *Providing their facilities are adapted for the task, side-chain*
> *equalizers provide great benefit over pre-compressor equalizers.*

Having said that, it pays to remember that post-compressor equalizers can damage the dynamic control achieved by the compressor. For instance, we might have balanced vocals out of the compressor, but then boosting around 1760 Hz would make all the A notes jump out (1760 Hz is A, and thus an harmonic of all A notes below it). There is no magic solution to such a problem. Sometimes the choice of pre- or post-compressor equalizer is a matter of experimentation.

Compressors on the mix bus

Engaging a compressor on the mix bus while mixing, also known as *bus compression*, is another source of a long debate in the mixing community. Some mixing engineers use bus compression passionately from very early stages of the mix. Often they have at their disposal a respectable compressor, like the SSL bus (or quad) compressor, and they drive it with rather subtle settings. The bus compressor might be used during the majority of the mix, but it is often switched out before the mix is printed – mastering engineers seem rather united against mixes that have already been stereo-compressed. On the other hand, some mixing engineers would rather be mixing on the edge of an airport runway than mix with a compressor on the mix bus – they are that much against it. Both types include some of the world's most creditable mixing engineers.

Perhaps the least practical reason for using bus compression is that it simply makes the mix more exciting, and excitement is an important part of mixing. Bus compression also lets us evaluate how the mix might translate to radio or in clubs – both involve heavy compression and/or limiting in their signal chain. Mostly in these situations we care how the low-end affects the rest of the mix – is the kick too boomy? Is it thin?

There are a few risks in using bus compression. The greatest of them is that it limits our ability to judge relative levels. If the bus compressor is set to anything but subtle, it should be no wonder if when it is bypassed the mix relative levels would turn chaotic. Also, by applying compression, signals that overshoot the threshold are reduced by a specific ratio. If a fader is brought up by 4 dB, its instrument might only raise by 1 dB due to 4:1 ratio on the bus compressor. In the way compressors work, pushing the kick up might result in more compression of low frequencies, which can actually make the kick quieter (there is no individual make-up gain when the stereo mix is compressed). So it is not only the precision of the faders that we can lose with bus compression – level changes might be opposite to fader movements. Another thing to ask is how exactly one should go about servicing the bus compression? Any change in the mix can also affect the compression. So should we revisit the bus compressor after every mixing move? It should be clear that only respectable compressors and subtle settings would be fit for the task.

Perhaps above all comes the idea that only trained ears can recognize the subtleties of bus compression. For the novice, bus compression can be like a translucent layer that only blurs what is behind it – the actual mix.

> *Unless you have a good compressor, subtle settings and trained*
> *ears, bus compression can be self-defeating.*

Dynamic convolution compressors

The concept of convolution, in simple terms at least, involves capturing the characteristics of a specific audio system so they can be emulated digitally. The remarkable thing about this process is that it is done without any knowledge of how the system actually works – it simply involves learning the system response (output) to input signals. While the concept might seem rather complex, it is a truly simple one – we can take a vintage analog unit, learn its characteristics, then build a plugin that emulates these characteristics. If all the science is done properly, the plugin would sound very much like the original unit.

Convolution reverbs are now widely known and will be discussed in Chapter 23. *Dynamic convolution* is a fairly new technology, developed by a company called Sintefex Audio. Dynamic convolution is far more complex than the convolution process of reverbs. In order to learn the characteristics of compressors (or equalizers), a tedious series of measurements has to be taken before any simulation can be crystallized. Theoretically, it involves studying the compressor response to any possible combination of input levels, frequencies and settings on the unit itself. In practice, not every possible combination is studied, but a large-scale step-wise approach is taken instead. For example, instead of studying every possible ratio setting, we might only learn integer-based ratios (1:1, 2:1, 3:1, 4:1, 5:1 and so forth). In theory, the emulating processor would only offer the settings that were studied – if only integer-based ratios were learned, the emulator would not offer a 2.5:1 ratio. This would be an annoying limitation. Luckily, the current technology allows dynamic convolution compressors to offer settings that were not studied, and smart algorithms produce an approximate result (a 2.5:1 ratio would produce something between 2:1 and 3:1).

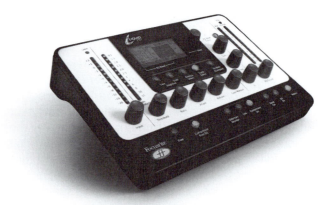

Figure 16.45 The Focusrite *Liquid Mix*.

Dynamic convolution is mostly known for its usage in the Focusrite Liquid Channel and Liquid Mix. The former is a hardware unit, the latter is a DSP platform for a DAW. The Liquid Mix, shown in Figure 16.45, includes emulations of 40 compressors (and 20 equalizers) that pretty much constitute a list of the most respectable units one can find in a professional studio.

17 Limiters

A true limiter has one critical task: to ensure that signals do not overshoot the threshold. May I add, *no matter what*. Often it is sudden transients and peaks that we limit. Limiters have many applications outside mixing: they are necessary to protect PA systems, to prevent one FM radio station from intruding on the frequency band of an adjacent station and in mastering, limiters help maximize loudness. In mixing, the task of limiting peaks is essential when the final mix is printed, where signals must not exceed 0 dB before being converted to digital or bounced to integer-based files (or any other digital media).

We already know from chapter 16 that a compressor can be configured to behave much like a limiter. To make a compressor a true limiter we must have peak-sensing, hard knee and zero attack time. Compression (well, limiting really) of this nature is the most aggressive and obvious that can be achieved with a compressor. Anything but subtle gain reduction can bring about some dreadful results.

A limiter takes a different approach. The idea is still to have no signals overshooting the threshold, but a limiter employs means to soften the drastic effect of hard limiting. A limiter's principle operation is really concerned with what happens below the threshold and how gain reduction can start before the signal overshoots. One of the ways of doing this is by introducing soft knee that grows progressively to reach a limiting ratio of $\infty:1$ at the threshold point. Also, a sophisticated attack function can observe levels below the threshold in order to adapt to the dynamics of the input signal. A look-ahead function on software limiters is standard. Some limiters make use of two stages, where the first stage provides more transparent limiting but no overshoot guarantee, then if any signals manage to overshoot they are trimmed down by a second, more aggressive stage.

Figure 17.1 shows a typical limiter with fixed threshold. The three pots on the left are standard for this type of limiter: input level, output level and release. Some limiters offer variable threshold, but often bringing it down results in respective gain being automatically applied, so the eventual result is the same as bringing up the input level. The input or threshold controls are similar to that of a compressor. The output level often denotes the highest possible output level, also known as *ceiling level*. Often 3 dB of mix headroom is reserved for mastering. Also, even if peaks are limited to 0 dBFS, they can exceed the

Figure 17.1 The Universal Audio Precision*Limiter* plugin.

equivalent analog reference during the digital-to-analog conversion. To prevent this, a little headroom (0.2 dB) is still exercised.

As opposed to compressors, limiters rarely give the operator any control over the attack, hold, ratio, knee or peak/RMS sensing. A user control over these parameters could defeat the limiting task. As elementary as it may sound, limiters *do not* provide a replacement for compressors – they give us far less control over the final outcome. The novice might be tempted to look at a limiter as a shortcut to compression. But limiters and compressors are worlds apart in the control they offer over the final outcome.

> *Limiters should not be used as a quick replacement for compressors.*

We can look at a limiter as a limited mutation of a compressor. As such, they have their own distinctive **effect** that can be appealing and have a place in the mix. They can also succeed a compressor in order to **perfect our compression aims**, like loudening or level balancing. In that sense they are used for their original task – **limiting peak**s. Occasional clipping can be handled with a limiter, yet regular clipping would normally call for level attenuation.

Finally, as well as being used for peak limiting, limiters can be used as loudness maximizers. As long as the limiter operation is unnoticeable, we can push the level against it to raise the loudness of the source material. Limiters are often judged based on how hard the input level can be pushed before non-musical artifacts appear. The extra loudness that is such a common part of the standard limiter operation is a risk since it makes A/B comparison unfair – a limiter might fool you into perceiving sonic improvement, even if the limiter is not limiting at all. In mixing scenarios it is worth using the output control in its conventional compressor role as make-up gain.

> *Limiters can easily fool you into perceiving a sonic improvement.*

Track 17.1: Drums Source
The source track to be limited in the following samples.

Track 17.2: Drums Limited
Mind your monitoring levels - this track is louder than the previous one. This is the result of dialing +9 dB of input gain on the limiter. This might appear to be a substantial improvement compared to the previous track.

Track 17.3: Drums Limited Compensated
This track involves the same settings as in the previous track, only that this time the post limiting signal level is attenuated to match the loudness of the source track. Compare this track to Track 17.1 to see that the difference between the two is not that great. Also, notice how in the limited version the snare and the kick have lost some attack and have been sent backward.

Track 17.4: Drums Limiting Effect
Setting the limiter's input to the maximum of +24 dB produced this distorting sound, which can be used in a creative context. The level of this track is compensated as well.

Plugin: Universal Audio *Precision Limiter*

18 Gates

Following compressors, gates are perhaps the second most common dynamic range processors in mixing. Gates are also called *noise-gates* – a name that implies their traditional usage as **noise** eliminators. In the past, tapes and analog equipment were the main contributors to added hiss. Nowadays, digital technology tends to allow much cleaner recordings. Still, in project studios unwanted noise on a recording might be the outcome of ground-loops, a computer fan or airborne traffic. Noise, however, is not the only thing gates are employed to reduce. The **rumble** produced by the floor tom when the rest of the drum kit is played is often gated. The hi-hats **spill** on a snare track and the headphone spill on a vocal track are just two more examples. In addition to their traditional role, gates are also used for more creative tasks, like **tightening drums**, **adding punch** or **applying dynamic movement**.

Figure 18.1 A gate plugin. The MOTU *MasterWorks Gate*.

Controls

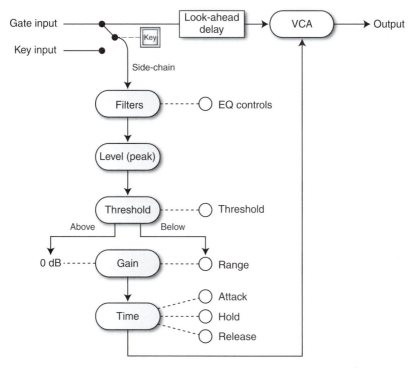

Figure 18.2 Inside a gate. The vertical chain shows the main side-chain stages, and the controls link to each stage.

Threshold

Gates affect signals *below* the threshold – these are either attenuated or muted. Signals above the threshold pass through unaffected, unless some attack is applied. A gate only cares whether the signal is above or below the threshold; a gate is said to be *closed* when the signal is below the threshold and *open* when the signal is above it. Figure 18.3 illustrates the function of a gate on a snare.

Threshold settings on a gate might seem straightforward – we set the threshold below the wanted signals and above the unwanted signals. In practice, this affair can be tricky since a certain threshold not always meets both criteria. For example, the snare in Figure 18.3 lost some of its attack and much of its decay. We could retain more of both by lowering the threshold, but this would cause false triggering by the hi-hats. Solutions to this common issue are discussed later on. In the meantime we should note:

> *Lower threshold settings on a gate are often sought as they help in retaining more of the natural attack and decay.*

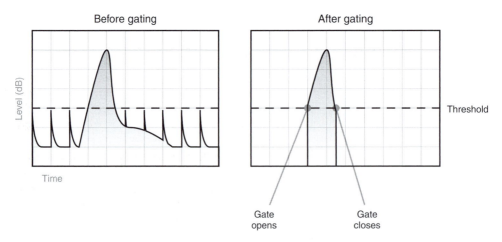

Figure 18.3 Gate threshold function. The original signal before gating involves a single snare hit and a few hi-hat spikes we wish to eliminate. The threshold is set above the hi-hat peaks, so the gate would only open once the snare hits. The gate would close as the snare hit drops below the threshold. After gating, only the loud portion of the snare remains.

Track 18.1: Snare Source
The source snare track used in the following tracks, where the aim is to gate all drums apart from the snare.

Track 18.2: Snare Threshold −40 dB
A −40 dB threshold is too low. The drums hardly drop below the gate's threshold, and when they do, the gate produces abrupt level changes.

Track 18.3: Snare Threshold −30 dB
The kick, toms and later the hi-hats still overshoot the −30 dB threshold. There are even more abrupt drops in level on this track.

Track 18.4: Snare Threshold −20 dB
Apart from a few tom hits, it is only the snare that overshoots the −20 dB threshold.

Track 18.5: Snare Threshold −10 dB
Only the snare triggers the gate opening here. However, due to the high threshold some of the snare's leading edge, and much of its decay are gated as well.

Plugin: McDSP *Channel G*

Hysteresis

To allow quick gate opening once the signal exceeds the threshold, the level detection on most gates is based on peak-sensing. Level fluctuations are more profound with peak-sensing than with RMS. While fluctuating in level, signals may cross the threshold in both directions many times over a short period of time. This causes rapid opening and closing of the gate, which produces a type of distortion called **chattering** (see Figure 18.4a).

(a)

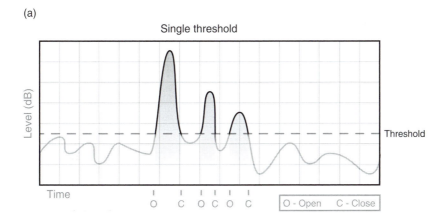

Single threshold

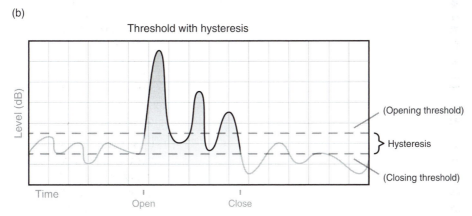

(b)

Threshold with hysteresis

Figure 18.4 Hysteresis control on a gate. (a) A gate with a single threshold and no hysteresis control. Chattering is introduced due to the rapid opening and closing of the gate, which are triggered by quick level changes. (b) A gate with hysteresis control. The gate only opens as the signal overshoots the opening threshold and only closes as the signal drops below the closing threshold. Level variations between the two thresholds do not cause toggling of the gate state, thus no chattering occurs.

One way to overcome this is by having two thresholds – one above which the gate opens, another below which the gate closes. Having two threshold controls would be cumbersome since adjustments to one will call for adjustments to the other. Instead of providing two controls, gates offer a single threshold and a control called hysteresis. The threshold is the opening threshold, and the hysteresis determines how many dB below it the closing threshold is set. For example, with the threshold set to −20 dB and hysteresis to 5 dB, the closing threshold would be at −25 dB. Figure 18.4b illustrates this.

Many gates do not offer hysteresis as an adjustable control, but have a built-in setting fixed between 4 and 10 dB. When hysteresis control is given, these figures provide a good starting point.

Track 18.6: Kickverb Source
A kick sent to a reverb, which is gated in the following tracks.

Track 18.7: Kickverb No Hysteresis
The chattering on this track was caused due to the reverb level fluctuating around the gate's threshold (−37 dB).

Track 18.8: Kickverb Hysteresis −6 dB
With −6 dB of hysteresis, the gate opens at −37 dB and closes at −43 dB. As can be heard, this reduces chattering, but does not eliminate it.

Track 18.9: Kickverb Hysteresis −12 dB
A −12 dB of hysteresis eliminates chattering.

Plugin: Logic *Noise Gate.*
(Reverb: Universal Audio *Plate 140.)*

Range

Range, or *depth*, defines the amount of gain applied on signals below the threshold. A range of −10 dB means that signals below the threshold are attenuated by 10 dB. Often signals below the threshold are considered muted, although in practice this perceived muting is due to heavy attenuation with the typical range of −80 dB. Figure 18.5 shows the transfer function of a gate, while Figure 18.6 demonstrates the effect of different range settings.

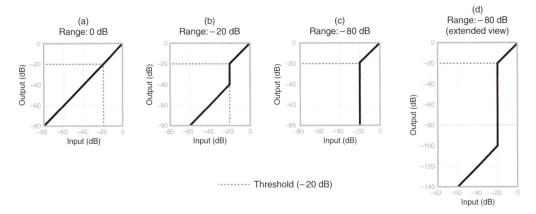

Figure 18.5 The transfer function of a gate. (a) A gate with 0 dB range: both below and above the threshold the input–output ratio is at unity gain, and the gate would have no effect on the signal. (b) A gate with −20 dB range: all signals below the threshold are simply attenuated by 20 dB. For example, an input signal at −40 dB will leave the gate at −60 dB. (c) A gate with −80 dB range, which effectively mutes all the signals below the threshold. (d) Again −80 dB range, but with an extended output scale that reveals what happens below the limits of the standard scale. We can see clearer here that an input signal at −40 dB will leave the gate at −120 dB.

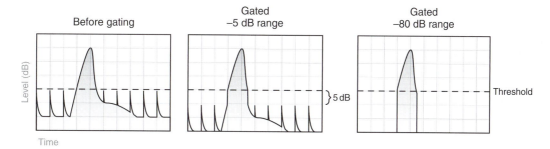

Figure 18.6 The range function of a gate. When the range is set to −5 dB, signals below the threshold are attenuated, but still heard. With large range, such as −80 dB, signals below the threshold become inaudible.

The range value can be expressed in either positive or negative values. Depending on the manufacturer, a gain attenuation of 40 dB might be expressed with a range of 40 or −40 dB. This book uses the negative notation. Small range denotes little effect (e.g., −5 dB), while large range denotes more effect (e.g., −80 dB).

Generally speaking, large range settings are more common in mixing. However, sometimes it is only a gentle attenuation we are after. One example would be gating vocals to reduce breath noises between phrases – removing breaths altogether is often perceived as artificial, so these are often only attenuated. Another example involves reducing the room reverb captured on a recording.

The following tracks demonstrate different range settings on a gate. The larger the range the less gated portions of the sound can be heard. Notice that it is only with a range of −20 dB or less that the gated material is easily audible.

Track 18.10: Snare Range −60 dB
Track 18.11: Snare Range −40 dB
Track 18.12: Snare Range −30 dB
Track 18.13: Snare Range −20 dB
Track 18.14: Snare Range −10 dB

Plugin: McDSP *Channel G*

Attack and release

Attack controls how quickly the gate opens, release controls how quickly the gate closes. For example, with the range set to −80 dB, a closed gate would apply −80 dB of gain on the input signal. The attack determines how quickly these −80 dB rise to 0 dB when the gate opens, while the release determines how quickly the gain returns to −80 dB when the gate closes.

The response times on a gate are normally set in milliseconds. Attack times usually span between 0.010 ms (10 μs) and 100 ms. Release times are often within the

5–3000 ms range. Like with compressors, both the attack and release times determine the *rate* of gain change. For instance, a gate might define that response times are referenced to 10 dB of gain change. With an attack of 1 ms and a range of −10 dB, it will take 1 ms for the gate to fully open; but with −80 dB of range, it will take the gate 8 ms to open. The practical outcome of this is that the range affects the overall time it takes a gate to open and close – where appropriate, we can achieve faster attack and release times by dialing a smaller range. In turn, this promotes smaller range settings than a gate might offer – on a busy mix, a range of −40 dB could suffice to make signals below the threshold inaudible.

It is worth noting that both the attack and release controls have the opposite effect on dynamic envelopes than compressors. A longer attack on a gate means that less of the natural attack is retained, a longer release on a gate means that *more* of the natural decay is retained. Figure 18.7 illustrates this on a gated snare.

| *On a gate: longer attack, less of the natural attack; longer release, more of the natural decay.* |

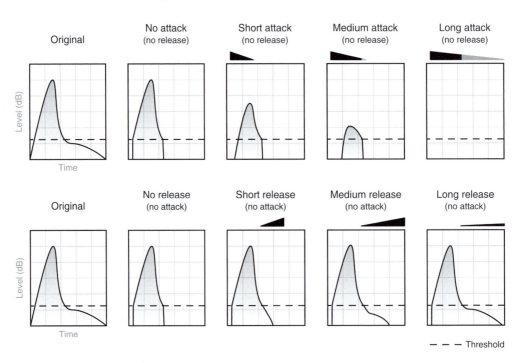

Figure 18.7 The effect of attack and release on a gated snare. The top row shows different attack times with no release; the drop in gain reduction is shown as a triangle above the graphs. We can see that the longer the attack is the longer it takes gain reduction to drop and the less of the snare's original attack is retained. Long attack setting causes gain reduction to drop so slowly that none of the signal is audible before the input drops below the threshold. The bottom row shows different release times with no attack; the rise in gain reduction is shown as a triangle above the graphs. We can see that the longer the release is, the slower the gate reduces the signal once it falls below the threshold, so more of the natural decay is audible.

> **Track 18.15: Snare TC Source**
> The source track used in the following tracks, that demonstrate different attack times on a snare. The threshold was set to −20 dB, the range to −60 dB, and the release to its minimum of 100 ms. Notice how the longer the gate's attack is, the less of the snare's natural attack is retained.
>
> **Track 18.16: Snare Attack 10 Microseconds**
> **Track 18.17: Snare Attack 1 ms**
> **Track 18.18: Snare Attack 5 ms**
> **Track 18.19: Snare Attack 10 ms**
> **Track 18.20: Snare Attack 50 ms**
> **Track 18.21: Snare Attack 100 ms**
>
> Another set of tracks with the same gate settings, only this time with attack fixed to its minimum of 10 µs and varying release times. The longer the release the more of the natural decay is retained:
>
> **Track 18.22: Snare Release 100 ms**
> **Track 18.23: Snare Release 300 ms**
> **Track 18.24: Snare Release 600 ms**
> **Track 18.25: Snare Release 900 ms**
>
> *Plugin:* McDSP *Channel G*
> *Snare:* Toontrack *EZdrummer*

On the same principle discussed in Chapter 16, very short attack and release times can cause audible **clicks** due to abrupt level changes. A steep level rise produced by fast attack **generates high frequencies**, the shorter the attack the higher the frequency (Figure 16.30 illustrates the reason for this). The same applies to release, although the gate operation on the quiet signals below the threshold tends to be slightly less noticeable. Depending on the gated signal, fast response times might also alter the low and mid frequencies, which essentially means that a gate can affect any part of the frequency spectrum – a side-effect we have to observe.

The attack setting has an extra weight when gating transients, especially those with low-frequency content like kicks. Most of the kick's character is in its very first cycle, with most of the impact gearing up at the beginning of this cycle. A gate is most likely to reshape this important part of the waveform – either with a short attack that produces sharp level changes or with a longer attack that reshapes the waveform and withholds the kick impact. One way or another, the kick's tonality is very likely to change, with the lows, mids and highs, all likely to suffer. Ideally, we would like the attack to retain the original shape of the signal's waveform, which is often only possible with look-ahead (discussed later). Figure 18.8 illustrates this.

Very short settings can also cause **low-frequency distortion** due to the gate operation within the cycles of a low-frequency waveform. Longer attack, release or hold can rectify this issue. In the case of percussive performance, we must consider how the release and hold settings, which affect the length of the sound, **lock to the rhythmical feel** of the track. In addition, we want the gate to fully close before the next hit arrives, or successive hits will be gated differently (due to variety in gain reduction during the attack phase).

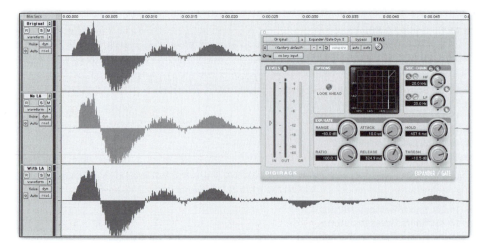

Figure 18.8 A gated kick. The top track shows the original kick, and it is hard not to notice the dominant first cycle. The middle track shows the kick after gating with the settings shown on the Digidesign DigiRack Expander/Gate on the right. The attack was set as short as possible and the look-ahead was disabled. You can see the partial loss of initial impact and the steep level climb caused by the gate opening, which generates high frequencies. The bottom track shows original kick after gating with the same settings, but with look-ahead enabled. You can see how the initial attack was not affected.

Track 18.26: Kick Source
The source kick track used in the next track.

Track 18.27: Kick Rising Attack
For the purpose of demonstration, the threshold of the gate in this track was set high at −15 dB. The attack was automated to rise from 10 μs to 24 ms. First, notice how the impact of the kick has completely disappeared. Then, notice how the click generated by the steep level change decrease in frequency as the attack is lengthened. Once the attack is long enough, it doesn't generate a click but yields a drop in level (this track was not faded out).

Plugin: Digidesign *DigiRack Expander/Gate Dyn 3*
Kick: Toontrack *EZdrummer*

* * *

Track 18.28: Kickverb Release Click
The click generated when the gate closes is caused by the steep drop in level – the result of 0 ms release.

* * *

Track 18.29: Bass Source
This is the source track for the following sample.

Track 18.30: Bass LF Distortion
The extremely unflattering distortion in this track is the result of threshold at −14 dB, range of −60 dB, and 0 ms for all time constants. Although exaggerated in this track, a distortion caused by super-fast time constants on a gate are mostly as unflattering as this.

Plugin: Logic *Noise Gate*

One issue with the principal operation of gates is that we often want short attack, so more of the signal above the threshold passes through, and short release, so the signal below the threshold is attenuated quickly. These typical short settings are more obstructive for the many reasons explained above. We have to remember that in many cases the attack and release are applied on large-scale gain changes, with a range of −60 dB or more. Compressors, on the other hand, often work on 20 dB or so, and moreover, gain changes are not as steep due to the gradual development of the input signal. A gate has no such softening mechanism – it is either open or closed, and there is often quite some gain involved in toggling between the two.

> *The attack and release on a gate tend to be obstructive due to their typical short settings.*

Short attack and release settings are not always appropriate though. Longer times are often used when the gated instrument has long natural attack and release, for example, a synthesized pad that rises and falls in a gradual fashion. Short attack and release will simply trim the parts of the signal that ascend or descend below the threshold. Long settings will keep some gradual sense for both the natural attack and decay.

Longer attack times might let us lower the threshold by a small amount. The reason for this is that false triggers will not be long enough to become audible. For example, a long attack in Figure 18.3 would allow the threshold to be slightly below the hi-hat peaks. Although the top of each hit would trigger the gate opening, the slow attack would mean the gate will not fully open by the time the hit drops below the threshold, potentially leaving these hits inaudible.

Hold

Once the signal has dropped below the threshold, the hold time determines how long the gate's gain reduction is held unaltered. For example, if 8 dB of gain reduction is applied as the signal undershoots the threshold, a hold period of 2 seconds would mean 2 seconds with 8 dB of gain reduction. Once the hold period is completed, the release phase starts. Gates typically offer a hold period of between 0 and 5 seconds.

Hold often replaces release in the task of **retaining the natural decay** of an instrument. There are two reasons for this. First, we can see from Figure 18.7 that quite a long release setting is needed in order to keep the natural decay. Such a long release is not always practical since it might not end before the next time the signal overshoots the threshold. Second, having the release starting right as the signal drops below the threshold causes an escalated decay – the natural decay of a snare, for example, will be superimposed by the artificial descent caused by the release function. Using hold instead of release lets us keep the fall rate of the natural decay. The hold time can be made to match the interval between two hits, while some release is still used in order for the gate to close without audible clicks (Figure 18.9).

> *Hold can be used to retain the character of the natural decay.*

Like with compressors, longer hold periods can reduce **low-frequency distortion**. Longer hold time can also reduce **chattering**, as the gate is held open while the signal rapidly

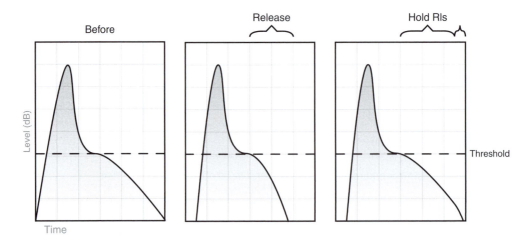

Figure 18.9 The difference between release and hold. With release, the natural decay of the instrument is reshaped, which is not the case with hold. The short release after the hold period is applied to prevent clicks.

leaps between the two sides of the threshold. In that sense, the hold facility provides a similar function to hysteresis, so when no hysteresis is offered, hold can be used instead. In addition, hold might be used to **compensate for any look-ahead** time that might be in force. For instance, if look-ahead is set to 10 ms, the gate will start closing 10 ms before the signal drops below the threshold. Setting a hold time of 10 ms will compensate for this early response.

The following two tracks demonstrate the use of hold to prevent chattering. The source track is identical to Track 18.7, with no hysteresis. A hold setting of 50 ms is nearly enough to prevent the chattering; 200 ms of hold rectify it completely:

Track 18.31: Kickverb hold 50 ms
Track 18.32: Kickverb hold 200 ms

* * *

Track 18.33: Tom Source
The source track used for the following samples.

Track 18.34: Tom Short Hold Long Release
The gate is set with 50 ms of hold and 710 ms of release. The release reinforces the dropping level of the tom, accelerating its decay.

Track 18.35: Tom Long Hold Short Release
The gate is set with 710 ms of hold and 50 ms of release. Once the tom drops below the threshold (−17 dB) the hold function maintains the gate open, allowing its decay to pass through unaffected. Then, the release quickly closes the gate. The tom's decay is retained here far better then in the previous track, but at the cost of some extra spill level.

Plugin: Logic *Noise Gate*

Look-ahead

This is perhaps the ultimate problem with gates: short attack results in clicks, long attack softens the natural attack and can even repress the beginning of words. Lower threshold? Cannot be done or spill will open the gate. Filter the side chain? Done, helps a little. But, for the ultimate problem there is an ultimate solution – look-ahead.

Like with a compressor, look-ahead lets the side chain examine the input signal slightly before processing takes place. This means that we can dial longer attack times, since the attack phase starts sooner. By the time the signal arrives to the gain stage, gain reduction has decreased to a degree that allows the leading edge of the signal to pass through the gate with no clicks or envelope reshaping. The very same principle applies to release – since the release phase starts shortly before the treated signal drops below the threshold, we can set longer release times. One of the prime benefits of look-ahead is that it allows longer attack and release times, which make the gate operation less obstructive.

| *Look-ahead makes the gate operation less obstructive.* |

A look-ahead function on an analog gate will introduce output delay. Software gates have no such issue, provided auto plugin delay compensation is active. Later, we will look at an extremely useful look-ahead trick that results in no delay with both analog and digital gates.

> **Track 18.36: Kick No Look Ahead**
> Using Track 18.26 as a source, the click is caused by a very short attack (10 µs, not automated in this track). The threshold is set to −20 dB and no look-ahead is involved.
>
> **Track 18.37: Kick with Look Ahead**
> By simply enabling look-ahead (of 2 ms) the click has disappeared and the full impact of the kick is maintained.
>
> *Plugin:* Digidesign *DigiRack Expander/Gate Dyn 3*

Side-chain filters

As with compressors, some gates let us equalize the side-chain signal that triggers the gating. A classic example of how extremely helpful this is, can be seen when gating a kick track in order to get rid of spillage like the hi-hats or snare. It goes without saying that both the hi-hats and snare have less energy on the lows than the kick. So we can filter the highs and mids content from the side-chain, leaving mostly the low-level energy of the kick. In turn, this lets us lower the threshold, so more of the original kick attack can be retained.

Key input

Gates, just like compressors, let us feed an external signal into the side-chain. On a compressor, the external source input is called *external side-chain*. On a gate, the same input is often called *key input* instead (to prevent confusion when a dynamic range processor offers both a compressor and a gate, and the side-chain of each can be fed from a different external source).

We can feed the key input with either a similar signal to the one being gated or a different signal altogether. In the case of the former, an example would involve a *gate microphone* – an additional microphone on a specific drum, which is later used in mixdown to feed a gate's key input. A gate microphone has one role – capturing the drum with as little spill as possible. Thus, gate microphones (often dynamic/cardioid) are placed as close as possible to the drum skin and are not concerned with faithfully capturing the timbre of the drum. When we gate a snare, we feed the snare microphone into the gate input and the snare's gate-mic into the key input. This way, we can dial lower threshold, achieve more accurate gating and have more freedom with the different control settings. Gate microphones aside, sometimes a specific track will benefit from being gated in respect to a similar track. For example, a kick-out track is better gated with respect to the kick-in microphone, which would normally capture less spill (Figure 18.10).

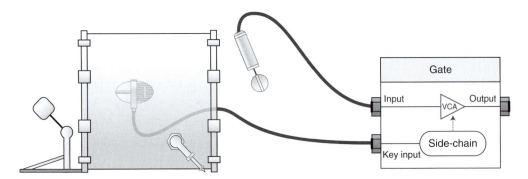

Figure 18.10 Gating kick-out with relation to kick-in. This illustration shows a recording setup, although we usually do this during mixdown with the recorded tracks. The kick-in track is likely to incorporate less spill than the kick-out. While gating the kick-out, feeding the kick-in to the gate's key input is likely to provide more musical gating. Also, the sound arrives at the kick-in microphone slightly before the kick-out microphone, providing a natural look-ahead. A gate-mic can replace the kick-in for other drums like the snare.

Output level

Although not often provided, the output level control offers a similar functionality to a compressor's make-up gain – boosting or attenuating the entire signal by a set amount of dB.

Stereo link

The stereo link function on a gate is similar to that on a compressor – it ensures that both left and right channels are gated identically, so no image shifting occurs.

Meters

Some gates provide a gain reduction meter just like compressors. Some have two indicators – one lights up when the gate is open, one when the gate is closed. Some gates add an additional indicator that lights up during the attack, hold and release phases, when the gate is neither fully open nor fully closed.

Applications

Removing noise

Noise can be introduced onto a recording in many ways. Tape media, microphones, microphone preamps, any analog component or an A/D converter are just a few examples of systems that superimpose their own inherent noise on the signal. When a recording is done in a domestic facility, background noise is also often an issue. Essentially, any signal that ever roamed the analog domain includes some noise. Even purely digital signals, e.g., a bass line generated with a software instrument can incorporate either quantization distortion or dither noise.

Luckily, modern technology enables cleaner recordings than in the past, so much of the noise on today's recordings is inaudible. Generally speaking, if high-quality equipment is used, and used correctly, a gate for noise removal might not be needed at any stage of the mix. If, for example, the signal-to-noise ratio on a digital recording is 60 dB (noise level at −60 dBFS), chances are that the noise will not be a problem. Even higher noise levels might not be an issue since in many cases the wanted signal masks it. Noise tends to be more noticeable with sparse arrangements and during the quieter sections of a production – in both cases there is less to mask it.

One thing worth considering is that noise can become more noticeable as the mix progresses. For example, after applying make-up gain on a compressor, the noise level would rise as well. Also, by boosting the highs of a specific track, the noise is likely to become more noticeable.

Even when the noise is audible, one must ask: **what is wrong with a little bit of noise?** Many people associate the synthetic sound of digital systems with the lack of noise and distortion. In fact, some engineers deliberately add noise or distortion to a digitally clean recording in order to replicate some of the familiar analog idiosyncrasies. The noise in these cases might be similar to a tape hiss or even crackling vinyl. Nevertheless, other noises like those generated by hard drives or washing machines are a problem – they are unlikely to remind anyone of the familiar analog sound.

> *Some judgment has to be made as to what noise needs gating and what noise can be tolerated.*

Another point is that our ears find varying noise levels (breathing) more disturbing than constant noise levels. This fact is taken into account in many noise reduction systems. We have to take this into account when gating a noisy track – varying noise level after gating might be more noticeable than the constant noise level before gating. Say for example we have a sparse arrangement with a noisy vocal track. The noise is likely to be masked or be less noticeable while the vocals are sung. The challenge is to make sure that the gate opening and closing does not cause noticeable noise variations. For example, once the vocal drops below the threshold and diminishes, a slow release can cause a noticeable descent in noise level. If the vocal is sent to a reverb emulator, this drop might be even more noticeable.

Exit Music (For a Film)
Radiohead. *OK Computer*. Parlophone, 1997.
Mixed by Nigel Godrich.

What is wrong with a little bit of noise? Judge for yourself. This track starts with what some consider a generous amount of noise. It is worth noting how the vocals mask the noise, and once the arrangement becomes denser the noise appears inaudible. It is also worth noting the drop in noise level as the guitar is ridden on the second bar. This is just one commercial track out of many that includes a noticeable amount of noise. The earth still spins.

Removing spill

Much of today's tracking is done in overdubs. Overdubs are less prone to spill, and mostly we deal with headphone leakage and the spill on the various drum tracks. Spill can result in four main problems:

- **Impairing separation** – ideally we want each track to represent a single instrument or drum (room-mics and overheads are obvious exceptions). A snare track that also contains the hi-hats would make it hard to separate the two. For example, such a hybrid track would restrict independent panning of each drum and in turn could cause a smeared stereo image to at least one of them.
- **Combfiltering** – the hi-hats on the snare track might not be phase-aligned with the hi-hats track or with the overheads. When the snare is mixed with either track, combfiltering is likely to impair the timbre of the hi-hats and give it a hollow, metallic or phasing sound. Any instrument might suffer from loss of focus, impact or timbre coloration if its intended track is mixed with its own spill on another track.
- **Adding dirt** – whenever an instrument is not playing, a spill on its track can produce unwanted sounds. Floor toms are notorious for producing rumble when the rest of the kit is played. Headphone spill might add unwanted noise during quiet sections.
- **Interfering with processing** – to give an example, a loud kick on a snare track might trigger compression and interfere with the snare compression. Brightening a tom track might also emphasize any hi-hats spill it includes.

This list suggests that spill should be removed whenever possible. On occasion, after removing the spill we may find that bypassing the gate actually has a *positive* effect on the mix. The reasons for this are mostly unpredictable. Regardless, we must remove the spill first in order to learn whether its removal actually improves the mix. It is also worth remembering that processors added later in the mix, like compressors and equalizers, could also have an effect on this. With this said, once the mix is in its final stages, it is worth trying to bypass spill gates, and see whether the mix changes for the better or worse.

One of the main challenges with gating is keeping the timbre of the gated instrument. It has already been mentioned that a lower threshold setting would help in doing so. This task is made harder when the wanted signal and the spill are relatively close in levels, especially if they share the same frequency regions. Snare and hi-hats, kick and toms are potentially problematic pairs, especially if spill was not considered during microphone selection and placement. We may employ any possible gate facility in order to improve our gating. Side-chain equalization lets us attenuate the spill by attenuating its dominant frequencies, look-ahead lets us retain more of the natural attack, and hysteresis lets us keep more of the instrument's natural decay. For exceptionally tricky gating tasks, it would be extremely beneficial to use at least one of these facilities, if not all of them.

> *One of the main challenges in gating is keeping the timbre of the instrument; mainly its natural attack and decay.*

When gating drums, there is often a trade-off between the length of the natural decay and the amount of spill – the longer we retain the natural decay the more spill will escape gating. The spill is often made louder by a compressor following the gate, and the instrument can become louder in the mix for short periods while the gate is open. For example, the hi-hats spill on a snare track might make them louder while the snare decays. One solution to this involves ducking the intended hi-hats track in an opposite fashion to the snare's gating, so while the open gate adds some hi-hats spill, the actual hi-hats track is attenuated. Another solution is **decay reconstruction** – the gate shortens much of the decay so no spill remains, and a reverb is employed to imitate the missing decay.

Reshaping dynamic envelopes

Just like compressors, gates are also employed to reshape the dynamic envelope of instruments, mainly of percussive ones. We can say that a compressor operates on a transient (once it overshoots the threshold) and on what comes shortly after it (due to the release function). A gate operates on both sides of the transient (essentially the signal portion below the threshold). Yet, a gate might also affect the transient itself due to the attack function.

As per our discussion in Chapter 16, a part of **adding punch** is achieved by shortening the length of percussive instruments. Typically, the natural decay that we try to shorten is below the threshold, which makes gates the best tool for the job. It is a very old and common practice – you gate a percussive instrument, and with the gate release (and hold) you determine how punchy you want it. Just like with compressors, we must consider the rhythmical feel of the outcome of gating. As gates constrict the length of each hit, the overall result of gating percussive instruments tends to make them **tighter** in appearance. Figure 18.11 demonstrates this practice. The threshold is set to the base of the natural attack and above the natural decay. The range would normally be set moderate to large. The release and hold settings determine how quickly the natural attack is attenuated (with longer settings resulting in longer natural decay). If we only want to **soften the natural decay**, not shorten it altogether, we can use smaller range settings.

> *Gates are often employed to add punch to percussive instruments.*

We can also **accent the natural attack**, **accent transients**, **revive transients** or **reconstruct lost dynamics** using a gate. This is normally done by setting the threshold somewhere along the transient and dialing a very small range so everything below the threshold is mildly attenuated. It is also possible to boost the output signal, so the transient is made louder than before, but everything below the threshold returns to its original level. Figure 18.12 illustrates this.

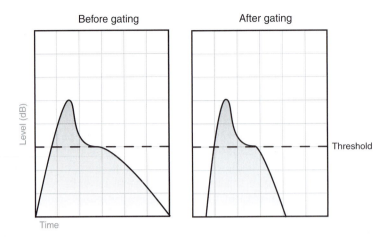

Figure 18.11 Adding punch to a percussive instrument using a gate. The threshold is set to the base of the natural attack and the range is set moderate to large. The release together with hold will determine how quickly the natural decay is attenuated and in turn the overall length of the hit.

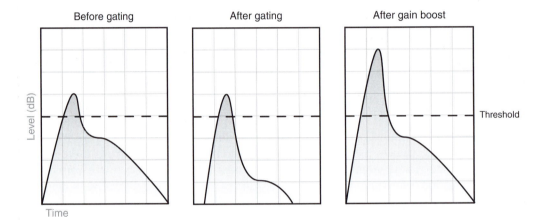

Figure 18.12 A gate emphasizing the attack of a snare. The gate threshold is set halfway through the natural attack. After gating, both sides of the transient have been reduced in level, including the snare's decay. If we boost the gate output, the material below the threshold returns to its original level, but the snare's attack ends up louder in level compared to the signal before gating.

It is worth noting that the threshold in Figure 18.12 was set higher than the base of the natural attack. Had the threshold been set lower than that, the gate operation would most likely alter the dynamic envelope of the snare in a drastic, unwanted way. Figure 18.13 illustrates why this specific application can be somewhat trickier to achieve with a gate than with a compressor. We already said that a gate's attack and release functions tend to be more obstructive due to their typical short settings. While on a gate we would dial both fast attack and release for this application, on a compressor we would dial longer, less obstructive settings. One outcome of this is that when gates are employed for this task, the results tend to be more jumpy than those of a compressor. For these reasons, the gate's range is often kept as small as possible, and look-ahead is employed.

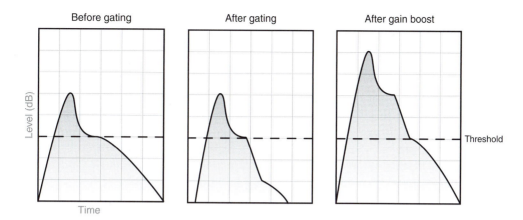

Figure 18.13 A gate reshaping destructively the dynamic envelope of a snare. Due to the low threshold setting, the release function has altered the dynamic envelope in a drastic way, resulting in a new shape that does not resemble the input signal.

Track 18.38: Kick Snare Source
The source track for the following samples, in which the snare is to be gated.

Track 18.39: Kick Snare Range −4 dB
The gate's threshold was set to −36 dB, the attack to the minimum of 5 μs, hold 30 ms, and release to 17 ms. The range was set to −4 dB, so the snare's decay is only attenuated a little. Despite no gain compensation, the snare in this track has more profound attack and punch.

Track 18.40: Kick Snare Range −30 dB
With the same settings as in the previous track but −30 dB of range, the snare's decay is removed. This can also translate into a more punchy sound, that despite sounding unnatural can be suitable for some genres.

Track 18.41: Kick Snare Range Longer Hold
We can lengthen the hold parameter to determine the length of each hit. The settings in this track are identical to those in the previous track, only the hold is lengthened to 90 ms. Compare this track to track 18.38.

* * *

The kick in the previous four tracks is also gated, but only in order to get rid of some low-frequency tail:

Track 18.42: Kick No Gate
Track 18.43: Kick Gated

Plugin: Sonnox *Oxford Dynamics*
Drums: Toontrack *EZdrummer*

In practice

The problem with gates

What is being gated, and how, is largely determined by the threshold setting. Ideally, we would like the gated material to show little gating inconsistencies. This can be achieved if the static threshold setting always affects identical parts of the signal, which is often not the case. For example, if two tom hits vary in level, a gate would attenuate more of the quiet hit's natural attack and decay. Snare hits might be roughly at the same level, but if at a specific section the drummer plays a crescendo, the first few quiet hits might be below the threshold and therefore muted completely (in this specific case, we often bypass the gate momentarily).

Normally, we loop a specific section while setting the gate. Once happy with our settings, it is important to perform a *through run* – playing the track while soloed, beginning to end, to ensure that the gate works as intended throughout the full track length. Even in the case of toms that only play occasionally, it is wise to listen from beginning to end (rather than jumping between the hits) to make sure that no false triggering occurs (for the most part of these through runs we will hopefully enjoy the silence).

Traditionally, gates are placed *before compressors* since the compressor's nature of reducing dynamic range makes gating a trickier affair. However, on a software mixer, it would not be considered odd practice to place a balancing compressor before the gate, so level inconsistencies are reduced before the signal is gated. The gate might still be followed by another compressor that has a different role, like adding some impact.

Gate alternatives – manual gating

Gating inconsistency is an outcome of varying levels, which are only a natural part of a real performance. If we take drums for example, it could be great having the ability to gate each hit with different settings, but automating a gate for such a purpose would be unwieldy. Fortunately, audio-sequencers offer a gating alternative that lets us do exactly this, a practice we will call hereafter *manual gating*. To be sure, manual gating would not produce a characteristic effect like many gates do and can sometimes be senseless – treating 128 kick hits individually can be tedious. But with instruments like toms, which involve fewer hits, individual treatment for each hit can be feasible. Manual gating can also work on vocals, electric guitars and many other instruments. It usually takes more time to perform than standard gating, but it gives us more control over the final results.

> *When sensible, manual gating provides a powerful alternative to standard gating.*

Essentially, manual gating is like offline gating that we can then tweak. It involves three main steps, that are demonstrated in Figure 18.14:

- Strip silence
- Boundary adjustments
- Fades

Strip silence is an offline process that removes quiet sections from audio regions. It works by splicing an audio region into smaller regions that contain the signals we wish to keep. Figure 18.15 shows the Detect Silence window of Cubase. Other audio sequencers provide parameters akin to the ones in this figure. The similarity to a gate is evident from the open and close threshold settings. A minimum time for 'gate open' and 'gate close' is also given, as well as pre- and post-roll times, which create pad margins before and

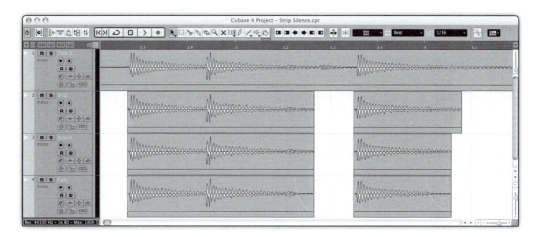

Figure 18.14 Manual gating using strip silence. From top to bottom: the original three tom hits, after strip silence, after boundary adjustments and after fades. Note that only the right boundary of the second region (third tom hit) was adjusted and the different fade lengths for each region. This screenshot makes use of four tracks to show each step at a time; normally, we perform all steps on the original track.

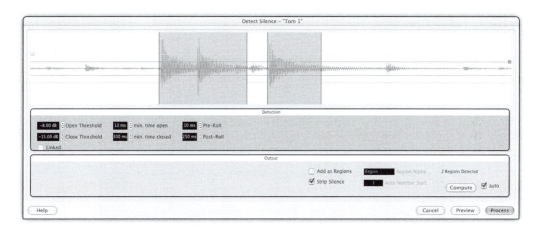

Figure 18.15 Cubase's Detect Silence window. In this screenshot, the process is applied on three tom hits. Due to the close proximity of the first two, they end up within the same region. It is also evident that the decay of the third hit is being trimmed (but also some spill that is not visible in this illustration). It is worth remembering that the low-level decay being trimmed here might not be audible unless the tom is played in isolation. One way or another, we can adjust the region boundaries afterwards.

after each splice. The pre-roll time is of great importance otherwise the process is likely to produce clicks, just like a normal gate with a very short attack would.

The main aim in the strip silence step is to divide the wanted signal into sensible regions. The exact boundaries of each region are less critical, since the next step involves *adjusting each region's boundaries* individually. If we set the strip silence threshold and pre-roll time appropriately, modifications to the start of each region might not be required. The region's end is adjusted to match the decay length we are after, while taking into account the level and amount of spill. The final step of manual gating involves fading in and out each region, which is similar in effect to the attack and release functions of a normal gate. We can set different fade lengths (attack and release) to each region and normally have control over the shape of each fade.

Gate alternatives – denoisers

Another alternative to gates, when used to remove hiss or other kinds of noise, is the digital denoiser (Figure 18.16). Most modern denoisers need to be taught the nature of the noise to be removed. This is done by feeding pure noise to the denoiser while in learn mode. Such pure noise is often available at the start and end of tracks. Feeding a denoiser with anything but pure noise can result in a wired digital effect that might be justified in some creative (somewhat experimental) context. Generally, denoisers do a better job than gates in removing noise and involve a quicker and easier setup. However, this is subject to the noise being of a consistent nature, like the static hiss of analog components. Denoisers are not so good at reducing dynamic noise like headphone spill or street noise.

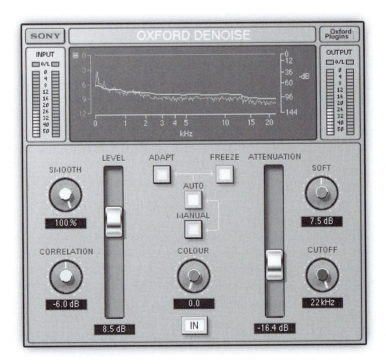

Figure 18.16 A Denoiser plugin. The Sonnox *Oxford Denoise*.

Tricks

Manual look-ahead

A look-ahead facility on a gate is invaluable. Most software gates provide it, but not all hardware units. Often the look-ahead time is fixed, and there might be a switch to bypass this feature altogether. Some gates also let us choose different look-ahead times, like the MOTU *MasterWorks gate* in Figure 18.1 (the control below the plot). Whatever gate you use, the trick presented here provides a definitive look-ahead with any gate and, in fact, is one of the more useful tricks in mixing.

The idea of manual look-ahead is fairly simple. Say we gate a kick, we duplicate the kick track, nudge it back a few milliseconds in time and instead of routing it to the mix bus we feed it to the key input of the kick's gate. Figure 18.17 illustrates this arrangement. One advantage of manual look-ahead is that we can set any look-ahead time we fancy. For example, by nudging the duplicate kick 20 ms backward, we get 20 ms of look-ahead time, and we can always nudge the duplicate back and forth (say in 5 ms steps) to see which time works best. Typically, gates have look-ahead times within the 0–10 ms range, but in specific circumstances even 40 ms could be appropriate. Since the duplicate track is early in time compared to the kick track, the gate's side chain gets to see the kick a few milliseconds before processing it. The gate, whether software or hardware, has no output delay whatsoever – it is the key input signal being pulled back in time, not the gated kick being delayed.

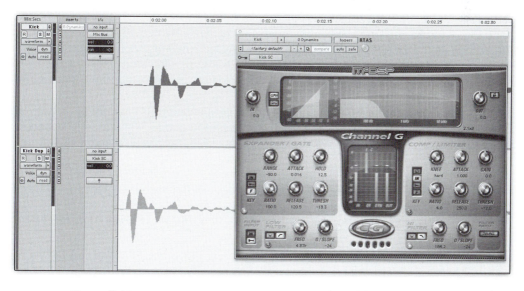

Figure 18.17 Manual look-ahead in Pro Tools. The top track is the gated kick, the bottom is a duplicate nudged 20 ms backward. The duplicate output is routed to a bus named 'Kick SC', which feeds the side-chain of the McDSP *Channel G*.

Note that the *Channel G* plugin in Figure 18.17 is set so the key input signal is filtered (where hi-cut filter is applied to reduce spill). Had the gate not provided side-chain equalization, we could have simply inserted an equalizer on the duplicate track to achieve the same effect.

Virtually all audio sequencers and digital recorders let us duplicate a track and nudge it backward in time. But nudging is not exactly a part of any tape machine. Therefore, the arrangement would be very similar to the way analog gates implement look-ahead – a track would feed both the gate input and its key input, but the copy sent to gating is delayed with a delay unit (obviously causing output delay).

> While manual look-ahead is extremely useful with gates, it can be used with any other dynamic range processor that has an external side-chain facility – a compressor can benefit from this trick all the same.

Kick-clicker

It has already been mentioned that short attack settings on a gate can produce an audible click, that the shorter the attack the higher the byproduct frequency, and that the steep level rise can impair the timbre of the gated instrument, especially that of kicks. However, the resultant click caused by fast attack can also have a positive effect on kicks – it adds some definition. Indeed, sometimes gating a kick with a short attack improves its overall presence in the mix, although in most cases we have to compromise with some loss of impact. A simple trick can come to our rescue, enabling us to both keep the impact and add definition.

The arrangement is as follows: the original kick is gated so its leading edge is kept intact (often using look-ahead), while a duplicate track of the kick is gated to produce the click. The two are layered, and we can use the duplicate's fader to determine how much click we want. We can also use the attack and threshold on the clicker-gate to alter the character of the click (instead of using an EQ). Normally, the hold and release are kept very short, so the clicker-gate only opens for a very short period, just to allow the click to pass through. The beauty of this arrangement is that along with the click some of the early impact also passes through the gate, resulting not only in improved definition, but also in some added power.

The kick-clicker trick works exceptionally well with sequenced music where all the hits are of equal or very similar velocity. The problem with a recorded kick is that level variations between hits will produce clicks with changing character. There is an elegant way around this – placing a compressor before the clicker-gate.

Track 18.44: Kick Clicker
This track is a gated version of Track 18.43. A threshold at −13 dB and the shortest release (5 µs) produce a click.

Track 18.45: Kick Snare with Kick Clicker
This track is the result of layering the previous track at −9 dB with Track 18.41.

Plugin: Sonnox *Oxford Dynamics*
Drums: Toontrack *EZdrummer*

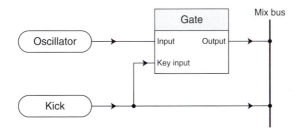

Figure 18.18 Adding sub-bass to a kick. The oscillator is gated with relation to the kick. Every time the kick hits, the gate opens and lets the oscillator signal pass through. Both the kick and the oscillator are mixed.

Adding sub-bass to a kick

This trick is used to add some low-energy power, or oomph, to kicks. Although some kicks already contain a hefty low-end, club music or genres like drum 'n' bass and reggae can all benefit from some extra force that some describe as chest pressure, others as LF wind. This trick is also sometimes used in rock mixes.

Figure 18.18 shows the layout of this trick. The idea is to add some low frequency whenever the kick hits. The low frequency is generated by an oscillator, which is gated with relation to the kick – whenever the kick hits the gate opens to let the oscillator signal through. Figure 18.19 shows this setup in Logic.

Since it is pure power we are after, we do not want the oscillator to produce a recognizable pitch. Thus, a sine wave is used (being the only waveform that generates a pure frequency, with no harmonics to contribute to pitch recognition). The oscillator frequency is often set to 40 or 50 Hz, where 40 Hz would be more powerful, but fewer systems will be able to produce it faithfully. The higher the frequency the more obvious its pitch becomes, thus frequencies above 50 Hz are rarely used.

To prevent a low-frequency hum that will constantly muddy the mix, the gate is set with maximum range so as to mute the oscillator whenever the kick is not playing. The attack is set short, but it is important to have it long enough to prevent clicks. The clicks produced by a gated low-frequency sine are most likely to vary in nature, as for each hit the gate opens at random points along the waveform cycle (Figure 18.20). A very short release will also produce clicks, which can be even more disturbing as they are often delayed in respect to the kick's attack. With the hold and release we can set the length of each sub-bass beat. A short beat will simply reinforce the kick, and the longer it is made, the stronger the impact. One creative effect involves making the sub-bass longer than the kick itself, while setting the hold and release to create some noticeable decay (an effect mostly associated with drum 'n' bass tracks). That said, if the sub-bass beat is made too long, we could end up with a constant low-frequency content that will muddy the mix. Another creative approach is to make the sub-bass on the downbeat longer than all other beats, thereby emphasizing the downbeats.

Figure 18.19 Adding sub-bass to a kick in Logic. The sub-bass track is an aux track on which an oscillator is inserted to generate a 45 Hz sine wave. The oscillator is followed by a gate, which has its side-chain sourced from the kick track (track 1).

Track 18.46: Kick Snare No Sub-bass
The source track before the addition of sub-bass.

Track 18.47: Kick Snare With Sub-bass
50 Hz sub-bass is added to this track. The gate is set with its threshold to −40 dB, range to −80 dB, attack to 4 ms, hold to 150 ms and release to 400 ms.

Track 18.48: Kick Snare Sub-bass only
Only the sub-bass from the previous track.

Track 18.49: Sub-bass 30 ms Hold
Slightly tighter effect is achieved by shortening the hold parameter to 30 ms.

Track 18.50: Sub-bass 300 ms Hold
And a different effect is achieved by lengthening the hold parameter to 300 ms.

Track 18.51: Sub-bass Varying Clicks
This track demonstrates the varying clicks issue. The gate's attack in this track was set to the minimum of 10 μs.

Plugin: Digidesign *DigiRack Expander/Gate Dyn 3*
Drums: Toontrack *EZdrummer*

Figure 18.20 Varying clicks on a sub-bass gated with fast attack. A 40 Hz sine wave was gated with relation to a kick. This illustration shows the post-gate outcome of four hits. The short attack produces a click, but as the gate opens at random points along the cycle, a different click is produced for each hit.

The great challenge with sub-bass beats is sensibly balancing them on the frequency spectrum of the mix. Full-range monitors (or a subwoofer) and a controlled acoustic environment are a requisite for this task. The process is usually this: we set the sub-bass level using the fader, then toggle between the full-range and the near-fields. As we toggle, we would like the relative instrument level to remain as similar as possible, and by switching to the full-range speakers we only want to gain some extra power – it is an extension we are after, not an intrusive addition.

The arrangement used for this trick is most often employed to add sub-bass to a kick. Another popular version of this trick involves adding filtered white noise to a snare in order to add some definition. We can just as easily add any signal to any other percussive instrument, whether for practical or creative reasons.

> **Track 18.52: Kick Snare with White Noise**
> This track is similar to Track 18.47 only with gated white noise triggered by the snare. The noise was treated with high-shelving attenuation and low-pass filtering. The gate settings are: Threshold −40 dB, range −80 dB, attack 4 ms, hold 12 ms, and release 260 ms.
>
> *Plugin:* Digidesign *DigiRack Expander/Gate Dyn 3*
> *Drums:* Toontrack *EZdrummer*

Tighten bass to a kick

One way of looking at the previous trick is this: when the kick gets louder, the sub-bass gets louder. We had to mute the sub-bass at all other times to prevent low-frequency hum. How about keeping the same setup, but exchanging the oscillator with a bass guitar or a bass line? The bass will be muted apart from when the kick hits. First, muting the bass between kick hits will result in low-end deficiency (one fundamental mixing concept is that the kick provides low-end rhythmical power, while the bass fills the low-end gaps between the hits). Second, we lose important musical information as the bass often provides the harmony root. So this idea is not that great. But how about setting a small range on the gate so the bass is only attenuated by a small amount? The bass guitar will still be audible, but whenever the kick hits it will become slightly louder. This can be very useful if the bass is not extremely tight to the kick. Making the bass louder whenever the kick hits can give the impression that the bassist played the accents tight to the beat. The one issue with this practice is that we invite masking between the already competing kick and bass. Making this trick effective requires a flawless equalization to both the kick and the bass (which, to be fair, is our aim anyway).

Gate as a ducker

A small setup lets us employ a gate for the purpose of ducking, just in case no duckers are at our disposal. Say for example we want to duck a guitar with relation to the vocals. We can create a duplicate of the guitar and invert its phase, then load a gate and source its key input from the vocals. Whenever the vocals overshoot the threshold, the gate opens, the phase-inverted duplicate will be mixed with the original guitar, causing phase cancellation and gain reduction. With the level of the duplicate we control the range of attenuation. In order for this setup to work, the original guitar and its duplicate must be identical – if we change the equalizer on the original, we must update the duplicate respectively. It is a cumbersome path to ducking, yet cumbersome paths are sometimes the only option.

19 Expanders

Expanders are very common in mixing environments, although not so common in mixes. Expanders are the opposite of compressors in that they expand dynamic range rather than compress it. Indeed, expanders are often used to expand what a compressor has compressed – a do-undo system called a *compander*. Apart from being a vital part of noise-reduction systems for tapes, a compander has little to do with mixing. Still, compressors and expanders work on a very similar principle – both are based around the same design concept and share identical controls, most notably, a ratio. The fundamental difference

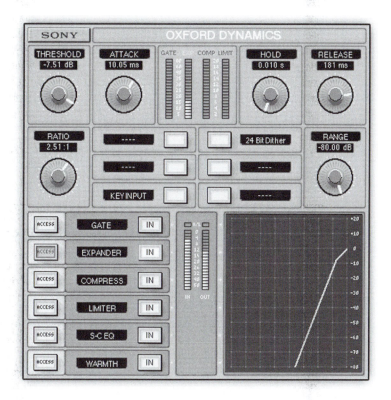

Figure 19.1 An expander plugin. The expander module, part of the Sonnox *Oxford Dynamics*.

between the two is that a compressor affects signals above the threshold, while an expander affects signals *below* the threshold. If one understands how a compressor works, there is not much to learn about expanders.

Controls

An expander is said to make quiet signals quieter. The ratio, just like with compressors, is expressed with the input:output notation. We can see from Figure 19.2 that with a 1:2 ratio, 20 dB fall below the threshold for input signals results in 40 dB fall for output signals. Visually speaking, the portion of the signal below the threshold is stretched downward to be made quieter. It is interesting to note that with the 1:100 ratio the transfer function of the expander looks like a gate. Indeed, a gate is often an expander with a large ratio, and most processors are an expander/gate rather than one or the other. Commonly, we get a

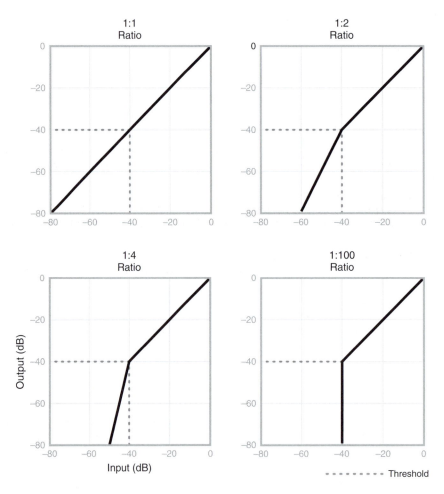

Figure 19.2 The transfer function of an expander with different ratios.

control that lets us dial ratios between 1:1 and 1:100. On some hardware units, a switch will toggle a gate into an expander with fixed ratio, often 1:2.

> An expander ratio is often noted 2:1 rather than 1:2. This is done for convenience, despite not complying with the standard "input:output" notation. In this book, an expander ratio of 1:8 is higher than an expander ratio of 1:2.

Track 19.1: Piano Source
The source track used in the following three tracks. The sound in this track rises and decays slowly.

The expander in the following track is set with its threshold at −39 dB, the release and attack set to fastest and the hold set to 1.1 seconds (in order to prevent chattering). Note how the higher the ratio is the quicker the leading sound rises and the closing sound falls.

Track 19.2: Low Expander Ratio (Piano)
1:2 ratio.

Track 19.3: Moderate Expander Ratio (Piano)
1:4 ratio.

Track 19.4: High Expander Ratio (Piano)
1:16 ratio.

- - - - - - - - - - - - - - -

Track 19.5: Snare Source
The source track used in the following three tracks.

The expander's threshold in the following tracks was set to −17 dB, so only the snare would overshoot it. The attack and release were both set to minimum, and the hold was set to 2.9 seconds to maintain some of the snare decay.

Track 19.6: Low Expander Ratio (Snare)
With 1:2 ratio, the spill is not only audible, but it also fluctuates quite noticeably in level. The audible spill is the outcome of insufficient attenuation; the noticeable fluctuations are caused by the gradual transfer curve.

Track 19.7: Moderate Expander Ratio (Snare)
1:4 ratio reduces the spill much better, yet some of it can still be heard (although it would probably be masked by the rest of the mix).

Track 19.8: High Expander Ratio (Snare)
The 1:16 ratio in this track yields very fast changes in gain reduction. Compared to the previous track, the function of the expander here is more aggressive, and the spill is reduced in full.

Plugin: Sonnox Oxford Dynamics

Expanders rarely provide a soft-knee option, which means that the transition between treatment and no-treatment at the threshold point can be quite harsh. Like with compressors, one way to reduce this artifact is using the attack and release. In order to understand how the attack and release work, it is worth remembering how a compressor works: the amount of gain reduction is determined by the threshold and ratio settings, then the attack and release slow down changes in gain reduction. An expander works on exactly the same principle, only the amount of gain reduction is determined by the signal level below the threshold, not above it (the undershoot, not the overshoot amount).

This is worth stressing: both the attack and release on an expander simply slow down changes in gain reduction. More specifically, a gain change from −40 to 0 dB will be slowed down by the attack, and a change from 0 to −40 dB will be slowed down by the release. Neither the attack nor the release has any relation to the threshold setting. If the signal level fluctuates below the threshold, the attack and release will affect respective changes in gain reduction. Once the signal overshoots the threshold, the applied gain reduction will diminish as fast as the attack allows, and once down to 0 dB the expansion stops. As the signal then drops back below the threshold, expansion starts again, and the release starts to take effect.

The very common hold function on an expander can be implemented in two different ways: the compressor style or the gate style. A compressor-style hold simply alters the release rate, so gain reduction starts very slowly. A gate-style hold freezes the gain reduction for the hold period once the signal has dropped below the threshold.

Most expanders also offer a range control. This parameter defines the maximum amount of gain that would be applied on the signal. For example, a range of −20 dB means that the signal will never be attenuated by more than 20 dB. Figure 19.3 illustrates this.

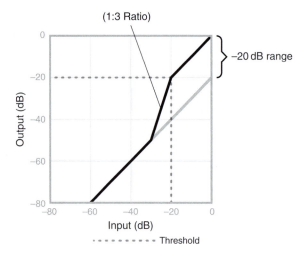

Figure 19.3 Range function on an expander. The gray line denotes the curve of a plain 20 dB attenuation, which sets the bottom limit for the expansion effect. With the resultant ratio curve (black line), the maximum amount of attenuation would never exceed 20 dB.

In practice

While the technical descriptions above might seem somewhat complex, the principal usage of expanders in mixing is very simple – a softer alternative for gates. It was stressed in the previous chapter that gates tend to be obstructive, mostly due to the sharp transition between treatment and no treatment. An expander, with its ratio, provides a more gradual transition between the two. Whether it is large or small attenuation we are after, an expander can give smoother results. The idea is that while a gated signal jumps from one level to another (as the threshold is crossed), an expanded signal slides. Figure 19.4 demonstrates these differences. There is an important outcome to this sliding nature of expanders. While on a gate the attack and release are often employed to soften the gate's operation, on an expander this responsibility is consigned to the ratio, allowing

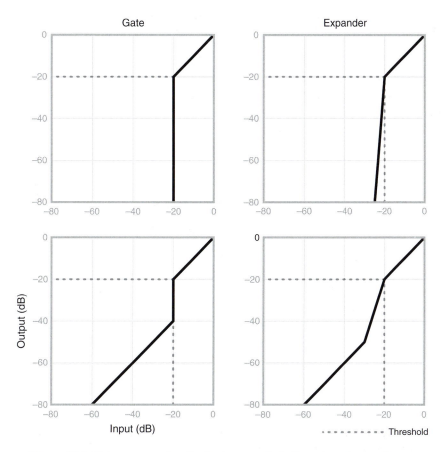

Figure 19.4 Gate vs. expander. On the top row, both the gate and expander are employed to drastically attenuate the input signal. While the gate toggles instantly between large attenuation to no attenuation, an expander would slide between the two. Even when small range is used, like in the bottom row, the same sliding nature of expanders results in less-obstructive effect.

the attack and release to concern themselves with the musical aspects of the signal dynamics.

| *Expanders are used as a softer, less obstructive alternative for gates.* |

The following tracks demonstrate how an expander can be used to reduce ambiance from a drum track, and how it differs from a gate.

Track 19.9: Compressed Drums
This is the source drum track from the following samples. The drums are compressed in a way that contains the attack of the kick and snare. Also, the loud ambiance makes the overall drum image appear backward.

Track 19.10: Reduce Ambiance Step 1
The initial expander/gate settings involve a range of −80, a ratio of 1:100 (essentially a gate), and the fastest attack, release and hold. The threshold was set to −11 dB, which determines what material won't be attenuated by the expander/gate. This material can be heard in this track.

Track 19.11: Reduce Ambiance Step 2
In this step the range is brought up to −13 dB. Sounds that were lost in the previous track can now be heard. However, like in the previous track, some clicks are evident. Although such treatment can be appropriate sometimes, it might also be considered too drastic.

Track 19.12: Reduce Ambiance Step 3
In this step the ratio is brought down to 1.5, effectively turning the gate into an expander. The treatment in this track sounds less obstructive than in the previous track. Compare this track to the source track to see how the ambiance is reduced and how the drums image shifts forward. This can also be described as attack accenting.

Track 19.13: Reduce Ambiance Less Ambiance
We can now control the amount of ambiance with the range control. In this track it is brought down to −40 dB.

Track 19.14: Reduce Ambiance More Ambiance
This track is the result of −6 dB of range. Compared to track 19.12, there's more ambiance here.

Plugin: McDSP *Channel G*
Drums: Toontrack *EZdrummer*

However, there might be some issues with expanders when large ranges are involved. If we assume that large ranges are used when we want to mute signals below the threshold, only gates can ensure such behavior. Expanders can make the very quiet signals inaudible, but signals right below the threshold might still be heard. In order to make these signals less audible we have to increase the ratio. The issue is that the higher the ratio the more the expander behaves like a gate – the ratio curve becomes steeper and level changes slide faster. In some situations, signals right below the threshold can only be made inaudible with very high ratios, which effectively turns the expander into a gate. In a way, expanders are more adequate for small range settings.

Another scenario where expanders have an advantage over gates is when we want to de-emphasize the quiet sounds in a gradual fashion. If a gate could talk, it would say: 'If it's above my threshold I keep it, if it is below I attenuate it'. If an expander could talk, it would say: 'Well, if it's below my threshold I attenuate it, and the quieter it is the more I attenuate it'. While on a gate all signals below the threshold are attenuated by the same amount, expanders let us keep more of the louder details (those right below the threshold) and less of the very low details (which might not be heard anyway). This is suitable in situations where it is actually what is above the threshold we are trying to emphasize,

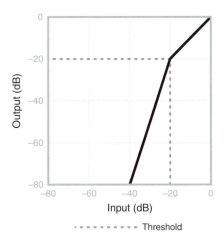

Figure 19.5 Gradual attenuation using an expander. This type of expansion lets us keep more of the loud details below the threshold, while less of the very low level details.

but we do so by attenuating what is below the threshold (which is still important). An example would be adding some dynamic life to a flat drum loop. Figure 19.5 shows a typical transfer function for these types of applications.

Upward expanders

Why upward expanders are so hard to find is baffling. Both with compressors and gates we can make the loud signals louder, but we have to take a back-door approach – make what is below the threshold quieter then bring the whole signal up. Upward expanders are designed to make loud signals louder, which makes them an ideal tool for **accenting the natural attack, accenting or reviving transients, reconstructing lost dynamics, adding liveliness, snap or punch**.

Figure 19.6 shows the transfer function of an upward expander with 0.5:1 ratio. The fraction denotes upward behavior (some manufacturers will write the same ratio as 1:2). It can be seen from Figure 19.6 that for a 20 dB input rise above the threshold, the output rises by 40 dB. With these specific settings, an input signal entering the expander at 0 dB would come out at +40 dB. Naturally, this would result in hard clipping or overload on most systems. In practice, the ratios used in upward compression are very gentle and do not often go below 0.8:1. Still, if the input signal ever hits 0 dB, the output will exceed the 0 dB limit. Thus, often the output level control is used to bring the overall output level down.

When it comes to upward expansion, we no longer talk about gain reduction, we talk about gain increase. In that sense, the attack controls how quickly the gain increase can rise, while the release determines how quickly it can fall. If a snare is being treated and the threshold is set to capture the attack, shorter attack means more of the natural attack, shorter release means less of what comes after it. Figure 19.7 demonstrates this.

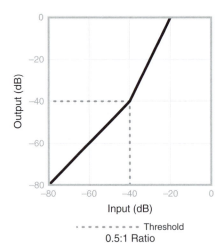

Figure 19.6 An upward expander. Signals below the threshold remain at the same level, while signals above the threshold become louder based on the ratio settings. With the settings shown in this illustration, any input signal above −20 dB will exceed the 0 dB at the output and might overload the system.

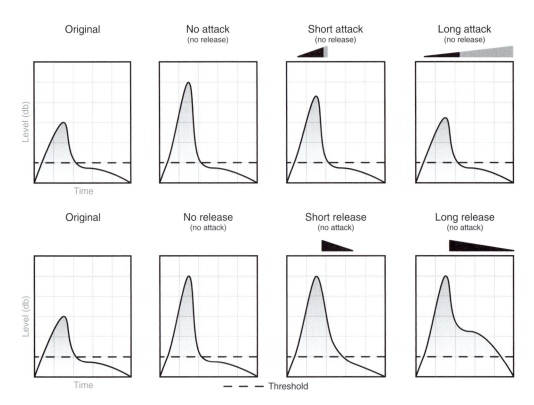

Figure 19.7 The effect of attack and release on upward expansion. The shorter the attack the quicker the signal above the threshold is boosted, resulting in a stronger effect. Long release can be used to boost the natural decay.

Both the attack and release functions on upward expanders tend to be far more transparent than on a compressor, since in most cases the expander operates *with* the signal direction, not against it. Despite the dominance of compressors and gates, upward expanders provide an advantageous choice when our goal is *making something louder*. Both compressors and gates are designed to reduce the gain; upward expanders are designed to increase it. If we want to emphasize the impact of a kick, upward expansion is likely to bring natural results, as it simply reinforces the already rising attack.

> *Upward expanders might be the most suitable tool when it comes to making things louder.*

Track 19.15: Upward Expander
An upward expanded version of the compressed drums from Track 19.9. The expansion ratio is 0.5:1.

Plugin: Logic *Expander*
Drums: Toontrack *EZdrummer*

20 Duckers

We are all familiar with duckers from their common broadcast application – when the DJ speaks, the music is turned down. Most hardware gates provide a switch that turns the gate into a ducker. Even the small dynamic sections on large-format consoles sometimes offer such a feature. For no apparent reason, duckers are hard to find in the software domain. At the time of writing, the *SV-719* in Figure 20.1 is one of the very few plugins that offers ducker functionality. However, ducking as an application is a common part of mixing that is arguably slightly underused.

Figure 20.1 The Sonalksis *SV-719*. This gate/expander plugin also offers a ducker mode.

Operation and controls

Looking back at the internal architecture of a gate in Figure 18.2, all that needs to be done to turn a gate into a ducker is to swap the threshold outputs (above and below). Indeed, duckers offer exactly the same controls as gates and work in a similar way. The sole difference between the two is that while a gate attenuates signals below the threshold, a ducker attenuates signals *above* the threshold. Put another way, once the signal overshoots the threshold, it is attenuated (ducked). The amount of attenuation is fixed and determined by the **range**. Figure 20.2 illustrates the transfer function of a ducker.

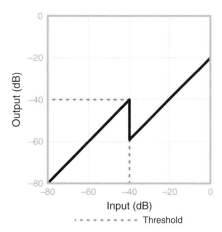

Figure 20.2 The transfer function of a ducker. Signals above the threshold are attenuated by a fixed amount determined by the range (20 dB in this illustration).

The rationale for attenuating a signal by a fixed amount once it exceeds a certain threshold is yet to be discovered. But there are many reasons for attenuating a signal once *a different* signal exceeds a certain threshold like in the radio example above, where we want the music attenuated when the DJ speaks. To achieve this we use the key input – the ducked signal (e.g., the music) is fed into the ducker's input, and the ducking signal (e.g., the voice) is fed into the key input. Whenever the voice exceeds the ducker's threshold, the music will be attenuated. Figure 20.3 illustrates this arrangement, and Figure 20.4 shows it in action. Ducking applications always make use of the key input in a ducked/ducking arrangement.

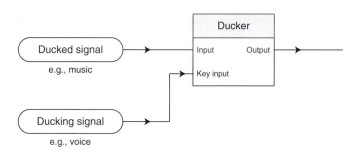

Figure 20.3 The ducked/ducking arrangement. The signal to be ducked is fed into the ducker's input, while the signal to trigger the ducking is fed into the key input. If music is the ducked signal and the voice is the ducking signal, every time the voice exceeds the ducker's threshold the music will be attenuated.

As can be seen in Figure 20.4, gain reduction on the ducked music was applied gradually. This is due to the **attack** and **release** times, which ensure that no clicks occur. In the case of a ducker, the attack determines how quickly the ducked signal is attenuated, while the release determines how quickly the signal returns to its normal level. Also in Figure 20.4, inspecting the music levels during the attack descent suggests that some music could still

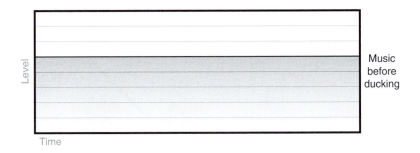

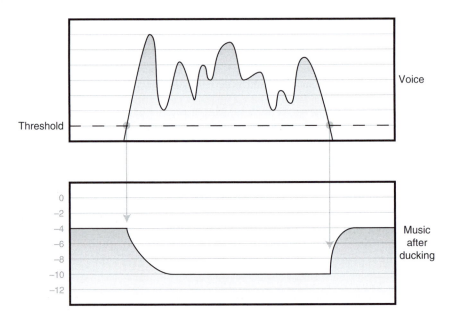

Figure 20.4 Ducking music with respect to voice. The top graph shows the music before ducking, which for simplicity shows no level variations. The middle graph shows the voice that triggers the ducking. As the ducker inspects the side-chain signal, the threshold function is applied to the voice. The bottom graph shows the music after ducking with −6 dB range. The arrowed lines denote the points where the voice crosses the threshold, which is where the ducker's gain reduction starts to rise and fall.

interfere with the voice. Also, there is a hint that once the voice faded away, the music rise during the release stage was audible. Shortening the attack and release might cause clicks and might not be musical. This is where **look-ahead** can be extremely useful – ducking starts and ends slightly beforehand, letting us dial more musical time constants. The **hold** parameter gives us added control over the ducker behavior once the side-chain signal drops below the threshold.

It has been mentioned that often compressors are utilized as duckers. It would be worth demonstrating the difference between a real ducker and a compressor used as a ducker. Both processors attenuate the treated signal once the side-chain signal overshoots the threshold. While on a ducker the amount of attenuation is fixed, on a compressor it varies with relation to the overshoot amount. This means that as the side-chain signal fluctuates in level, the amount of gain reduction fluctuates and so does the compressed signal. These differences are illustrated in Figure 20.5.

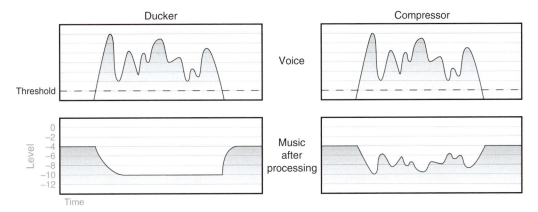

Figure 20.5 A compressor vs. a ducker in the ducking challenge. The amount of gain reduction on a ducker is fixed; on a compressor it is dependent on the overshoot amount. As the side-chain signal fluctuates in level above the threshold, the ducked signal also fluctuates in level. This is not the case with a ducker.

The following tracks demonstrate the difference between a ducker and a compressor-ducker. In both tracks, the uncompressed vocal ducks the guitars. With the ducker, the gain reduction is constant, whereas with the compressor-ducker the gain reduction varies with relation to the level of the voice:

Track 20.1: Ducker
Track 20.2: Compressor Ducker

Plugins: Logic *Ducker*, Logic *Compressor*

Applications

In a mixing context, duckers make one instrument quieter when another gets louder – an incredibly useful way to combat **masking**. Say for example we want the full slap of the kick, but the distorted guitars mask much of it. We can reduce this masking interaction by ducking the distorted guitars whenever the kick hits. It can be surprising how much attenuation can be applied before it becomes noticeable, especially when it comes to normal listeners. Whether done for a very brief moment, or for longer periods, ducking is an extremely effective way to clear some space for the really *important* instruments. Indeed, it is mostly the important instruments serving as ducking triggers and the least important instruments being the ducking prey. While the options are endless, here are a few common places where this can be done:

- **Vocals ducking their own reverb** – one of the oldest tricks in mixing involves attenuating the vocal reverb whenever the vocalist sings. Vocal reverbs are typically added as an effect and less for any spatial purposes. In fact, the spatial nature of reverbs can be a great threat to vocals, since it tends to send the vocals back in the depth field (which normally is not desired). In addition to the ability to mask the vocals, reverbs might blur the intelligibility of the sung words. Ducking the reverb while the vocals are sung makes the vocals more focused and prominent, but we still get the reverb effect during vocal pauses. There are endless accounts of such ducking application in commercial releases.

- **Snares ducking their own reverb** – for the very same reasons as vocals.
- **Kicks and snares ducking their opponents** – it is a typical aim in many contemporary mixes to give an added emphasis to the kick and snare. In dance genres, a potent kick can be one of the prime objectives in the mix. In busy mixes in particular, many instruments can mask the kick and snare, reducing their brunt. By way of analogy, ducking lets the kick and snare push aside any instrument that stands in their way for presence. Both the kick and snare are impulse sounds, so the brief drop they cause on the ducked instruments does not normally result in great loss of musical information. On some mixes, it is even appropriate to duck the bass with respect to the kick. In addition to kicks and snares, toms can also benefit from the same technique, and if the majority of the tom sound derives from the close-mics, it is not unthinkable to have them ducking the overheads.
- **A phrasal power guitar ducking a steady power guitar** – some of the heavier rock or metal arrangements involve a power guitar that plays continuously and another guitar that bursts every bar or so. Ducking the steady guitar with the phrasal one would cause the steady guitar to drop back (on some mixes even disappear) every time the phrasal guitar has something to say.
- **Ducked effects** – time-based effects occupy some space in the mix over time. Reverbs can cloud the mix, and delays can cause some confusion or draw unnecessary attention. Ducking interfering effects can result in increased clarity and tidiness. What instrument triggers the ducking is dependent on the arrangement.

> *Remember ducking when you fight masking, clouding or you are after prominence.*

Track 20.3: Vocal Reverb Normal
A normal arrangement where a vocal is sent to a reverb, but the reverb is not ducked. The dense reverb sends the vocal backward, clouds it and reduces its intelligibility.

Track 20.4: Vocal Reverb Ducked
The vocal in this track ducks its own reverb. Notice how during vocal pauses the reverb gets louder. Compared to the previous track, the vocal here is more focused and forward.

Plugins: Sonnox Oxford Reverb, Sonnox Oxford Dynamics

There is a way to make ducking less intrusive yet more effective, but it involves giving the ducker a rest and using a compressor instead, more specifically a multiband compressor. Back to the earlier example of distorted guitars masking the kick, instead of ducking the entire guitar tracks with the kick, we can feed them into a multiband compressor that is side-chained to the kick. On the multiband compressor we utilize the high-mids band, so the kick's attack will compress the guitars every time the kick hits. Since no other frequency range will be affected, we can set a deeper ducking to the high-mids of the guitars.

An important goal in mixing is controlled dynamics. Compressors are often employed to contain the level fluctuation of various instruments. However, once levels are contained the result can be rather vapid – all the instruments play at the same level throughout the song. Duckers can be utilized to insert some **dynamic motion**. Since we have full control over how this is done, we are kept in the zone of *controlled* dynamics. The idea is to add sensible level variations that reflect the action in the mix. Drums are often a good source of ducking triggers; vocals as well. We can, for example, duck the hi-hats with relation to the snare and we can even set a long release on the ducker so the hi-hats rise gradually after each snare hit. Or we can duck the organ in relation to the vocals or the pad with respect to the lead. This is not done to resolve masking or add prominence, it is simply done to break a static impression and add some dynamism.

| *Ducking can infuse motion into a dynamically insipid mix.* |

Then there is always ducking for creative purposes and sheer effect. We can make drum loops and drum beats far more interesting by incorporating ducking into them. We can feed a delay line into a ducker to create rhythmical level pulses. We can also duck the left channel of a stereo track with relation to the right and vice versa, to create intentional image shifting.

Track 20.5: Drums No Ducking
This is the source track used in the following track. No ducking is applied on this track.

Track 20.6: Drums with Ducking
The hats and tambourine are ducked by the kick, with relatively long release. Notice how this track has an added dynamic motion compared to the previous one.

Plugin: Digidesign *DigiRack Compressor/Limiter Dyn 3*
Drums: Toontrack *EZdrummer*

Polly
Nirvana. *Nevermind.* **Geffen Records, 1991.**

During the verse of this song the vocals appear to duck the guitar. However, the inconsistency of the level variations reveals that this specific ducking-like effect is the outcome of gain-riding.

Biscuit
Portishead. *Dummy.* **Go! Discs, 1994.**

The notable drum beat in this song involves a creative ducking setup, which entails more than one ducking trigger, including some 'hidden triggers'.

21 Delays

Delay basics

Before the emergence of digital units, engineers used tapes to generate delays (often termed *tape-echo*). Not only is it interesting to see how these earliest delay arrangements worked, but it can also be easier to understand the basics of delays using tapes rather than the less straightforward digital adaptation. Famous units like the *Echoplex* and the Roland *RE-201 Space Echo* were designed around the principles explained below. Despite many digital tape delay emulations, these vintage units have not yet vanished – many mixing engineers still use them today. Delays are also used to create effects like chorusing and flanging, which are covered in the next chapter.

Figure 21.1 The Universal Audio Roland *RE-201 Space Echo*. This plugin for the UAD platform emulates the sound of the vintage tape loop machine. The interface looks very similar to that of the original unit.

Delay time

The most basic requirement of a delay unit is to delay the input signal by a set amount of time. This can be achieved using the basic arrangement shown in Figure 21.2. A magnetic tape loop rolls within the unit, where a device called a capstan controls its speed. The tape passes through three heads: the erase head, the record head and the playback head (also known as replay head). After previous content has been erased by the erase head, the input signal is recorded via the record head. Then it takes a period of time for the recorded material to travel to the replay head, which feeds the output of the unit. The time it takes the tape to travel from the record to the replay head determines the delay time.

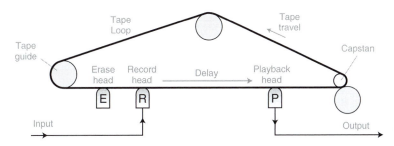

Figure 21.2 Simple tape loop.

There are two ways to control the delay time: either changing the distance between the record and replay head or changing the tape speed. Moving any head is a mechanical challenge, and it is much easier to change the tape speed. There is a limitation, however, to how slowly the tape can travel, so some units employ a number of replay heads at different distances, and the rough selection of delay time is made by selecting which head feeds the output. The tape speed is then used for fine adjustments.

For readers completely new to delays, here are a few tracks demonstrating different delay times. The original hi-hats are panned hard left, the delay is panned hard right. Notice that both 1 and 25 ms fall into the Haas window, therefore individual echoes are not perceived.

Track 21.1: HH Delay 1 ms
Track 21.2: HH Delay 25 ms
Track 21.3: HH Delay 50 ms
Track 21.4: HH Delay 75 ms
Track 21.5: HH Delay 100 ms
Track 21.6: HH Delay 250 ms
Track 21.7: HH Delay 500 ms
Track 21.8: HH Delay 1000 ms

Plugin: PSP *Lexicon 42*

Modulating the delay time

The delay time can be modulated – shortened and lengthened in a cyclic fashion. For example, within each cycle, a nominal delay time of 100 ms might shift to 90 ms, then back to 100 ms, then to 110 ms and back to the starting point – 100 ms. This pattern will repeat in each cycle. How quickly a cycle is completed is determined by the *modulation rate* (or *modulation speed*). How far below and above the nominal value the delay time reaches is known as the modulation depth.

If heads could easily be moved, modulating the delay time would involve moving the replay head closer and further to the record head. But since it is mostly the tape speed that grants the delay time, a special circuit slows down and speeds up the capstan. There is an important effect to this varying tape speed – just like the pitch drops as we slow a vinyl, the delayed signal drops in pitch as the capstan slows down; and similarly, as the capstan speeds up, the delayed signal rises in pitch. These changes in pitch are highly noticeable with both high modulation-rate or depth settings.

Track 21.9: ePiano Not Modulated
This track involves an electric piano panned hard left and its 500 ms delay hard right. In the following tracks, the higher the depth, the lower and higher in pitch the echo reaches; the higher the rate the faster the shift from low to high pitch. The depth in the track names is in percentages:

Track 21.10: ePiano Modulated Depth 24 Rate 1 Hz
Track 21.11: ePiano Modulated Depth 100 Rate 1 Hz
Track 21.12: ePiano Modulated Depth 24 Rate 10 Hz
Track 21.13: ePiano Modulated Depth 100 Rate 10 Hz

Plugin: PSP 84

Feedback

Using the arrangement discussed so far, with no feedback applied, the delayed copy would produce a signal echo when mixed with the original signal. It is often our aim to produce repeating echoes, whether in order to simulate the reflections in large spaces or as a creative effect. There is a very easy way to achieve this, which simply involves routing the replay signal back to the record head (Figure 21.3). This creates a feedback loop where each delayed echo returns to the record head, then is delayed again. Providing

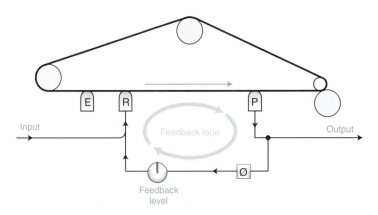

Figure 21.3 Tape delay with feedback loop.

a fixed delay time of 100 ms, the echo pattern will involve echoes at 100, 200, 300, 400, 500 ms and so forth.

A feedback control determines the amount of attenuation (or sometimes even boost) applied on repeating echoes. If this control attenuates, say by 6 dB, each echo will be 6 dB quieter than its preceding echo, and the echoes will diminish over time. If the feedback control is set to 0 dB or higher, the echoes will become progressively louder until the tape saturates (even 0 dB produces such an effect as either tape hiss or the input signal causes increasing level when mixed with the echoes). Such rising echoes are used for creative effect, but at some point the feedback has to be pulled back to attenuation or the echoes will continue forever, causing increasingly distorted sound.

You probably noticed the phase switch before the feedback level control. This switch inverts the phase of the signal passing through the feedback loop (on some units this is achieved using *negative feedback*). The first echo is phase inverted compared to the input signal. But once this echo is delayed again and travels through the loop as the second echo, its phase is inverted again to make it in-phase with the input signal. Essentially, the phase flips per echo, so odd echoes are always phase-inverted compared to the input signal, while even echoes are in-phase. Very short delay times cause combfiltering and can alter the harmonic content of the material quite noticeably. Inverting the phase of signals traveling through the feedback loop alters this combfiltering interaction – sometimes for better, sometimes for worse.

Track 21.14: dGtr No Feedback
This track involves a distortion guitar panned hard left and 500 ms delay panned hard right. No feedback on the delay results in a single echo. The following tracks demonstrate varying feedback gains, where essentially the more attenuation occurs the less echoes are heard.

Track 21.15: dGtr Feedback −18 dB
A −18 dB of feedback gain corresponds to 12.5%.

Track 21.16: dGtr Feedback −12 dB
A −12 dB of feedback gain corresponds to 25%.

Track 21.17: dGtr Feedback −6 dB
A −6 dB of feedback gain corresponds to 50%.

Track 21.18: dGtr Feedback 0 dB
With 0 dB gain on the feedback loop (100%), the echoes are trapped in the loop and their level remains consistent. This track was faded out manually after 5 seconds.

- - - - - - - - - - - - - - -

Track 21.19: Feedback No Phase Inverse
This is the result of 10 ms delay with feedback gain set to −2 dB. This robotic effect is due to the combfiltering caused by the extremely short echoes.

Track 21.20: Feedback Phase Inverse
By inverting the phase of the signal passing through the feedback loop, the robotic effect is reduced.

Plugin: PSP *84*

Wet/dry mix and input/output level controls

Like most other effects, most delays let us set the ratio between the wet and dry signals. There might also be input and output level controls. Adding these controls, Figure 21.4 shows the final layout of our imaginary tape-echo machine.

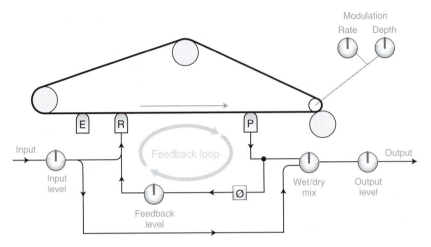

Figure 21.4 A tape-echo machine with all common controls.

The tape as a filter

Tapes do not have a flat frequency response. This is one of the most important characteristics that distinguishes them from precise digital media, for recordings in general and with relation to delays in particular. The most noticeable aspect of this uneven frequency response is the high-frequency roll-off, which becomes more profound at lower tape speeds. Essentially, a tape can be looked at as having a LPF with its cut-off frequency dependent on the tape speed (and quality). With relation to tape delay, each echo experiences some degree of high frequencies softening. Repeating echoes become progressively darker, which in our perception makes them appear slightly further away.

Track 21.21: dGtr Feedback Filtering LPF
The arrangement in this track is similar to that in Track 21.17, only with LPF applied on the feedback loop. For the purpose of demonstration, the filter cut-off frequency is set to 1.5 kHz (which is much lower than what a tape would filter). The result is echoes that get progressively darker.

Plugin: PSP *84*

Types

As already mentioned, tape-based delay units are still one of the most revered tools in the mixing arsenal. Around the late 1960s, nonmechanical analog delays emerged, making use of shift register designs, but these were quickly replaced by today's rulers – digital delays. A digital delay, often referred to as *digital delay line* or *DDL* for short, works with a striking similarity to tape delays. Instead of storing the information on magnetic tape, the incoming audio is stored in a memory buffer. Conceptually, this buffer is cyclic, just like the tape is arranged in a loop. There are record and playback memory pointers, very much like the record and playback heads. To achieve smooth modulation of the delay time, sample rate converters are used, providing functionality similar to slowing down and speeding up the tape. Since digital algorithms have inherently flat frequency response (up to the Nyquist frequency), digital delays often provide a LPF to simulate the high-frequency roll-off typical to tapes. Perhaps one of the greatest limitations of digital delays is that their high memory consumption restricts the maximum delay time. However, we can always cascade two delays to achieve longer times.

Although most digital delays are based on similar principles, there are a few main types:

The simple delay

The simple delay (Figure 21.5) has a single feature – delay time, which is set in either samples or milliseconds. Often this type of process is used for manual plugin delay compensation but we will see later that even such a simple delay can be beneficial in a mix.

Figure 21.5 A simple delay plugin.

Modulated/feedback delay

Figure 21.6 shows the Digidesign *Mod Delay II* in its mono form. The plugin offers the standard functionality of a DDL: input gain, wet/dry mix, a LPF, a delay time set in milliseconds, modulation depth and rate, and feedback control. The LPF is placed within the feedback loop to simulate the high-frequency roll-off typical to tapes. In the case of the *Mod Delay II,* the phase switch simply inverts the input signal phase, not that of the feedback signal.

The depth parameter is often given in a percentage and determines the delay time alternation range. For example, with a delay time of 600 ms, 100% depth denotes 600 ms difference between the shortest and longest delays, which would result in a delay time within the range of 300–900 ms. It is worth mentioning that the higher the delay time is, the more effect the depth will have – while 50% depth on 6 ms means 3 ms difference, the same depth on 600 ms would yield 300 ms difference. The modulation rate can be in

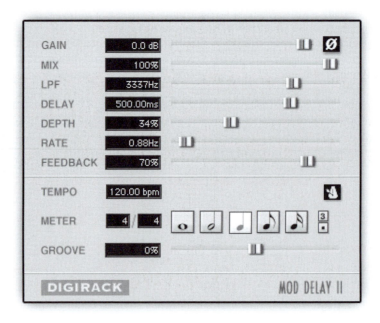

Figure 21.6 The Digidesign *DigiRack Mod Delay II*.

Hz, but can also be synced to the tempo of the song. There is some link between the modulation depth and rate settings, where the higher the depth is the slower the rate needs to be for the effect not to become peculiar.

It is worth noting the tempo-related controls in Figure 21.6, which are a very common part of digital delays. In its core, instead of setting the delay time directly in milliseconds, it is set based on a specific note value. In addition to the common values, like a quarter-note, there might also be triplets and dotted options. Often a dotted quarter-note will appear as *1/4D*, and a half-note triplet will appear as *1/2T*. But the truly useful feature of DDLs is their ability to tempo-sync – given a set note duration, the delay time will alter with relation to the changing tempo of the song. This way, as the song slows down, the delay time is lengthened. On a digital delay unit this often requires MIDI clock-based synchronization. Plugins, on the other hand, natively sync to the changing tempo of the session. In cases where the musicians did not record to a click, a *tempo-mapping* process can map tempo changes to the tempo ruler.

 Appendix B includes delay-time charts for common note values and BPM figures.

From the basic features shown in Figure 21.6, delays can involve many more features and controls. One of the more advanced designs – the PSP *84* – is shown in Figure 21.7. Among its features are: separate left and right delay lines, a modulation source with selectable waveforms and relative left/right modulation phases, a resonant filter of three selectable natures, a drive control to simulate the saturation of analog tapes, and even an integral reverb. In addition, the various building blocks can be configured in a modular fashion typical to synthesizers. For example, the modulation section can affect both the buffer-playback speed and the filter cut-off. The comprehensive features on the PSP *84* make it capable of producing a wider variety of time-based effects than just plain delays. But at least some of these features are often also part of other delay lines.

Figure 21.7 The PSP *84*.

Ping-pong delay

The term ping-pong delay describes a delay where echoes bounce between the left and right channels. If both channels' feedback gets sent to the opposite channel, the resultant signal flow would behave the same as in Figure 21.8.

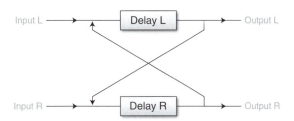

Figure 21.8 A pure ping-pong delay.

Figure 21.9 The Logic *Stereo Delay*.

Although there are some dedicated ping-pong delays (mostly programs within a unit or a plugin), any stereo delay that lets us pan the feedback from one channel to the other can achieve such behavior. Instead of offering feedback pan control per channel, some plugins offer two controls per channel: feedback and feedback crossfeed to the opposite channel. Such a plugin is shown in Figure 21.9.

> **Track 21.22: Ping-Pong Delay**
> This track already appeared in Chapter 13 on panning but, is reproduced here for convenience.
>
> *Plugin:* PSP *84*

Multi-tap delay

The feedback mechanism, despite its wide use, has some limitations: the echoes are spaced at regular intervals, their level drops in a predictable way and their frequency content is correlated. We say that each echo is dependent on the echo preceding it. Sometimes, we have a very clear idea of the exact number of echoes we would like, what their level should be, where they should be panned, and we wish to have full control over their spacing and frequency content. A multitap delay is designed to provide such precise control. Converting a standard tape delay into a multitap delay would involve adding, say, seven playback heads (each considered as a tap), giving us the ability to position each within a set distance from the record head and providing a separate output for each head. These are rather challenging requirements for a mechanical device, but there is no problem implementing multitap delays in the digital domain. Figure 21.10 shows the PSP *608* (an eight-tap delay). We have individual control over each echo's timing (horizontal sliders), gain, pan position, feedback send and filter.

| *Multitap delay allows precise control over each echo.* |

> **Track 21.23: Multitap Source**
> The source arrangement used in the following tracks, with no delay involved.
>
> **Track 21.24: Multitap I**
> The delay here demonstrates echoes of unequal spacing. Five taps are configured so there is an echo on the second, third, fifth, sixth and seventh 8th-note divisions. The taps are panned left, right, right, left, right, and the different tap gains make the echo sequence drop then rise again.
>
> **Track 21.25: Multitap II**
> Seven taps are used in this track, arranged in 8th-note intervals. The taps are panned from left to right, and drop in level towards the center. This creates some cyclic motion.
>
> **Track 21.26: Multitap III**
> In this track, seven taps are configured so the echoes open up from the center outwards. An LPF is applied on each tap, with increasing cut-off frequency and resonance per tap.
>
> *Plugin:* PSP *608*
> *Hi-Hats:* Toontrack *EZdrummer*

Figure 21.10 The PSP *608* multi-tap delay.

Phrase sampler mode

This is an added feature on some delay units (also called *freeze looping*) enabling the sampling of a specific phrase then playing it over and over again. A unit might provide freeze and trigger switches or a single switch that simply enters the mode by replaying part of the currently stored buffer. Many units can tempo-sync this feature so all sampling and triggering lock to the tempo of the song. This feature is different from setting the feedback to 0 dB, since with phrase samplers each repeat is identical to the original sampled material (whereas in feedback mode each echo is a modified version of the previous echo).

In practice

Delay times

Different delay times are perceived differently by our ears and thus play a different role in the mix. We can generalize and say that short delays are used very differently than longer delays. What differs between the two is whether or not we can discern distinct echoes. Yet, we can categorize delay times even further into five main groups. The times shown here are rough and vary depending on the instrument in question:

- **0–20 ms** – very short delays of this sort, provided the dry and wet signals are mixed in mono, produce **combfiltering** and alter the timbre of instruments. Applications for such short delays often involve modulation of the delay time and are discussed in the next chapter. If the dry and wet signals are panned each to a different extreme, the result complies with the **Haas effect**.
- **20–60 ms** – this range of time is often perceived as **doubling**. However, it is the cheapest-sounding approach to doubling, which can be improved by modulating the delay time.
- **60–100 ms** – a delay longer than approximately 60 ms is already perceived by our ears as a distinct echo. Nevertheless, the short time gap between the dry and wet sounds results in what is known as **slapback** delay – quick successive echoes, often associated with reflections bouncing between parallel walls. Slapback delays are used (more in the past than today) as a creative effect. Since slapback delay could be easily produced using a tape machine, it was extremely popular before digital delays emerged and can be recognized in many mixes from the 1960s.
- **100 ms quarter note** – such delay times are what most people associate with plain **delay** or **echo**, where echoes are spaced apart enough to be easily recognized. For a tempo of 120 BPM at common time a quarter-note would be 500 ms. Delay times within this range are very common.
- **Quarter note and above** – long delay times as such are perceived as **grand-canyon echoes**.

Track 21.27: Vocal Dry
The source track used in the following samples. All the delays in the following tracks involve 30% of feedback and are mixed at −6 dB. A 414 ms roughly corresponds to a quarter-note, 828 ms to a half-note. Notice that the effect caused by 40 ms of delay can easily be mistaken for a reverb.

Track 21.28: Vocal 15 ms Delay
Track 21.29: Vocal 40 ms Delay
Track 21.30: Vocal 80 ms Delay
Track 21.31: Vocal 200 ms Delay
Track 21.32: Vocal 414 ms Delay
Track 21.33: Vocal 828 ms Delay

Plugin: Digidesign *DigiRack Mod Delay 2*

Panning delays

The delay output can be either mono or stereo. In the case of stereo output, where there is a clear difference between the left and right channels, the echoes are often mirrored around the center of the stereo panorama or around the center position of the source instrument. As an example, if vocals are panned center, the echoes might be panned to 10:00 and 14:00. Mono delays can be panned to the same position as the source instrument, but often engineers pan them to a different position, so they do not clash with the source instrument. This not only makes the echoes more defined, but also widens the stereo image of the source instrument. Very often if the instrument is panned to one side of the panorama, its echoes will be mirrored on the other side. Instruments panned center would normally benefit from a stereo delay, so the effect can be balanced across the sound stage. But there are quite a few examples of an instrument-panned center, with its mono delay panned only to one side.

Useless
Kruder Dorfmeister. The K&D Session. !K7 Records, 1998.

At the very beginning of this track, the echoes of the word 'echoing' are panned right. Shortly after, the echoes of the word 'mind' are panned center.

The following tracks demonstrate various delay panning techniques:

Track 21.34: Delay Panning Crossed
In this track, the hats are panned hard right and their mono delay hard left. The bass line is panned hard left and its delay hard right.

Track 21.35: Delay Panning Crossed Narrowed
The hats are panned to around 15:00 and their delay is mirrored to 9:00. The bass line is panned 9:00 and its delay 15:00.

Track 21.36: Delay Panning Outwards
Both the hats and the bass line are panned center. The hats' delay is panned hard left and the bass line's delay hard right.

Track 21.37: Delay Panning Mono
The hats and their delay are panned hard left. The bass line and its delay are panned hard right.

- - - - - - - -

Track 21.38: Vocal Delay Wide
A stereo delay (with no feedback) is panned hard to the extreme. This creates a loose relationship between the echoes and the dry signal.

Track 21.39: Vocal Delay Narrow
With the stereo delay narrowed between 10:00 and 14:00, the echoes are less distinguishable from the dry signal, yet there is a fine sense of stereo width.

Track 21.40: Vocal Delay Mono
Panning the stereo delay hard center results in a retro effect that lacks size. Also, the dry signal clouds the echoes.

Plugin: PSP *84*
Hi-hats: Toontrack *EZdrummer*

Tempo-sync

It would seem very reasonable to sync the echoes to the tempo of the song, especially in the various dance genres. Since delay is the most distinct timing-related effect, off-beat echoes can clash with the rhythm of the song and yield some timing confusion. Delays longer than a quarter-note are especially prone to such errors. However, there are situations where it actually makes sense to have the echoes out-of-sync. First, very short delay times, say those shorter than 100 ms, seldom play a rhythmical role or are perceived as having a strict rhythmical sense. Second, for some instruments tempo-synced echoes might be masked by other rhythmical instruments. For example, if we delay hi-hats playing eight-notes and sync the delay time, some of the echoes will overlap with the original hits and be masked. Triplets, three-sixteenth-notes and other anomalous durations can often prevent these issues while still maintaining some rhythmical sense. Perhaps the advantage of out-of-sync delays is that they can draw more attention, being contrary to the rhythm of the song. This might work better with occasional delay appearances, where these off-beat echoes create some tension that is later resolved.

It is worth remembering that before audio sequencers became widespread, engineers had to calculate delay times in order to sync to the tempo of the song. Despite the knowledge of how to do this, often delay times were set based on experiment rather than calculations, and indeed, many times the result was not strict note values.

| *The delay time does not always have to be tempo-synced.* |

The following tracks demonstrate various delay times. All apart from Track 21.47 are tempo-synced to a specific note duration:

Track 21.41: Half-Note Delay
Track 21.42: Quarter-Note Delay
Track 21.43: Quarter-Note Dotted Delay
Track 21.44: Quarter-Note Triplet Delay
Track 21.45: Eight Note Delay
Track 21.46: Eight Note Dotted Delay

An 8th-note dotted is equal in duration to three 16th-notes. This type of duration can be very useful since the echoes hardly ever fall on the main beat, yet bear a strong relationship to the tempo.

Track 21.47: 400 ms Delay
Despite having no direct relationship with the tempo of the track (120 BPM), 400 ms delay does not sound so odd.

* * *

Track 21.48: Sines No Delay
The source track used for the following two samples, involving 16th-note sequence.

Track 21.49: Sines Tempo Synced
The delay on this track involves 16th-note delay on the left channel and 8th-note delay on the right channel. Although the delay can be heard, the dry sequence notes and the delay echoes are all tight to 16th-note divisions, therefore the dry notes always obscure the echoes.

Track 21.50: Sines Not Tempo Synced
Having the delays out of sync with the tempo (being 100 ms for the left channel and 300 for the right) produces more noticeable delay.

Plugin: Digidesign *DigiRack Mod Delay 2*
Drums: Toontrack *EZdrummer*

Filters

Echoes identical to the original source can be quite boring. In addition, if these are quite loud compared to the original, they can be mistaken for additional notes rather than echoes. Although the complexity of tape delay is greater than what simple filtering can achieve, the feedback filters let us incorporate some change and movement into the otherwise statically repeating echoes and distinguish them from the original sound. A LPF makes each echo slightly darker than the previous one, which also makes the echoes progressively undefined and distant. A HPF has the opposite effect – it makes each echo more defined, so despite their decreasing level, late echoes might be perceived better. Yet, they are still distinguishable from the original sound.

> **Track 21.51: dGtr Feedback Filtering HPF**
> The arrangement in this track is similar to that in Track 21.21, only with HPF in the feedback loop instead of LPF. The cut-off frequency is 500 Hz.
>
> **Track 21.52: dGtr Feedback Filtering BPF**
> Some delays also offer a band-pass filter. In this track, its center frequency is 1 kHz.
>
> *Plugin:* PSP *84*

Modulation

The modulation feature of a delay line is mostly used with very short delay times. This is done in order to enhance effects like chorusing or create effects like flanging. It can be used with longer delay times to add some dimension, movement or size to a simple delay line, but with anything more than subtle settings, the change in pitch becomes evident. This might be appropriate in a creative context, but less so in a practical context.

The automated delay line

One very early and still extremely popular mix practice is automating a delay line. Done mostly on vocals or other lead elements of the mix, we often want the delay to catch the very last word or note of a musical phrase. To achieve this, we automate the send level to the delay unit, bringing it up before the word we wish to catch, then down again. Depending on the exact effect we are after, the feedback is set to moderate so the echoes diminish over time, or we set it high to create echoes that barely decline, then we cut the effect by turning the feedback all the way down.

> **Track 21.53: Automated Delay Send**
> Demonstrated here on the vocal, the delay send level was brought up on 'I' and then on the closing word 'Time'.
> *Plugin:* Digidesign *DigiRack Mod Delay 2*

Applications

Add sense of space and depth

A detailed exploration into our perception of depth is given later in Chapter 23 on reverbs. For now, it would suffice to say that the very early reflections from nearby walls and other boundaries give our brain much of the information regarding spatial properties and the front/back position of the source instrument in a room. To be frank, reverb emulators are designed to recreate a faithful pattern of such early reflections, and they do so much better than any delay line. However, even the simplest echo pattern caused by a stereo delay can create a vague impression of depth – it is unlikely to sound as natural as what most reverbs produce, but it would still have some effect.

One advantage that delays have over reverbs is that reverbs provide far denser sounds that can consist of thousands of echoes. So many echoes might not only produce an excellent space simulation, but also cloud the mix and cause some masking. Delays, on the other hand, are far more sparse and often only involve a few echoes. When natural simulation is not our prime concern or in cases where vague space is what we are after, delays may be suitable for the task.

Delays can be used to create a vague sense of space and depth,
with a relatively small increase in sound density.

There is another important advantage in the vague simulation that space delays produce – they do not tend to send instruments to the back of the mix as much as reverbs do. This lets us add some sense of space, while still keeping instruments in the front of the mix.

Delays do not tend to send instruments to the back of the mix as
much as reverbs.

Compare Track 21.48 to Track 21.49 to see that the sines appear further back with the delay; the image shifts even further back in Track 21.50, where the delay is more noticeable. Also, compare Track 21.27 to Tracks 21.29 and 21.30 for a similar demonstration of depth.

Add life

This use, to some extent, is very similar to the previous one – by adding some echoes to a dry instrument, we add some dimension that makes its sound more realistic. Again, the idea that any depth and distance are vague is often an advantage here. Perhaps the main beneficiary of this type of treatment is the vocals – we can add some vitality and elegance, without sending them backward in the mix.

Track 21.54: Vocal with Life
Some listeners will agree that this track sounds slightly more natural when compared to the dry track 21.27. This track involves a two-tap delay mixed at −30 dB.

The comparison between Tracks 21.48 and 21.49 can serve as another demonstration of how the sines – despite being synthesized – sound more natural and 'alive' with the delay.

Plugin: PSP Multidelay 608

Natural distance

If we imagine a large live room, where all the musicians are positioned in front of a stereo pair, the sound arriving from musicians right at the back would have to travel a longer time than the sound arriving from musicians close to the microphones. As sound takes approximately 1 ms to travel 1 ft, the sound from a drummer placed 25 ft (approximately 8 m) from the microphones would take around 25 ms to travel. It will take around 3 ms for the voice to travel from a singer a meter away. In many recordings done today using overdubs there is no such delay – apart from the overhead, all other instruments might be recorded with close-mics, often no further than 2 ft away. Essentially, when all these overdubs are combined, it would create the impression that all the musicians were playing equidistant from the microphone, in other words the very same line in space. Although it is primarily reverbs that give us the impression of depth, this very sense can be enhanced if we introduce some short delay on instruments we want further away in the mix. In this specific case, we only mix the delayed signal without the original one and make no use of feedback or modulation.

Fill time gaps

The echoes produced by a delay can fill empty moments in the mix where nothing much develops otherwise. For this task, a generous feedback is often set along with relatively long delay times, and a filter is used to add some variety to the echoes. We usually want the delay to fill a very specific gap and diminish once the song moves on, so this application often involves an automated delay line with ridden feedback.

Fill stereo gaps

Delays can solve stereo imbalance problems where one instrument is panned to one side of the panorama with no similar instrument filling the other side. The arrangement is simple – the original instrument is sent to a delay unit, and the wet signal is panned opposite the dry sound. For instruments of less rhythmical nature, a short delay is probably more suitable. For more rhythmical instruments, say a shaker, the delay time can be longer and tempo-synced. For sheer stereo balance purposes, a single echo (no feedback) would normally do. We can also attenuate the wet signal to make the effect less noticeable.

Pseudo-stereo

It is sometimes our wish to make a mono instrument stereophonic, for example, a mono output of an old synthesizer. One way to achieve this involves the same arrangement as above – the mono instrument is panned to one channel, its delay to the other. It is worth mentioning that this can sometimes also work when applied on one channel of a stereo recording. For example, an acoustic guitar that was recorded with body and neck microphones; if the neck microphone has obvious deficiencies, we can use the body-mic only and create a stereo impression by delaying it and mirroring the delay opposite the dry signal. Applying filter on the delay would normally make the effect more natural.

Make sounds bigger/wider

Similarly to the filling stereo gaps application above, delays can also be used to make sounds bigger. Often we use a single echo with short delay time, which is neither too short to cause combfiltering nor too long to be clearly perceived as a distinct echo. This way, we simply stretch the image of a monophonic sound across a wider area on the stereo stage and create an altogether bigger impression. Using this effect may result in loss of focus, but in specific cases it is still a highly suitable effect.

Let us not forget the Haas trick that was explained in Chapter 11 – a common way to enrich and widen sounds by panning a mono copy to one channel and a delayed duplicate to the other channel. The delay time is usually smaller than 30 ms and the effect can give some extra size if the delay time is modulated.

As a key tool in dance music

Delays are one of the most common tools in dance music production. They are used with virtually any combination of settings to enhance many aspects of the production and for a few good reasons. For one, dance music has a profound rhythmical backbone, and tempo-sync delays can easily enhance rhythmical elements – no other tool has such a strong timing link as delays. Then, sequenced dance productions call for little or no natural sound stage, so delays can easily replace the role of the more natural sounding reverbs. This is not to say that reverbs are not used, but the upbeat tempo of most dance productions creates a very dense arrangement where reverbs might not have enough space to deliver full effect. In a way, reverbs are more evident in mixes with slower tempos, whereas delays are more evident in mixes with fast tempos. Although we have already established that many recorded music mixes nowadays do not provide a natural sound stage, there are still some limits to how unnatural things can be made before the mix turns purely creative (say for example, delaying bass guitar or overheads). In dance music there are hardly any such limitations – delays can be applied on nearly every track, and it takes a while before things start to sound too weird.

22 Other modulation tools

Modulated delay lines are at the heart of a few other effects. Although neither phasers nor tremolos are based on delay lines, their operation is based on modulation. This chapter covers these tools.

Vibrato

Vibrato is the effect caused by rapid and subtle alteration of pitch. Such an effect can be part of a vocal performance, is an extremely common playing technique with string instruments, a known technique for guitars and can also be the byproduct of the Doppler effect as produced by a Leslie cabinet (which was originally called a 'Vibratone'). Vibrato can be achieved by modulating a delay line – without mixing the source signal. Subtle settings create similar effects to those that occur in nature. The evident pitch shifting caused by more aggressive settings will make the effect very obvious and unnatural (although this can still be used to creative effect in certain circumstances). The actual settings involve:

- **Dry/wet:** fully wet, dry signal not mixed.
- **Delay time:** between 3 and 8 ms.
- **Modulation depth:** around 25% (2 ms). This control is used for more or less effect.
- **Modulation rate:** around 3 Hz.
- **Feedback:** none.

Note that some delay time is still needed. If the delay time is set to 0, it is like having the record and playback heads at the same position. Although mechanically impossible, no matter how much the tape speed was modulated with such an arrangement, the pitch would never alter.

Track 22.1: Original Vocal

Track 22.2: Vibrato (Vocal, Light)
Settings: 100% wet, 8 ms delay time, 50% modulation depth, 1.5 Hz modulation rate, no feedback.

Track 22.3: Vibrato (Vocal, Deep)
Settings: 100% wet, 8 ms delay time, 75% modulation depth, 2.2 Hz modulation rate, no feedback.

Track 22.4: Original Guitar

Track 22.5: Vibrato (Guitar)
Settings: 100% wet, 8 ms delay time, 40% modulation depth, 1 Hz modulation rate, no feedback.

Track 22.6: Original Snare

Track 22.7: Vibrato (Snare)
Settings: 100% wet, 8 ms delay time, 65% modulation depth, 1 Hz modulation rate, no feedback. In this track, the slow vibrato applied on the snare results in subtle pitch changes. This can be used to make snare samples more realistic.

Plugin: PSP *84*
Drums: Toontrack *EZdrummer*

ADT

Not all singers are fans of the effort involved with real double-tracking. John Lennon was one of them. In search of an artificial alternative, Ken Townshend, then an engineer for The Beatles at Abbey Road Studios in London, came up with a solution. He connected two tape machines, so the delayed vocals from the second are rerecorded onto the first. The modulation on the second machine was achieved using a purpose-installed voltage-controlled oscillator. Having the original voice mixed with its own delayed-modulated version created an effect similar to double-tracking. The technique was termed *artificial double-tracking* (or *automatic double-tracking*) and was used liberally on Revolver and subsequent Beatles' albums. Townshend's actual arrangement also helped give birth to chorus and flanging effects.

Although the general consensus is that properly performed double-tracking is better than ADT, having the ability to create this effect artificially during mixdown is a handy one – it gives an impression of fullness, whether enriching vocals or other instruments. A few artists are associated with the double-tracking effect because they made generous use of it (the late Elliott Smith was one such exponent). But mostly, double-tracking is introduced during specific sections of a song to add some interest and breadth.

The execution of the ADT effect is similar to that of vibrato, except that we mix the dry signal as well:

- **Dry/wet:** possibly at the same level or with dry slightly louder than wet.
- **Delay time:** between 20 and 50 ms.
- **Modulation depth:** 40% as a starting point.
- **Modulation rate:** less than 1 Hz.
- **Feedback:** none.

> **Track 22.8: ADT (Vocal)**
> Settings: 50% wet, 20 ms delay time, 21% modulation depth, 0.775 Hz modulation rate, no feedback.
>
> *Plugin:* PSP *84*

Chorus

The actual idea of multiple singers singing in unison was established well before harmony was introduced into western compositions. Mostly, this was done solely for the purposes of loudness – even today the size of an orchestra is dependent on the size of the venue and the audience within it. But multiple unison performance is also known to produce an extremely appealing effect called *chorus*. When 20 cellos play the same note, as each player produces a slightly different frequency (within the cents range) and as each cello is either slightly closer to or further from the listener, the combined effect is mesmerizing. Synthesizers offer a similar feature called *unison*, where the same patch is layered with identical patches slightly detuned.

Perhaps of all the effects mentioned in this chapter, chorus is the most useful for mixing purposes. We might use it in the traditional sense of creating the impression of multiple performance with slightly different pitch and timing. It can, for example, enrich backing vocals and strings. It might also be used to add some richness and polish to plain sounds, like tame organ presets or synthesized strings. Also, it is used to simply make other instruments, like an acoustic guitar, slightly bigger and wider, and using more subtle settings, to improve fluidity, on a bass guitar for example. It can even mask slight pitch detunes in a vocal performance. The denser the mix the heavier the chorus would need to be to have an effect, but once the settings result in clear pitch alteration, the effect can lose its practical impact and sound gimmicky. In sparse mixes a touch of chorus will often suffice. One of the qualities of chorus (which can also be regarded as a disadvantage), is that it tends to soften sounds and send them slightly further away in the mix.

Chorus can be considered a form of ADT with slightly different settings. Chorus tends to sound richer with stereo delays, where slightly different delay times are set on each channel:

- **Dry/wet:** wet set between 20% and 70%, the more the wet signal the clearer the effect.
- **Delay time:** between 20 and 80 ms.
- **Modulation depth:** around 10%.
- **Modulation rate:** can be anywhere between 0.1 and 10 Hz.
- **Feedback:** none or very little.

Figure 22.1 The Universal Audio *DM-1* in chorus mode. This screenshot shows the settings of the Chorus Dimension preset. Essentially, very short delay on both channels, negative feedback of 25% and feedback filter set to 4 kHz; slow rate, little depth and only 19.7% of wet signal, which is phase inverted.

Track 22.9: Chorus 1 (Vocal)
Settings: 30% wet, 21 ms (L), 26 ms (R), 10% modulation depth, 0.5 Hz modulation rate, 20% feedback. The effect return is panned fully to the extremes.

Track 22.10: Chorus 2 (Vocal)
Settings: 50% wet, 40 ms (L), 35 ms (R), 11% modulation depth, 1 Hz modulation rate, −25% feedback. The effect return is panned fully to the extremes. This track has more noticeable chorus than the subtle chorus in Track 22.9. Note how compared to the original vocal (Track 22.1) the chorused version appears both deeper and wider.

- - - - - - - - - -

Track 22.11: Chorus 1 (Guitar)
Settings: Same as Chorus 1 (Vocal)

Track 22.12: Chorus 2 (Guitar)
Settings: Same as Chorus 2 (Vocal)

- - - - - - - - - -

Track 22.13: Original Distorted Guitar
Track 22.14: Chorus 1 (Distorted Guitar)
Settings: Same as Chorus 1 (Vocal)

- - - - - - - - - -

Track 22.15: Original Bass
Track 22.16: Chorus 3 (Bass)
Settings: Same as Chorus 1 (Vocal), with two exceptions: Wet is set to 10%, and the effect is panned 50% to each extreme.

Plugin: UAD *DM-1*

Flanging

Just like with guitars, flanging is mostly used as a distinct effect in the mix. Flanging is based on combfiltering created due to very short delay times. It yields a series of peaks and dips along the frequency spectrum, which are organized in harmonic series (see Figure 22.2). The actual character of the combfilter is determined by the delay time, the feedback phase switch and the amount of feedback. The modulation sweeps the comb up and down in frequencies, producing a whooshing sound. Modulation rates are usually kept low and might be synced to a half-note or a one-bar duration. If any waveform option is given, either a sine or a triangle would normally be used.

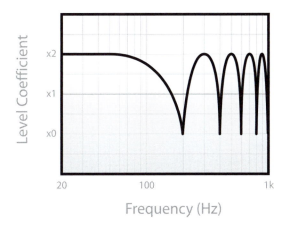

Figure 22.2 The combfilter response caused by a flanger. The frequencies of the peaks and dips are harmonically related. As the effect is modulated, this response pattern expands and contracts, but the harmonic relationship remains.

By way of analogy, the flanging effect is like a siren where the depth determines how low and high the siren goes and the rate determines how quickly it goes. The feedback can be thought of as a resonance controller, the higher the feedback the more the resonance. Phase-inverting the feedback will normally produce a bolder effect:

- **Dry/wet:** 50/50.
- **Delay time:** between 1 and 20 ms.
- **Modulation depth:** often higher than 25%.
- **Modulation rate:** slow, often tempo-synced to a half note or a bar.
- **Feedback:** the more feedback the more resonance.

Figure 22.3 The PSP *Lexicon 42* in deep flanger mode. This screenshot shows
the setting of the deep flanger preset – high feedback, 1 ms of delay, high depth, slow
rate and both the feedback signal and the wet signal are phase inverted.

The following tracks were produced using both the light and deep flanger presets of the PSP 42. The deep
flanger preset can be seen in Figure 22.3. The light flanger preset is only different from the deep one in that
its feedback is set to 0%.

Track 22.17: Flanger (Vocal, Light)

Track 22.18: Flanger (Distorted Guitar, Light)
Track 22.19: Flanger (Distorted Guitar, Deep)

- - - - - - - - - -

Track 22.20: Original Drums
Track 22.21: Flanger (Drums, Light)
Track 22.22: Flanger (Drums, Deep)

- - - - - - - - - -

Track 22.23: Original Shaker
Track 22.24: Flanger (Shaker, Light)

Plugin: PSP *Lexicon 42*
Drums: Toontrack *EZdrummer*

Phasing

Flanging and phasing are twin effects, sounding very similar, yet different. A true phasing
effect cannot be achieved using a delay. Instead, phasers employ a series of all-pass
filters. These filters, instead of altering the amplitude of specific frequencies, change their
phase. When the original and treated signals are combined, the result is a series of peaks
and dips along the frequency spectrum. But unlike flanging, the peaks and dips are not
organized in harmonic series (Figure 22.4).

Generally, two all-pass filters are required to create one dip. Often phasers use an even
number of filters, where fourth order would create two dips, sixth order would create
three and so forth. The higher the order the stronger the effect. A feedback loop might
also be included, which, like a flanger, would produce a more resonant, bolder effect.

Like flanging, phasing is an easily recognizable effect, although it is generally subtler. In
addition to electric guitars, it can be applied on various instruments just to create some
cyclic movement or be introduced on specific sections of the song just to add interest.

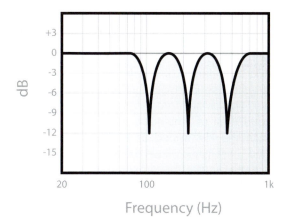

Figure 22.4 The response caused by a sixth order phaser. The peaks and dips are not harmonically related. As the effect is modulated this pattern would not expand or contract, it would simply shift lower and higher in frequencies.

Figure 22.5 The Logic *Phaser*.

The settings for all these tracks are identical: fourth order, 0.5 Hz modulation, sweep floor 40 Hz, sweep ceiling 10 kHz. It is interesting to compare these tracks to the flanger tracks and learn to recognize the differences between the two effects.

Track 22.25: Phaser (Vocal)
Track 22.26: Phaser (Distorted Guitar)
Track 22.27: Phaser (Drums)
Track 22.28: Phaser (Shaker)

Plugin: Logic *Phaser*

Tremolo

A tremolo produces cyclic modulation of amplitude, which makes instruments passing through it periodically fall and rise in level. This effect can be traced back to the Leslie cabinet (mentioned earlier), which also produces a vibrato. While vibrato is a modulation of pitch, tremolo is a modulation of amplitude. The Leslie cabinet (and similar products) is famous for its use with the Hammond organ. Tremolos were later introduced into many guitar amplifiers, and there are endless examples of songs with guitar tremolos. Till this day, tremolo is frequently applied on organs and electric guitars, but apart from its legacy, there is nothing to prevent it from being applied on any other instrument. Being a distinctive effect in its own right, tremolo can be used either subtly or more evidently to add a sense of motion to sounds.

Tremolo controls are mostly modulation-related, we often get:

- **Rate:** the modulation rate. Can be expressed in Hz, or if it can be tempo-synced it might be given as note value.
- **Depth:** the intensity of attenuation. Expressed in percentages, where 0% means no effect, 100% means the signal level will fall down to complete attenuation.

Some tremolos provide a choice of modulation waveform (sine, sawtooth, square). Also some stereo tremolos employ a separate tremolo for each channel, where the modulation phase between the two can be altered. For example, 180° on a sine modulator would mean that as one channel is at the highest amplitude the other is at the lowest. This can be used to create interesting panning effects.

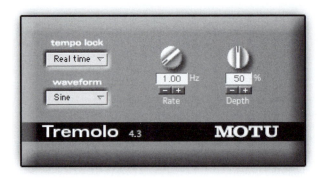

Figure 22.6 The MOTU *Tremolo plugin.*

Track 22.29: Tremolo 1 (Distorted Guitar)
For the most obvious effect, the tremolo depth in this track is set to 100%; the rate was set to 2 Hz.

Track 22.30: Tremolo 1 (Distorted Guitar)
The more musical settings in this track are the outcome of 30% depth and 5 Hz rate.

Track 22.31: Tremolo (Vocal)
40% of depth, 5 Hz rate, applied on vocal.

Plugin: UAD *Tremolo*

23 Reverbs

With the exception of pure orchestral recordings, reverbs are used in nearly every mix. The common practice of close-miking and the dry nature of some sounds produced by synthesizers and samplers result in initial mixes that lack both ambiance and depth (such mixes are often described as '2D mixes'). Reverbs allow us to add and build up these missing elements during mixdown, but they can also be beneficial for many other mixing tasks.

What is reverb?

In nature, reverb is observed mostly within enclosed spaces, such as rooms. Reverbs are easier to understand if we imagine an impulse sound, like a hand clap, emitted from a sound source in an empty room. Such a sound will propagate in a spherical fashion and for simplicity we should regard it as traveling in all directions. The emitted sound will travel in a direct path to a listener (or a microphone) followed by reflections that bounced from the walls, floor and ceiling. These will be gradually followed by denser reflections that have bounced many times from many surfaces. As sound both diminishes when traveling through air and being absorbed by surface materials, the reflections will slowly decay in amplitude. Reverb is the collective name given to the sound created by bounced reflections from room boundaries (which we consider to be the main reverb contributors, although in a room there might be many surfaces). In mixing, we use reverb emulators, either hardware units or software plugins, to simulate this natural phenomenon.

Applications

Simulating natural or creating imaginary ambiance

Due to the inflexible nature of ambiance recordings and the poor reverb characteristics of many live rooms and home studios, many engineers choose to record most instruments dry and apply artificial ambiance during mixdown. While mixing, reverb emulators give us more options and control over the final ambiance of the mix.

Crafting the ambiance of a production is considered one of the more important and challenging practices in mixing – ambiance transforms a collection of dry recordings into an inspiring spatial arrangement that gives our mix much of its character. The reverb controls let us mold the ambiance and make creative decisions that can make our mix more cohesive and vigorous.

> *Many regard the creative process of crafting the ambiance of the mix as a vital and most exciting process.*

Here again, we have a choice between a natural or unnatural (or imaginary) outcome. A natural ambiance is expected in some mixes and for some genres; jazz, bossa nova or productions that include an orchestral performance for example. For such projects we might choose to utilize natural reverb simulations so the listener can conceive the performance as if being played in a very familiar type of space. Realism is an important factor in natural simulations – not only is the reverb choice important, but also the way we configure it to support the depth of individual instruments and the different properties of the simulated space.

While we can imagine the typical ambiance of a jazz performance, what ambiance should we employ for a trance production? How about electro, hip-hop, trip-hop, techno or drum 'n' bass? There is a loose link between sequenced productions and acoustic spaces. For such projects, the ambiance is realized based on the creativity and vision of the mixing engineer. Furthermore, sometimes an imaginary ambiance is applied even when a natural one is expected. This is done in order to achieve something refreshing, different and contemporary. Such implementation is evident in some recent jazz albums that, in addition to unnatural ambiance, utilize many modern recording and mixing techniques.

An important point to remember is that while selecting an ambiance reverb we might ask ourselves **'How attractive is this reverb?'** and **'How well does it fit into the mix?'** rather than dealing with the question of **'How natural?'** Such thinking promotes less natural reverbs and we have already discussed how a non-natural mix can bring about a fantastic result – an imaginary space can excite us more than the familiar natural ones.

> *While a natural choice might be suitable for some productions, an imaginary ambiance can sometimes be more rousing and effective.*

Gel the instruments in the mix

After combining the various tracks, each instrument in the mix might have a distinctive appearance, but heard together they might feel a little foreign to one another – there would be nothing to link them. A reverb, even one that can hardly be perceived, can gel the various instruments, making them all appear as a natural part of the mix. Similar to ambiance reverbs, one emulator might have many tracks sent to it but in contrast to ambiance reverbs, the gelling reverb is more subliminal, it might not be audible or have a clear spatial sense and will be felt rather than heard.

Increase the distinction of instruments

Although we have just advocated many instruments sent to the same reverb, different reverbs on different instruments can sometimes be beneficial. For example, having the guitars feeding a specific reverb and the vocals feeding another, despite creating a less natural space, would increase the distinction between the two instruments, resulting in increased separation.

Depth

In nature, many factors contribute to our perception of depth. The natural reverb in a room is a very important one. During mixdown, reverbs are the main tool we use when positioning sounds in the depth field. This is extremely desirable in many situations – it adds another dimension to the mix, lets us position instruments more naturally, establishes hierarchy and can also help with masking. But the bond between depth and reverbs does not come without a price – the perceived depth caused by the addition of reverb is sometimes unwanted, for example, when adding reverb to vocals while still wanting to keep them in the front of the mix.

> *Although sometimes unwanted, reverb addition often increases the perceived source-to-listener distance and therefore reverbs are regularly utilized for applying depth.*

Enhance a mood

Different productions have different moods – happy, mellow, dark and aggressive are just a few examples. Reverbs are often associated with tenderness, romance, mystery, intimacy and many other sensations and atmospheres. We can enhance, and even establish, a production mood by using reverbs creatively.

Some genres, like chillout and ambient, embrace reverbs as a production means. But while reverbs are usually more evident in slow productions, it is perhaps the spatial disorientation caused by certain drugs that fosters reverb use in some upbeat dance genres (and psychedelic music in general).

Livening sounds

Those who have heard a recording that took place in an anechoic chamber or those who attended an unamplified performance in an open field know how unsatisfying such a reverb-free performance can be. Reverb is such a native part of our life as it accompanies most of the sounds we are exposed to. A reverb-free recording of specific instruments, as good as it may be, can sound cold and unnatural. By adding reverb we take a dry artificial recording and breathe 'life' into it – adding the reverb we are so familiar with in our daily life.

However, it is not only the existence of a reverb that matters – it is also the amount of reverb. Ask yourself, do you like your singing better when performing in your bedroom or in your bathroom? Most people will choose the latter since mostly it has a louder and longer reverb. Reverbs can make sounds more impressive and more present; when we say 'livening' we also mean taking a stodgy performance and turning it into a captivating one.

Another issue that comes to mind here is the use of mono samples in sequenced production. For instance, a mono ride sample can sound odd compared to the familiar sound of a ride captured by stereo overheads. Adding reverb to mono samples can produce a more natural and less mechanical feel to sequenced instruments.

Filling time gaps

Empty time gaps can break the balance and integrity of our mix. Reverbs can make mix elements longer and fill empty time gaps, an application more common in slow, sparse productions.

Filling the stereo panorama

Consider a slow production consisting of an acoustic guitar and vocal recordings only. How can one craft a full stereo panorama with just two instruments? Stereo reverbs can be utilized in mixes to create an imaginary sound curtain behind the performers. Often reverbs of such type span the stereo panorama. But we can also use reverbs to fill gaps in the stereo panorama by simply padding specific instruments with a stereo reverb, making their stereo image wider.

Track 23.1: Guitar Dry
The source guitar track, dry, with its focused and thin image at the center of the panorama.

Track 23.2: Guitar with Reverb
The added reverb fills the stereo panorama. Also, notice that the guitar appears further back although the level of the dry signal has not changed.

Plugin: Universal Audio *DreamVerb*

Changing the timbre of instruments

In many situations, added reverbs modify the timbre of the original sound, whether it be our desire or not. A loud reverb on a percussive instrument can smoothen its attack, slow its decay or create a completely different decay envelope. The friction between a violin string and a bow can produce a harsh noise that can be blurred and softened by a reverb. On the contrary, it is exactly that friction noise which can give a double bass added expression in the mix. A reverb can be used to constructively alter timbre, but it can also deform it (a possibility we must be aware of). In most situations, the right reverb choice along with appropriate control adjustment will render the desired result.

> *When adding reverb we listen closely to timbre changes and, when needed, tweak the reverb parameters.*

Reconstruct decays and natural ambiance

Reverbs can also be used to replace the natural decay of an instrument. Such an application might be required if a tom drum recording contains exaggerated spill. Toms are commonly compressed in order to give them more punch and bolder decay, but doing so will also amplify the spill, which can cause some level fluctuations in the mix. To solve this we gate the toms so the decay is shortened, then feed the gated tom into a reverb unit, which produces an artificial decay. We have much more control over artificial decays and we can, for example, make them longer and bigger, so as to make the toms prominent.

A similar application is the reconstruction of room ambiance. A classic example is the ambiance captured by a pair of overheads. It can be very tricky to control such ambiance and fine-tune it into the mix, mainly with regards to frequencies and size. It can become even more problematic after these overheads tracks are compressed, since this will emphasize the ambiance even more. The low and mid frequencies, at which most of the presence and size of the ambiance are present, can be filtered and reconstructed using a reverb.

Track 23.3: Drums Before Reconstruction
These are the original drums, unprocessed.

Track 23.4: Drums After Filtering
Only the room and the overheads are filtered in this track. A 12 dB/oct HPF at 270 Hz removes much of the ambiance. The drums now sound thinner and their image appears to be closer.

Track 23.5: Drums After Reconstruction
The close-mics are sent to a reverb emulator, which is preceded by a LPF. The result is ambiance different in character from that in Track 23.3.

Plugin: Audio Ease *Altiverb*
Drums: Toontrack *EZdrummer*

Resolve masking

Although a potential masker in its own right, a reverb can be used to resolve masking. The common usage of reverbs for the purpose of depth gives us the ability to position sounds in a front–back fashion in addition to the conventional left–right panning. This gives another dimension and much more space in which we can place our instruments.

A stereo reverb, which spreads across the sound stage, has more chance of escaping masking, and its length means that it will have more time to do so. Say we have two instruments playing the same note and one completely masks the other. By adding reverb to the masked instrument, we make it longer and are able to identify it more because of its reverb.

More realistic stereo localisation

Sounds that are not accompanied by reflections are very rare in nature. Panning a mono source to a discrete location in the stereo panorama can therefore produce an artificial result. Reverbs can be added to mono signals in order to produce more convincing stereo localization. The reverb will also extend the stereo width of the mono signal, which can make it clearer and more defined. The reverbs that are used for such tasks must be stereophonic, rely on a faithful early reflection pattern and are usually very short.

A distinctive effect

Sometimes reverbs are added just to spice things up and add some interest. Snare reverb, reversed reverb, gated reverb or many other reverbs are simply added based on our creative vision and not necessarily based on a practical mixing need. Reverbs can be automated between sections of the production in order to create some movement or be introduced occasionally to mark transitions between different sections.

Impair a mix

Although not exactly a constructive application, it is worth summarizing the common problems that reverbs might cause or otherwise what we should watch out for:

- **Definition** – reverbs tend to smear sounds, make them unfocused, distant, and can decrease both intelligibility and localization. A proper configuration of a reverb emulator can usually prevent these problems.
- **Masking** – usually being a long, dense and wide sound, reverbs can mask other important sounds. As with any case of masking, mute comparison can reveal what and how reverbs are masking.
- **Clutter** – if used unwisely (which usually stands for a reverb too loud and too long), reverbs can clutter the mix and give a 'spongy' impression. Part of the challenge is making reverbs effective yet not over-dominant.
- **Timing** – as a time-based effect, a reverb can affect the timing of a performance, especially a percussive one.
- **Change timbre** – discussed previously.

Types

Stereo pair in a room

The earliest way to incorporate reverbs into a mix (long before the invention of reverb emulators) was by capturing the natural reverb using a set of microphones in a room. Producers used to position musicians in a room in relation to how far away and how loud they should appear in the mix. Back then and even today, different studios are chosen based on their natural reverb characteristics. In some studios, special installations, like moving acoustic panels, are made to enable a certain degree of control over the reverb.

In most situations where a small sound source is recorded, say less than ten musicians, the ambiance and depth of the room is captured by two microphones positioned using an established stereo-miking technique. As already discussed, such a stereo recording results in what can be described as a 'submix' which can be altered only slightly during mixdown. However, the celebrated advantage of this method is that it captures the highest extent of complexity that a natural reverb offers. No reverb emulator has the processing means to produce such a faithful, accurate and natural simulation, especially not in realtime. As acoustic instruments radiate different frequencies in different directions, a microphone in close proximity can only capture parts of the timbre. But all radiated frequencies reflect and combine into a rich and faithful reverb, which is then captured by the stereo pair.

> *The reverb and depth captured by a pair of microphones in a room are the most natural and accurate ones, but they can be seriously limiting if they do not fit into the mix.*

Reverb chambers (echo chambers)

A reverb chamber is an enclosed space in which a number of microphones are placed in order to capture the sound emitted from a number of speakers and the reverb caused by surface reflections. Either purposely built or improvised, these spaces can be of any size and shape. Even a staircase has been utilized for such a purpose. A send from the control room feeds the speakers in the reverb chamber and the microphone signals serve as an effect return. Whilst reverb chambers are generally not an accessible mixing tool for most engineers, every studio can be used as a reverb chamber – all studios (live rooms) have microphone lines and many have loudspeakers installed with a permanent feed from the control room.

The advantage of using reverb chambers is that the captured reverb is, again, of a very high quality being produced by a real room. As opposed to a stereo recording into which most mixing elements are immovable, reverb chambers are used during mixdown and provide more control over the final sound. This is achieved either by different amounts of send and return or by physical modifications applied in the chamber itself, for example, changing the distance between the microphones and the speakers or adding absorbent (or reflective) materials. However, such alterations result in a limited degree of control and most of the reverb characteristics will only vary by insignificant amounts, most notably the perceived room size. We consider each reverb chamber as having a very distinctive sound that can only be 'flavored' by alterations. Moreover, the loudspeakers used in a reverb chamber differ substantially from a real instrument in the way they produce and radiate sound, so reverb chambers are just a simple alternative to placing a real instrument in a real room. Another issue with reverb chambers is their size – a general rule in acoustics is that larger rooms have better acoustic characteristics than smaller ones. A small reverb chamber can produce a reverb with broken frequency response and reflections that can color the sound. A large reverb chamber is expensive to build and only big studios can afford it.

Track 23.6: Vocal Chamber Convolution
A chamber reverb simulation produced by a convolution emulator.

Plugin: Trillium Lane Labs *TL Space*

Spring reverb

Since reverb is essentially a collection of delayed sound clones, it is no wonder that one of the first artificial reverb to be invented had its origins in a spring-based delay device. The original device was conceived by Bell Labs researchers who tried to simulate the delays occurring over long telephone lines. The development of the spring reverb, starting as early as 1939, is credited to engineers from Hammond who tried to put life into the dry sound of the organ. During the early 1960s, Leo Fender added Hammond's spring reverb to his guitar combo and was followed by manufacturers such as Marshall and Peavey.

A spring reverb is an electromechanical device that uses a system of transducers and steel springs to simulate reflections. The principle of operation is simple: an input transducer vibrates with respect to the input signal. Attached to the transducer is a coiled spring on which the vibrations are transmitted. When vibrations hit the output transducer on the other end of the spring, they are converted back into output signal. In addition, parts of the vibrations bounce back onto the spring, then bounce back and forth between the ends. Such reflections are identical to those bouncing between 'surfaces in a one-dimensional space' (technically speaking, a line, with its boundaries being its end points).

Figure 23.1 The Accutronics Spring Reverb (courtesy of Sound Enhancement Products, Inc.).

In reality, the science of a spring reverb is more complex than explained above, but these units are relatively cheap to manufacture and still shipped today with many guitar amplifiers. Although spring reverbs do a pretty bad job in simulating a natural reverb, listeners have grown to like their sound. Furthermore, many digital reverb emulators fail to match the sound of a true electromechanical spring reverb, which explains why standalone rack-units are still in production. Such units will be fitted with very few controls and in most cases will have fixed decay and pre-delay times. (It is worth noting that pre-delay could always be achieved if a delay was connected before the reverb; however, it was not until the early 1970 s that digital delays started replacing tape delays for such purposes.) Quiet operation and flat frequency response were never assets of true electromechanical spring reverbs, and they are known to produce an unpleasant sound when fed with transients. Therefore, spring reverbs are usually applied on the likes of pads and vocals, normally while only mixing a small amount of the wet signal.

> *While spring reverbs are far from spectacular in simulating a natural reverb, the sound created by a true electromechanical unit can serve as an identifiable retro effect, especially if applied on guitars and other non-percussive sounds.*

Track 23.7: Vocal Spring Convolution
A spring reverb simulation produced by a convolution emulator.

Track 23.8: Snare Spring Convolution
The snare on this track is sent to the same emulator and preset as in the previous track.

Plugin: Trillium Lane Labs *TL Space*

Plate reverb

If a spring reverb is considered a 'one-dimensional' reverb simulator, the logical enhancement to such a primitive design was going two-dimensional. The operation of a plate reverb is very similar to that of a spring reverb, only that the actual vibrations are transmitted over a thin metal plate suspended in a wooden box. An input transducer excites the plate and output transducers are used to pick up the vibrations. In the case of plates, these vibrations propagate on its two-dimensional surface and bounce from its edges.

In 1957, a pioneering German company called EMT invented and built the first plate reverb – the EMT 140. This model, which is still a great favorite, had the impressive dimensions of 8 ft × 4 ft (2.4 m × 1.3 m) and weighed around 600 pounds (more than quarter of a ton). These original units were very expensive but still much cheaper than building a reverb chamber. They can be heard on countless records from the 1960s and 1970s and on many other productions that try to reproduce the sound of these decades.

Whilst mobility was not one of its main features, it had better sonic qualities than the spring reverb. In addition to a damping mechanism that enabled control over the decay time, its frequency response was more musical. The reverb it produced, although still not resembling a true natural reverb and being slightly metallic, blended well with virtually every instrument, especially with vocals. The bright, dense and smooth character of plate reverbs also made them a likely choice for drums, which explains why they are so frequently added to snares.

> *Although being a serious artificial reverb simulator, a plate reverb does not produce a truly natural-sounding reverb. Hence it is used more as an effect to complement instruments like snares and vocals.*

Track 23.9: Vocal Plate Convolution
A plate reverb simulation of the EMT 140 produced by a convolution emulator.

Track 23.10: Snare Plate Convolution
The same emulator and preset as in the previous track, only this time being fed with a snare.

Plugin: Trillium Lane Labs *TL Space*

Digital emulators

So far we have discussed natural and mechanical reverbs. The invention of digital reverberation is credited to Manfred Schroeder, then a researcher at Bell Laboratories, who back in 1961 demonstrated a simple digital reverberation system (Figure 23.2a). It took a while to turn his digital reverberation ideas into a tangible commercial machine, but with the performance rise of DSP chips and the fall in their price, digital emulators were destined to take over. It was EMT again, with help from an American company called Dynatron, that in 1976 revealed the world's first commercial reverb unit – the EMT 250. It had very basic controls such as pre-delay and decay time and could also produce effects like delay and chorus.

(a)

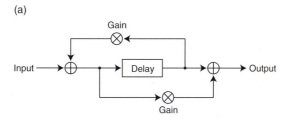

(b)

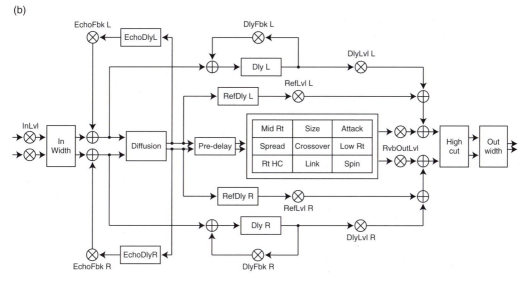

Figure 23.2 Block diagram of a reverb. (a) The original Schroeder reverb design (1961). Later digital designs are far more complex as can be seen in (b) the plate program of the Lexicon *PCM91* (1998).

Track 23.11: Vocal EMT 250 Convolution
A plate reverb simulation of the EMT 250 produced by a convolution emulator. Notice that compared to Track 23.9 this reverb has better spatial characteristics.

Plugin: Trillium Lane Labs *TL Space*

A digital reverb is not a dense collection of simple delays, since these would cause much coloration to the direct sound and produce a broken frequency response. There are many internal designs for digital reverbs and they vary in their building blocks and in the way these blocks are connected. At the lowest level, digital emulators are implemented using mathematical functions that are executed on internal DSP chips. To differentiate from digital convolution reverbs (discussed later), these type of reverbs are now referred to as *algorithmic reverbs*.

No digital emulator will ever be able to produce a reverb completely identical to the one created in a real room. This is mainly due to the complexity of such a reverb – there are thousands of reflections to account for, different frequencies propagate and diffract in a different fashion, different materials diffuse and absorb sound in a different way, even the

Figure 23.3 An algorithmic reverb plugin – the Eventide *Reverb* TDM.

air itself changes the sound as it travels through it. Manufactures must take shortcuts, and the more processing power is at their disposal the fewer shortcuts they have to take, thus the more realistic the reverb is likely to be. It is worth noting that although DSP chips were far less powerful in the 1970s and 1980s, many units from these decades are still regarded today as state-of-the-art, which shows just how important the design of such units is. Since each respectable manufacturer has their secret architecture, each has a distinct sound.

Back in the 1990s, when realtime plugins emerged, CPUs had less than a tenth of the processing power compared to modern processors and could only handle a few plugins at a time. Today, more and more people rely solely on their computer CPU for all mixing tasks. Yet, a high-quality reverb plugin can consume tenths of the CPU processing capacity, so many still find the need for additional processing power – either in the form of external hardware units or as internal DSP expansion cards.

As algorithmic reverbs have no physical or mechanical limitations, they provide a multitude of controls that let us tweak nearly every property of the reverb. This makes them an extremely flexible and versatile tool in many mixing scenarios, and it is no wonder that they are the most common reverbs in our mixing arsenal. An issue with high-quality emulators is that they are expensive both in terms of price and processing-power consumption.

> *Digital reverbs are the most common type in use and give us great control over reverb parameters, but they can consume large amounts of processing power depending on their quality.*

Figure 23.4 The 'larc' (lexicon alphanumeric remote control) for the Lexicon *480L*; the reverb unit itself is rack-mounted and often unseen in studios. Despite being released in 1986, the 480L is still considered by many as one of the best reverb emulators (courtesy of SAE Institute, London).

Convolution (sampling) reverbs

Even during the 1970s various people toyed with the possibility of capturing the reverb characteristics of an acoustic space, so it could be applied later to any kind of recording. What might sound like a wild dream has become a reality with the new millennium, when DSP technology became fast enough to accommodate the realtime amount of calculations required for such a process.

Reverb sampling is normally done by placing a pair of microphones in a room, then recording the room response (i.e., the reverb) to a very short impulse, like that of a starter pistol. Since it is very hard to generate a perfect impulse, an alternative method involves playing a sine sweep through speakers instead. The original sound might be removed from the recording, leaving the reverb only. The recorded impulse response (IR) is then loaded into a convolution reverb, which analyzes it and constructs a huge mathematical matrix that is later used by the convolution formula. With every sound fed through the unit, a reverb very similar to that of the original space is produced.

An emulator can be based on one of two types of convolution, either one that is done in the time domain (pure convolution) or one that is based on the frequency domain (convolution or Fourier based) – each generates the same result, only in some situations one will be faster than the other. If pure convolution is used, an impulse response of 6 seconds at 44.1 kHz would require around 23 billion mathematical operations per second – an equivalent to the processing power offered by a 2.2 GHz processor. It can be easily seen how such a process might be unwieldy in some situations. As a general rule, the shorter the original impulse response is, the less the processing needed would be.

Convolution reverbs experience increased popularity nowadays. They let us incorporate into our mix the reverb of many exotic venues and spaces. For the film industry this is a truly revolutionary tool – engineers can record the reverb of any location and then apply it during post-production. For mixing, however, it is doubtful how much the reverb characteristics of the Taj Mahal can contribute to a modern rock production. Yet, the impulse recordings shipped with emulators include less exotic spaces that can be used in every mix. Many impulse responses can also be downloaded from the Internet, many are free. The quality of the impulse recording is determined by the quality of the equipment used, which is a vital factor if natural results are sought. It is generally agreed that a good impulse recording produces an extremely believable reverb simulation that matches (if not exceeds) the quality of the best algorithmic emulators.

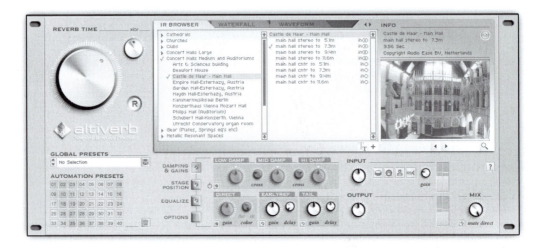

Figure 23.5 A convolution reverb plugin. The Audio Ease *Altiverb*.

Convolution reverbs are not only used to reproduce the reverb of acoustic spaces. Impulse responses of virtually any vintage unit like the EMT 140 or 250 are also available. Fascinating results can also be achieved if instead of loading a real impulse response, one loads just a normal short recording.

One problem with convolution reverbs is that all the properties of the reverb, as captured during the recording process, are imprinted into the impulse recording and can be altered very little later on without some quality penalty. An example of this is the inability to change the distance between the sound source and the listener or the different settings used while a hardware unit was sampled. Many convolution reverbs include only limited amounts of controls that make use of simple envelopes and filters in order to give some degree of control over the reverb characteristics. But for more natural results, many reverbs ship with a variety of impulse recordings of the same space, each based on a different recording setup. Generally speaking, convolution reverbs produce the best simulations provided the original impulse response is unaltered by the artificial envelope and filter controls. This fact makes convolution reverbs, to some extent, a hit-and-miss affair – if the pure impulse response is right, it should sound great; if it isn't, additional tweaking might make it more suitable for the mix, but not as great. Like with algorithmic reverb, the initial preset choice is crucial.

> *Good convolution reverbs can produce exceptionally natural results,*
> *but they consume processing power and are normally inflexible.*

Perhaps the irony of convolution reverbs is that now after having them at our disposal, many find the old algorithmic reverbs far more interesting and suitable for both creative and practical tasks. Despite the fact that convolution reverbs are considered superior in simulating natural spaces, algorithmic reverbs are still favored for many other mixing applications.

The following tracks demonstrate various IR presets from the Logic *Space Designer* convolution reverb. It is for readers to judge how realistic and appealing each of these reverbs is. It is perhaps most interesting to observe how. Despite hardly adding a noticeable reverb, the Wine Cellar track has extra sense of space and depth compared to the dry track:

Track 23.12: Conv Source
The dry source track used in the following samples.

Track 23.13: Conv Wine Cellar
Track 23.14: Conv Wooden Room
Track 23.15: Conv Drum Chamber Less
Track 23.16: Conv Wet Bathroom
Track 23.17: Conv Small Cave
Track 23.18: Conv Music Club
Track 23.19: Conv Concert Hall
Track 23.20: Conv Canyon
Track 23.21: Conv Roman Cathedral

Here are two sets of samples demonstrating the use of just a normal sample as an IR. The results are unnatural, but can be used in a creative context:

Track 23.22: Conga IR
Track 23.23: Conv Conga

Track 23.24: FM Noise IR
Track 23.25: Conv FM Noise

Plugin: Logic *Space Designer*

Reverb programs

Digital emulators can simulate a number of spaces, each will exist as a loadable program (preset). As there can be hundreds of programs, these are organized in categories.

A high-quality reverb emulator might implement a different algorithm for each category. The main categories are:

Category	Description	Application
Halls	Large, natural sounding, live spaces	Natural
Chambers	Simulate reverb chambers or spaces that have slightly less natural reverb behavior and a less defined size	Natural
Rooms	Normal rooms of different sizes	Natural
Ambiance	Concerned more with placing the sound naturally in a virtual space, caring less about the actual reverberation. Most often an ambiance preset involves early reflections only	Natural
Plate	Plate reverbs	Effect

The following categories are also common:

Category	Description	Application
Studio	Simulate the reverb in a recording studio live room	Natural
Church/ cathedral	These types of spaces might produce a highly impressive reverb for certain types of instruments, like organs, but they generally result in poor intelligibility	Natural
Spring	Spring reverbs	Effect
Gated	Nonlinear reverb (explained later)	Effect
Reversed	Rising reverb instead of decaying one (explained later)	Effect

Some mixes are expected to present a more natural space than others. If one seeks a truly natural space in the mix, hall or room presets can produce good results. If less natural reverb is required, either chamber or ambiance might be used. Most ambiance programs are essentially the shortest and most transparent of all reverb programs. They excel at blending mono instruments into a spatial arrangement, without adding a noticeable reverb.

The choice of category is usually determined by two main factors. First, the size of the room should co-exist with the type of music. A classical recording or a chill-out production might use a large space such as a hall; a bossanova or jazz production can benefit from a moderate-size hall, while a heavy metal or trance productions might make use of very small space simulations. Second, the decay time (or length) of the reverb should be relative to the mix density – a dense mix will suffer from long reverb tails that will cause masking and possibly clutter; in a sparse mix, longer decays can fill empty spaces.

For a truly natural space, hall and room programs can work better than chambers and ambiance. The room size and the decay time are the two main factors when selecting a reverb program.

Within each category various programs reside. For example, halls category might include small hall, large hall, vocal hall and drum hall. As opposed to EQs and compressors, it is uncommon for a mixing engineer to program a reverb from scratch – in most cases a preset selection is made and then altered if needed. Initially, choosing the right program for the mix is based on experiment – many mixing engineers will focus on one or two categories and will try all the different programs within them; if different units are available each might be tested as well. It is time worth spent, especially when selecting one of the main reverbs in the mix. A reverb that sounds good when soloed might not interact well with the rest of the mix – final reverb selection is better done in mix context. With experiment comes experience, and after a while we learn which emulator and which program works best for a specific task. The factory settings of each program are optimized by the manufacture based on studies and experts' opinions, therefore drastic changes to these settings can theoretically degrade the quality of the simulation. This makes the initial selection of an appropriate program even more important.

The selection of an appropriate reverb program is highly important and is a process worth spending time on.

When a reverb is used as an effect, an artistic judgment is required since there are no golden rules. Many use plate programs on vocals and drums; snares are commonly treated with a plate reverb or a gated one. For more impressive effect, chambers or halls can work. Halls can also be suitable for orchestral instruments such as strings, flutes, brass and the less orchestral instrument, the saxophone. Bass guitars are usually kept dry, and distorted electric guitars can benefit from a subtle amount of small-room reverb that will add a touch of shine and space. As the association between synthesized sounds and natural acoustic spaces is loose, chamber and ambiance programs might be more suitable in a sequenced production.

The above should merely serve as guidelines – the choice of a reverb program is truly subject to experiment and varies for each individual mix.

The following tracks demonstrate various reverb programs. Reverb programs can vary noticeably between one emulator and another, so readers are advised to treat the following tracks as a rough sampler.

Track 23.26: Vocal Dry
Track 23.27: Drums Dry
The source tracks used in the following samples.

Track 23.28: Program Hall (Vocal)
Track 23.29: Program Hall (Drums)
Track 23.30: Program Chamber (Vocal)
Track 23.31: Program Chamber (Drums)
Track 23.32: Program Room (Vocal)

Track 23.33: Program Room (Drums)

Track 23.34: Program Ambiance (Vocal)
Track 23.35: Program Ambiance (Drums)
Track 23.36: Program Plate (Vocal)
Track 23.37: Program Plate (Drums)
Track 23.38: Program Studio (Vocal)
Track 23.39: Program Studio (Drums)
Track 23.40: Program Church (Vocal)
Track 23.41: Program Church (Drums)
Track 23.42: Program Cathedral (Vocal)
Track 23.43: Program Cathedral (Drums)
Track 23.44: Program Spring (Vocal)
Track 23.45: Program Spring (Drums)

Plugin: Audio Ease *Altiverb*

Reverb properties and parameters

The parameters found on most digital reverb emulators are tightly related to the properties of the natural reverb produced in acoustic spaces. The understanding of these parameters is important not only for more natural reverb simulations, but also for incorporating reverbs more effectively and musically into mixes.

Reverb emulators vary in their internal design, and some are intended for specific applications more than others. The controls offered by reverb emulators can vary substantially between one emulator and another, and it is impractical to discuss all the possible controls offered by each individual emulator. Moreover, identical control on two different emulators can be implemented in a different way and can yield different results. Reading the manual will reveal the purpose of each control, and some manuals include useful tips on optimal utilization of the emulator or reverbs in general. This section covers the most important parameters that are typical with many emulators.

Direct sound

Direct sound is not part of a reverb. It is the sound that travels the shortest distance between the sound source and the listener, and in most cases it does so while traveling

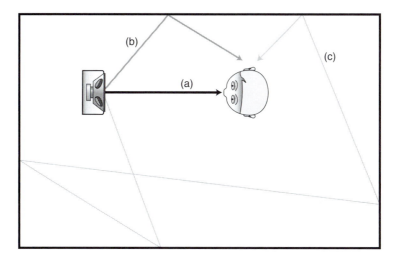

Figure 23.6 Direct and reflected sounds. (a) The direct sound reaches our ears without encountering any boundaries. (b, c) Reflected sound. Note that the path of reflection (c) is valid considering the propagation characteristics of low-frequencies.

in a direct line between the two (path a in Figure 23.6). It is the first instance of the sound arriving at the listener and hence provides an important psychoacoustic cue. As discussed in previous chapters, both the level and the high-frequency content of the direct sound contribute to our perception of depth.

The direct sound is the dry signal that we feed into a reverb emulator so it can produce a simulation of a reverb. A reverb emulator might have a parameter to determine whether this dry signal is mixed with the generated reverb (dry/wet mix). If the reverb is connected via an aux send, the original dry signal is mixed anyway, and the copy sent to the emulator would become redundant once mixed with the reverb. This is not the case when a reverb is connected as an insert, where the dry signal would only be heard if blended with the reverb at the emulator output.

> *If reverbs are connected via an aux send, make sure to toggle off the dry signal on the emulator. However, toggle it on if reverbs are connected as an insert.*

Pre-delay

Pre-delay is the time difference between the arrival of the direct sound and that of the very first reflection. However, in a few reverb emulators this parameter stands for the time gap between the dry signal and the later reverberation.

Pre-delay gives us a certain clue regarding the **size** of the room, where in larger rooms the pre-delay is longer as it takes more time for reflections to travel to the boundaries and back to the listener. Pre-delay also gives us a certain clue regarding the **distance** between the source and the listener, but here an opposite effect takes place to what might initially seem: the closer the source to the listener the longer the pre-delay. This is due to the fact that the *relative* distance between the direct and the reflected sounds is getting

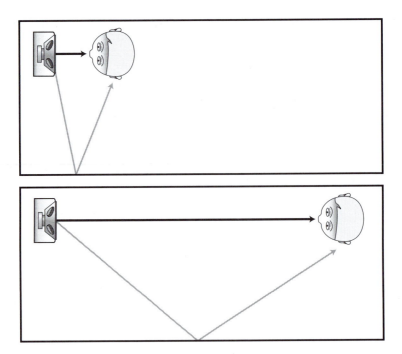

Figure 23.7 Distance and pre-delay. When the listener is closer to the sound source the relative travel distance between the direct sound and the first reflection is bigger, which results in a longer pre-delay.

smaller the further away the source is from the listener (Figure 23.7). This phenomenon is commonly put into practice when a reverb is required but not the depth that comes with it – we simply lengthen the pre-delay. Pre-delay is normally expressed in milliseconds, and for natural results our brain requires that it is kept below 50 ms. However, longer pre-delay times are still used, for example, when trying to keep things at the front of the mix.

> *Long pre-delay time can help keep an instrument at the front of the mix.*

The pre-delay time also determines when reflections start to mix with the direct sound. Reflections caused by a real room are far more complex than those produced by an emulator with its limited building blocks. Needless to say, high-quality emulators do a better job in that respect, but we still say that digital reflections are more likely to color the original sound than those created by a real room. Thus, the reflections generated by a digital emulator might cause combfiltering and other side effects when mixed with the original signal. The sooner they are mixed, the more profound this effect will be. It is worth remembering that even the sound of a snare hit can easily last for 80 ms before entering its decay stage. If the early reflections are mixed with the direct sound within the very first milliseconds (like they mostly do in nature) the timbre might distort. We lengthen the pre-delay time in order to minimize this effect. On the same basis, early reflections can be masked by the original signal. Since the early reflections give us most of the information regarding the properties of the space, lengthening the pre-delay can nudge these reflections outside the masking zone and result in clearer perception of important psychoacoustic cues.

Another issue related to short pre-delay settings is intelligibility – if the reverb is mixed with direct sound very early, it might blur and harm both clarity and definition. This is extremely relevant when adding reverb to vocals – if the words cannot be understood, we might just as well call it a day and look for a job in a scuba club.

A very long pre-delay can cause an audible gap between the original and the reverb sounds. This will separate the reverb from the dry sound, cause the reverb to appear behind the dry material and usually produce an unnatural effect along with possible rhythmical disorder. We usually aim at a pre-delay time that will produce minimal timbre distortion without audible time gaps. Nevertheless, audible pre-delay gaps have been used before in order to achieve interesting effects, including rhythmical ones.

> *A too short pre-delay can distort timbre, while too long can cause audible time gaps. We usually aim for settings that exhibit none of these issues while taking into account the intelligibility of the source material.*

Track 23.46: Pre-Delay 0 ms Snare
When the pre-delay is set to 0 ms, both the dry snare and its reverb appear as one sound.

Track 23.47: Pre-Delay 50 ms Snare
The actual delay between the dry snare and the reverb can be heard here, although the two still appear to be connected.

Track 23.48: Pre-Delay 100 ms Snare
A 100 ms of pre-delay causes two noticeable impulse beats. The obvious separation between the two sounds yields a non-musical result.

- - - - - - - - - - - - - - - - -

Track 23.49: Pre-Delay 0 ms Vocal
As with the snare sample, 0 ms of delay unifies the dry voice and the reverb into a smooth sound.

Track 23.50: Pre-Delay 25 ms Vocal
Despite a hardly noticeable delay, the intelligibility of the vocal is slightly increased here. This is due to the fact that the delayed reverb density leaves some extra space for the leading edge of the voice.

Track 23.51: Pre-Delay 50 ms Vocal
The 50 ms delay can be discerned in this track, and it adds a hint of a slap-delay impression. Similar to the previous track, the voice is slightly clearer here as there is more space for the leading edge of the voice. However, the delay makes the overall impression busier and less smooth, resulting in slightly more clutter.

Track 23.52: Pre-Delay 100 ms Vocal
A 100 ms of pre-delay clearly results in a combination of a delay and a reverb. This is a distinctive effect in its own right, which might work in the right context. Notice how compared to the 0 ms version, the reverb on this track appears clearer and louder.

Plugin: t.c. electronic *ClassicVerb*

Early reflections (ER)

Shortly after the direct sound, bounced reflections start arriving at the listener (path b in Figure 23.6). Most of the early reflections only bounce from one or two surfaces, and they arrive at relatively long intervals (a few milliseconds). Therefore, our brain identifies them as discrete sounds that are correlated to the original signal. The early reflections are indispensable in providing our brain with information regarding the space characteristics and the distance between the source and the listener. A faithful early reflection pattern is obligatory if a natural simulation is required, and alterations to this parameter can greatly enhance or damage the realism of the reverb. Neither spring nor plate reverbs have distinct early reflections due to their small size and therefore they do not excel at delivering spatial information.

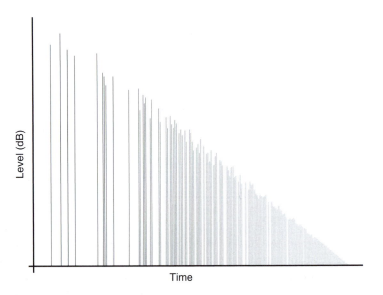

Figure 23.8 Possible reflection pattern in a real room.

Gradually, more and more reflections arrive at the listener, and their density increases until they cannot be distinguished as discrete echoes. With dependence on different room properties, early reflections might arrive within the first 100 ms following the direct sound. It is worth remembering that early reflections within the first 35 ms fall into the Haas zone, hence our brain discerns them in a slightly different manner. In addition, these very early reflections are readily masked by the direct sound. As discussed earlier, we can increase the pre-delay time in order to make the early reflections clearer.

> *Early reflections provide our brain with most of the information regarding space properties and will contribute greatly to the realism of depth.*

The **level** of the early reflections suggests how big the room is – a bigger room will have its boundaries further away from the listener, thus bounced reflections will travel longer distances and will be quieter. **Surface materials** also affect the level of reflections. For example, reflections from a concrete floor would be louder than reflections from a carpet.

With relation to **depth**, the level of early reflections might again have the opposite effect to what we might expect. Although the further away the listener is from the sound source the longer distance the reflected sound travels, it is the difference in distances travelled between the direct and the reflected sounds that matters here – a close sound source will have a very short direct path but a long reflected path. The further away the source is from the listener the smaller the difference in distance between the two paths. In practice, the further away the source and the listener are, the closer the direct and reflected sounds will be or in relative phrasing: the louder will be the early reflections.

> *Louder early reflections denote greater distance between the source and the listener.*

Early reflections are the closest sound to the dry sound, hence they are the main cause for timbre distortion and combfiltering. One of the biggest challenges in designing a reverb emulator involves the production of early reflections that do not color the dry sound. Sometimes attenuating or removing the early reflections altogether can yield better results and a more healthy timbre. The same practice can also work with a multi-performance such as a choir – in a real church each singer produces a different early reflection pattern, and the resultant early reflection pattern is often dense and indistinct. Nevertheless, removing the early reflections can result in poor localization and vague depth, so it may be unsuitable when trying to achieve a natural sound.

Finally, one trick involves adding the early reflections alone to a dry signal. This can enliven dry recordings in a very transparent way and with fewer side effects. The explanation for this phenomenon goes back to the use of delays to open up sounds and create some spaciousness. Good early reflections engines create more complex delays that will not color the sound as much while giving a greater sense of space. This trick can work on any type of material, but it does an exceptional job on distorted guitars.

> *Early reflections can distort the dry signal timbre. For less natural applications, concealing the early reflections can sometimes be beneficial. On the contrary, sometimes the early reflections alone can work magic.*

> **Track 23.53: Organ Dry**
> The dry signal used in the following tracks.
>
> **Track 23.54: Organ Dry and ER**
> The dry signal and the early reflections with no reverberation. Note the addition of a spatial cue and the wider, more distant image.
>
> **Track 23.55: Organ ER Only**
> The early reflections in isolation without the dry signal or reverberation.
>
> **Track 23.56: Organ Dry ER and Reverberation**
> The dry signal along with early reflections and reverberation.
>
> **Track 23.57: Organ Dry and Reverberation**
> The dry signal and reverberation together, excluding the early reflections. Notice the distinction between the dry sound and its reverb, whereas the previous track has much more unified impression and firmer spatial localization.
>
> *Plugin:* Sonnox *Oxford Reverb*

Reverberation (late reflections)

The reason that the term *reverberation* is used in this text and not *late reflections* is that the term 'reflection' suggests something distinct like an echo. However, the reflection pattern succeeding the early reflections is so dense that it can be regarded as one bulk of sound. Sometimes, reverberation is referred to as the *reverb tail*. The reverberation consists of reflections that bounced from many surfaces many times (path c in Figure 23.6). As sound is absorbed every time it encounters a surface, later reflections are absorbed more as they encounter a growing amount of surfaces. This results in reverberation with decaying amplitude. The **level** of the reverberation is an important factor in our perception of depth and will be explained shortly.

> **Track 23.58: Organ Reverberation Only**
> The isolated reverberation used in the previous tracks.
>
> *Plugin:* Sonnox *Oxford Reverb*

Pure reflections

Early studies on reverbs have shown that in comparison to the reverberation, the spaced nature of the early reflections gives our brain different psychoacoustic information regarding the space. Many reverb designers have based their designs on this observation with many units having different engines (or algorithms) for each of these alleged components. Later studies have shown that such an assumption is incorrect and that it is wrong to separate any reverb into any discrete components. Another argument commonly put forth is that a room is a single reverb generator and any distinction made within a reverb emulator might yield less natural results. Such an approach was adopted by respectable companies like Lexicon and Quantec and can be witnessed in the form of superb reverb simulations that both of these companies are known for.

Reverb ratios and depth

Reverbs are the main tool we employ when crafting depth into the mix. In order to understand how this can be best employed, it is important to first understand what happens in nature.

The inverse square law defines how sound drops in amplitude with relation to the distance it travels. For example, if the sound 1 m away from a speaker is 60 dBSPL, the sound 5 m away from the speaker will drop by 14 dB (to 46 dBSPL), and the sound 10 m away from a speaker will drop by 20 dB (to 40 dBSPL). It should be clear that the further away the listener is from the sound source the lower in level the direct sound will be. But the direct sound is only one instance of the original sound emitted from the source. The reverberation is a highly dense collection of the reflections, and although each reflection has dropped in level, summing all of them together results in a relatively loud sound. In a simplified scenario, a specific room might have reverberation at a constant level of 46 dBSPL.

If a listener in such a room stands 1 m away from the speaker he/she will hear the direct sound at 60 dBSPL and the reverberation at 43 dBSPL. The further away the listener gets from the speaker the quieter becomes the direct sound, but the reverberation level remains the same. Put another way, the further away the listener is the lower the ratio is between the direct sound and the reverberation. At 5 m away from the speaker the listener will hear the direct sound and the reverberation at equal levels. We call the distance at which such equality happens *critical distance*. Beyond the critical distance the reverberation will be louder than the direct sound and will damage intelligibility and clarity. As the direct sound will still be the first instance to arrive at the listener, it will still be used by the brain to decode the distance from the source, but up to a limited distance.

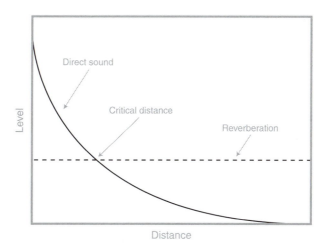

Figure 23.9 Critical distance. The relative levels of the direct sound and the reverberation with relation to the distance between the source and the listener. The distance at which the direct and reflected reverberations are equal in level is known as critical distance.

The direct-to-reverberant ratio is commonly used by recording engineers, especially orchestral ones, to determine the perceived depth of the sound source captured by a stereo microphone pair. For mixing purposes, we decrease the level ratio between the dry signal and the reverberation in order to place instruments further away in the mix. As many emulators do not have a separate level control for reverberation, we more often compromise and adjust the ratio between the dry signal and the wet signal (which contains both early reflections and reverberation). Just like in nature, intelligibility and definition can suffer if the reverb is louder than that of the dry signal. It should be added that the perceived loudness of the reverb is also dependent on its decay time and its density.

> *Depth positioning during mixdown is commonly achieved by different*
> *level ratios between the dry sound and the reverb.*

Track 23.59: Depth Demo
This track is similar to Track 6.11, where the lead is foremost, the congas are behind it and the flute-like synth is further back. The lead is dry, while the congas and the flute are sent at different levels to the same reverb.

Track 23.60: Depth Demo Lead Down
Compared to the previous track, the lead is lower in level here. This creates a very confusing depth field: On the one hand, the dry lead suggests a front positioning. On the other, the louder congas, despite their reverb, suggest that they are at the front. When being asked which one is foremost, the lead or the congas, some people would find it hard to decide.

Track 23.61: Depth Demo Flute Up
Compared to the previous track, the flute is higher in level. As the reverb send is post-fader, the flute's reverb level rises as well. The random ratio between the dry tracks and their reverb creates a highly distorted depth image.

Plugin: Audioease *Altiverb*
Percussion: Toontrack *EZdrummer*

Decay time

How long does it take a reverb to disappear? In acoustics, a measurement called RT60 is used, which measures the time it takes reverb in a room to decay by 60 dB. In practical terms, 60 dB is the difference between a very loud sound and one that is barely audible. This measurement is also used in reverb emulators to determine the 'length' of the reverb. Scientifically speaking, the decay time should be measured in relation to the direct sound, but some emulators reference it to the level of the first early reflection. In nature, a small absorbent room will have a decay time of around 200 ms, while a very large arena can have a decay time of approximately 4 seconds.

Decay time gives us a hint regarding the **size** of a room – bigger rooms have longer decay times as the distance between surfaces is bigger and it takes more time for the reflections to diminish. Decay time also gives us an idea of how reflective the **materials** in the room are – painted tiles reflect more sound energy compared to glass wool.

The decay time in a digital emulator is largely determined by the size of the room. One should expect a longer decay time in a church program than in a small room. Longer decay creates a heavier ambiance, while a shorter decay is associated with tightness. Longer decay also means a louder reverb that will cause more **masking** and possible intelligibility issues. With vocals, there is a chance that the reverb of one word will overlap the next word. This is more profound in busy-fast mixes where there are little time gaps between sounds. If too much reverb is applied in a mix and if decays are too long, we say that the mix is 'washed' with reverb.

In many cases, especially when reverbs are used on percussive instruments, the decay time should make **rhythmical sense**. For example, it might make sense to have a snare reverb dying out before the next kick. This will 'clear the stage' for the kick and also reduce possible masking. There is very little point delving into time calculations here – the ear is a better musical judge than any calculator, especially when it comes to rhythmical feel. In addition, reverbs may become inaudible long before they truly diminish.

Snare reverbs are commonly automated so they correspond to the mix density, a common trick is to have a shorter snare reverb during the chorus, where the mix calls for more power. As reverbs tend to soften sounds, it might be appropriate to have a longer reverb during the verse. Some engineers have also automated snare reverbs with relation to the note values played – a longer decay for quarter-notes, a shorter decay (or no reverb at all) for sixteenth-notes.

Track 23.62: Decay Source
The untreated track used in the following samples.

Track 23.63: Decay Long
A long decay which results in a very present effect that can shift the attention from the beat (this can be desirable sometimes).

Track 23.64: Decay Beat
A shorter decay time that becomes inaudible with the next kick. Note how in comparison to the previous track, the kick has more presence.

Track 23.65: Decay Lag
Slightly shorter decay than in the previous track; however, this decay makes less sense rhythmically, as the audible gap between the reverb tail and the kick can create an impression that the kick is late and therefore slightly off-beat.

Track 23.66: Decay Short
A short decay results in an overall tighter and punchier impression, leaving the spotlight on the beat itself.

Plugin: Universal Audio *Plate 140*
Drums: Toontrack *EZdrummer*

It is worth remembering that a longer pre-delay would normally result in a later reverb decay – lengthening the pre-delay and shortening the decay can result in an overall more present reverb. Level-wise, many find that a short, loud decay is more effective than a long and quiet one.

Long decay times can increase masking and damage clarity. When used as a special effect, mainly on percussive instruments, rhythmical judgment should be made.

Room size

The room size parameter determines the dimensions of the simulated room, and in most cases it is linked to the decay time and the early reflection pattern. Coarse changes to this parameter distinguish between small rooms like bathrooms and large spaces like basketball arenas. Generally speaking, the smaller the room is the more coloration occurs. Increasing the room size can result in a more vigorous early reflections pattern and longer pre-delay – combined with shorter decay time, the resultant reverb could be more pronounced.

Density

A density parameter on a reverb emulator can exist for the early reflections alone, for the reverberation alone or as a unified control for both.

The density of the early reflections gives us a hint regarding the size of the room, where denser reflections suggest a smaller room (as sound quickly reflects and re-reflects from nearby surfaces). The density parameter determines how many discrete echoes constitute the early reflections pattern and with low values discrete echoes can be clearly discerned. Reducing the early reflections density can minimize the combfiltering caused by phase interaction with the direct sound.

With both early reflections and reverberation densities, higher settings result in a thicker sound that can smooth the sharp transients of percussive instruments. Low density for percussive instruments can cause an unwanted metallic effect similar to flutter echo. But the same low-density settings can retain clarity when applied to sustained sounds such as pads or vocals (which naturally fill the gaps between the sparse reverb echoes). The density setting also relates to the masking effect – a denser reverb will mask more brutally than a sparse one.

Low-density settings can reduce both timbre distortion and masking.
High density is usually favorable for percussive sounds.

Track 23.67: Density Low
With low density, discrete echoes can easily be heard and the sound hardly resembles a reverb.

Track 23.68: Density High
Higher density results in a smooth reverb tail.

Plugin: Logic *PlatinumVerb*
Snare: Toontrack *EZdrummer*

Diffusion

The term diffusion is used to describe the scattering of sound. A properly diffused sound field will benefit from more uniform frequency response and other acoustic qualities that make reverbs more pleasant. Diffusers are commonly used in control and live rooms in order to achieve such behavior. Diffusion is determined by many factors, for instance, some **materials** like bricks diffuse sound more than other materials such as flat metal. An **irregularly shaped** room also creates more diffused sound field compared to a simple cubical room. When diffusion occurs, the reflection pattern becomes more complex, both in terms of spacing and levels.

Different manufactures try to imitate a diffused behavior in different ways. In most cases the implementation of this parameter is very basic compared to what happens in nature. Many link the diffusion control to density, sometimes in a way that high diffusion settings result in growing density over time or produce less regular reflection spacing. Density and diffusion are commonly confused as their effect can be identical. The variety of implementations makes it essential to refer to the emulator's manual in order to see what this parameter exactly does. Listening to what it does can teach even more.

Track 23.69: Diffusion Low
With low diffusion, the reverb decay is solid and consistent.

Track 23.70: Diffusion High
High diffusion results in a decay that shows some variations in level and stronger differences between the left and right channels.

Plugin: Cubase *Roomworks*
Snare: Toontrack *EZdrummer*

Frequencies and damping

Frequency treatment can happen at three points along the reverb signal path:

1. **Pre-reverb** – where we usually remove unwanted frequencies that can impair the reverb output.
2. **Damping** – frequency treatment within the reverb algorithm, which relates to the natural properties of the simulated space.
3. **Post-reverb** – where we usually EQ the reverb output in order to fit it into the mix.

Some emulators enable frequency treatment at all three points. Others offer damping only, which can be used as an alternative to post-reverb EQ. If neither pre- nor post-EQ processing is available, a dedicated EQ can always be inserted manually into the signal path – within an audio sequencer this simply involves inserting an EQ plugin before or after the reverb (on the aux track). In the analog domain this requires patching an EQ between the aux send output and the reverb. Patching an EQ after the reverb is seldom required as most effect returns include some form of frequency control.

Pre-reverb equalization usually involves pass or shelving filters. Low-frequency content can produce a long, boomy reverb sound that can clutter and muddy the mix. A HPF placed before the reverb can prevent this by filtering low-frequencies content – like those of a kick in a drum mix. Some high-frequency content might produce a luminous and unpleasant reverb tail – a pre-reverb shelving EQ can resolve such a problem.

Damping is concerned with the frequency behavior over time. High frequencies are easily absorbed: It takes 3" (76 mm) of acoustic foam to eliminate all frequencies above 940 Hz; it takes a bass trap 3' deep (1 m) to treat a frequency of 94 Hz. High frequencies are also absorbed by the air itself, especially when traveling long distances in large spaces. The natural reverb of an absorbent space has its high frequencies decaying much faster than low frequencies, resulting in less high-frequency content over time.

One could be mistaken for thinking that a pre- or post-shelving EQ could achieve an identical effect, but such a static treatment is found mostly within cheap reverb emulators. A better simulation is achieved by placing filters within the emulator's delay network, which results in the desired attenuation of frequencies over time.

The damping parameter usually represents the ratio between the reverb decay time and the frequency decay time. More specifically, a standard decay time of 4 seconds and a HF damping ratio of 0.5 means that the high frequencies will decay within 2 seconds. Damping ratio can also be higher than 1, in which case specific frequencies will decay more slowly over time. Further control over damping is sometimes offered in the form of an adjustable crossover frequency. If an emulator has a single control labeled 'damping', it is most likely to be a HF damping control (as these are more commonly used than LF damping controls).

Many digital emulators produce brighter reverbs than those produced by real spaces. **Damping HF** can attain more natural results or help in simulating a room with many absorptive materials. But HF damping can be useful in other scenarios – the sibilance of vocals can linger on a reverb tail making it more obvious and even disturbing (although often mixing engineers intentionally leave a controlled amount of these lingering high frequencies). Noises caused by synthesized sounds that include FM or ring modulation, recordings that capture air movements like those of a trumpet blow, distortion of any kind, harsh cymbals or even the attack of an acoustic guitar can all linger on a long reverb tail and add unwanted dirt to the mix. In such cases, which many novice mixing engineers tend to overlook, HF damping can serve as a remedy.

> *HF damping can result in more natural simulation and also prevent unwanted noises from lingering on the reverb tail.*

LF damping can be applied in order to simulate rooms with materials that absorb more low frequencies than high frequencies, like wood. Sometimes low-frequency reverberation is required but only in order to give an initial impact. Employing LF damping in such cases will thin the reverb over time and will prevent continuous muddying of the mix.

Post-reverb equalization helps to tune the reverb into the mix. After all, a reverb is a long sound that occupies frequency space that might interfere with other signals. Just as relevant is a discussion about how different frequencies can modify our reverb perception. High frequencies give the reverb a spell of presence and sparkle that many choose to retain, especially when reverbs are used as an unnatural effect. On the contrary, a more transparent, even hidden reverb can be obtained if its high frequencies are softened. Very often high frequencies are attenuated in order to create a warm or dark effect – such a reverb can be heard on many mellow guitar solos. Attenuating high frequencies can also result in an apparent increase in distance. Low frequencies make reverbs bigger, more impressive and warmer. A boost with a low-shelving EQ can accent these effects. The more low frequencies a reverb has the bigger the space will appear. LF attenuation (or filtering) will thin the reverb and make it less imposing.

> *The high-frequency content of a reverb contributes to its presence,*
> *while the low frequencies contribute to its thrill.*

Track 23.71: RevEQ Untreated
The source track for the following samples involves both the dry and wet signals. Note the sibilance lingering on the reverb tail.

Track 23.72: RevEQ Pre-Shelving Down
A shelving EQ at 1.25 kHz with −6 dB of gain is placed before the reverb. This eliminates much of the lingering sibilance, but results in a muddy reverb that lacks spark.

Track 23.73: RevEQ Damping HF Down
The reverb's damping band Mid-Hi crossover frequency was set to 1.25 kHz, the damping factor was 0.5. The resultant effect is brighter, and the lingering sibilance has been reduced drastically. Compared to the untreated track, the high frequencies on this track decay much faster.

Track 23.74: RevEQ Post-Shelving Down
The same shelving EQ as in Track 23.72 is placed after the reverb. Despite sounding dark like Track 23.72, it is possible to hear the lingering sibilance being attenuated by the post-reverb filter.

Track 23.75: RevEQ Damping HF Up
With the damping factor set to 1.5, it is easy to hear how high frequencies take much longer to decay.

Track 23.76: RevEQ Damping LF LMF Down
By damping the low and low-mid frequencies with a factor of 0.5, the reverb retains its warmth despite losing some sustained lows.

Plugins: t.c. electronic *VSS3*, PSP *MasterQ*

Reverbs and stereo

Mono reverbs

A certain practice in drum recording involves a single room microphone in addition to the stereo overheads. Although the room microphone is positioned further away from the drum kit and therefore captures more reverb compared to the overheads, it is interesting to hear how the stereo recording from the overheads can sound more reverberant, spacious and natural.

There are a few aspects that contribute to the limited realism of mono reverbs. One of them is the masking effect, where the direct sound masks the early reflections. An early trick to solve this involved panning the reverb away from the direct sound. But while this can yield an interesting effect, it contributes very little to realism. The main issue with a mono reverb is that it does not reassemble the directional behavior of a natural reverb, which arrives at our ears from all directions. A reflection pattern that hits the listener from different angles will sound more realistic than one that arrives from a single position in space. Although our pair of monitors can usually only simulate one sixth of the space around us (and even this happens on one dimension only), a substantial improvement is achieved by using a stereo reverb.

> *Stereophonic behavior is essential for the reproduction of a lifelike reverb.*

Track 23.77: Percussion (Dry)
Track 23.78: Percussion (Stereo Reverb)
Track 23.79: Percussion (Mono Reverb)
The dry source is monophonic. Note how the stereo reverb produces a realistic sense of space while the mono reverb is more reminiscent of a retro effect.

Plugin: Trillium Lane Labs *TL Space*
Percussion: Toontrack *EZdrummer*

Having said that, it might be surprising to learn that mono reverbs are nothing rare in mixing. For one, they are used as a distinctive effect. Then, they are used when some properties of a stereo reverb are unwanted, notably its width. To give an example, say we have two distorted guitars each panned hard to a different extreme, and we want to add a sense of depth to these guitars or just add a bit of life. Sending both guitars to a stereo reverb will fill the stereo panorama with harmonically rich material that might decrease the definition of other instruments. In addition, the localization of the two guitars could blur. Instead of sending both guitars to a stereo reverb, we can send each to a mono reverb and pan both the dry and wet signals to the same position. The mono reverbs might not produce a natural space simulation, but they would still provide some sense of depth and life without filling the stereo panorama. It is worth remembering that usually using a single channel from a stereo reverb will translate better than panning both channels to the same position; this is due to the phase differences that many reverbs have between the two channels.

> *Mono reverbs are used either as a special effect or when the width of a stereo reverb is unwanted.*

Track 23.80: Guitars (Dry)
The dry guitars panned hard left and right.

Track 23.81: Guitars (Stereo Reverb)
Each guitar sent to a different extreme of the same stereo reverb.

Track 23.82: Guitars (Two Mono Reverbs)
Each guitar is sent to a mono reverb, which is panned hard to the respective extreme. It is worth noting that there is more realistic sense of ambiance with the stereo reverb, although the two mono reverbs still provide strong spatial impression. Also, compared to the stereo reverb, the two mono reverbs can appear somewhat tidier.

The following tracks are identical to the previous three, but only involve one guitar:

Track 23.83: Guitar (Dry)
Track 23.84: Guitar (Stereo Reverb)
Track 23.85: Guitar (Mono Reverb)

Plugin: Trillium Lane Labs *TL Space*

Stereo reverbs and true stereo reverbs

The original digital reverb proposed by Schroeder was capable of mono-in/mono-out. A decade later Michael Gerzon designed a stereo-in/stereo-out reverb. Needless to say, stereo-in reverbs support the common configuration of mono-in/stereo-out, but as many hardware units will use different algorithms for mono and stereo inputs, an exact selection of input mode needs to be made. Software plugins are self-configuring.

Not all the reverbs that accept a stereo input process the left and right channels individually. Some reverb emulators sum a stereo signal to mono either at the input or at a specific stage such as after the early reflections engine. If a reverb sums to mono internally, panning a stereo send might not result in the respective position imaging – the generated reverb can remain identical no matter how the feeding signal is panned. An emulator that maintains stereo processing throughout is known as a *true stereo reverb*.

The implementation of a stereo reverb requires different reflection content on the left and right output channels. It is very common that the two channels are highly phase-shifted, which is why soloing a reverb will often make a phase meter jump to its negative side. These phase differences result in a spacious reverb that spans beyond the physical speakers and can seem to be coming from around us. The drawback of these phase differences is their flimsy mono-compatibility. Some reverbs give different controls that dictate the overall stereo strength – the more stereophonic a reverb is the more spacious it will appear.

Panning stereo reverbs

The practice of panning a stereo reverb return to the extremes is not always justified, since the reflections will occupy the full width of the stereo panorama. The potential masking that a reverb might cause can be reduced if a narrower panning tactic is used. There are many cases where narrowing a stereo reverb is desirable. For example, a strong stereo reverb on a snare can sound foreign to the mix if panned to the extremes. More cohesive results can be achieved if the reverb is panned around the source instrument so it only occupies a slice of the stereo panorama. Identical panning schemes can be applied to nearly every reverb which is used as an effect, including vocal reverbs.

In less natural mixes even the ambiance can be narrowed down. This creates a less spacious impression, but more intense effect. It also keeps the stereo extremes clear for other sounds. As narrowing a reverb makes it more monophonic, it can result in a decrease of perceived depth, which can work well in a powerful mix. However, if the ambiance is narrowed too much, it can give the impression that the instruments are located in a long tunnel.

Although the narrowing of the reverb output is commonly done using pan pots, these can cause unwanted phase interaction between the left and right channels. It is worth checking if the emulator has a stereo spread control (sometimes called stereo width). With some emulators, using the stereo spread control instead of panning the reverb return can minimize the phase interaction happening between the two channels and produce a healthier reverb altogether.

If reverbs are used as a special effect or to fill stereo gaps, they do not have to be panned mirrored around the center. For instance, if a snare is panned to 11:00, it is reasonable to pan the snare reverb between 10:00 and 12:00 (Figure 23.10).

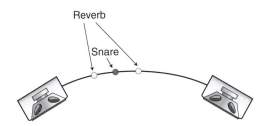

Figure 23.10 Possible snare Reverb Panning.

A stereo reverb need not be panned to the stereo extremes or be placed symmetrically around the center.

Track 23.86: Reverb Panning Full Width
The stereo reverb in this track is panned to the extremes, therefore it spans the full stereo panorama.

Track 23.87: Reverb Panning 6 Hours
The reverb is panned between 9:00 and 15:00. Notice that there is still a good sense of space while the extremes are clear.

Track 23.88: Reverb Panning 3 Hours
The reverb is panned between 10:30 and 13:30. Its narrow image is reminiscent of the sound at a the opening of a tunnel.

Track 23.89: Reverb Panning Mono
Both reverb channels are panned center resulting in mono reverb. This yields an unnatural impression, and the dry signal masks some of the reverb qualities.

Plugin: t.c. electronic *MegaReverb*

- - - - - - - - - - - - - -

Track 23.90: Panning SnareVerb Source
The source track used in the following samples.

Track 23.91: Panning SnareVerb Full Width
Although sometimes desirable, such a wide panning of the reverb results in an unnatural spatial image. It feels as if the drums are in one room and the snare is in another.

Track 23.92: Panning SnareVerb Half Width
Narrowing the reverb around the snare makes it sound more like an effect than a spatial reverb.

Track 23.93: Reverb Panning Mono
Panning both extremes of the stereo reverb to the center results in a mono reverb. Even more than in the previous track, the reverb appears as an effect here, but the lost stereo width has some consequences on its impression.

Plugin: Universal Audio *Plate 140*
Drums: Toontrack *EZdrummer*

Other reverb types

Fixed source reverb emulators

Most digital reverb emulators can be seen as a virtual room in which artificial reverb is simulated. Despite this, most reverb emulators do not offer the ability to position the sound source (or the listener) in the virtual room. This can be down to, among other issues, processing power limitations. Many of today's emulators are still based on the principles of the original Schroeder design. For simplicity, we shall regard such emulators as having a fixed position for both the sound source and the listener, and we will therefore title them 'fixed-source' reverbs.

Figure 23.11 illustrates the reflections of a real room where a violin and a trumpet are placed in front of a stereo microphone pair. The very first reflection to arrive at the left microphone will be that of the trumpet, and it will be immediately followed by the first reflection from the violin.

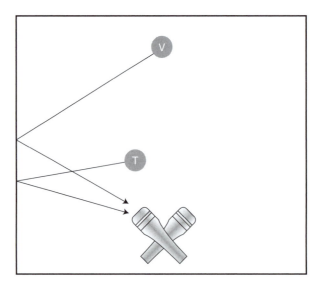

Figure 23.11 The very first early reflections in a real room from a violin (V) and a trumpet (T).

When we aim to simulate this recording scenario with a fixed-source emulator, we feed it with a stereo mix of both the trumpet and the violin. To reflect their position in Figure 23.11, the trumpet will be panned slightly to the left and the violin will be panned center. Both channels of the stereo mix will contain both the trumpet and the violin, although at different level ratios.

Figure 23.12 illustrates what will happen within the reverb emulator, where the speakers represent the stereo input signal and the microphones represent the stereo output. In this case, the first early reflection will consist of a mix between the trumpet and the violin; in fact, all reflections will be made of such a mix between the two instruments.

In a scenario like the one in Figure 23.12, which reminds us very much of the way sound is recorded in reverb chambers, lies one of the great limitations in feeding any reverb emulator with a composite stereo mix. Reverb emulators are incapable of separating the instruments in a stereo mix and therefore cannot produce an individual reflection pattern per source. When all reflections originate from such a stereo mix, our perception links the relative position of the different instruments. An accurate localization and depth can be achieved if each of the instruments has a distinct reflection pattern that is not correlated to any other instruments. Even tiny reflection variations between the different instruments can make a big difference.

One way to overcome this, usually only possible in the digital domain, is to send each instrument to a single emulator. Each of the emulators has to be loaded with an identical program, with slight variations in parameters between the emulators. The small variations are likely to create a distinct reflection pattern for each instrument, but unlikely to cause ambiance collision. Nevertheless, the fixed position of the source in the virtual room can still limit the extent to which realistic localization can be achieved.

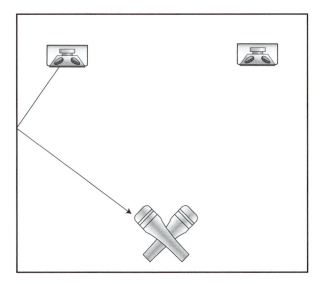

Figure 23.12 Early reflections reproduction within a fixed-source reverb emulator.

Moveable source reverb emulators

Movable source reverbs (sometimes just *source reverbs*) give us control over the position of the sound source in the virtual room. Control over stereo sources is achieved by front–back positioning of either the imaginary speakers or listener; a mono signal can also be moved left and right. The Sonnox *Oxford Reverb* (Figure 23.13) is such a plugin, enabling source positioning via its position control.

A positioning of multiple instruments in one virtual space requires sending each of the instruments to a different emulator with an identical program loaded. The only parameter that will change between each emulator is the position of the source (or the listener). Such a setup will bring about an individual early reflection pattern for each instrument and therefore much more faithful localization and more defined depth.

While some digital reverb emulators enable flexible positioning, some convolution reverbs simply involve different impulse responses that were taken at different positions within the room. The distance is often mentioned in the preset name (for example, 1 m away, 3 m away and so forth). Some convolution reverbs even provide a graphic display of the room and show the different positions at which the impulse response was recorded.

Gated and nonlinear reverbs

The reflections in a real room can build up and decay in amplitude very differently from those produced by a reverb emulator. While most emulators loudest reflection is the first one, the reflection pattern in some spaces can build up for a while before reaching maximum amplitude – an equivalent to the attack stage on a dynamic envelope. In addition, the reflections in a real room fluctuate in level, unlike the *linearly* decreasing reflection levels in some reverb emulators. The shape of the reverb envelope is important for natural simulation, and some emulators provide related controls. Mixing engineers can employ

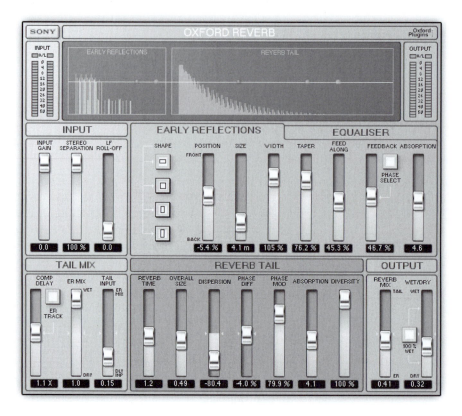

Figure 23.13 The Sonnox *Oxford Reverb*. The position of the source within the space algorithm is adjustable via the position control.

various tools to make the decay of reverbs far less natural. Despite potentially damaging its natural impression, such a nonlinear reverb can have more impact.

The decay of a reverb is what we most commonly shape. The simplest way of shaping it is by inserting a compressor after the reverb. Often the compressor is set to moderate threshold, moderate ratio and fast attack. For truly wild decay shapes the release can be set to minimum so some pumping takes place; for more subtle results the release time can be lengthened.

The most famous nonlinear reverb is the *gated reverb*. Rumor has it that it was discovered by accident during a recording session of Peter Gabriel's third album. It is mostly recognized for its use in Phil Collins' *In the Air Tonight*. In its most simple form, gated reverb is achieved by inserting a gate after the reverb and setting the gate threshold so it opens with the initial reverb burst. Usually a fast attack and release are used and hold settings that make rhythmical sense. This configuration does not alter the shape of the decay curve until its later stages – during the gate hold period the reverb decay remains as before (Figure 23.14b). One rationale for gating a reverb in such a way has to do with the fact that with sustained dry signals the reverb is likely to be masked by the direct sound as soon as it drops by 15 dB. In a busy mix the reverb tail can be masked by many other sounds – a gated reverb can sharpen the sonic image by cutting these inaudible parts of the reverb.

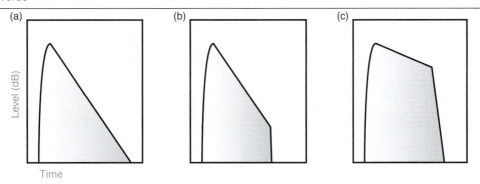

Figure 23.14 Reverb decay. (a) A linear reverb decay. (b) A simple gated reverb decay. (c) A decay that first falls slowly and then dives quickly.

If instead of decaying linearly, the decay shape is altered so it sustains at maximum level (or falls slowly) and then decays abruptly, the reverb becomes more prominent. To achieve this, a compressor is inserted after the reverb. The compressor is set to flatten the decay so it hovers longer at higher amplitudes. But this will result in a reverb that might not drop below the gate's threshold. Therefore, the original sound is sent to the gate's key input. Figure 23.15 demonstrates such a setup.

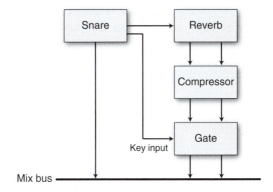

Figure 23.15 Gated snare setup. A schematic signal flow for an 'improved' gated reverb.

There are endless amounts of productions in which gated reverb has been applied on snares and toms, but it can be applied to many other instruments, even those of a non-percussive nature. Shaping the reverb's decay envelope can give the reverb more punch or more defined rhythmical characteristics (which must be observed while tweaking the gate). Gated reverbs are so popular that many digital reverb emulators include such a program. Still, better control is usually achieved by setting up the independent building blocks.

> *While gated reverbs are often used on drums, nonlinear reverbs in general can give the reverb more entity and a stronger rhythmic form.*

> **Track 23.94: SnareTom Source**
> The source drum sequence before the addition of reverb.
>
> **Track 23.95: SnareTom Reverb Only**
> Both the snare and the toms are sent to the same reverb, which on this track is neither compressed nor gated.
>
> **Track 23.96: SnareTom Compressed Reverb**
> The reverb in this track has been compressed but not gated.
>
> **Track 23.97: SnareTom Compressed Gated Reverb**
> This is essentially the common gated reverb effect. The reverb is both compressed and gated.
>
> **Track 23.98: SnareTom Uncompressed Gated Reverb**
> Removing the compressor preceding the gate results in the effect heard on this track.
>
> *Plugins:* Sonnox *Oxford Reverb*, Digidesign *DigiRack Expander/Gate Dyn 3*
> *Drums:* Toontrack *EZdrummer*

Reverse reverbs and preverb

A reversed reverb does not really reverse any audio, but it does reverse the amplitude envelope of the reverb so it starts quietly, grows gradually and then cuts abruptly (Figure 23.16a). This effect gives a certain backward impression that can be used as a special effect.

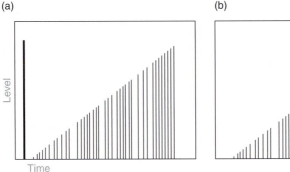
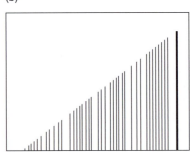

Figure 23.16 (a) A reversed reverb. (b) Preverb. The bold line denotes the direct sound.

A more common practice in mixing is called preverb. It is done by recording the reverb and then playing it backward so that it rises before the actual sound (Figure 23.16b). In order for this to work, the reverb has to be fed with a backward version of the original material. With tapes, this is easily done by reversing the reels, recording the reverb on to an available track and then reversing the reels again. In the digital domain, the original material has to be reversed, the reverb has to be bounced and then both the original sound and the reverb have to be reversed again.

> **Track 23.99: Reversed Reverb Mono**
> Both the dry voice and the reverb are panned hard center in this track.
>
> **Track 23.100: Reversed Reverb Stereo**
> The voice is panned hard left and the reverb to hard right. This makes the reversed development of the reverb easier to discern.
>
> *Plugin:* Logic *EnVerb*
>
> - - - - - - - - - - - - - -
>
> As with the previous two tracks, the following two demonstrate preverb:
>
> **Track 23.101: Preverb Mono**
> **Track 23.102: Preverb Stereo**
>
> **Track 23.103: Preverb Before Reversing**
> This is the voice and the reverb, before reversal.
>
> *Plugin:* Universal Audio *RealVerb Pro*

Reverbs in practice

Reverb quantity and allocation

A common mixing discussion relates to the amount of reverbs one should use in a mix. To start the debate, one should imagine oneself in a standard professional studio back in 1971. Such a studio had mostly analog equipment, maybe up to three high-range dedicated reverb units, perhaps along with a couple of less expensive units and a few multi-effect units that had reverb programs as well. It would be fair to say that for most mixes in such a setup the high-range units were allocated to the ambiance and vocals, while other reverbs were allocated based on their quality and the importance of the instruments sent to them. Any mix that involved more than eight reverbs would be considered as reverb-lavish, with many mixes completed with no more than five reverbs and some with as little as two. Previously, there was little choice other than to feed the same reverb unit with more than one track using an aux send. While there are mixes from that era that involve more than a dozen reverb emulators, it is worth remembering that many beloved mixes from the past were mixed with just a few reverb units. A practice that still remains.

The inherent problem with hardware units is that they cannot be multiplied with a click of a mouse like plugins can. Even in 2001, many mixing engineers had to limit the amount of reverb plugins due to processing power limitations, and reverbs had to be connected in parallel and in many cases were fed with a mix of many tracks using an aux send.

It is only in recent years that computers have become powerful enough to accommodate even 100 reverb plugins in a single mix. A different reverb plugin can be used as an insert on each individual track and one can also feed a single vocal track into five reverb plugins connected in series. Nowadays, for many of those who software-mix, having eight reverbs in a mix does not seem lavish at all.

If mixing engineers back in 1971 had had an array of 20 reverb units, they might have used them all. But two issues come to mind. First is the fact that it is easier to grow a tree than a forest. Put another way, a mixing engineer might be tempted to use many reverb plugins while compromising on quality and giving less attention to each reverb. Selecting and configuring a reverb can be a truly time-consuming business, especially for the less experienced engineers who are not yet sonically familiar with their tools. Second, the use of too many reverbs, especially if a different program is loaded on to each, can cause what is known as *ambiance collision* – a blend of too many spaces, which does not make any spatial sense. A certain objective when using reverbs is the creation of a firm, convincing ambiance even when an imaginary ambiance is involved. Although many reverbs can work together, a certain risk exists if insufficient attention is given to how they interact. Generally speaking, when using more than one ambiance reverb, softening the highs of the short reverbs can help conceal them and reduce collision.

Track 23.104: MultiAmb Dry
The dry signal used in the following samples.

Track 23.105: MultiAmb One Reverb
The woodblock, tambourine and congas are all sent to the same reverb emulator.

Track 23.106: MultiAmb Three Reverbs
The three instruments are each sent to a different reverb emulator loaded with a different preset. We can still localize the instruments in the depth field, but the simulated space is less coherent than in the previous track.

Plugin: Universal Audio *DreamVerb*
Percussion: Toontrack *EZdrummer*

While there is almost total agreement that expensive emulators are superior in producing a natural reverb, it is sometimes the cheapest and most basic emulators that can produce a distinctive effect. One advantage that comes with software plugins is an ever-growing selection – some of today's plugins offer unique controls and sound that might not be found in expensive emulators.

Aux send or insert?

Convention has it that reverbs are fed via auxiliary sends. The most obvious advantage in using a send is that we can route more than one signal to the reverb emulator and therefore save processing power or share high-quality emulators. Connecting reverbs as a send on an analog desk requires just a few more easy steps compared to an insert connection. However, in the digital domain the same task usually requires a little extra work, like the creation of auxiliary tracks and bus routing. If many reverbs are used, but each is fed from one single track, the multitude of auxiliary tracks can clutter the mixer, and many buses are employed purely for routing purposes (rather than summing). Therefore, some argue that in specific situations it is sensible to use an insert rather than a send, for instance, when a reverb is added to a kick for two bars only; or when the reverb only adds a touch of early reflections that are meant to enrich a mono signal.

Another important point worth considering is that both in the analog and digital signal flows the inserts are pre-fader. Thus, connecting reverbs as an insert forces a pre-fader feed that leaves all dry/wet adjustment to the reverb emulator. Accessing the emulator every time we want to change the dry/wet balance is somewhat less convenient compared to accessing the send controls.

| *In all but very specific situations, reverbs are added using sends.* |

Pre- or post-fader?

As just discussed, reverbs connected as an insert are pre-fader (with rare exceptions). When connected as an aux send, we have a choice between either a pre- or post-fader send. A post-fader send is also post-cut. Similarly, a pre-fader send is also pre-cut.

The basic principle behind using post-fader send is that the level of the dry signal and the reverb level are linked – when the dry signal level increases, the reverb level increases as well. In practice, sometimes a fader adjustment might not lead to the desired reverb level change, and some additional adjustment to the reverb level will be needed. A post-fader send is necessary if any level or mute automation should take place – when we fade out or mute the dry signal we want the reverb to fade out or mute as well.

Another point relates to the common use of destructive solos during mixdown. Say we have vocal and guitar both sent to an ambiance reverb. The reverb is solo safe to prevent it from being cut when other instruments are soloed. If the guitar send is pre-fader and the vocal is soloed, the guitar reverb will still be heard along with the vocal and its reverb. This can be highly annoying. Post-fader sends ensure that such situations do not occur.

| *Post-fader reverb send is most common in mixing, and a requisite if* |
| *any level or mute automation is to take place on the source signal.* |

In some situations we need to bounce the reverb separately from its dry signal. In many software applications this can be achieved by soloing the reverb auxiliary track, but as software solos are often destructive, this will cut the dry signal track as well. In order to still have the dry signal feeding the auxiliary bus and consequently the reverb unit, a pre-fader send can be used. Another situation where a pre-fader send can be beneficial is when the original track is mixed at a very low level and therefore the post-fader send will not provide enough level to produce an audible reverb.

Pre-compressor reverb send

Some interesting results can be achieved if the dry signal is compressed but the signal sent to the reverb emulator is taken before the compression takes place. In a performance with vibrant dynamics, an increased expression is conveyed by louder phrases. Such a performance might be a problem in a mix and require compression. But if sent to a reverb unit before being compressed, the dynamics of the performance will trigger more reverberation during the louder moments. This 'manual automation' effect is most suitable on a performance with quick dynamic changes like those of some vocal performances.

While on an analog desk a copy of the uncompressed signal can always be taken from the multitrack return or a pre-processing insertion point, software users might scratch their head wondering how this can be done – a software send (the reverb) is always post-insert (the compressor). Figure 23.17 illustrates how this can be done in Logic – the dry signal is sent to a bus, on which the reverb is inserted. Note that the vocal track output is not set to the main mix bus, but to another bus on which the compressor is inserted.

Figure 23.17 A pre-compressor reverb send in Logic.

Connecting reverbs for depth applications

The ratio between the direct sound and the reverb is commonly used for depth applications. It is important to discuss the different connection options for depth reverbs.

If a reverb is connected using an insert, any dry/wet ratio adjustments must take place on the reverb emulator. Some reverb emulators offer separate controls for the dry and wet signal levels, but some offer a single control that simply defines the ratio between the wet and dry signals. The problem with the latter is that modifying the ratio will affect not only the perceived depth, but also the perceived level of the track, requiring further fader adjustments.

In both the digital and analog domains, the ratio between the dry and wet signals can be more instantly altered if reverbs are connected using an aux send. With a post-fader send, boosting the dry signal in order to bring sound forward will also result in a reverb boost that will send the sound backward. This affair does not happen with a pre-fader send, which therefore can be useful for depth applications.

While the dry signal is controlled using the channel fader, the reverb level can be altered in two ways: either using the local aux send level or the reverb return level. If more than one source is feeding the reverb emulator, changing the reverb return level will affect all sources, so it is the local aux send level that is used. If only one source is feeding the reverb, better gain structure calls for setting optimum levels into the reverb unit, the aux send level should be set to optimum, while the reverb return level might be attenuated.

> *Depth reverbs might benefit from a pre-fader feed. The reverb level will be controlled by the local aux send level if more than one source is feeding the reverb or by the reverb return level otherwise.*

How loud, how much?

Back in the 1980s, when the use of digital reverb units became widespread, much excitement surrounded this new tool and reverbs manifested in many mixes. As time went by, mixing engineers started to make more relaxed use of reverbs and the general tendency today is that reverbs serve the mix rather than being a sound in their own right. In some of today's mixes, a trained ear will be able to identify many more reverbs than initially seem present – reverbs nowadays tend to be more transparent and subtler.

This might go against the intuition of many novice mixing engineers who wish to hear the reverb in the mix, having introduced it. But the beauty of a reverb can persist even if it is not very obvious. One way to check the effectiveness of reverbs, even if hidden, is by muting them and seeing how they contribute to the mix. Yet, in specific situations having a bold, well-defined reverb is our aim; solo saxophone in a mellow production could be one of these situations.

Tuning ambiance

The standard way to create ambiance involves sending various instruments to one or more ambiance reverbs. However, the resultant ambiance can suffer from various issues due to the frequency content of the various instruments sent to the reverb. Being of such potent nature, the ambiance in a mix can be enhanced using a specific advanced technique. It involves sending a modified version of the instruments sent to the mix bus to the ambiance reverbs. On an audio sequencer, this can be achieved by duplicating a specific track and disabling its standard output routing to the main mix. Yet, this track is sent to the ambiance reverb instead of the original. We can then process (mainly equalize) the ghost track in different ways, so as to tune it to the ambiance created by the reverb it is sent to.

Delay and reverb coupling

A very common mixing technique involves a blend between a delay and a reverb. Most often this technique is used on vocals, but it can be applied on other instruments just as well. Blending a delay with a reverb is known to result in more impressive effect than having only one of them. One way to look at it is as if the delay enhances the reverb. Multitap delays with two active taps (each panned to a different side of the panorama) can be highly suitable for the task. Most often the delay and the reverb are connected in parallel, i.e., neither feeds the other. On an audio sequencer this can be achieved by sending a track to a bus, then feeding the bus into two different aux tracks – on one a reverb is loaded, on the other a delay. It also pays to try to connect a delay before a reverb – such an arrangement can also produce an appealing effect.

> **Track 23.110: Vocal Reverb No Delay**
> This track involves a reverb, but no delay.
>
> **Track 23.111: Vocal Reverb With Delay**
> The delay in this track is added in parallel to the reverb. Notice how, compared to the previous track, the effect here sounds slightly livelier and richer.
>
> *Plugins:* Audio Ease *Altiverb*, PSP *82*

Reverb coupling

Sometimes we combine two reverbs in order to achieve a specific task. However, this has to be done with caution or some undesirable problems might arise. Coupled reverb is the term used to describe a 'reverb of a reverb'. In real life, such phenomenon can happen when a reverb from one space travels to an adjacent space that adds its own reverb. Coupled reverbs are mainly a concern for natural reverb simulations – they place a space within a space, they increase the reverb time and introduce reverbs that can have broken decay slopes. In mixing, coupled reverbs can be introduced with the addition of a reverb to a recording that already contains a reverb or when connecting two reverb emulators in series.

Recordings that include a reverb, unless initially intended, can sometimes create great difficulties for the mixing engineer and in extreme situations be unusable. Such recordings are sometimes done by inexperienced recording engineers who may choose to record a guitar amp or an electronic keyboard with their internal (sometimes monophonic) reverb effect. Sometimes a recorded reverb is simply the result of a very reverberant recording environment. One way or another, such reverbs are then imprinted into the recording, often sound cheap and leave the mixing engineer between a rock and a hard place. It is worth remembering that reverbs often become louder after being compressed. Compressing a vocal recording that contains reverb can result in solid vocal levels but fluctuating reverb. While it is only possible to achieve a completely reverb-free recording in an anechoic chamber, recording engineers can achieve a direct-to-reverb ratio that will present no limitations during mixdown.

With the exception of a recording that was intended to capture ambiance, a reverb on a recording is like oil on water – they are likely to not mix well.

Nonetheless, a combination of two reverb emulators during mixdown can be useful. A valid practice is to send the same signal to two reverb emulators that are connected in parallel. This creates a composed reverb that can have richer texture. Another possible experiment that can yield an interesting effect is using two mono reverbs and panning each to an opposite extreme.

Yet another trick relates to the way emulators work – many of them have separate engines for the early and late reflections. Some emulators feed the early reflections into the late reflection engine, while others have complete separation between the two. Often we can choose the balance between the early and late reflections or simply switch off one engine or the other. This yields some interesting possibilities where we can combine two emulators, each producing either the early or later reflections. If we have two emulators and one is better at producing early reflections while the other is better at producing reverberation, we can mute the weak engine in each emulator to achieve a reverb that has the best of both worlds. This can work exceptionally well when a convolution reverb is the generator of the early reflections. Convolution reverbs tend to produce a highly believable early reflection pattern, but might be very CPU consuming if the long reverberation is produced as well. By assigning the convolution reverb with the task of producing the early reflections only, and the less critical task of late reflections to an algorithmic reverb, we can save CPU power, yet end up with more convincing results. The emulators can be connected either in parallel or in series, depending on whether we want to feed the early reflections into the late ones (Figure 23.18).

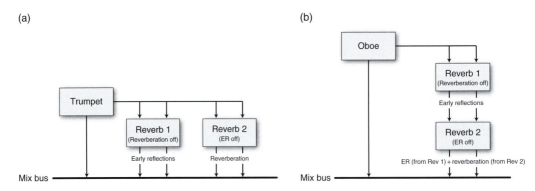

Figure 23.18 Reverb coupling. (a) Parallel reverb setup where each emulator produces a different reverb component. (b) Serial reverb setup where the early reflection from the first emulator feeds the second emulator, that only generates reverberation.

Kruder Dorfmeister. *The K&D Session*. !K7 Records, 1998.

This downbeat electronica remix album is a master-class in reverbs, delays and mixing in general. It demonstrates how reverbs can be used to craft an extremely powerful ambiance and put forth a bold mood. Both the depth and stereo domains are utilized immaculately to create a rich and inspiring sound stage.

24 Distortion

Mixing is a dynamic art. Mixing trends are changing with years, and throughout past decades the extent and the way we use different tools has changed as well. There is little argument that distortion is the tool that has gained the most popularity in recent years. Distortion, in that specific mixing context, is not exactly the screaming pedal we use with guitars, but the more subtle harmonic distortion, in its various forms.

There are two main reasons why distortion has become so widely used. First, the majority of pop mixes nowadays tend to be more aggressive than in the past. This applies in both analog and digital domains, where distortion is added to various degrees depending on the production. Strictly speaking, the use of distortion is not limited to rock or metal productions – even genres like dance, chart-pop or trip-hop might involve a certain degree of distortion on more than a few tracks. Second, distortion is used to compensate for what can be described as the 'boringly precise' digital sound – the inherent distortion of analog equipment and tape media is not an asset of digital audio. Engineers use distortion to add some edge to digital audio by introducing some degree of 'appealing error'.

Distortion basics

Harmonic distortion and inter-modulation

In the recognition of sounds, our ears use both the fundamental and its harmonics. Harmonics, as already explained, are integer multiples of the fundamental, and thus the two are closely related. By emphasizing harmonics we increase the definition of sounds. Lower harmonics are louder than higher harmonics. For example, the second harmonic is louder than the 20th harmonic. Due to their relatively loud level, the few low-order harmonics have some extra importance. The second harmonic is an octave above the fundamental, and the fourth harmonic is two octaves above the fundamental. Being octave-related to the fundamental, both harmonics are considered musical – emphasizing them tends to give appealing results. The third and fifth harmonics are also important, but neither is octave-related to the fundamental – both tend to produce a slightly more colorful sound.

A linear device is one that has perfect proportionality between input and output. The transfer curve of such a device is one straight line. An example of a nonlinear device is a compressor with any ratio other than 1:1. When signal passes through a nonlinear system, different kinds of distortion are produced. The less linear the system the more profound the distortion. One type of distortion is *harmonic distortion*, essentially added harmonics. Analog components are incapable of being perfectly linear. A specification called *total harmonic distortion* (THD) measures the harmonic distortion content produced by an analog device under standard test conditions. There are different flavors to analog distortion. The ratio between the low-order harmonics produced by a tube is different than those produced by a transistor. This is a major reason for the different sounds that a tube and solid-state equipment produce. Although technically speaking the lower the distortion the better, harmonic distortion is an intimate part of the analog sound in general and the characteristics of analog gear in particular. Digital systems are capable of being perfectly linear and thus might not produce harmonic distortion. Although being technically superior, many find the digital sound lifeless and pale.

Another type of distortion is *inter-modulation*. Like total harmonic distortion it can be measured and the specification given is simply called inter-modulation (IMD). Unlike harmonic distortion, inter-modulation distortion involves additional frequencies, but unlike harmonic distortion these are not necessarily harmonically related to the sound. Therefore, inter-modulation is often considered harsh and it is mostly unwanted. Yet, it is an inseparable part of any nonlinear system.

The problem with distortion in the digital domain

Often distortion produced within an analog system is more musical than that produced within a digital system. The reason for this originates from the fact that analog systems do not have a frequency limit like digital systems do. The highest frequency a digital system can accommodate is called the *Nyquist frequency*, and it is always half the sample rate. The problem with digital systems is that any content produced within the system that exceeds the Nyquist frequency mirrors around it. For example, say we have a system running at a sample rate of 44,100 Hz. The highest frequency such a system can accommodate is 22,050 Hz (the Nyquist frequency). Then say an 8 kHz sine wave is distorted. The resultant harmonic distortion would include a second harmonic at 16 kHz and a fourth harmonic at 32 kHz – both have an octave relationship to the fundamental. However, since the 32 kHz harmonic is higher than the Nyquist frequency, it would mirror around it, resulting in an alias frequency at 12.1 kHz. This frequency is not harmonically related to the 8 kHz fundamental and would produce a harsh sound. On the same basis, any distortion content that exceeds the Nyquist frequency creates aliasing frequencies below it. On an analog system no such thing happens – any distortion content above the 20 kHz is simply unheard.

We can minimize the harsh aliasing phenomenon of digital processing by using higher sample rates. If, for example, the sample rate of the system above was 88.2 kHz, the Nyquist frequency would be 44.1 kHz. This means that the 32 kHz harmonic would not alias – it simply would not be heard. Even content that does alias might do so into the inaudible range between 20 and 44.1 kHz. With 88.2 kHz, only content above 68.2 kHz would alias back into the audible frequency range. The problem with using higher sample rate is that more samples have to be processed. Compared to a sample rate of 44.1 kHz, an 88.2 kHz audio would require twice the samples processed, which essentially halves the processing capabilities of the CPU (or any other digital processor for that matter). In practice, the actual processing overhead in using higher sample rates can be more than initially seen – some plugins would need to perform substantially more calculations in order to operate at higher sampling rates.

High-quality plugin developers take this aliasing problem into account in their designs by implementing internal upsampling and then downsampling. Essentially, even if the project's sample rate is 44.1 kHz, the plugin upsamples the audio to, say, 176.4 kHz (4×). This results in a new Nyquist frequency of 88.2 kHz, and only content above 150.4 kHz would alias back into the audible frequency range. The plugin then performs its processing, which might produce content above the original Nyquist frequency – 22.05 kHz. Then any content above 20 kHz is removed using an anti-aliasing filter (essentially a high-quality LPF) and the audio is downsampled back to the original project's sample rate. This way, any content that would otherwise alias back into the audible range is simply removed.

Track 24.1: Hats Upsampling
These highly distorted hi-hats were produced while the plugin ×16 upsampling option is enabled.

Track 24.2: Hats No Upsampling
The added frequencies in this track are the result of aliasing frequencies. This track is produced while the upsampling option is disabled.

Plugin: Bram @ SmartElectronix.com *Cyanide 2*

It should be mentioned that the aliasing distortion produced within a digital system is often inaudible. The problem is not as bad as one might think. Some manufacturers do not perform upsampling for this reason. Still, accumulating distortion (a distortion of distortion) could become audible.

Parallel distortion

Just like the parallel compression technique, where a compressed version is layered underneath the original, signals are sometimes distorted and then layered below the original. This gives us added control over the amount of distortion being added. Consequently, this lets us drive the distortion harder to produce a stronger effect, but then layer it underneath at lower levels so it is not too obvious or crude.

The following tracks demonstrate parallel distortion on vocal and drums:

Track 24.3: Vocal Source
The unprocessed source track.

Track 24.4: Vocal Distortion Layer
The vocal in the previous track distorted. In the following tracks, the unprocessed track and this distorted layer are mixed, with the distorted layer a few dB below the unprocessed track.

Track 24.5: Vocal Distortion −18 dB
Track 24.6: Vocal Distortion −12 dB
Track 24.7: Vocal Distortion −6 dB
Track 24.8: Vocal Distortion 0 dB

- - - - - - - - -

The same principle as in the previous tracks, a similar demonstration on drums:

Track 24.9: Drums Source
Track 24.10: Drums Distortion Layer
Track 24.11: Drums Distortion −26 dB
Track 24.12: Drums Distortion −18 dB

Plugin: SansAmp *PSA-1*
Drums: Toontrack *EZdrummer*

Ways to generate distortion

Gain controls

As already described in Chapter 9, a boost on the gain control, affecting the signal early in the signal path, can overload successive components. In the analog domain we usually speak of *overloading* first, which mostly adds subtle distortion and only then saturation that can produce drastic distortion. The nature of solid-state components is that they provide increasing distortion with gain. Tube equipment tends to produce a very appealing distortion up to a specific point where the system seems to 'break' and produce very undesired clicks. Tapes can be overloaded on the very same principle by boosting the input signal. Bipolar transistors, FET, tube and tape distortion all have different qualities as they all produce different harmonic content. Also, the more each system overloads the more compression occurs, which makes the overall effect even more appealing.

Digital clipping

Of all types of distortion, probably digital clipping is the least appealing. The harsh limiting of audio exceeding the 0 dB threshold on a digital system can produce extremely unpleasant sound, especially if the added content aliases. There are also different types of digital clipping, where differences are dependent on what happens to signals that exceed the system limit. On most floating-point systems these signals are trimmed to the highest possible value (simply hard-limiting). However, different integer notations can produce different results, where exceeding signals can alias around the highest sample value; they can be displaced to the bottom of the value range or displaced across the zero line.

Figure 24.1 The Logic *Clip Distortion* Plugin.

Generally speaking, clipping tends to produce the strongest type of distortion. Some plugins try to imitate the clipping sound of analog devices by introducing additional processing along with the basic digital clipping. The clip distortion in Figure 24.1 is one of them. It is worth noting the Mix control, which lets us blend the undistorted and distorted sounds.

The following tracks demonstrate clip distortion. The dB in the track names denotes the approximate amount of dB by which the signal exceeded the 0 dBr limit of the system.

Track 24.13: Vocal 6 dB Clip Distortion
Track 24.14: Vocal 18 dB Clip Distortion
Track 24.15: Drums 6 dB Clip Distortion
Track 24.16: Drums 18 dB Clip Distortion

These tracks are produced by overloading the mix bus in Pro Tools LE. (After bouncing, the 18 dB versions are attenuated as a measure of ear safety.)

- - - - - - - - -

The following tracks are produced using Logic Clip Distortion. The drive is set to 30 dB, the symmetry to 0 or 60% (versions 1 and 2, respectively):

Track 24.17: Vocal Clip Distortion 1
Track 24.18: Vocal Clip Distortion 2
Track 24.19: Drums Clip Distortion 1
Track 24.20: Drums Clip Distortion 2

Plugin: Logic *Clip Distortion*
Drums: Toontrack *EZdrummer*

Short attack and release on dynamic range processors

In Chapter 16 that covered compressors, we saw that a short attack and release can act within the cycle of low frequencies, thus altering the shape of the waveform and producing distortion (refer to Figure 16.29 for a demonstration of this). This type of distortion is not reserved solely for compressors and can be produced by short time constants on any dynamic range processor. We can use this type of distortion to add some warmth and definition to instruments with low-frequency content, notably basses.

Wave shapers

A wave shaper is a pure implementation of a transfer curve. Essentially, the signal passes through a transfer characteristics function over which we have some control. This is somewhat similar to a dynamic range processor with no time constants and unrestricted transfer curve. Wave shapers are not very common and might be a sub-facility within a different effect. However, they can produce extremely interesting results that can be used in subtle amounts for gentle enhancements or in a drastic form for creative effect.

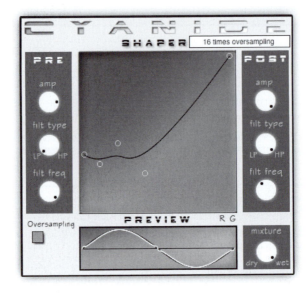

Figure 24.2 The Smartelectronix *Cyanide 2* plugin (donationware).

A few tracks demonstrating the impressive distortion capabilities of wave shapers:

Track 24.21: Wave Shaper 1 (Drums)
Track 24.22: Wave Shaper 1 (Vocal)
Track 24.23: Wave Shaper 2 (Drums)
Track 24.24: Wave Shaper 2 (Vocal)
Track 24.25: Wave Shaper 3 (Drums)
Track 24.26: Wave Shaper 3 (Vocal)
Track 24.27: Wave Shaper 4 (Drums)
Track 24.28: Wave Shaper 4 (Vocal)
Track 24.29: Wave Shaper 5 (Drums)
Track 24.30: Wave Shaper 5 (Vocal)

Plugin: Bram@SmartElectronix.com *Cyanide 2*
Drums: Toontrack *EZdrummer*

Bit reduction

Early digital samplers that emerged around the early 1980s had low specs compared to those used today – in addition to low sample rates, these were designed around 8-bit (and later 12-bit) samples. Even now, the 8-bit sound is sought after in genres such as hip-hop, where many find the lo-fi sound of drum samples appealing.

When we discussed software mixers in Chapter 10 we discussed the importance of dither, which meant to rectify the distortion caused by bit reduction. Unsurprisingly, like many other areas in mixing where we are after the 'technically wrong' and the less precise sound, bit reduction can also have an appealing effect. Essentially, a bit reduction process simply reduces the bit depth of digital audio. Although within the audio sequencer the audio is still represented by 32-bit float numbers, sample values are quantized to the steps of the target bit depth. For example, reduction to 1 bit would produce 32-bit float values of 1.0 and −1.0 only (in practice, however, most processors also output 0.0 as a possible sample value). This process produces quantization distortion, where the lower the target bit depth the more the distortion. Clearly, we wish to keep this distortion, so no dither is applied. Just like with drum samples, bit reduction can be used to give a lo-fi sense to various instruments. It can also be used as a creative effect.

Figure 24.3 The Logic *Bitcrusher*. This plugin combines bit reduction, three-mode clipping distortion and digital downsampling. In this screenshot, only the reduction to 3-bits is utilized. The eight quantization steps that would affect a sine waveform are visible on the display.

Track 24.31: Moroder Source
The source track, before applying bit reduction.

Track 24.32: Moroder 5 Bits
All instruments but the kick are reduced to 5-bits in this track. While the lows of the bass line become wimpy, the effect on both the hats and the snare is applicable.

Plugin: Logic *BitCrusher*

Amp simulators

A famous mixing practice in the analog domain is called *re-amping* – feeding a guitar recording during mixdown back into a guitar amp so a different sound texture and distortion can be created. Mostly, the signal sent to the amps during mixdown is the direct recording (before any pedals or amp). Many plugins nowadays are designed to imitate the sound of classic guitar cabinets, and most often these plugins also include other guitar processors like tremolos, compressors, gates, echo engines and guitar reverbs. When there is a direct recording of a guitar, we can easily choose the final guitar sound during mixdown (perhaps along the cabinet recording; perhaps replacing it).

Even before the digital age, re-amping was not limited to guitars. Sometimes engineers sent vocals, drums or other tracks into a guitar amplifier (or guitar pedals), using it as a distortion or effect unit. Using the same principle, an amp simulator plugin can be used to distort any type of material. Like any other type of distortion, subtle settings can be useful for gentle enhancements and more trashy settings for more drastic results.

Figure 24.4 The Universal Audio *Nigel*. The center blocks on this plugin are the amp simulator where different amps and cabinets can be chosen. The side blocks provide additional effects like tremolo and delay.

Track 24.33: dGtr SM57
A microphone recording of a distorted guitar.

Track 24.34: dGtr Direct
This is the direct recording (before amp or pedals) of the same recording in the previous track.

Each set in the following samples contains three tracks: The first is the direct recording from the previous track processed with an amp simulator. The second and third are vocal and drums being processed with the same preset:

Track 24.35: Amp Simulator 1 (Guitar)
Track 24.36: Amp Simulator 1 (Vocal)
Track 24.37: Amp Simulator 1 (Drums)

Track 24.38: Amp Simulator 2 (Guitar)
Track 24.39: Amp Simulator 2 (Vocal)
Track 24.40: Amp Simulator 2 (Drums)

Track 24.41: Amp Simulator 3 (Guitar)
Track 24.42: Amp Simulator 3 (Vocal)
Track 24.43: Amp Simulator 3 (Drums)

Track 24.44: Amp Simulator 4 (Guitar)
Track 24.45: Amp Simulator 4 (Vocal)
Track 24.46: Amp Simulator 4 (Drums)

Track 24.47: Amp Simulator 5 (Guitar)
Track 24.48: Amp Simulator 5 (Vocal)
Track 24.49: Amp Simulator 5 (Drums)

Plugin: McDSP *Chrome Amp*
Drums: Toontrack *EZdrummer*

25 Drum triggering

A short story: a band goes to a studio to record a song. The producer and the engineer spend some time miking up the drum kit. They try different microphones in different positions and mic each drum with two microphones. They experiment with different snares of different sizes and woods, go through the process of tuning the toms as they wish, build the famous kick tunnel and damp all the drums. After perhaps seven hours of preparation they cut the first drum take. Two weeks later the track arrives to the mixing engineer, who in his own studio, at night, replaces all the drum recordings with samples – perhaps 20 minutes work. None of the original drum tracks are left in the mix. He also filters some ambiance from the overheads and reconstructs it from the drum samples. Three days later, the clients come around to hear the mix. The drums sound amazing. The band is happy, the producer is happy, the record company is happy and four months later the listeners are happy – everyone is happy. Many tracks have similar stories; one of them is *Smells Like Teen Spirit*.

The practice of using samples to add to or replace real drums started somewhere around the early 1980s. It was used in pop and soon became an extremely common practice in the more energetic metal genres (where it is still prevalent). Throughout the 1990s, engineers started using it more and more in virtually any genre that involved recorded drums. This technique has always had its critics, who dismissed it as being disloyal to the true essence of performance and recording. Nonetheless, drum triggering is an *extremely* popular practice nowadays, and many recorded productions involve some degree of drum samples.

One approach to generating the samples is programming a drum sequence that matches the recorded performance, then using a sampler to trigger the drums. Before audio sequencers existed, this involved syncing a tape machine to a hardware sequencer. Unless the recorded drums were tight to a metronome, this involved (and still does) the generation of tempo maps. Altogether, this method can be rather tedious. Today, the most common approach to drum triggering involves feeding a drum track, say the snare, to a machine or a plugin that triggers the samples in realtime. The samples can be those of a drum module, taken from a sample library, or just clean individual hits that were recorded during the session. The samples either **replace** or are just **layered** with the original recording. Although most commonly used with drums, we sometimes use samples to reinforce cymbals.

There are various benefits of drum replacement: First, we have total freedom over the sample we use, so if we do not like the **sound** of, say, the toms, we replace them with tom samples we do like. When layered, samples might be used to add missing timbre components. Second, the resultant samples track is **spill-free**, which makes gating, compression and the addition of reverb both easier and more consistent. Third, most often as part of the process of drum triggering, we have ultimate control over the level of the hits, which can make the **performance** more consistent (and therefore possibly more powerful).

Drum replacement is ideal for budget drum recordings made in problematic rooms. The good news is that you can get away with horrendous close-mic recordings. The bad news is that you still need good overheads recording, which out of all drum recordings is the hardest one to get in problematic rooms – the small dimensions of the room can cause combfiltering, profound room modes can alter the captured frequency content and, above all, the reverb in such rooms is often not the greatest. Still, drum replacement is a blessing in situations where the drum recordings are flawed. Similar to guitar re-amping, it allows us to select the sound of our drums during mixdown.

The drums in the following demonstration are from the *Hero* production:

Track 25.1: Kick Original
A blend between the recorded kick in and out microphones.

Track 25.2: Kick Trigger
The original audio kick track is converted to MIDI using Logic's Audio to Score function, then Toontrack *EZdrummer* is used to trigger this kick sample. Note the consistent velocity, unlike in the previous track.

Track 25.3: Snare Original
A blend between the recorded snare top and bottom microphones, and a snare reverb.

Track 25.4: Snare Trigger
The snare trigger, which is a product of Trillium Lane Labs *Drum Rehab*. This trigger is not intended to replace the original snare, just to blend with it.

Track 25.5: Snare Original and Trigger
A blend between the original snare and its trigger. Compare this track to Track 25.3.

Track 25.6: Drums Without Triggers
The recorded drum kit without triggers.

Track 25.7: Drums With triggers
In this track, the original kick is replaced by the kick trigger, while the snare trigger is only layered with the original snare.

Plugin: Trillium Lane Labs *Drum Rehab*
Kick: Toontrack *EZdrummer*

Methods of drum triggering

Delay units with sample and hold

One early method of triggering drum samples was using a delay unit that offered a sample and hold facility. A single drum hit was sampled and then could be triggered by feeding the unit with the performance itself. The main disadvantages of this method are that it required sampling a drum hit first (rather than choosing between different prerecorded samples) and that the triggered sample was always the same, including its level.

Drum module triggers

Another early method, still used today, involves the drum modules used with electronic drum kits. The drum module of these kits has trigger inputs that are fed from the drum pads via a standard audio connection. Instead of feeding these inputs with the pads, we can feed a line-level signal from the multitrack or perhaps a gated version from the desk. Drum modules let us choose different samples via their interface. Most modules are velocity-sensitive, meaning that the triggered sample level corresponds to the level of the input hit. Some modules also provide dynamic control where we can set the ratio between input and output velocities. For example, we might configure the unit, so wild variations in the level of input hits will only result in small (or no) level variations of the output samples.

Footswitch on a synchronizer

Some synchronizers have an input footswitch. The synchronizer can generate a specific MIDI note every time a signal exists at the footswitch input. By routing a drum track into the footswitch input, the synchronizer generates a MIDI note with every hit. We can connect the MIDI output of the synchronizer to a drum module or a sampler.

Drum replacement plugins

Early drum replacement plugins, like Digidesign's *Sound Replacer*, worked offline. Recent products like *TL Drum Rehab* (Figure 25.1) or *Drumagog* perform drum replacement in realtime. Essentially, these plugins are loaded as an insert on a drum track, various detection parameters are configured and we get to choose from various drum samples on our hard drive. We can also set the mix between the original track and the triggered samples. Most replacement plugins let us set the ratio between input and output velocities. Also, triggering might be based on velocity layers, where different input velocities trigger

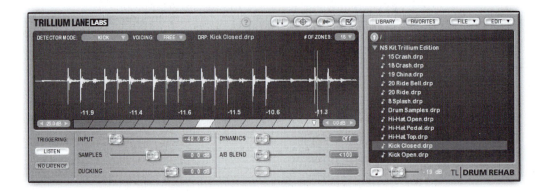

Figure 25.1 The Trillium Lane Labs *Drum Rehab*.

different samples, each of the same drum played at different velocities. This gives more realistic results.

Audio to MIDI

Some applications let us convert audio into MIDI data. In Logic, the single audio track is first analyzed offline, and after configuring various settings a new MIDI region is created with notes representing each drum hit (Figure 25.2). The advantage of audio-to-MIDI conversion is that we can then alter more easily the various hits, whether in level or in timing. Digital Performer offers a very similar conversion that takes place in realtime via a plugin called *Trigger* (Figure 25.3).

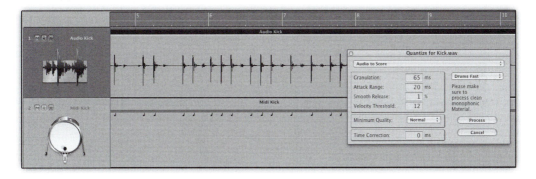

Figure 25.2 Logic Audio to Score conversion. The top track shows the original kick track, the floating window shows the Audio to score dialog with its various configuration options and the bottom track shows the resultant MIDI track after all notes have been transformed to C1.

Figure 25.3 The MOTU *Trigger* plugin.

After the audio is converted into MIDI, the track can feed either a sampler or a software instrument. Various software instruments are designed specifically for drums and can become extremely handy for the purpose of drum triggering. One of them is the Toontrack *EZdrummer* shown in Figure 25.4.

Figure 25.4 The Toontrack *EZdrummer*. This multilayer drum sampler has various drum sets; within each, different drums can be selected. These types of products are highly useful during drum triggering.

Manually pasting samples onto a track

Although not exactly triggering, on an audio sequencer we can copy a specific drum sample, then go from hit to hit and paste the sample to replace original drum hits. This is achievable with toms or crashes, where there are not many hits to replace. However, it can be very tedious with snares and kicks of which there might be more than a hundred hits. Locating the beginning of each hit can also be a difficult task. Features like Pro Tools' tab-to-transient or beat-detective can be of real help in these situations as they enable you to precisely locate each hit. Then keyboard macros like those offered by applications such as Quick Keys can automate the process so that hits can be replaced in a matter of seconds.

26 other tools

To cover each and every processor and effect available for mixing engineers would be exhausting. Combined with the huge amount of plugins that emerge each year, a whole new book could be written. This chapter covers a few mixing tools that may come in handy and are therefore worth knowing about. Since these are less common than the tools covered so far, the information is kept short.

MS

Background

Mono vinyls utilized lateral (horizontal) groove excursions to encode amplitude variations. When stereo records were introduced in 1958 there needed to be a way to encode both the left and right channels. One proposal was to encode the left channel as lateral excursions and the right channel as vertical ones. The problem was that had a stereo record been played on a mono player, only the left channel would be heard (at that time this could mean losing the drums, vocals or bass, since due to the infancy of pan pots many records had different instruments panned hard to the extremes). As part of his stereo invention, Alan Blumlein taught us that any left and right information can be converted into middle and side and vice versa. Based on this, EMI cut the first stereo record where the middle is encoded as lateral groove excursions, and the side as vertical. This way, even if a stereo record was played on a mono player, the player would only decode the lateral motion, which represents the mid (the mono sum of left and right). A stereo player would decode both lateral and vertical motions (mid and side) and would convert these into left and right. This mid-side system is used in vinyl pressing to this day. Stereo transmissions of FM radio and television are also encoded in MS.

MS (or *mid-side* or *mono/stereo*) is the general name given to stereo information existing in the mid-side form. It is a well-known stereo recording technique that was already discussed in Chapter 13 (also see Figure 13.15). Many mixing engineers come across MS at least once – the famous *Fairchild 670* could work in either left/right mode or mid/side mode (termed Lat/Vert on the 670 for its relation to records). Some plugins provide MS operation modes in addition to the standard left/right.

Figure 26.1 The Universal Audio *Fairchild* plugin. This plugin looks and works very similar to the original unit (the price of which can easily exceed the $25 000 mark due to its vintage status). Like the original unit, the plugin can work in either left/right or mid/side modes (determined by the control between the two channels). The mid/side mode is labeled Lat/Vert (lateral/vertical) as per the stereo record cutting method.

MS and LR conversion

The conversion from LR to MS is straightforward: mid is half the sum of left and right, while side is half the difference between left and right. When such conversion takes place, the mid is sent to the left channel, and the side to the right. The conversion from MS to LR is even simpler: left is the sum of mid and side, while right is the difference between them. Expressed in equations:

- $M = (L + R)/2$
- $S = (L - R)/2$
- $L = M + S$
- $R = M - S$

We can achieve such conversion with any desk or audio sequencer. To get M, for example, we have to mix (sum) the left and right channels, then attenuate 6 dB (which is the halving of voltage or sample value). To get S, we do exactly the same, only that we invert the phase of the right channel. Figure 26.2 shows one way how LR to MS, and MS to LR can be done in Pro Tools, and it does not take much to see how unwieldy this process can be.

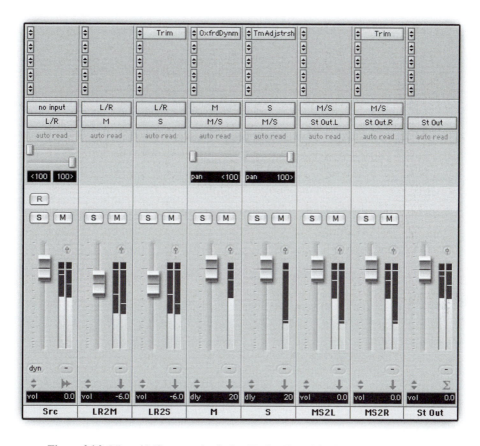

Figure 26.2 LR and MS conversion in Pro Tools. The original track is routed to the L/R bus, which feeds both the LR2M and LR2S auxiliaries. Since the output of LR2M is mono, the left and right channels are summed. There is also a 6 dB attenuation corresponding to the required halving. LR2S does exactly the same, only that using the Trim plugin the right channel is phase inverted, resulting in mono output which is the difference between left and right (attenuated by 6 dB). The M and S buses are each fed into a mono auxiliary, where each can be processed individually. The *TimeAdjuster* plugin is only inserted on S to manually compensate for the 20 samples plugin delay the Sonnox *Oxford* on M involves. Both M and S are panned to the extremes and routed to a different bus called M/S, which feeds both the MS2L and MS2R auxiliaries. These two auxiliaries feed the respective left and right channels of the St Out bus. While MS2L simply sums M and S, the Trim plugin on MS2R inverts the phase of S, thus the mono output is the difference between M and S.

There are a few ways we can shorten the overhead this conversion involves. Ideally, we would like to have a plugin that does the conversion for us. Both Digital Performer and Logic provide such a plugin. Looking at the equations above, it is clear that the conversion is very similar, only that during LR to MS conversion we have to attenuate by 6 dB. Indeed, the same plugin can perform both conversions, and we might need to manually reduce the MS output by 6 dB. In addition, we can use a plugin that works on only one of two channels. This way, we could first convert from LR to MS, process either the M or S channels, then convert back to LR. Figure 26.3 demonstrates this.

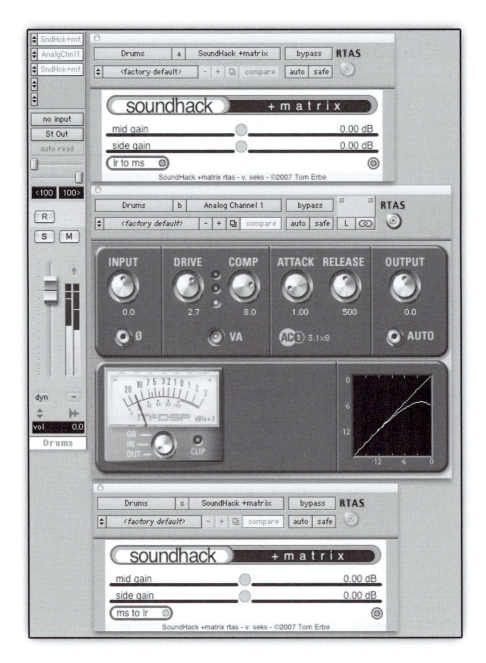

Figure 26.3 Optimized MS processing. The *+ Matrix* plugin by Soundhack (freeware) was loaded on the first insert slot and converts LR to MS. Pro Tools provides a multi-mono mode, where a plugin can process the left and right channels individually. The McDSP *Analog Channel 1* was loaded in such a mode on the second insert slot and only affects the left channel, which represents M. Then on the third insert slot another *+ Matrix* plugin is loaded, this time converting the MS back to LR.

Applications of MS

There are two principal ways in which MS can be used in mixing. We are discussing stereo tracks here, and it can be easier to understand the whole concept if we imagine overheads panned hard to the extremes. The first use of MS involves changing the ratio between M and S. There is little point attenuating S, since this produces similar results to narrowing the stereo image using the pan pots. But something very interesting happens when we attenuate M. Basically this results in a wider, more spacious stereo image, which can be useful sometimes, for example, if we want to widen a stereo overheads recording that presents a narrow image. What effectively happens as we attenuate M is this: instruments panned dead center drop in level, and instruments panned slightly off-center start shifting toward the extremes. Again, this can be practical if the snare on the overheads is only slightly off-center, and we want to shift it further sideways to clear more space for the vocals, bass or kick. But a really interesting thing happens with instruments panned to the extremes. As we attenuate M, such instruments start appearing on the opposite extreme out-of-phase. Essentially, this creates a similar effect to the out-of-speakers trick. But when discussing the out-of-speakers trick we said that it works best with low frequencies; high frequencies simply tend to appear around the center. So as M is attenuated, low-frequency instruments panned to the extremes tend to appear as if coming out-of-speakers, while instruments with high-frequency content tend to shift inward toward the center. An absorbing effect indeed.

The second beneficial way of using MS involves situations where we want to process the mid or side individually. For example, say we want to attenuate the kick on the overhead, since it does not blend well with the close-mic. We can use the close-mic to duck the overheads, but this would mean that any hi-hats, toms or crashes would be ducked as well. Since on overheads the kick is often found around the center, whereas hi-hat, toms and crashes are panned off-center, it makes sense to only duck the overheads in the center (mid) and not both the left and right channels. Another example would be a stereo choir recording fighting with the lead vocal. We can achieve better separation by rolling-off around, say, 3 kHz on the mids of the choir. Other, more creative applications might involve an instrument ducking the S of its own ping-pong delay; this way, while the instrument is playing, its echoes would appear monophonic, and when the instrument is not playing the slow release of the ducker would slowly make the echoes open into stereo. Many more practical and creative examples exist, but the principle should be clear by now.

Track 26.1: Drums Original
These drums are intentionally programmed to have a narrow stereo width. The kick is panned hard center. The snare is panned slightly to the left, the hi-hats slightly further to the left from the snare.

Track 26.2: Drums MS Processed
The drums from Track 26.1 are converted to MS, then the M channel is attenuated by 2 dB and the S channel boosted by 4 dB, then converted to LR format again. The result is drums with wider stereo image, and both the snare and the hi-hats evidently shifted further to the left.

- - - - - - - - -

Track 26.3: Loop Original LR
The original loop, with the kick panned hard right, the snare hard center and the hi-hats hard right.

Track 26.4: Loop MS
The loop after MS encoding. The snare which was previously panned hard center now only appears on the M channel (left).

Track 26.5: Loop M Only
Only the M channel of the encoded loop.

Track 26.6: Loop S Only
Only the S channel of the encoded loop.

Track 26.7: Attenuating M
In this track, the M channel changes in level prior to decoding back to LR. This track is subdivided into four-bar chunks. Bars 1–4: The original loop. Bars 5–8: The M channel is gradually attenuated to $-\infty$. Bars 9–12: involve no M, only S (which is essentially the same S on both left and right channels with one channel out of phase). Bars 13–16: The M is gradually boosted back to 0 dB. Bars 17–20: The original loop again.

Plugin: Soundhack *+Matrix*
Drums: Toontrack *EZdrummer*

Pitch shifters and harmonizers

A pitch shifter alters the pitch of the input signal without altering its duration. The way it works is very interesting and provides the basis for pitch correction and granular synthesis. If we zoom into the recording of a human voice we would identify repeating patterns that only vary over time. For example, when one sings a long 'Ahhhh', one produces many repeating patterns that vary slightly over time. In practice, nearly everything we sing is essentially made up of small repeating patterns. The sound of most instruments is produced in exactly the same manner. A pitch shifter identifies these patterns, then shrinks or stretches them into the same time span. For example, shrinking a pattern to half of its length, then fitting two shrunken halves into the same space as the original pattern would raise the pitch by an octave. Doing so pattern by pattern would raise the overall input signal by an octave. To pitch down something, half of the original pattern is placed into the space of the original one. The actual process is more complex, but at its core is this rudiment. Also, percussive instruments do not present clear patterns, so pitch-shifting them involves slightly different algorithms.

Pitch shifters provide transposition values in the form of semitones, and often cents. Some provide wet/dry control. They can be used to improve double-tracking or ADT. But their real power is in their ability to artificially produce harmonics. Perhaps most commonly, pitch shifters are beneficial with bass guitars. By layering a bass guitar with an artificially produced second harmonic (an octave above), we can increase the guitar definition. We can also add the third (an octave and perfect fifth or 19 semitones) and fourth (two octaves above) harmonics and alter the level between them to shape the color of the guitar. We can also blend the bass guitar with a version of itself an octave-down. This would enhance the low-energy power of the guitar. Indeed, this process is used in many bass enhancers.

Harmonizers differ from pitch shifters in that they produce more than one transposed version of the input signal and are configured around musical intervals and chords. As one would expect, harmonizers can enrich a vocal performance, but can also be used with guitars and keyboards.

Track 26.8: Vocal Delay Only
This track involves a blend of the dry voice and a 85 ms delayed version.

Track 26.9: Vocal Delay with Pitch Shift
The delayed version in this track is pitch-shifted by 30 cents. This track sounds more like double-tracking whereas the previous track sounds more like a plain delay.

Plugin: Logic *Pitch Shifter 2*

Exciters and enhancers

Exciters and enhancers are two terms that can be used interchangeably. Both exciters and enhancers have the role of making instruments, mixes and masters sound better. Better, for that matter, is often described as a combination of cleaner, brighter, closer, wider, livelier and more defined. There are two questions to ask: How do they do it? And, why *shouldn't* we always use them?

The first aural exciter was produced by Aphex in 1975. Legend has it that the whole concept was discovered by accident, when one channel of a stereo amplifier faultily produced distortion. When added to the properly working channel, this improved the sound. Today, many companies manufacture exciters, and we also have a multitude of dedicated plugins or sometimes an extra facility within other types of plugins. Although the exact way in which each exciter works is a guarded secret often patented, the core principles of aural enhancement are widely known:

- **Addition of harmonics** – by adding a controlled amount of harmonics, sounds tend to sound brighter and clearer. This is somewhat similar to distorting or saturating signals, only that within an exciter it is done in a more controlled fashion.
- **Level enhancements** – an enhancer might employ a process that increases the perceived loudness of signals. Essentially, part of the device involves some degree of loudness maximizing.
- **Dynamic equalization** – by dynamically equalizing the input signal, it can appear more solid and less prone to frequency/level imbalance caused due to varying notes.
- **Stereo enhancement** – by introducing subtle variations between the left and right channels, instruments might appear to widen, have more life and become clearer. We have seen this previously when discussing stereo equalization.
- **Phase enhancement** – frequency-dependent phase shifts are said to occur acoustically and are a proven part of signal processing and reproduction. Enhancers employ a process that delays different frequency ranges by different times, and by doing so rectify phase misalignments. This can increase the definition and clarity of the signal in question.

Another look at this list reveals that all, apart from the last technique, have already been described in this book. Exciters and enhancers often involve very few controls. For example, the more basic enhancers only have an amount pot. While this means they can be employed very quickly, this can also be limiting to some extent – we might be able to achieve similar enhancements using standard tools like equalizers and distortion, with which we have further control over the final result.

Figure 26.4 The Noveltech *Character* plugin. The enhancement process involves just three controls: character, target and one of three modes.

In analog studios there would normally have been no more than one or two enhancers. Mixing engineers had to choose which instrument (if any) called for enhancement the most. In the plugin age, we can enhance every single instrument. So why wouldn't we just enhance all of them? The answer goes back to the basic mixing question: What is wrong with it? If nothing is wrong with an instrument, why should we enhance it? Even if something is wrong, enhancers might not be the solution – perhaps compression is the answer.

If we enhance each instrument in the mix, they will each sound better but not with respect to one another. From a mix perspective, they might all sound equally exciting. Only enhancing one or two instruments creates some contrast that can be beneficial to the mix. In addition, enhancing one instrument might reduce the clarity of another instrument. This is partly due to the fact that enhancers add frequency content in the form of harmonics that can mask other instruments.

Finally, there is another risk in using enhancers called *frying* or *overcooking*. We find the immediate effect of exciters and enhancers very appealing, and we can easily be tempted to drive an enhancer harder than necessary. This can result in sound that is fatiguing to the ear, and sometimes it is only later that we discover how overuse of exciters makes the mix sound brittle and undefined.

Without a doubt, exciters and enhancers can produce beautiful and magical results, but the points above must be considered when deciding when they should be used and to what extent. Like many other mixing tools, using them sparingly can often yield the most effective results.

Track 26.10: Bass Source
The source bass track used in the following sample, before enhancement.

Track 26.11: Bass Enhanced
Apart from being louder, this track is also slightly brighter and overall presents more vitality.

Plugin: The Enhance facility of the Sonnox *Oxford Limiter*

- - - - - - - - -

Track 26.12: Drums Source
The source drums to be enhanced in the following samples.

Track 26.13: Drum Enhancement 1
Compared to the source track, there is added warmth and low power in this track.

Track 26.14: Drum Enhancement 2
Compared to the source track, this track is brighter and sounds more alive.

Track 26.15: Drum Enhancement 3
This is frying – this track presents distorted frequency content favoring the highs. Listening to this track too loud or for a long time can easily cause ear fatigue.

Plugin: Novaltech *Character*

Transient designers

Transient designers, or *transient enhancers*, are designed specifically to accent or contain transients. Essentially, most of them are a hybrid of upward expanders (for accenting) and downward compressors (for containing). However, most transient designers examine differences between peak and RMS variation of the input signal in order to determine when transients really happen. Transients rise quickly enough to yield differences between the peak and RMS readings; a signal that rises slowly would not. It is for this very mechanism that transient designers have an advantage over upward expanders or compressors – they tend to handle transients more uniformly regardless of at which level these happen. Put another way, a quiet snare hit that does not overshoot the threshold would not trigger compression; a transient designer can be set so both quiet and loud hits would be treated.

Like enhancers, transient designers are something of an automatic tool that can easily add punch to percussive instruments. They can also add life and accent or revive the dynamics of various instruments. Like with compressors, we can use them to reduce ambiance and decay. Altogether, they tend to produce a very appealing effect, especially if used with more conservative settings.

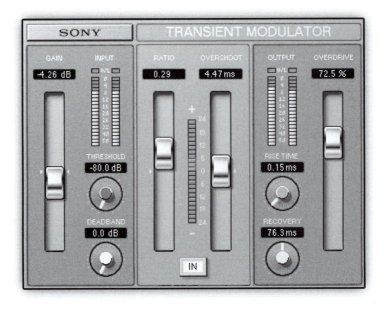

Figure 26.5 The Sonnox *Oxford Transient Modulator* plugin.

Track 26.16: Compressed Bass
This is the source track for the following samples. The bass in this track is compressed so as to suppress its attack.

Track 26.17: Bass Transient Revival
After applying a transient modulator, the bass gained some extra attack.

Track 26.18: Drum Compressed
These are the source drums used in the following sample.

Track 26.19: Drum Enhancement 3
After applying a transient modulator, the snare, hats and the kick have more attack. Note that the ambiance has been reduced and the drum image moved forwards.

Plugin: Sonnox *Oxford Transient Modulator*
Drums: Toontrack *EZdrummer*

27 Automation

It is as simple as this – commercial mixes are full of automation. This process, which is performed naturally by professional mixing engineers, is somewhat overlooked by the novice. Even before the invention of multitrack recorders, engineers used to ride levels of different microphones during recording. With the introduction of multitrack recorders this practice didn't stop, but some of it was postponed to the later mixing stages. Before automation computers were integrated into consoles, the engineer, the assistant, the producer or even the band members (as many hands as were needed) used to gather around the console to **perform** automation passes that were printed straight onto the final two-track. Each person knew exactly what needed to be moved, how and when. For the more complex mixes an 'action score' was written. It was, practically, a performance, and the console was the instrument. If someone botched, the whole performance had to start all over again. Automated consoles were introduced around the late 1970s. Even today, many of them can only write the automation of channel faders, mutes and solos, and studio engineers and their companions continue to perform live automation on analog desks.

Nowadays, audio sequencers let us write the automation for virtually every control in the mix and, moreover, provide graphical editors in which automation can be corrected or even drawn from scratch. There is no need for 12 hands since we can automate different controls during different passes. Never in the history of mixing was writing and editing automation as easy as it is today, with audio sequencers. It is surprising that many DAW users fail to comprehend the benefits of this powerful facility and the potent effect automation can have on the mix.

We say that each song is a story. In the case of classical music, progressive rock and many jazz pieces, the stories might be epics. In modern pop productions, the story is often squeezed into around three minutes, and changes happen very quickly. There is a lot happening in each story – the music develops, the arrangement changes, different sections should have different impacts and the importance of different instruments varies. It would be unfair to have so much action streaming through a static mix. It is our responsibility as mixing engineers to accommodate the dynamic movement of the music and the structural elements of the production. Then, automation can always be used to create some interest or add some extra movement. The options are endless, and virtually any process or effect can be automated. It is possible to regard this late phase in the mix as playtime, where creativity and experiment might replace practical needs. Finally, level automation, mostly on vocals, is sometimes done before a compressor for a more musical result.

479

Any list of possible automation examples would be partial – the options are truly endless. It should be stressed that the most common mix automation (and often the most practical one) involves levels, so fader rides should probably precede any other type of automation. Here are just a few things we can automate, some of which have been mentioned in this book already, while others can be heard on some commercial mixes:

- Raise the level of a specific instrument during the chorus, then bring it down again during the verse. Likely candidates are vocals, kicks, snares and guitars.
- Mute some instruments early in the song, then introduce them later.
- Make an instrument brighter or darker during specific sections.
- Ride the level of overheads during crash hits.
- Ride the level of overheads or any other instrument up and down with relation to the tempo.
- Pan something to one place during one section, then pan it somewhere else during another.
- Introduce some interesting vocal effect momentarily.
- Change the timbre of the kick during some sections.
- Mute the double-tracked vocals at points.
- Change the reverb of the snare between sections or with relation to the note values being played or just alternate it between hits.
- Bring down the level of some instruments to clear some space for another instrument.
- Introduce distortion on the bass during the chorus only.
- Increase the compression on the drums as the song progresses.
- Widen and narrow the stereo width of an instrument during various sections.

Automation engines

Automation engines work on the principle of storing the position of controls using automation events. An automation event typically includes which control has been automated and its position at specific timestamps (very similar to the way MIDI systems handle MIDI control messages). Before any automation has been written, each control is free to move. Once even a single automation pass has been performed, the control position is often bound to the calculated position between two automation events or the position of the latest automation event.

> It is worth knowing that some automation systems store events with SMPTE time code, which advocates the use of 30 fps for higher-resolution automation. This is subject to the project not involving any visuals that might dictate a different frame rate.

The automation process

Performing vs. drawing

When *performing* automation we move controls during playback. Audio sequencers (and some digital recorders) provide a graphical display of automation events and also let us draw them on screen. Sometimes, drawing automation is quicker. For example, if we want to mute a specific instrument for a minute, it should take less time to draw two mute events than performing mute automation and, having to wait for a minute. Drawing automation can also be beneficial when we want events to be quantized to the tempo grid. However, some argue that performing automation yields more musical results as there is always an interaction between what we hear and what we do – we respond to the music rather than approximating the effect of drawn events (which effect we mostly hear after drawing). Depending on the situation, different methods may be more appropriate, but it is worth remembering that between performing and drawing, the former is more likely to involve feel.

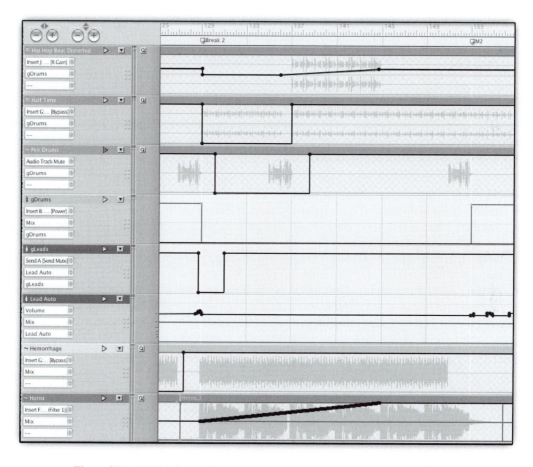

Figure 27.1 This Digital Performer screenshot shows all the automation events happening during the second break of *The Hustle.* Some of these events have been drawn (like those in the top track), while others performed (like those in the Lead Auto track).

Performing automation

Automation is said to be written rather than recorded. We do not have to press a record button in order to write automation, although sometimes we have to tell the system which control is to be automated and sometimes we have to assign (arm) a specific channel of interest to the automation engine (more common on digital desks).

When writing automation on an audio sequencer, we either use a control surface or the mouse to alter the position of controls during playback. Ideally, automation systems would like to know when a control is touched and when it is released – often automation is only written between these two events. Some control surfaces feature *touch-sensitive controls*; these are either faders or (less commonly) rotary knobs that by way of varying capacitance detect a finger touch. If a control is not touch-sensitive, the automation engine would start writing automation either with the first control movement or as soon as the control position matches the existing automation value. In most cases, such control is considered released after a certain period has passed with no position changes (a period often called *touch timeout*). Controls on a computer screen are regarded as touch-sensitive – a control is considered touched as soon as the mouse button is pressed, and released as soon as the mouse button is released.

Automation modes

Automation engines may vary in their modes, features and response to user action. We can, however, generalize a few automation modes typical in many systems. The modes in this list are illustrated in Figure 27.2:

- **Off** – automation data is neither read nor written.
- **Read** – previously written automation is read, but new control changes are not written.
- **Touch** – new automation data is written as long as a control is touched, otherwise previous automation is read.
- **Latch** – automation is written from the moment a control is touched until playback stops.
- **Write** – new automation information is written, as long as the playback is running, overriding previously recorded automation. To prevent unwanted automation overrides, automation engines often switch to a different mode after each pass in write mode.

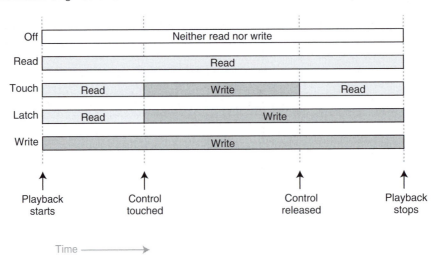

Figure 27.2 The five typical automation modes.

Apart from when the playback stops, the writing of automation might also stop if the mode is set back to Read or if the specific channel automation assignment is disarmed. Two things can happen when automation writing stops during playback: either the control jumps instantly to the previous automation position (which could generate clicks) or it slides to that position. Often the time it takes a control to slide between the two positions is called the *match period*.

Another mode known as **Trim** or *Relative* mode usually applies to levels only (faders and sends). It is useful when we want to adjust the level of previously written automation. The idea is that instead of writing the absolute level values, trim mode simply offsets the existing automation levels by the amount of dB we move the fader by. For example, if the fader is brought down from −6 to −12 dB, all automation events during the writing pass would drop by 6 dB. Systems often provide the functionality to apply the relative change throughout the song.

Automation is often one of the very last things we do in a mix before printing it. After recording automation for a specific control, any adjustments can be something of an effort since we have to adjust the full automation information, rather than just move the control position. This can be especially annoying if we want to alter the level of tracks after writing level automation. This is exactly what trim mode came to solve, but even trim mode for global level adjustments can be cumbersome at times. An elegant solution to this problem was mentioned earlier – we could insert a gain plugin and use it to perform any level automation. The track's fader would then be free for global level adjustments. Another solution is to send different instruments to an audio group and automate the level of the group instead of that of the original tracks.

Automation alternatives

Duplicates

In some situations, we might want to apply more than a few changes to a specific instrument. For example, during the chorus we might want to make the overheads louder, compress them more, narrow their stereo width and alter their equalization. We can automate all the related controls, but a different approach could be quicker: most audio sequencers allow more audio tracks than any project requires. In scenarios where serious changes are needed, it pays sometimes to duplicate the track in question, trim or mute the respective sections on the two tracks (for example, having the choruses on the duplicate only, but removing it from the original track) and mix the duplicate differently, with all the involved changes (Figure 27.3).

Fades

Mix-fades are another often overlooked practice – instruments very often fade in or out rather than instantly starting or stopping. The risk of clicks exists with any mute automation or region-trimming, whereas fades are click-proof. Transitions between sections are gelled using fades, whereas both mute automation and region-trimming can come across as very

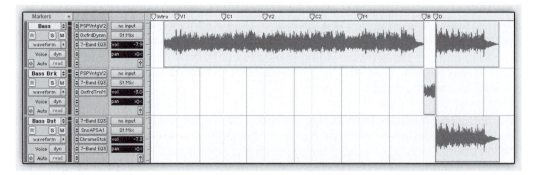

Figure 27.3 Duplicates 'automation'. This screenshot shows three bass tracks. The top track plays throughout most of the song and provides the main bass sound in the mix. During the break, a different sound was sought, involving a variation of tonal characteristic, level and some additional processing. To achieve this, the bass was moved onto a new track with different processing and level (Bass Brk) for the length of the break only. The bottom track is a distorted layer that is only mixed with the main bass track during the outro.

unnatural. While crossfades are more of an editing affair, fade ins and outs are a powerful mixing tool. We can achieve fades using level automation, but automation makes level adjustment a longer procedure, and where fades are needed is often realized very early in the mixing process. The fade tools provided by every audio sequencer are used for this task.

Control surfaces

When Digidesign launched the *Icon D-Control* (Figure 27.4) back in 2004, eyebrows were raised as to whether there was a place for a pure control surface as big as a large-format console. But for many engineers that used large-format analog consoles this seemed all too justified.

The analog vs. digital debate goes beyond audio quality to the realm of human interaction. Analog consoles provide the highest level of accessibility – all the controls are laid out in front of you and any sonic action you'd like to perform is within reach. In most cases, 2 or 3 seconds is all it takes to translate your sonic vision into sound. Then there are the layer-based digital desks, where usually a third or half of the mix might be readily accessible, although you might have to press the select button in order to equalize a specific track or navigate some on-screen menus in order to tweak a certain effect. On the bottom of the accessibility ladder comes a computer system where the whole mix has to be channeled through a rather primitive device called a mouse. Compared to large-format desks, even a 21-inch screen is tiny. The more tracks a project involves the slower the mixer navigation becomes. Plugin windows have to be opened and closed. Things can take time. Had it only been a matter of more or less time, perhaps control surfaces as

Figure 27.4 The Digidesign *Icon D-Control*. This product, which looks like a large-format console, is a pure control surface for Pro Tools. Products of this kind enable a mixing experience that was once reserved for users of large-format consoles, and all their technical and creative benefits (courtesy of Digidesign, Bill Schwob Photo).

large as the *Icon* would not exist. But the creative flow can easily be restrained while the brain is busy operating the computer. Audio sequencers provide a few features that let us shorten the time between our vision and its implementation. But there is no argument that mouse-mixing will never be as fast as mixing on a large-format analog console or on large control surfaces like the *Icon*.

It goes even beyond that. Most people find sliding faders and turning knobs much more natural than dragging a mouse on the screen, let alone when it comes to automation. Then there is also the fact that a mouse is a serial interface – rarely can we change more than one control at a time, although many times in mixing we need such ability. We might, for example, alter the ratio and threshold of a compressor simultaneously, we might fancy boosting the highs on one channel while attenuating them on another, or we might want to bring one fader down while bringing another up during an automation pass. There are many examples.

Control surfaces come in various forms and sizes: from the large Icon, to the moderate-size designs like the 9-fader *Mackie Control*, to compact designs like the *AlphaTrack* in

Figure 27.5 The Frontier Design Group *AlphaTrack*. Despite its compact size, a control surface like this gives automation in particular and mixing in general a far more natural feel.

Figure 27.5. Many of them can be cascaded, and the larger the work surface we are using the more accessible our mix becomes and the faster we can realize our sonic vision. But, even compact surfaces like the *AlphaTrack* can make automation and others aspects of mixing a far more natural experience.

Part III
Sample Mixes

Regarding the naming conventions of tracks on the DVD:

Many of the audio tracks in this part involve both soloed and non-soloed versions of the instrument and treatment in question. Audio tracks with '(i)' denote the soloed (isolated) version, while '(m)' denotes the non-soloed (mix) version.

28 Hero (Rock)

Performed by AutoZero (*www.autozero.co.uk*)
Lyrics by Dan Bradley, music by AutoZero.
Dan Bradley: vocals, guitars.
Lish Lee: bass, vocals.
Lee Ray Smith: drums.

Produced by AutoZero.
Recorded at Soho Studios, London.
Engineered by Guy Katsav.
Edited by Luca Barassi.
Mixed (in Pro Tools) by Roey Izhaki.
Mastered by Mandy Parnell (*www.mandyparnell.com*) at Electric Mastering, London.

AutoZero is a three-piece indie band from London. *Hero* is a rock song based on classic rock instrumentation – drums, bass, rhythm/power guitars and vocals. The song was recorded overnight at Soho Studios, London. Although offering state-of-the-art mixing facilities, Studio 1 (where *Hero* was recorded), has a very small live room – just big enough for a drum kit. This small room is evident on the drum recordings. The limited time – 7 hours – in which this production was recorded is another contributor to the fact that the raw material was less than ideal. As with many other recordings done in limited time, much of the mix efforts focused on correcting rather than elevating.

Track 28.1 is a home-recorded demo of the song, involving an electronic drum kit. Track 28.2 is a mix-ready version involving rough levels and panning, but no processors or effects. Track 28.3 is my own rough mix of the pre-edited multitrack. Neither the toms nor the backing vocals are mixed in this rough mix, and the work on it stopped once I felt familiar enough with the tracks and had a firm plan for the real mix. Perhaps the greatest lesson offered by the rough mix was that some distinction had to be made between the verses and the choruses – the original recordings involved little arrangement changes between the sections. Track 28.4 is a mix-in-progress version, Track 28.5 is the final mix, while Track 28.6 is the mastered final version.

Drums

Overheads

Inserts: Sonnox *Oxford EQ*, PSP *Vintage Warmer*.

The small live room in which the drums were recorded resulted in an overheads recording that had a few flaws (Track 28.7). The small-room ambiance captured on the recording was unappealing. In addition, the individual kick, snare and tom tracks all suffered from ill interaction with the overheads. For these reasons, the cymbals-only approach was chosen for the overheads. First, they go through the *Oxford EQ*, where a steep HPF [*400 Hz, 30 db/oct*] goes as high as it can to roll-off all the lows (Track 28.8). In addition, a parametric filter [*702 Hz, −8.37 dB, Q 3.08*] further attenuates the snare and removes an annoying mid-range noise (Track 28.9). The EQ is followed by the *Vintage Warmer* [*drive +12.9 dB, knee 41.1%, speed 35.8%, auto release, mix 78%*], where parallel compression tightens the sound (Track 28.10). The level of the overheads was automated to rise during the last drum bar. Also, their stereo image was narrowed to give a tighter impression and to clear space on the extremes for the distorted guitars.

Kick

Kick 1 Inserts: Toontrack *EZdrummer*, Digidesign *Smack!*, Sonnox *Oxford EQ*.
Kick 1 Sends: Ambiance Reverb *(*UAD *DreamVerb)*.

Kick 2 Inserts: Toontrack *EZdrummer*, Digidesign *Smack!*, Sonnox *Oxford EQ*.

The multitrack contained both kick-in and kick-out tracks (Tracks 28.11–28.12). Having auditioned these tracks while working on the rough mix, it quickly become apparent that replacing the recorded kick with triggered samples would not only provide easier material to work with, but would also yield a better sound altogether. The kick on the rough mix is already triggered. The triggering process involved importing the kick-in track into Logic and converting it to MIDI using the *Audio to Score* feature. The resultant MIDI track was then loaded on an instrument track in Pro Tools, with *EZdrummer* as a virtual instrument. The kick sound used is the default 22" GMS Felt Beater from the Pop/Rock kit.

In the final mix, there are two kick tracks. The two never play together and the toggling between them sharpens the more powerful moments in the song, as well as adding some interest. Both kick samples are identical, both are processed using the same plugins, but the settings on these plugins are different.

Kick 1 (Track 28.13) is the more rounded, settled kick and it plays through most of the verses and during the break. It is first compressed using *Smack!* [*Norm mode, input 5, ratio 2:1, attack 4.1, release 0.8*], which adds some punch and mass (Track 28.14). Then, it goes through the *Oxford EQ*, where four parametric filters shape the sound of the kick: the first filter [*83.3 Hz, 3.24 dB, Q 2.83*] adds some thud (Track 28.15), the second filter [*220 Hz, −5.26 dB, Q 2.57*] attenuates tapping low-mids (Track 28.16), the third filter [*1003 Hz, 5.6 dB, Q 2.83*] adds attack (Track 28.17), and the fourth filter [*5 kHz, −2.87 dB, Q 1.93*] reduces click (Track 28.18). This kick is also sent at a low level to the ambiance reverb, partly to soften it, partly to blend it into the ambiance, partly to move it further back in the mix and partly to add to its timbre (Track 28.19).

Kick 2 (Track 28.20) is the more powerful kick between the two, having more snap and presence. It plays during the choruses, bridge, outro and some verse parts. It is first compressed using *Smack!* [*Norm mode, input 6.4, ratio 6:1, attack 10, release 0*], which adds tightness and punch (Track 28.21). Also, the distortion facility in *Smack!* [*Odd + Even*] was used to add some grit (Track 28.22). The compressor is followed by the *Oxford EQ*, where four parametric filters shape the sound of the kick: the first filter [*110.5 Hz, 3 dB, Q 2.83*] adds thud (Track 28.23), the second filter [*260 Hz, −5.78 dB, Q 1.6*] reduces unnecessary mids (Track 28.24), the third filter [*938 Hz, 11.45 dB, Q 4.83*] adds attack (Track 28.25), and the fourth filter [*5.1 kHz, 2.4 dB, Q 3.9*] adds click (Track 28.26).

Snare top

Inserts: PSP *Neon*, Digidesign *DigiRack EQ 3.*
Sends: Ambiance Reverb (UAD DreamVerb).

Having tried gating, compression and triggers, I came to the conclusion that equalization alone was the most suitable approach for the snare top. A healthy amount of hi-hats and ride spill on this track made the cymbals unstable once compression or gating had been applied, and a fully triggered snare sounded too mechanical. Not gating the snare meant that spill on the track limited equalization possibilities, mainly with regard to the highs, where the hi-hats and the ride roam. On the other hand, there was some benefit in the support the hi-hats got from the snare track.

Snare top (Track 28.27) first goes through *Neon*, which in linear-phase mode shapes its sound. A low-shelving filter [*1.1 kHz, −5.9 dB*] attenuates an excess of lows and low-mids (Track 28.28), a parametric filter [*312 Hz, −4 dB, Q 1.8*] attenuates a resonant tone (Track 28.29), and another parametric filter [*2.5 kHz, −5 dB, Q 2.9*] attenuates some cymbals spill without harming the snare timbre too much (Track 28.30). On the *DigiRack EQ*, a high-shelving filter [*4.6 kHz, −1.8 dB*] softens the snare and reduces the cymbals spill (Track 28.31). The high-shelving filter was automated to only take effect during the verses − in other sections the snare becomes brighter and the cymbals slightly louder. The sole reason for the addition of another EQ was that automating it was more straightforward than automating the *Neon*.

Snare top is sent to the ambiance reverb, which gels it to the sound stage and shifts it slightly backward (Track 28.32). The level of this track was automated to balance specific hits that were either too loud or too quiet. The track is panned to approximately 10:30.

Snare bottom

Inserts: Digidesign *DigiRack EQ 3,* Sonnox *Oxford Dynamics.*
Sends: Snare Reverb (Audio Ease Altiverb, McDSP Channel G Dynamics).

Snare bottom has somewhat more importance in this track since snare top did not deliver a complete snare sound. Snare bottom complements snare top and contributes both presence and crispiness (Tracks 28.33–28.34). The raw track (Track 28.35) first goes through the *DigiRack EQ*, where a HPF [*313 Hz, 24 dB/oct*] rolls off much of the kick spill and the snares body, which is already contributed by the snare top (Track 28.36); a parametric filter [*1 kHz, 5.1 dB, Q 1*] adds some presence and attack (Track 28.37), and a high-shelving filter [*6 kHz, 4 dB*] brightens the snare and accentuates its crispness (Track 28.38). Following the EQ the *Oxford Dynamics* was inserted, where both the gate and the compressor are employed. The gate [*threshold −19.2 dB, range −40 dB, attack 0.01 ms, hold 30 ms, release 11.2 ms*] removes spill, notably the snare rattle caused by kick hits (Track 28.39). The compressor [*threshold −18 dB, ratio 2:1, soft knee 40 dB span, attack 5.2 ms, release 130 ms*] adds some weight and density (Track 28.40). Snare bottom is panned to the same position as snare top.

Snare bottom is sent to *Altiverb*, and the IR used is the *Plate Short* from the *EMT 140 (Wendy Carlos)* category (Track 28.41). The only modification on the reverb is a 9.2 dB boost on the equalizer's treble (Track 28.42), which gives the reverb more shine. The reverb is gated by the *Channel G dynamics*, with its key input set to the dry snare (Track 28.43). This gated reverb was added as an effect, but also in order to send the snare further back.

Snare triggers

Main track Inserts: Digidesign *Smack!* Digidesign *DigiRack EQ 3.*
Support track Inserts: Digidesign *DigiRack EQ 3*, McDSP *Compressor Bank CB2.*

A triggered snare track was generated using Trillium Lane Labs *Drum Rehab*. From its integrated library, the sample chosen was the *DW1* snare from the *Perfect Drums* collection. Although eventually the snare sound is based on the recorded snare-top track, the triggered snare is still mixed, but only as a layer underneath the recorded snare. The triggered snare is mixed in all sections of the song apart from the verses. Its main role is to add brightness and definition to the recorded snare sound (Tracks 28.44–28.45). Also, since the velocities of the triggered hits are nearly identical, they balance out to some extent the overall snare hits.

The raw triggered snare (Track 28.46) had a decay that was too long and yielded a fluffy overall snare sound. So it is first treated by *Smack!* [*Norm mode, input 6.3, ratio 3:1, attack 5.8, release 4.7*], which adds punch using moderate attack but also attenuates the decay using moderate release (Track 28.47). The compressor is followed by the *DigiRack EQ*, where a combination between a HPF [*262 Hz, 12 dB/oct*] and a deep dip on a parametric filter [*427 Hz, −14 dB, Q 1*] removes all the snare body and tone (Tracks 28.48–28.49); a high-shelving filter [*6 kHz, 1.7 dB*] adds a touch of highs as part of tonality shaping (Track 28.50).

From the triggered snare track a few hits were copied onto a new track. These provide an extra support during the final sections of the song, mainly boosting the overall snare's level but also causing a subliminal change to the snare tonality. These few hits appear right before the break, right after it, halfway through the outro and at the very end. These hits (Track 28.51) are first treated with the *DigiRack EQ*, where a HPF [*476 Hz, 6 dB/oct*] rolls off much of the lows (Track 28.52); a high-shelving filter [*4.61 kHz, −3.4 dB*] attenuates the highs so these hits would not be over-present (Track 28.53). The equalizer is followed by the *CB2* [*threshold −30.6 dB, comp 7.6, knee 15, attack 70 ms, release 273 ms*], which adds attack (Track 28.54).

Hi-hats

Inserts: Digidesign *DigiRack EQ 3.*

Having the hi-hats sufficiently present on the overheads and snare top meant that there was little point mixing the hi-hats track. Still, the hats could use a little push during the break, so this was the only place where this track was mixed (Track 28.55). The hi-hats only go through the *DigiRack EQ*, where a HPF [*668 Hz, 18 dB/oct*] removes kick spill and low-mids that contributed little to the sound (Track 28.56), and a LPF [*6.8 kHz, 6 dB/oct*] eases brittle highs (Track 28.57).

Ride

Having sufficient presence on the overheads, this track was omitted.

Toms

First Layer Inserts: McDSP *Compressor Bank CB2*, Digidesign *DigiRack EQ 3.*
First Layer Sends: Ambiance Reverb *(UAD DreamVerb).*

Second Layer Inserts: McDSP *FilterBank E2*, Digidesign *Smack!*

Like the kick, the toms in the mix are triggered. But at one specific section, just before the break (Tracks 28.58–28.59), the recorded toms did an exceptional job. The tom sound in this section is the outcome of two layers (Tracks 28.60–28.62), both involve the original tom recording (Track 28.63) but each processed differently. It is the level balance between these two layers that dictated the final toms sound.

The first layer first goes through the *CB2* [*threshold –37.3 dB, comp 3.4, knee 13.7, attack 70 ms, release 40.2 ms*], which was employed to condense the natural-sounding dynamics (Track 28.64). It is followed by the *DigiRack EQ*, where a parametric filter [*3.5 kHz, 5 dB, Q 0.81*] adds attack and highs that let the toms push more through the mix (Track 28.65). This layer is also sent to the ambiance reverb – although the reverb does not handle the toms very well, once mixed it adds a rich sense of warm space and a touch of chaos, which works well (Track 28.66).

The second tom layer first goes through the *FilterBank E2*, where a low-shelving filter [*968 Hz, –12 dB*] drys the lows and low-mids, leaving only the presence and attack portions of the toms (Track 28.67). The EQ is followed by *Smack!* [*input 7, ratio 6:1, attack 8.5, release 6.1*], which accents the attack and sharpens the tom's dynamics (Track 28.68).

Tom triggers

Tom 1 Inserts: Digidesign *Smack!*, McDSP *FilterBank E6.*
Tom 1 Sends: Ambiance Reverb *(UAD DreamVerb).*

Floor Tom Inserts: Digidesign *Smack!*, McDSP *FilterBank E6.*
Floor Tom Sends: Ambiance Reverb *(UAD DreamVerb).*

With the exception of the short section above, all other tom hits are triggered. This allowed the toms to project better and provided ultimate control over their sound.

Tom 1 is being used several times during the song. The tom hit (Track 28.69) was extracted from isolated hits recorded in the studio after the drums were tracked. These hits were pasted into the session manually using the tab-to-transient feature. They first go through *Smack!* [*input 8.3, ratio 6:1, attack 3.9, release 0*], where a moderate attack adds some punch while zero release lengthens the decay (Track 28.70). The distortion facility on *Smack!* [*Odd*] was also utilized to add a touch of grit (Track 28.71). The compressor is followed by the *FilterBank E6*, where three bands are used: a HPF [*135 dB, 12 dB/oct*] sets the low frequency limit of the tom, essentially preventing it from overloading the lows, including those of the reverb it is sent to (Track 28.72); a parametric filter [*648 Hz, −2.6 dB, Q 0.8*] shapes the tonality of the tom while removing dispensable mids (Track 28.73); another parametric filter [*4.35 kHz, 2 dB, Q 1*] accents the attack (Track 28.74). Tom 1 is sent to the ambiance reverb, mainly in order to detach it from the front of the mix (Track 28.75).

The floor tom (Track 28.76) only plays once just before the bridge. Like tom 1, it was also extracted from the isolated hits recorded in the studio. It is compressed quite similarly to tom 1 by *Smack!* [*input 8.3, ratio 6:1, attack 6, release 0*], which adds attack and lengthens the decay (Track 28.77). On the following *FilterBank E6*, a single parametric filter [*4.9 kHz, 6.9 dB, Q 0.9*] accents the attack and definition of the tom (Track 28.78). The floor tom is sent to the ambiance reverb, intentionally creating some distant thunderous thud (Track 28.79).

Tambourine

Inserts: Toontrack *EZdrummer*, Digidesign *DigiRack Dyn 3*.
Sends: Ambiance Reverb *(UAD DreamVerb)*.

Break Tambourine Inserts: Toontrack *EZdrummer*, Sonnox *Oxford Reverb*.

The tambourine in *Hero* was never recorded – instead, it was generated using *EZdrummer* during mixdown. It plays through most of the song except the verses and was added mainly to sharpen the contrast between the verses and the choruses. It was mixed as a support instrument, meaning it is not clearly defined (Tracks 28.80–28.81). The untreated tambourine (Track 28.82) was compressed by the *DigiRack Dyn 3* [*threshold −12.6 dB, ratio 5.3:1, soft knee 24.7 dB span, attack 40.6 µs, release 5.7 ms*], which balances out the loud strokes with the quiet shakes (Track 28.83). It is sent to the ambiance reverb, which sends it far back into the depth field (Track 28.84). Before the verses, the tambourine is panned left opposite the ride, which plays along with it. From the bridge onward, the tambourine plays with the hi-hats, which appear left. To prevent the high-hats from masking the tambourine and in order to create more balanced stereo panorama, the panning of the tambourine was automated to shift to the right from the bridge onward.

During the break, the tambourine's nature alters, and because of this a new track was created. The only plugin inserted on this track is the *Oxford Reverb* [*67% wet*], where a modified version of the *concert room* preset adds a long and shiny tail (Tracks 28.85–28.86). This track, which is panned center, was automated in level to rise toward the end of the break, building up to the explosive outro.

Bass

Main bass

Inserts: PSP *VintageWarmer,* Sonnox *Oxford Dynamics,* Digidesign *DigiRack EQ 3,* SoundHack *+chebyshev.*

Growing to like the bass sound in the rough mix, most of the plugins and their settings were transported from the rough mix to the final mix. The raw bass (Track 28.87) first goes through the *VintageWarmer* [*mix 68.4%, drive 11.6 dB, knee 31.9%, speed 91%, auto release*], which has two roles: it condenses the bass dynamics using parallel compression and adds warm saturation. This saturation yields added harmonics that sharpen the bass definition (Track 28.88). Following the *VintageWarmer* is the *Oxford Dynamics,* but the only facility used on it is the Warmth facility [*89%*], which adds some clarity while also enlivening the sound (Track 28.89). Then comes the *DigiRack EQ,* where three bands are utilized: a HPF [*72.4 Hz, 24 dB/oct*] was employed to remove low frequencies that jumped with very low notes, making the bass tonality inconsistent (Track 28.90); a parametric filter [*206.5 Hz, −1.6 dB, Q 1*] was added late in the mixing stage to ease the mids on the bass as part of frequency tuning (Track 28.91); another parametric filter [*636 Hz, 3.8 dB, Q 1*] simply shapes the sound of the bass by accenting low-order harmonics (Track 28.92). Following the EQ comes the *+chebyshev,* which adds even more distortion and a very distinct size (Track 28.93).

Break bass

Inserts: PSP *Vintage Warmer,* Digidesign *DigiRack EQ 3,* SoundHack *+chebyshev,* Sonnox *Transient Modulator.*

During the second part of the break, the bass plays an important solo role, which builds up to the outro. This called for a deviation from its normal sound. So the bass during the break was moved to a new track, which is processed differently from the main bass. This track was first treated with the *Vintage Warmer* [*mix 100%, drive 20.9 dB, knee 31.9%, speed 91%, auto release*], which tightens the sound while also adding some warmth (Tracks 28.94–28.95). Then comes the *DigiRack EQ* where: a HPF [*124 dB, 18 dB/oct*] intentionally dries the lows from the bass, with the idea that these will return with the outro (Track 28.96); a parametric filter [*340 Hz, −7.5 dB, Q 2.24*] attenuates disturbing mids (Track 28.97); another parametric filter [*497 Hz, 8.8 dB, Q 10*] opens up the flat sound and adds some resonance as an effect (Track 28.98); yet another parametric filter [*2.46 kHz, 10.8 dB, Q 4.06*] draws some presence (Track 28.99). Next comes the *+chebyshev* which adds an extra dimension to the sound (Track 28.100). Last comes the *Transient Modulator* [*ratio 0.25, overdrive 90%*], which accentuates attack and adds a subtle amount of harmonic content (Track 28.101).

Bass FX

Inserts: Digidesign *DigiRack EQ 3,* SansAmp *PSA-1,* McDSP *Chrome Stack,* Digidesign *DigiRack EQ 3.*

As part of early tryouts, I experimented with the idea of having a heavily distorted bass – an idea that did not quite work well in the context of this song (Track 28.102). Yet, I had the distorted layer (a duplicated track of the bass going through the following plugins) and it worked quite well during the powerful outro of the song – adding extra power to the bass and to the overall mix (Tracks 28.103–28.104). The first plugin in the chain is the *DigiRack EQ*, which acts as a tuner for the subsequent *SansAmp* (Tracks 28.105–28.107). Two bands are operational on the EQ: a HPF [*194.2 Hz, 24 dB/oct*] and a high-shelving filter [*8.81 kHz, −6 dB*]. The *SansAmp* is followed by the *Chrome Stack*, which contributes the main effect (Track 28.108). As part of final frequency tuning, another *DigiRack EQ* was inserted, where a parametric filter [*829.6 Hz, −2.9 dB, Q1*] pulls some mids (Track 28.109).

Rhythm guitar

Three takes of the rhythm guitar were recorded each using three sources: a DI recording, an on-the-grill Shure *SM57* and a distant Neumann *U87* (Tracks 28.110–28.112). Out of the three, the DI and *SM57* were used in the mix, while the distance captured on the *U87* recording proved too limiting and wasn't used.

The rhythm guitar plays a more dominant role during the intro and break compared to the rest of the song. So a dedicated track was allocated for these two sections.

Intro/break rhythm guitar

Inserts: Digidesign *DigiRack Dyn 3*, SansAmp *PSA-1*, Digidesign *DigiRack EQ 3*, Sonnox *Oxford Reverb*.
Sends: Ambiance Reverb (UAD *DreamVerb*).

The intro/break guitar, a DI recording (Track 28.113), first goes through the *DigiRack Dyn 3*, where the compressor [*threshold −26.4 dB, ratio 9.7:1, soft knee 30 dB span, attack 1 ms, release 15.1 ms*] evens out level fluctuations (Track 28.114). It is followed by the *SansAmp*, which adds some color to the pale DI sound while also adding some crunch (Track 28.115). Next comes the *DigiRack EQ*, where a low-shelving filter [*230.6 Hz, −7.2 dB, Q 1*] attenuates an excess of lows (Track 28.116), then a wide parametric filter [*2 kHz, 7.4 dB, Q 1*] adds essential highs to fabricate the final guitar sound (Track 28.117). The *Oxford Reverb* was inserted in series to add some stereo size to the mono recording. The reverb was programmed to only output early reflections with very little (8%) of the wet signal mixed (Track 28.118). In addition, this track was sent to the ambiance reverb in order to place it in the mix space (Track 28.119).

The intro guitar plays for four bars alone before other instruments are introduced – an introduction that in my opinion did not have sufficient power. So the level of the intro guitar was set intentionally at a level low enough to allow some level impact once other instruments are introduced (compare Tracks 28.120–28.121).

Main rhythm guitar

Inserts: Digidesign *DigiRack EQ 3*, Digidesign *DigiRack Dyn 3*.
Sends: Guitar Chorus (UAD *Delay Modulator DM-1*), Ambiance Reverb (UAD *DreamVerb*).

By the time all the tracks were brought up and coarsely mixed, there was very little space for the rhythm guitar. Its definition is in the high-mids – an area already occupied by virtually all other instruments but the bass. Luckily, a rhythm guitar need not be in the limelight as it plays a rather supportive role in the mix.

The raw track (Track 28.122), a DI recording, goes through aggressive equalization using the *DigiRack EQ*. A HPF [*648.7 Hz, 18 dB/oct*] rolls off a good portion of the lows and low-mids, which contributed little to the guitar sound yet cluttered a valuable range of frequencies (Track 28.123). A parametric filter [*1.1 kHz, −8.3 dB, Q1*] in combination with another parametric filter [*1.83 kHz, −9.5 dB, Q 3.79*] tunes the guitar into the busy high-mids (Tracks 28.124–28.125). Then a high-shelving filter [*6 kHz, −2.7 dB*] treats exaggerated highs (Track 28.126). The EQ is followed by the *DigiRack Dyn*, where a compressor evens out level fluctuations (Track 28.127). The guitar is sent to the *DM-1*, which by adding chorus effect sends the guitar back in the depth field and also creates wide curtain of sound (Track 28.128). The guitar is also sent to the ambiance reverb, which gels it into the sound stage (Track 28.129).

Support rhythm guitar

Inserts: McDSP *FilterBank E6.*
Sends: Ambiance Reverb (UAD DreamVerb).

The main rhythm guitar was panned to the right, yet despite the stereo effects added, something was still missing on the left. To combat this, a different take – an *SM57* recording – was mixed panned left. Combined with the main rhythm guitar, this additional track creates an even wider and richer impression (Tracks 28.130–28.131). The raw track (Track 28.132) was only equalized by the *FilterBank E6*, where a HPF [*1.23 kHz, 6 dB/oct*] rolls off dispensable lows and low-mids (Track 28.133). It is sent to the ambiance reverb, which places it on the same line as the main rhythm guitar (Tracks 28.134).

Distorted guitars

When it came to mixing the distorted guitars, a few things had to be considered. First, during the verses there are intensity changes between sections that involve the vocal and those involving the lead guitar. To support these changes the distorted guitars are split into two pairs: one pair, which I term 'curtain guitars', play during the relaxed sections of the verses; the other pair, simply termed 'power guitars', play during all other sections. The second consideration was that the power guitars in the choruses, bridge and outro should stand out more than those during the verses.

Like the rhythm guitar, the distorted guitars were recorded using the same DI, *SM57* and *U87* setup, and here again the *U87* was not used. The DI recordings that were used were processed by amp simulators – a process that allowed ultimate control over their final sound.

Curtain guitars

Guitar 1 Inserts: UAD *Preflex.*
Guitar 2 Inserts: Digidesign *DigiRack Mod Delay II,* Digidesign *DigiRack EQ 3.*
Sends: Ambiance Reverb (UAD DreamVerb).

Curtain guitar 1 is a DI recording (Track 28.135) processed by the *UAD Preflex* with a modified version of the *Foxy Gravy* preset (Track 28.136). This track was panned hard left.

Curtain guitar 2 is an SM57 recording of the same take as guitar 1. It is panned hard right. The raw track (Track 28.137) goes through the *DigiRack Mod Delay* [*mix 50%, 24 ms delay, no feedback or modulation*] for the sole purpose of sending it backward without adding stereo width (Track 28.138). On the succeeding *DigiRack EQ,* a low-shelving filter [*175.5 Hz, −8 dB*] attenuates an excess of lows (Track 28.139).

Both tracks are routed to an audio group, which is sent to the ambiance reverb. The reverb shifts the guitars further back in the mix and also detaches them from the extremes (Track 28.140). This creates a nice contrast between the curtain guitars and the power guitars – the curtain guitars appear further back and more centered; the power guitars appear further in front and their image is bound to the extremes. The level of the curtain guitars was automated so they only appear in the relaxed section of the verses.

Power guitars

Guitar 1 Inserts: SansAmp *PSA-1,* SansAmp *PSA-1,* Digidesign *DigiRack Mod Delay II.*
Guitar 2 Inserts: Digidesign *DigiRack EQ 3,* Digidesign *DigiRack Mod Delay II.*

Power guitar 1 is a DI recording (Track 28.141). The first plugin in the inserts chain is the *SansAmp,* and is bypassed during the verses (Track 28.142). It is followed by another *SansAmp,* which operates throughout (Track 28.143). While the second *SansAmp* plays alone during the verse, in all other sections it is combined with the first *SansAmp* – a combination that creates a richer, brighter and more powerful sound (Track 28.144). Next comes the *Mod Delay* [*36 ms delay on right channel only*], which by applying the Haas trick turns the mono track into stereo while adding a noticeable size (Track 28.145).

Power guitar 2 is an *SM57* recording of a different take than power guitar 1. The raw recording (Track 28.146) goes through the *DigiRack EQ,* where a low-shelving filter [*142.4 Hz, −4.4 dB*] attenuates overemphasized lows (Track 28.147), and a parametric filter [*2.5 kHz, 2 dB, Q 1*] adds a touch of presence (Track 28.148). The guitar then goes through the *Mod Delay* [*57.23 ms delay on the left channel only*], which turns the mono track into a wide stereo one; modulation on the delay [*depth 3%, rate 1.26 Hz*] adds further dimension and size (Track 28.149).

Distorted guitar group

Inserts: Sonnox *Oxford EQ.*

All distorted guitar tracks were routed to an audio group. On the group track, the *Oxford EQ* is responsible for tuning the guitars to the frequency spectrum. On the EQ four bands are employed: a HPF [*288.2 Hz, 18 dB/oct*] sets the low frequency-limit of the guitars (Tracks 28.150–28.151); a parametric filter [*600 Hz, −2.75 dB, Q 6.25*] attenuates some mids that the guitars could live without and other instruments could use (Track 28.152); for the same reason another parametric filter [*3.7 kHz, −2.75, Q 2.22*] clears some highs (Track 28.153); then a high-shelving filter [*9.4 kHz, −6.28 dB*] pulls some harsh highs (Track 28.154).

Lead guitar

Lead guitar

Inserts: Digidesign *DigiRack Dyn 3,* Digidesign *DigiRack EQ 3,* PSP *Nitro.*

The selected source for the lead guitar was the *U87* recording (Track 28.155). It is first compressed using the *DigiRack Dyn* [*threshold –48.9 dB, ratio 3:1, hard knee, attack 272.7 μs, release 13.6 ms*] in order to balance out level fluctuations and also to increase sustain (Track 28.156). The compressor is followed by the *DigiRack EQ,* where a steep HPF [*448.6 Hz, 24 dB/oct*] rolls off dispensable lows and low-mids (Track 28.157); a parametric filter [*2.09 kHz, 4.5 dB, Q 1.38*] adds some flesh (Track 28.158); and another parametric filter [*5.96 kHz, −4.1 dB, Q 3.23*] eases some harshness, resulting in an overall smoother, more rounded sound (Track 28.159). The sound of the lead guitar is mostly the outcome of the *Nitro* and its *No-Fi* preset – essentially a combination of bit reduction and sample rate reduction followed by a LPF (Tracks 28.160–28.161). The lead guitar is sent to the ambiance reverb, mainly in order to shift it backward in the depth field and to blend it into the sound stage (Track 28.162). The lead track level was automated at various sections of the song, for example, to rise during the intro.

Lead guitar FX

Inserts: Digidesign *DigiRack EQ 3,* Digidesign *Smack!,* PSP *84.*

Another element that was transported from the rough mix was the lead guitar effect. To prevent correlation to the lead track, the original lead track (Track 28.163) was duplicated and processed as follows: First, it is equalized by the *DigiRack EQ,* where a HPF [*623.3 Hz, 6 dB/oct*] rolls off dispensable lows (Track 28.164) and a high-shelving filter [*6 kHz, 3.6 dB*] adds some spark (Track 28.165). The succeeding *Smack!* [*norm mode, input 5, ratio 4:1, attack 2.9, release 5*] is in charge of balancing level variations and lengthening the sustain (Track 28.166). The wah-wah-like effect is the outcome of a modified *Sarawak* preset on the PSP *84* (Track 28.167). This track is also sent to the ambiance reverb (Track 28.168).

Vocals

Lead vocal

Inserts: Pultec *EQH-2,* McDSP *MC2000 MC2,* SoundHack *+chebyshev,* Sonnox *Oxford Dynamics.*
Sends: Ambiance Reverb (UAD *DreamVerb*), Vocal Reverb (UAD *DreamVerb*).

There was a conceptual issue with mixing the lead vocal. On the one hand, *Hero* has its share of aggressiveness; on the other, it has its pop characteristics and involves a catchy melody. So the main question was: Should the lead vocal be sweet and beautiful, or should it be bad and dirty? It ended up somewhere in between.

The first plugin in the inserts chain, the *EQH-2,* was employed to shape the basic tonality of the vocal. It removes dispensable lows [*100 Hz, attenuation 4.9*] and adds a touch of spark [*10 kHz, boost 1.5*] (Tracks 28.169–28.171).

Succeeding the EQ is the *MC2000* – a multiband compressor. I tried two different compressors (*Smack!* and the *Oxford Dynamics*) prior to trying the *MC2000*. *Smack!* gave an obvious character that I have found was too much in the context of the song; the *Oxford Dynamics* did a fair job, but not a perfect one. Dan's voice fluctuates noticeably at some sections of this song. Whatever settings I tried, the level of the vocal never seemed stable throughout the song. The fundamental problem was that the frequency content of the recording changes quite noticeably with relation to the pitch sung. Lower-pitch notes seemed to disappear in the mix – making the vocal come and go. This issue is not uncommon, but on this production it seemed extremely tricky. For this reason, I picked the *MC2000*. Perhaps the most important setting on the compressor was the crossover frequency, which ended up at 1.6 kHz. Both bands are heavily compressed, with the threshold of the low band set to –24 dB, and the threshold of the high band to –44 dB. In addition to their standards use, the independent gain controls of each band were used to further shape the tonality of the vocal (Tracks 28.172–28.174). The *MC2000* added quite some character to the vocal, which I liked.

Following the *MC2000*, the vocal sounded balanced but still did not stand out as much as it should have. This was mostly the outcome of missing vitality on the mids. The *+chebyshev* did a perfect job of filling this missing range. It also adds some aggression in the form of distortion that at points becomes rather obvious – an effect that I thought was for the better (Track 28.175). The last plugin in the chain is the *Oxford Dynamics* [*threshold –12 dB, ratio 2:1, soft knee 40 dB span, attack 0.52 ms, release 5 ms*], acting as a catch-all compressor to perfect the balancing of the distorted voice (Track 28.176).

The lead vocal is sent to both the ambiance reverb and to a dedicated vocal reverb. The ambiance reverb blends the vocal into the sound stage, while the vocal reverb – a slightly modified *PLATE 140 Vocal Plate* preset – adds a rich effect, life and size (Tracks 28.177–28.179). The send level to the vocal reverb was automated to rise for the duration of a few sustained vowels (Track 28.180).

Vocal FX

The lead vocal involves a few effects. First, there is a narrowed ping-pong delay after the first chorus. This delay was not produced by a plugin, but by crossed bus routing in Pro Tools – left bus sent to right bus, right bus sent to left bus. On each bus there is 830 ms of delay, which despite not being perfectly synced to the tempo sat quite right. In addition, the echoes were treated by the *DigiRack EQ 3*, which was inserted on each bus with a combination of a LPF and a parametric filter that creates a resonant filter. The send level on these buses was automated so only three echoes are produced – a fourth echo would clash with the verse vocal. The echoes are blended into the depth field by the ambiance reverb to which they are sent (Track 28.181).

The ping-pong delay closing the first chorus fills some space between the chorus and the second verse. The second chorus is followed by additional vocal phrase and then the lead guitar immediately takes the stage, so there was little place for the same ping-pong effect. The only processing on the vocal phrase after the second chorus is the *SansAmp PSA-1* (Track 28.182).

Halfway through the bridge the phrase 'deep within us' was also processed, this time using four plugins. The first plugin, which is the main effect contributor, is the PSP *Master Q*. The settings involve a HPF at 96 Hz, a LPF at 3.5 kHz and a mad boost of 24 dB at 981 Hz. The actual effect is also the outcome of the hard saturation on the saturation facility (Tracks 28.183–28.184). The *Master Q* is followed by the Lexicon PSP *42*, which adds 172 ms delay with feedback (Track 28.185). Then comes the *Digidesign D-Verb* set to large hall and 19% wet (Track 28.186). Finally, the *McDSP Chrome Tremolo* adds a subtle tremolo effect (Track 28.187). This specific phrase is sent to the ambiance reverb (Track 28.188).

Shortly after the 'deep within us' phrase, the same phrase repeats. This second instance could have been processed in the same manner, but to create some variation and contrast a different effect was used: the lead vocal sends to the ambiance, and vocal reverbs were muted, causing forward shifting of the voice (Track 28.189).

Fake second voice

Inserts: Lexicon PSP *42*, Digidesign *DigiRack EQ 3*.
Sends: Vocal Reverb (UAD DreamVerb).

One of the last overdubs to be recorded, at around 6 am, was the second voice. By that point, Dan's vocal cords were a few takes away from snapping. Second voice harmony was recorded for all but the verses, but the performance during the choruses and the outro was pretty unconvincing. As a replacement, the lead vocal track was duplicated, trimmed in the right places and processed with ADT settings on the PSP *42* [*15 ms modulated delay, no feedback*] (Tracks 28.190–28.192). Following the PSP *42* comes the *DigiRack EQ*, where a HPF [*239.4 Hz, 18 dB/oct*] rolls off dispensable lows (Track 28.193) and a parametric filter [*1.79 kHz, 7.6 dB, Q 2.81*] boosts some presence (Track 28.194). The track is sent to the vocal reverb, which sends it backward in the depth field and also correlates it to the lead vocal (Track 28.195). To distinguish it from the lead vocal, it was panned left to around 9:00. As part of a gradual build-up, the second voice is only introduced during the second verse.

Real second voice

Inserts: PSP *84*, Sonnox *Oxford Dynamics*.
Sends: Vocal Reverb (UAD DreamVerb).

The only section in which the recorded second voice is featured is the bridge. The original recording (Track 28.196) was first processed with the PSP *84*, but no delay is actually applied. Instead, the filter, drive and plate reverb were used to create a unique effect (Track 28.197). Following the PSP *84* is the *Oxford Dynamics*, where a compressor [*threshold –26.5, ratio 3.48:1, soft knee 30 dB span, attack 0.52 ms, release 5 ms*] contains level fluctuations (Track 28.198). The vocal is sent to the vocal reverb (Tracks 28.199–28.200).

Ahh BV

Inserts: Digidesign *DigiRack EQ 3*, McDSP *Compressor Bank CB1*, PSP *Nitro*.
Audio Group Inserts: Digidesign *DigiRack Mod Delay II*, McDSP *FilterBank E2*, SoundHack *+chebyshev*.

Two backing vocals tracks were recorded for the bridge (Track 28.201) and like the second voice performance, they are not the greatest. A conscious decision was made to bury these in the mix. The two tracks are processed identically and then routed to an audio group for additional processing. The first plugin is the *DigiRack EQ*, where a HPF [*260.6 Hz, 12 dB/oct*] rolls off dispensable lows (Track 28.202). It is followed by the *CB1* [*threshold –40 dB, comp 3.2, knee 4.7, attack 5 μs, release 204 ms*], which flattens the levels (Track 28.203). The succeeding PSP *Nitro* was employed to create an autopan effect (Track 28.204).

On the audio group, the *Mod Delay* [*23.75 ms delay of the right channel only, 44% depth and 0.74 Hz modulation*] produces the Haas trick (Track 28.205). A low-shelving filter [*1.7 kHz, −5.1 dB*] on the *FilterBank* tunes the vocals into the mix (Track 28.206). Then the +*chebyshev* adds some edge to what was a somewhat numb sound (Track 28.207).

Break BV

Inserts: Digidesign *DigiRack EQ 3*, Digidesign *DigiRack Dyn 3*.
Audio Group Inserts: PSP *84*.

Three backing vocal tracks were recorded from the break onward, and these also conclude the production. Two of the tracks were performed by Dan and one by Lish. Lish's voice was mixed to be the leading voice out of the three, panned a notch to the left; Dan's vocals were each panned nearly fully to an opposite extreme. The three voices were first equalized using the *DigiRack EQ 3*, then compressed using the *Dyn 3* before being routed to a bus for additional processing. Perhaps the most interesting processing is the EQ on Lish's voice, where a HPF [*505.6 Hz, 12 dB/oct*] rolls off an excess of lows (Tracks 28.208–28.209), a parametric filter [*2.93 kHz, −4 dB, Q 1*] softens some edge (Track 28.210) and a high-shelving filter [*3.67 kHz, 7.4 dB*] boosts a generous quality of air and definition (Track 28.211).

The effect applied on the audio group is a slightly modified version of the PSP *84's Gasherbrum Four* preset, set to 50% wet (Tracks 28.212–28.213). The group on which the PSP *84* was loaded is sent to the ambiance reverb (Track 28.214).

29

It's Temps Pt. II (Hip Hop/Urban/ Grime)

Music and production: Brendon Octave Harding *(www.myspace.com/octaveproductions)*.
Lyrics and rap: Temps *(www.myspace.com/temps14)*.
Mixed (in Cubase) by Roey Izhaki.
Mastered by Mandy Parnell (*www.mandyparnell.com*) at Electric Mastering, London.

'Ain't hip hop, ain't urban, ain't grime . . . it's temps . . .' say the lyrics. Indeed, *It's Temps Pt. II* does not fall directly into these genres, but it involves a fresh approach that has its roots in all three. Without a doubt, the most distinguishing element in this production is the complex beat. I have taken the liberty of mixing this distinctive beat in a distinctive way, trying to craft a less-than-usual sound. As typical of many hip hop mixes, the beat and the vocals are the life and soul of the production, with all other elements tailored behind them.

Track 29.1 is the producer's rough mix. Track 29.2 is a mix-ready version involving rough levels and panning, but no processing or effects. Track 29.3 is a snapshot of the mix in progress. Track 29.4 is the final mix, while Track 29.5 is the mastered mix.

Beat

Main kick

Inserts: UAD *Cambridge EQ,* Cubase *Compressor.*

The main kick (Track 29.6) first goes through the *Cambridge EQ*, where a single parametric filter [*310 Hz, –6.8 dB, Q 1.05*] attenuates low-mids that contributed little to the overall kick sound – with less mids the kick appears to have more oomph and more attack (Track 29.7). The EQ is followed by the *Cubase Compressor* [*threshold –20 dB, ratio 2:1, peak sensing, attack 67.4 ms, release 10 ms*], which adds some punch (Track 29.8).

Second and Bat Kick

Neither the second kick (Tracks 29.9–29.11) nor the Bat Kick (Tracks 29.12–29.14) were treated. They are both simply layered underneath the main kick.

Main snare

Inserts: Sonnox *Oxford Dynamics*, t.c. electronic *EQSat*, SoundHack *+Decimate*.
Sends: Snare Reverb (Cubase *RoomWorks*).

The main snare (Track 29.15) first goes through the *Oxford Dynamics*, where a compressor [*threshold –16.8 dB, ratio 2:1, hard knee, attack 2.78 ms, release 127 ms*] adds some attack (Track 29.16). The compressor is followed by the *EQSat*, where three bands are employed: a parametric filter [*126.7 Hz, 3.5 dB, bandwidth 0.63 oct*] adds some body (Track 29.17), another parametric filter [*902.1 Hz, 4.5 dB, bandwidth 1.6 oct*] adds some presence and attack (Track 29.18) and a high-shelving filter [*1.45 kHz, 1 dB*] adds some highs as part of frequency tuning (Track 29.19). Following the EQ comes *+ decimate*, which reduces the bit depth to 8.04 bits. This adds some definition-noise and to some extent gives a touch of the famous 8-bit sample sound (Track 29.20). The level of the snare was automated to rise during the set of choruses after the second break. The main snare is also sent to the *RoomWorks* reverb, where a very short reverb sends it backward from its front position, which I found was too much in-your-face (Track 29.21).

Second snare

Inserts: PSP *MasterQ*.
Sends: Snare Reverb (Cubase *RoomWorks*).

The second snare (Track 29.22) is not a layer of the main snare, but an independent instrument that plays on different beats (Track 29.23). This track only goes through the *MasterQ*, where a parametric filter [*306 Hz, –8.41 dB, Q 0.53*] pulls a wide range of mids that only muddled the sound while covering an important range for other instruments (Track 29.24). This track is also sent to the snare reverb (Track 29.25).

Claps

Inserts: Sonnox *Oxford Dynamics*, Cubase *StudioEQ*, SoundHack *+Decimate*.

The original claps (Track 29.26) sounded too light and too natural. The initial burst of claps is usually sufficient for the listener's perception. There is little damage in shortening claps (especially in a busy mix), a practice that can also make claps snappier. The gate [*threshold –8.4 dB, range –41 dB, attack 5 µs, hold 35 ms, release 9.2 ms*] of the *Oxford Dynamics* was employed for this task (Track 29.27). It is followed by the *StudioEQ*, where three bands are used: a HPF [*357 Hz*] rolls off dispensable low-frequency content (Track 29.28), a parametric filter [*2057 Hz, –3.1 dB, Q 0.5*] further reduces high-mids that contribute little to the mix (Track 29.29), a high-shelving filter enhances definition and tunes the claps into the frequency spectrum (Track 29.30). The EQ is followed by the *+Decimate* [*8.04 bits*], which similar to its function on the snare adds some grit and 8-bit sound (Track 29.31).

Hi-hats

The hi-hats track (Track 29.32), which plays through most of the production, was left untreated. The track already contained some spatial information that sent it backward in the depth field. Being an additional element to the beat, it was panned left to around 10:00.

Wood

Another beat element that was left untreated is the wood (Track 29.33). It was panned around 14:00.

Broken bells

The broken bells (Track 29.34) suffer from exaggerated harshness and lack of grace. I found this track somewhat disturbing and in my view the mix was better-off without it. It was omitted.

Cajon

Intro/break Inserts: t.c. electronic *Filteroid*, PSP *608 MultiDelay*.
Other Sections Inserts: PSP *MasterQ*.

One important percussion element in this tune is the Cajon. It plays throughout most of the song, but has special importance during the intro and the break. This led me to split the source track into two, having a dedicated track for the intro and break, and another for the rest of the song.

The intro/break Cajon (Track 29.35) first goes through the *Filteroid*, where a resonant LPF is bar-sync modulated (Track 29.36). Then it flows through the PSP *608*. On the multitap delay the mix control is set to 18% wet. Two taps are operational – one set to eighth note, another to quarter note. Both taps have a feedback set to around 50% and both have a HPF in their feedback loop (Track 29.37). Just before the first main beat drops in, the amount, gain and balance of the first tap are automated. This creates a sustained feedback that sweeps from left to right, and could also appear to sweep up and down (Track 29.38). The level of the Cajon track was automated to rise during the break.

The Cajon that plays throughout the rest of the song is layered underneath the main beat. It is only treated with the *MasterQ*, where a HPF [*97 Hz*] clears the lows (Track 29.39).

Beat group

Inserts: PSP *VintageWarmer*, UAD *Fairchild*, UAD *1176SE*.

All beat tracks but the main kick are routed to an audio group. The reason for the kick's exclusion is that I felt it would benefit from independent processing – not being affected by the heavy compression taking place on the beat group. A quick look at the inserts chain reveals that the beat is processed by three compressors connected in series. Here is what happened: The first compressor in the chain was the *VintageWarmer*, configured with parallel compression settings [*Mix 50%, Drive 0 dB, knee 100%, speed 100%, auto release*]. Its main role is to condense the beat and add some punch (Tracks 29.40–29.41). The *Fairchild* [*input gain 7 dB, threshold 3, fastest time constants*] was loaded for the purpose of enhancing the compression effect and to give some unique character (Track 29.42). The *1176* was loaded as a tryout alternative for the *Fairchild*, but when it was loaded with the *Fairchild* still on, the result of the two working in series was infused with character (Track 29.43). The *1176*, however, is not operational throughout the song since it made the beat somewhat slim; instead, it was automated to take effect during the last four bars of the first verse and throughout the duration of the second verse.

Break beat

Inserts: Sonnox *Oxford Dynamic.*
Sends: Break Reverb *(t.c. electronic ClassicVerb, Sonnox Oxford Dynamic, PSP Nitro, Sonnox Oxford Dynamic).*

The break beat (i.e., the beat during the break) called for a change of theme and more drama. I have chopped the kick, snare and hi-hats from their main tracks and moved each to a new track. The kick was equalized with the *PSP Neon* (Tracks 29.44–29.45), the snare was processed with the *PSP MasterQ* (Tracks 29.46–29.47) and the hats were left untreated. These three tracks were then routed to an audio group, where the *Oxford Dynamics* adds some compression in linear mode (Tracks 29.48–29.49).

But the powerful effect on the break beat is the outcome of an inserts chain that succeeds the reverb to which all three tracks were sent. The reverb is based on a modified *Small Hall* preset from the *ClassicVerb* (Track 29.50). The reverb is followed by the *Oxford Dynamics*, where the gate gates the reverb and creates chaotic level variations (Track 29.51). The combination of bit reduction and a wave shaping from the *PumpItUp* preset of the *Nitro* adds a generous amount of distortion (Track 29.52). Then another *Oxford Dynamics* compresses the sound, using the more hectic linear mode (Track 29.53). Finally, the second plugin in the chain – the gate – was bypassed shortly before the end of the break, to leave an audible reverb decay just before the chorus kicks in (Tracks 29.54–29.55).

Bass

Main bass

Inserts: Noveltech *Character,* PSP *Neon,* Sonnox *Transient Modulator.*

A single bass track was submitted with the multitrack. The original track (Track 29.56) had a generous amount of mids and highs that were meant to increase the bass definition. But I have found these specific frequency ranges unappealing, so I decided to filter them. The first plugin the bass goes through is the Noveltech *Character*. The enhancer was inserted to fatten the lows of the bass, although as a consequence the highs were also enhanced (Track 29.57). Next comes the *Neon*, where in linear-phase mode three bands are used: First to be applied is a LPF [*380 Hz, 12 dB/oct*] that filter the mids and highs of the bass. This makes the bass a pure bass, without the mids and highs (Track 29.58). A low-shelving filter [*100 Hz, –1.06 dB*] that controls the subs was added later in the mix as part of frequency tuning (Track 29.59). Also, a parametric filter with a narrow Q [*99 Hz, –5.18 dB, Q 11.25*] was employed to shape the sound of the bass by removing some tone from around 100 Hz (Track 29.60). Following the EQ comes the *Transient Modulator*, which adds some attack to the bass – this accents the initial level burst and lets the bass stand out better in the mix (Track 29.61).

Bass support

Virtual Instrument: Access *Virus TI.*

Filtering the mids and highs from the bass removed some problematic frequencies, but then the bass lacked definition. So an email was sent to the producer who replied with the MIDI bass track file. The file was loaded onto an instrument track, with the Access *Virus TI* as a virtual instrument. The level of this support track was automated to have various degrees of effect at different sections. For example, it was attenuated in level during the intro and second verse, where the sparser arrangement made the added distortion highly noticeable.

Two, slightly modified *Virus* presets were used in this arrangement: *Dukbass+BC* is layered with the bass throughout the production (Tracks 29.62–29.63). Rob Papen's *Vacin RP* is essentially a distorted lead, wide in image, which only plays during the intro and the choruses. Its addition made the bass bigger and more defined during these important sections (Tracks 29.64–29.65).

Other tracks

Steel Drums

Inserts: PSP *MasterQ.*
Finale Inserts: PSP *MasterQ,* Cubase *MonoToStereo,* PSP *Nitro.*

Steel Drums play a key role during the intro and break sections during which the track level was automated to rise. Two versions were submitted with the multitrack – one with and one without delay (Tracks 29.66–29.67). I chose to use the one with delay. The original image of the raw stereo track was too wide for the mix and I wanted the instrument to be panned left. To achieve this, the image was narrowed by moving the right pan pot to the center (Track 29.68).

The track was equalized by the *MasterQ*, where three bands are used: a HPF [*367 Hz, 12 dB/oct*] rolls off dispensable lows (Track 29.69), a parametric filter [*2.3 kHz, –6.6 dB, Q 0.26*] results in a more rounded sound (Track 29.70), and a shelving filter [*9 kHz, –5 dB*] tunes this track to the frequency spectrum and helps send it slightly backward (Track 29.71).

The Steel Drums also conclude the production and I wanted the effect during the final seconds to be more unique than in the raw track. So on a duplicate track, which only plays once the music drops out, the same *MasterQ* was inserted with the same settings as above; it was then followed by the Cubase *MonoToStereo*, which actually sums the stereo track to mono (Tracks 29.72–29.73); then it is followed by the *Nitro*, which applies an autopan effect synced to the tempo of the song (Track 29.74).

Gtr Pizz

Inserts: PSP Nitro, PSP MasterQ.
Sends: Gtr Chorus (UAD DM-1 Delay Modulator).

Gtr Pizz (Track 29.75) was first processed by the *Nitro* and its *PumpItUp* preset – a combination of a wave shaper, bit reduction [*10.22 bits*], sample rate reduction [*19.5 kHz*] and a HPF [*748 Hz*]. This combination adds some harmonics that fatten the sound (Track 29.76). The *Nitro* was followed by the *MasterQ*, where a HPF [*277 Hz, 12 dB/oct*] rolls off an excess of lows (Track 29.77). The original mono track sounded small and panning it to either extreme could create a stereo imbalance. Thus, it is sent to the *DM-1*, where a chorus [*left delay 23.3 ms, right delay 68.1 ms, modulation rate 0.67 Hz, modulation depth 10%, feedback –29%, damping 3.7 kHz, 100% wet*] adds a noticeable stereo size (Track 29.78).

French Horns

IInserts: PSP MasterQ, PSP Nitro, PSP 84.

French Horns (Track 29.79) play alongside Gtr Pizz on quite a few occasions. The track was first processed by the *MasterQ*, where a HPF [*234 Hz, 24 dB/oct*] rolls off an excess of lows (Track 29.80), a parametric filter [*754 Hz, –3.2 dB, Q 2.8*] lightly cleans the tonality of the instrument (Track 29.81) and another parametric filter [*4.4 kHz, 3.6 dB, Q 2.8*] adds some definition that lets the instrument to stand out more in the mix (Track 29.82). Following the equalizer comes the *Nitro* with the same *PumpItUp* preset that adds distortion and some flesh (Track 29.83). The original stereo track included a stereo effect that did not translate well when summed to mono (Track 29.84). The PSP *84* was employed to add 41 ms of delay to the right channel only. This not only resolved the mono-summing issue, but also gave a more distinct stereo effect bound to the extremes (Tracks 29.85–29.86). The image of Gtr Pizz is narrower than that of French Horns, so when the two play together the stereo image appears to widen. It is then narrowed when Gtr Pizz plays alone (Track 29.87).

Big Stack

Inserts: PSP MasterQ.
Sends: Big Stack Delay (Cubase MonoDelay).

Big Stack arrived on the multitrack as a mono track (Track 29.88). It was processed by the *MasterQ*, where a single parametric filter with a wide Q [*6.2 kHz, 1.5 dB, Q 0.41*] adds some definition and tunes the instrument to the frequency spectrum (Track 29.89). Quite clearly panning this track center would make it clash with more important tracks like the beat and vocals. Panning it to either side could cause stereo imbalance. A stereo effect had to be added, and a delay was chosen [*8th note delay, feedback 18%, feedback LPF at 15 kHz*]. The original track was panned to the left around 10:00 and although the delay could have been mirrored to 14:00, experimentation resulted in it sounding better panned further right to around 15:00 (Track 29.90).

To add some variation and create a better stereo balance, the original track was duplicated and the duplicate was panned to the right around 14:00, with another delay panned to 9:00. But the original and the duplicate never play together – separating each hit on both and muting each hit on either meant that either the original played with its right delay or the duplicate with its left delay. The mute arrangement was not based on an alternating left-right pattern, but on a more experimental arrangement (Track 29.91). The level of the duplicate was automated to make its perceived level more consistent. In addition, both the original and the duplicate were sent to an audio group for the sole purpose of level control and automation.

Dist Bells

Sends: Dist Bells Reverb (UAD *RealVerb Pro*).

The Dist Bells track was left untreated, but the dry track (Track 29.92) needed to be sent backward in the mix and called for a nice long reverb. It was awarded with its own reverb – the *RealVerb Pro*. The *Big Bright Hall* preset was selected and its decay time was lengthened to 14.5 seconds (Track 29.93). When heard in isolation, this long decay seems to take forever to diminish, but not so when the mix is played along. Also, the noise component of the distorted bells clearly lingers on the reverb tail when heard in isolation. Although this could have been rectified, I found this addition of sustained noise as some sort of high-frequency halo to the mix, and decided to keep it. The level of Dist Bells was automated between various sections.

LPTwin

Inserts: Sonnox *Oxford Dynamic*, PSP *MasterQ*.
Sends: Ambiance Reverb (UAD *DreamVerb*), *LPT Delay* (PSP *Lexicon 42*).

LPTwin is essentially a synthesized guitar (Track 29.94). The original track presented me with a common problem where low notes got lost in the mix while high notes stood out. The compressor on the *Oxford Dynamics* [*threshold –9.9 dB, ratio 4:1, hard knee, attack 5.2 ms, release 127 ms*] was employed to level out the various note levels (Track 29.95); this was done with help from side-chain equalization involving a drastic low-shelving filter [*1 kHz, –20 dB*] and a drastic parametric filter [*3.2 kHz, 17 dB, Q 1.287*]. The compressor is followed by the *MasterQ*, where a HPF [*229 Hz, 12 dB/oct*] rolls off lows that only muddy the mix (Track 29.96) and a parametric filter [*4 kHz, 1.31 dB, Q 0.85*] adds some definition (Track 29.97).

LPTwin is sent at low level to the ambiance reverb just to blend it into the mix ambiance (Track 29.98). The ambiance reverb is the UAD *DreamVerb* and its slightly modified *Hall Cathedral* preset. LPTwin is also sent to the PSP *Lexicon 42*, which adds a distinguished wild-west-like effect. While LPTwin is panned around 14:30, its delay is panned nearly center. This creates a wide enough image and a fine sense of depth (Track 29.99).

Trumpet 1

Inserts: UAD *DM-1 Delay Modulator*.
Sends: Ambiance Reverb (UAD *DreamVerb*).

On the submitted multitrack there was only one trumpet track, but I decided to split it into two tracks and process each differently. Trumpet 1 involves the staccato sections, while Trumpet 2 provides two legato sections. Trumpet 1 (Track 29.100) is first processed using the *DM-1*, where a chorus [*left delay 44 ms, right delay 44.9 ms, modulation rate 0.67 Hz, modulation depth 10%, feedback 20%, damping 7 kHz, 25% wet*] adds some dimension and sends the instrument backward (Track 29.101). The track is also sent to the ambiance reverb for depth positioning (Track 29.102).

Trumpet 2

Inserts: Sonnox *Oxford Transient Modulator*, Sonnox *Oxford EQ.*
Sends: Ambiance Reverb (UAD *DreamVerb*), TrumpetVerb (UAD *Plate 140*).

Trumpet 2 (Track 29.103) first goes through the *Transient Modulator* [*ratio 0.58*], which was employed to add attack so each note could cut through the mix better (Track 29.104). It is followed by the *Oxford EQ*, where a HPF [*188 Hz, 6 dB/oct*] rolls off dispensable lows (Track 29.105). Trumpet 1 is sent to the ambiance reverb for depth positioning (Track 29.106). I wanted a nice reverb with a nice tail on the trumpet, but increasing the send level to the ambiance reverb would also position the trumpet further into the depth field. So while the ambiance reverb is in-charge of the front–back position, the *Plate 140* was employed to add a nice dense tail (Track 29.107).

Screamin B4

Inserts: PSP *Nitro,* Cubase *VSTDynamics.*
Sends: Ambiance Reverb (UAD *DreamVerb*).

Screamin B4 (Track 29.108), a Hammond emulation, plays during most choruses and during the break. It first goes through the *Nitro*, where the *Phat Flutter* preset adds some size and distortion (Track 29.109). One issue with this track was that the last notes were too loud. The compressor of the *VSTDynamics* [*threshold –12.5 dB, ratio 1.8:1, attack 1 ms, release 228 ms*] was employed to contain the level of these last notes (Track 29.110). This track was sent to the ambiance reverb in order to blend it into the mix space (Track 29.111). The level of this track was automated to drop in level during the break, rise after it, then rise again during the final two choruses.

Analog King

Inserts: PSP *MasterQ.*
Sends: Ambiance Reverb (UAD *DreamVerb*).

Analog King (Track 29.112) plays in the last choruses of the outro and becomes one of the main instruments once the vocals drop out. This track was treated with the *MasterQ*, where three bands are used: a HPF [*304 Hz, 12 dB/oct*] rolls off dispensable lows (Track 29.113), a parametric filter [*2 kHz, –5 dB, Q 1.69*] is in command of tone shaping (Track 29.114) and a high-shelving filter [*8 kHz, 4.5 dB*] adds some definition and tunes the track to the frequency spectrum (Track 29.115). This track is also sent to the ambiance reverb, which blends it into the sound stage (Track 29.116).

Imposcar

Inserts: Cubase *Studio EQ.*

Imposcar only plays during the verses. A section of this track that played during the first chorus was muted due to arrangement density. Being one of the least important elements in the production, it was mixed underneath everything else. Although a mono version was also submitted, I have chosen to use the stereo version of this track, which was already imprinted with autopan (Track 29.117). Imposcar only goes through the *Studio EQ,* which shapes the tonality of this track to have more emphasis on the low-mids and less on the high-mids. Two bands are operational on the EQ: a parametric filter [*735 Hz, 5.3 dB, Q 0.5*] boosts the low-mids (Track 29.118) and another parametric filter [*3.3 kHz, –2.9, Q 0.5*] attenuates the highs (Track 29.119).

Vocals

The submitted multitrack consisted of 13 vocal tracks. Since some would benefit from a different treatment at different sections, I split certain tracks into two. The final mix involves 21 vocal tracks in this arrangement:

- Intro:
  - vIntro1
  - vIntro2
  - vIntro3
- Verse:
  - vVerseLead
  - vVerse1 (also plays during the intro)
  - vVerse2
  - vKsh
- Chorus:
  - vChorusLead
  - vChs1
  - vChs2
  - vChsAns1
  - vChsAns2
  - vChsAns3
  - vChsAns4
  - vTemps1
  - vTemps2
- Break
  - vBreak1
  - vBreak2
  - vFX
- Outro
  - vOutro
  - vOutroDelay

Having to compress 21 vocal tracks individually is not an easy affair. Instead, I dialed rough settings on the lead vocal track and, copied the compressor to other vocal tracks, then adjusted each compressor as needed. Nearly all the vocal tracks were compressed using the PSP *MasterComp* – it offered the right balance between character and control. On a few tracks extra control was needed, so the Sonnox *Oxford Dynamics* was used instead. The general compression approach was to contain level fluctuations but retain healthy dynamics.

In addition to splitting some vocal tracks, specific words, sometimes phrases, were muted at points where too many vocals obscured one another. One of the main challenges in mixing these 21 vocal tracks was getting the right panning and depth relationship between the different voices.

Vocals reverb

Inserts: t.c. electronic *ClassicVerb.*

Different vocal tracks needed to be sent to a reverb for subtle depth positioning. The reverb had to be rather transparent and also suit the vocals as a sheer effect. The ambiance reverb was not fit for this task, so a dedicated reverb aux was allocated for the vocals. The *ClassicVerb* was chosen and its *Small Hall* preset highly modified. Tracks 29.120–29.121 demonstrate the function of this reverb.

vIntro1

Inserts: Sonnox *Oxford Dynamics.*
Sends: Vox Reverb *(t.c. electronic ClassicVerb).*

vIntro1 (Track 29.122) was compressed using the *Oxford Dynamics* [*threshold –30.3 dB, ratio 1.55:1, soft knee 40 dB span, attack 0.52 ms, release 5 ms*] with the intention of containing level fluctuations (Track 29.123). The track was also sent to the vocal reverb for depth positioning (Track 29.124). The track was panned around 11:00, and its level was automated to drop once the bass is introduced (a point at which vIntro2 becomes the main voice).

vIntro2

Inserts: Sonnox *Oxford Dynamics.*
Sends: Vox Reverb *(t.c. electronic ClassicVerb).*

vIntro2 (Track 29.125) was also compressed using the *Oxford Dynamics* with exactly the same settings as vIntro1 (Track 29.126). The track was also sent to the vocals reverb (Track 29.127). vIntro2 was panned center, and its level was automated to drop during the very last sentence of the intro.

vIntro3

Inserts: PSP *MasterComp.*
Sends: Vox Reverb *(t.c. electronic ClassicVerb).*

vIntro3 (Track 29.128) is essentially chopped from vVerse2. It was compressed using the *MasterComp* [*threshold –24 dB, ratio 3.36:1, hard knee, peak sensing, attack 0.78 ms, release 100 ms*] (Track 29.129). It was also sent to the vocal reverb (Track 29.130) and is panned slightly to the left around 11:00.

vVerseLead

Inserts: PSP *MasterComp,* Sonnox *Oxford EQ.*
Sends: Vox Reverb *(t.c. electronic* ClassicVerb*)*.

vVerseLead (Track 29.131) was compressed using the *MasterComp* [*threshold –14.8 dB, ratio 4:1, soft knee, RMS sensing, attack 0.79 ms, release 227 ms*] (Track 29.132). After the compressor, an *Oxford EQ*'s, HPF [*141 Hz, 6 dB/oct*] rolls off some dispensable lows (Track 29.133). This track was also sent to the vocal reverb (Track 29.134). vVerseLead was panned center and its level was automated to rise during the very first words of the first chorus.

vVerse1

Inserts: PSP *MasterComp,* Sonnox *Oxford EQ.*
Sends: Vox Reverb *(t.c. electronic* ClassicVerb*)*.

vVerse1 (Track 29.135) also constitutes the fourth voice during the late intro moments. Together with vVerse2 it provides the accompanying voice to vVerseLead. It was compressed using the *MasterComp* [*threshold –27.7 dB, ratio 2.8:1, soft knee, RMS sensing, attack 0.01 ms, release 100 ms*] (Track 29.136). Following the compressor the *Oxford EQ* was inserted, where three bands are operational: a HPF [*229 Hz, 6 dB/oct*] rolls off some lows (Track 29.137), a parametric filter [*775 Hz, 1.8 dB, Q 2.83*] was employed as part of tonality shaping and frequency tuning (Track 29.138), and another parametric filter [*3.2 kHz, –1.65 dB, Q 2.83*] helps enhance the back panning of the voice (Track 29.139). Altogether, the equalization of vVerse1 also helps distinguish it from vVerseLead. This specific voice was panned to the right around 14:00. It was also sent to the vocal reverb (Track 29.140).

vVerse2

Inserts: PSP *MasterComp,* Sonnox *Oxford EQ.*
Sends: Vox Reverb *(t.c. electronic* ClassicVerb*)*.

vVerse2 (Track 29.141) was compressed using the *MasterComp* [*threshold –14.3 dB, ratio 8:1, hard knee, peak sensing, attack 0.015 ms, release 457 ms*] (Track 29.142). It was equalized using the *Oxford EQ* similar to vVerse1, only without the 775 Hz boost (Track 29.143). It was panned to the left around 10:30 and its level was automated as an additional measure to balance its level. It was also sent to the vocal reverb (Track 29.144).

vKsh

Inserts: PSP *MasterComp.*
Sends: Vox Reverb *(t.c. electronic* ClassicVerb*)*.

vKsh was not submitted on the multitrack. At a certain point during mixing, I noticed a percussive 'Ksh'on the second verse. It turned out that this was the tail of the word 'splash' from vVerse2 (Track 29.145). I then copied this 'Ksh' and pasted it into a new track on the downbeat of every other bar, for the duration of the second verse only. On this new track – vKsh – a *MasterComp* with the same settings as vVerse2 compresses the vocal. To send it further into the depth field, this track was sent at high level to the vocal reverb (Track 29.146).

The final verse vocal arrangement can be heard in Track 29.147.

vChorusLead

Inserts: Sonnox *Oxford Dynamics.*
Sends: Vox Reverb (t.c. electronic *ClassicVerb*).

vChorusLead (Track 29.148) was compressed using the *Oxford Dynamics* [*threshold –7.5, ratio 50:1, soft knee 40 dB span, attack 0.52 ms, release 10 ms*] (Track 29.149). It was also sent to the vocal reverb (Track 29.150). It was panned center.

vChs1

Inserts: PSP *MasterComp,* Sonnox *Oxford EQ.*
Sends: Vox Reverb (t.c. electronic *ClassicVerb*).

Both vChs1 and vChs2 are double-tracked versions of vChorusLead. vChs1 (Track 29.151) was compressed using the *MasterComp* [*threshold –14 dB, ratio 8:1, soft knee, RMS sensing, attack 0.010 ms, release 457 ms*] (Track 29.152). The compressor is followed by the *Oxford EQ*, where three bands are employed: a HPF [*229 Hz, 6 dB/oct*] rolls off dispensable lows (Track 29.153), a parametric filter [*2.7 kHz, –2.56 dB, Q 2.83*] helps pan the voice slightly further away (Track 29.154) and another parametric filter [*9.4 kHz, –1.5 dB*] reduces some sibilance (Track 29.155). As with the additional verse voices, the EQ helps to distinguish the second voice from the lead. This track was sent to the vocal reverb, with the intention of placing it slightly further behind the lead chorus voice (Track 29.156). It is panned to around 11:00.

vChs2

Inserts: PSP *MasterComp,* Sonnox *Oxford EQ.*
Sends: Vox Reverb (t.c. electronic *ClassicVerb*).

The vChs2 track (Track 29.157) was processed very similarly to vChs1. The settings on the compressor are all the same apart from the threshold which is set to –12.26 dB and the release to *Auto* (Track 29.158). The settings on the successive *Oxford EQ* vary only slightly: a HPF [*229 Hz, 6 dB/oct*], a parametric filter [*4.6 kHz, –2.15 dB, Q 1.67*] and another parametric filter [*10.2 kHz, –1.38 dB, Q 2.83*] (Track 29.159). This track is panned to 13:00 – mirroring vChs1. It is also sent to the vocal reverb (Track 29.160).

vAns1 and vAns2

Inserts: PSP *MasterComp.*
Sends: Vox Reverb (t.c. electronic *ClassicVerb*).

Both vAns1 and vAns2 (Track 29.161) were compressed very similarly using the *MasterComp.* The compressor settings on vAns1 are [*threshold –15.63 dB, ratio 8:1, soft knee, RMS sensing, attack 0.017 ms, release 100 ms*], while the following differences applied on vAns2 [*threshold –17.25 dB, release 161 ms*] (Track 29.162). Both tracks are sent to the vocal reverb (Track 29.163). vAns1 was panned to around 10:30, while vAns2 was panned to around 13:30. This panning position means that the answering vocals are not too far from the center, yet their image is wider than the lead chorus vocal and its doubles. This creates a centrally panned question and an answer that appears further to the sides.

vAns3 and vAns4

Inserts: PSP *MasterComp.*
Sends: Vox Reverb (t.c. electronic ClassicVerb).

Both vAns3 and vAns4 (Track 29.164) are chopped from vAns1 and vAns2, respectively. Both tracks are compressed exactly like their origin tracks. The only difference is that vAns3 and vAnd4 are panned a notch further outward and their send to the vocal reverb is lower, thus, they appear closer (Track 29.165).

vTemps1 and vTemps2

Both vTemps1 and vTemps2 contain five instances of the word 'Temps', which emphasizes the other vocals at the very end of each chorus. These two tracks, which were neither processed nor sent to the vocal reverb, were panned around 9:00 and 15:00, respectively (Tracks 29.166–29.167).

The final chorus vocal arrangement can be heard in Track 29.168.

vBreak1 and vBreak2

Inserts: PSP *MasterComp.*
Sends: Ambiance Reverb (UAD DreamVerb).

Both vBreak1 and vBreak2, as their name suggests, only appear during the break (Track 29.169). Both are compressed using the same setting on the *MasterComp* [*threshold –8.82 dB, ratio 8:1, soft knee, RMS sensing, attack 0.017 ms, release 457 ms*] (Track 29.170). The general idea with mixing these two voices was to have them further away in the mix, to create a contrast with the main vocals. Thus, instead of sending these to the vocal reverb, they were sent to the ambiance reverb (Track 29.171). vBreak1 and vBreak2 are panned around 10:00 and 14:00, respectively.

vFX

Inserts: PSP *MasterComp.*
Sends: Ambiance Reverb (UAD DreamVerb).

This track, which on the multitrack was already distorted and imprinted with delay (Track 29.172), was compressed using the *MasterComp* in order to contain level fluctuations, and sent to the ambiance reverb to help it gel with the mix ambiance (Track 29.173).

vOutro and vOutroDelay

Inserts: PSP *608 MultiDelay.*

Based on a rough mix submitted by the producer, instances of 'it's Temps' were added to the otherwise vocal-free ending of the song. This was done by cutting the final instance of this phrase from vTemps2 and pasting it into a new track called vOutro. Instead of automating a send to a delay, I duplicated vOutro and inserted it on the new track (vOutroDelay) the *PSP 608* [*100% wet*]. All the eight taps of the delay were utilized with a quarter-note spacing between each tap. The echoes were programmed to gradually drop in level and to shift outward between left and right (Track 29.174).

Vocals group

Inserts: t.c. electronic *Dynamic EQ,* Sonnox *Oxford EQ.*

All vocal tracks, apart from vOutroDelay, were routed to an audio group. The audio group provides a collective level control and processing. The first insert on this audio group is the *Dynamic EQ.* One issue with all vocal tracks was overemphasis in the 10 kHz area. This was made even more profound by the *MasterComp* loaded on many vocal tracks. The *Dynamic EQ* was configured with a single parametric band to the frequency of 10.3 kHz, −7.2 dB of gain and a Q factor of 1. The equalization was set to fully dynamic, with the threshold set to −21.7 dB, attack of 1 ms and release of 81 ms (Tracks 29.175–29.176). Essentially, the *Dynamic EQ* was utilized as a de-esser in the 10 kHz area. Following the *Dynamic EQ* came the *Oxford EQ,* which was added at a later mixing stage to tune all vocals into the frequency spectrum. Two bands are used on the EQ: a HPF [*67.1 Hz, 6 dB/oct*] controls the low limit of the vocals (Track 29.177) and a parametric filter [*2 kHz, 2.5 dB, Q 1.73*] adds some highs that were lost due to the function of the *Dynamic EQ* (Track 29.178). The vocals group level was automated in two places to balance level fluctuations.

30 Donna Pomini (Techno)

Written and produced by TheSwine (*www.theswine.co.uk*).
Mixed (in Logic) by Roey Izhaki.
Mastered by Mandy Parnell (*www.mandyparnell.com*) at Electric Mastering, London.

It happens, sometimes, that you get a production to mix, raise the faders to audition the multitrack, and realize that there is not much for you to do – all the tracks seem to combine pretty well and the core of the mix is already there. This lets you spend more time elevating the mix, focusing more on the general feel and nuances. Such was the case with *Donna Pomini*. In all honesty, the most challenging aspect of mixing this production was getting the levels right, especially between the varying sections.

The multitrack of this nearly 8-minute tune consisted of 70 tracks, but more than a few tracks involved dry and wet versions of the same instrument. The final mix consisted of 59 audio tracks, some of which only play for a few seconds. It is worth mentioning the sections in this production, as marked by the producer: intro, lead-in, break 1, main 1, straight, ocean, break 2, main 2, outro.

Track 30.1 is the mix-ready version. The mix took so little time to complete that there is only one snapshot of the mix-in-progress, presented on Track 30.2. There are not many changes between this mix-in-progress and the final mix presented on Track 30.3, but these changes are noteworthy. Track 30.4 is the mastered mix.

Ambiance reverb

t.c. electronic *ClassicVerb*.

The mix of *Donna Pomini* is a classic example of a mix where one reverb satisfies nearly all the mix reverb needs. There are very few other reverbs in the mix and most of them only appear occasionally. Productions of this type call for a reverb that creates a sense of space while also governing depth positioning, but the reverb itself does not need to be too evident since most of the action occurs at the front of the mix. The chosen ambiance reverb was the t.c. electronic *ClassicVerb*. The reverb itself was based on the *Classic Hall* algorithm and configured to have 0 ms of pre-delay. Additionally, the low- and high-color parameters were set to maximum and minimum, respectively – providing extra warmth and size but little definition. The tail of this reverb can be heard on Track 30.5, while a mix with and without the reverb can be heard on Tracks 30.6–30.7.

Beat

Out of the 59 audio tracks this mix involves, 32 are dedicated to the beat. In addition to the main beat that plays during the main sections, there are also dedicated beat tracks for the lead-in and the straight sections.

Main kick

Inserts: Logic *Compressor*, UAD *Cambridge EQ.*

The main kick (Track 30.8) was compressed using the Logic *Compressor* [*threshold – 11 dB, ratio 2.4:1, peak sensing, knee 0.5, attack 10.5 ms, release 92 ms*], which adds punch and power (Track 30.9). On the subsequent *Cambridge EQ* there is one parametric filter [*4.4 kHz, 2.8 dB, Q 1.12*], which adds attack and makes the kick even more prominent in the mix – a prominence customary in such productions (Track 30.10).

The compression and the added attack on what was already a powerful kick resulted in a massive sound. But during the first few bars of the production this made the kick appear completely detached from the sparse and relaxed arrangement. One way to combat this would be automating the level of the kick, but changing its tonality provided a more creative alternative. So two more bands on the *Cambridge* were employed that are only operational during the first few bars. On one band, a parametric filter with a wide Q [*991 Hz, –6.8 dB, Q 0.52*] reduces much of the kick's impact (Track 30.11), this filter is bypassed with the introduction of the hats. The second band involves a HPF [*34.7 Hz, 24 dB/oct*], which effectively starts rolling-off frequencies below 250 Hz (Track 30.12). Its role is to attenuate much of the powerful lows and subs. It is bypassed just before the introduction of the snare. The bars involving these two EQ automation events are presented in Track 30.13. The level of the main kick was automated to rise just before the first main section.

Pillow kick

Inserts: Sonnox *Oxford EQ.*

Pillow kick only appears in the two closing bars of the intro. The raw track (Track 30.14) was treated with the *Oxford EQ*, where a single parametric filter [*4.3 kHz, –4.4 dB, Q 2.83*] attenuates some click to result in a more pillow-like sound (Track 30.15).

One hit snare

Inserts: PSP *MasterQ*, Logic *Noise Gate*, Logic *PlatinumVerb.*

One hit snare (Track 30.16), as the name suggests, only involves one hit, which concludes the intro section. This track was first treated with the *MasterQ*, where three bands shape the tonality of the snare: a HPF [*73 Hz, 12 dB/oct*] removes dispensable lows (Track 30.17), a parametric filter [*470 Hz, –3.43 dB, Q 0.26*] attenuates some body (Track 30.18) and a high-shelving filter [*10 kHz, 2 dB*] adds a touch of definition (Track 30.19). Following the EQ comes the *Noise Gate* [*threshold –17 dB, –100 dB reduction, attack 3 ms, hold 130 ms, release 10 ms, hysteresis –3 dB*], which shortens the length of the snare so it fits better into the rhythm (Track 30.20). This gate was configured with the next *PlatinumVerb* already in place. On the reverb, which was added as an insert rather than a send, the dry control is set to 100% and the wet to 45%. The reverb adds an audible effect and its tails smoothes the transition to the lead-in section (Track 30.21).

LiSnare 1

Inserts: UAD *Cambridge EQ.*

LiSnare 1 (Track 30.22) plays during the first part of the lead-in section and during the last four bars before the outro. This track was treated using the *Cambridge EQ* where a HPF [*168 Hz, 6 dB/oct*] rolls off muddling lows (Track 30.23). Also, a high-shelving filter [*6.32 kHz, 4 dB*] brightens the snare and tunes it to the frequency spectrum (Track 30.24).

SnareB

Inserts: PSP *MasterQ.*

SnareB (Track 30.25) accompanies every other hit of LinSnare 1 during the first part of the lead-in section. It was treated with the *MasterQ*, where a HPF [*74 Hz, 12 dB/oct*] rolls off dispensable lows (Track 30.26) and a high-shelving filter [*10 kHz, 4 dB*] adds some extra definition (Track 30.27).

Claps

Inserts: t.c. electronic *ClassicVerb.*

There are five clap hits during the first part of the lead-in section, two of which conclude this first part. Only a reverb was added to these claps and the plugin was added as an insert. The wet/dry control was set to 50/50, the pre-delay to 91 ms and the early reflections were muted (Tracks 30.28–30.29).

LiMidSnare

Inserts: Logic *Channel EQ.*

Two hits of LiMidSnare play right before the second part of the lead-in section (along with two claps). The raw track (Track 30.30) was treated with the *Channel EQ*, where a HPF [*130 Hz, 24 dB/oct*] rolls off dispensable lows (Track 30.31) and a parametric filter [*7.7 kHz, 5.5 dB, Q 1.9*] adds brightness and definition (Track 30.32).

LiSnare 2

Inserts: UAD *Cambridge EQ.*

LiSnare 2 (Track 30.33) plays during the second part of the lead-in section. Its tonality was altered quite noticeably by the *Cambridge EQ*, mainly due to the HPF [*428 Hz, 12 dB/oct*], which dries out all the lows (Track 30.34). As part of frequency tuning, a parametric filter [*900 Hz, 4.2 dB, Q 2*] was also employed (Track 30.35) along with a high-shelving filter

[*6.32 kHz, 1.6 dB*] (Track 30.36). At some point during the mix, a compressor was also loaded onto this track, but a punchy snare didn't work quite well in the context of the relaxed lead-in section. What's more, the beat is not the most important element of this section – the real action happens later.

Kick 2

Inserts: Sonnox *Oxford EQ.*

Kick 2 plays during the main and the straight sections of the production and it provides a layer to the main kick (Tracks 30.37–30.38). The only plugin inserted on this track is the *Oxford EQ*, where a parametric filter [*95 Hz, –2.88, Q 1.4*] attenuates disturbing content around 95 Hz (Tracks 30.39–30.40).

Kick WoodKnock

Kick WoodKnock only plays during the first main and the straight sections. It is a complementary layer for the kick and it was untreated (Tracks 30.41–30.43).

Main Snare

Inserts: UAD *Cambridge EQ.*

Main Snare appears during the main sections. In many productions such as *Donna Pomini*, the kick is the mighty beat element and the snare gets much less stage space. To some extent, the Main Snare was mixed with slightly more entity than what some consider normal. The raw track (Track 30.44) was only treated with the *Cambridge EQ*, where a high-shelving filter [*4 kHz, 1 dB*] adds a touch of presence (Track 30.45). The level of this track was automated to rise during the second main section, but falls again in its second part.

Rvs Snare

Inserts: PSP *MasterComp.*

Rvs Snare plays during the first part of the first main section and throughout most of the second main section. It accompanies the main snare as a supportive beat element (Tracks 30.46–30.48). Rvs Snare is only treated with the *MasterComp* [*threshold –16.1 dB, ratio 1.4:1, RMS sensing, hard knee, attack 3.16 ms, release 1 sec*], which condenses the snare's dynamics (Track 30.49).

Straight Snare

Inserts: Sonnox *Oxford Dynamics*, Sonnox *Oxford EQ*, Logic *BitCrusher.*
Sends: SnareVerb *(*Sonnox *Oxford Reverb).*

Straight Snare (Track 30.50), which plays during the straight section, was first treated with the *Oxford Dynamics*, where both the compressor [*threshold –22.5 dB, ratio 2:1, hard knee, attack 15.3 ms, release 5 ms*] and the gate [*threshold –30 dB, range –40 dB, attack 5 μs, hold 550 ms, release 19.6 ms*] are employed to add weight and punch (Tracks 30.51–30.52). Then, the *Oxford EQ* shapes quite drastically the sound of the snare using four bands: a HPF [*271 Hz, 24 dB/oct*] rolls off the body and dispensable lows (Track 30.53), a parametric filter [*487 Hz, –8.44 dB, Q 2.83*] attenuates disturbing low-mids (Track 30.54), another parametric filter [*3.13 kHz, 7.71 dB, Q 2*] adds snap (Track 30.55) and a high-shelving filter [*3.32 kHz, 2.18 dB*] adds some brightness (Track 30.56). The EQ is followed by the *BitCrusher*, which adds its own touch of grit [*8 bits, drive 4.5 dB*] (Track 30.57).

The snare was sent to a dedicated reverb – a modified version of the *Hall Full* preset on the *Oxford Reverb*. The send level to the reverb was automated so as to open on every other hit (with one hit being an exception). Apart from making the effect appear on every other hit, this also makes one hit to appear in back, the other in front (Track 30.58). There is no practical purpose for this automation and it is hardly creative, but we do these things sometimes.

Whip Snare

Inserts: Sonnox *Oxford Dynamics,* Sonnox *Oxford EQ.*
Sends: Ambiance Reverb *(*t.c. electronic *ClassicVerb).*

Whip Snare (Track 30.59) plays during the last four bars before the outro. It first goes through the *Oxford Dynamics*, where much like with the Straight Snare, both the compressor [*threshold –17.7 dB, ratio 5:1, soft knee, attack 16.8 ms, release 130 ms*] and the gate [*–17.2 dB, range –40 dB, attack 5 μs, hold 120 ms, release 20.9 ms*] add snap and punch (Tracks 30.60–30.61). On the subsequent *Oxford EQ* a HPF [*210 Hz, 12 dB/oct*] rolls off dispensable lows (Track 30.62) and a parametric filter [*3.29 kHz, 3.2 dB, Q 1.96*] emphasizes some edge (Track 30.63). Whip Snare is sent at high level to the ambiance reverb, which places it quite deep in the depth field and adds an apparent tail (Track 30.64).

March Snare

Inserts: SoundHack +*chebyshev,* Logic *Gain,* UAD *GateComp,* t.c. electronic *Filteroid.*

March Snare is part of the buildup during the ocean section. The raw track already contained some room sound (Track 30.65). It is first treated with the +*chebyshev*, which adds some harmonics that fatten the sound (Track 30.66). Then the *Gain* plugin was inserted for level automation, where the snare progressively rises in level (Track 30.67). The succeeding *GateComp* adds weight to the snare with its compressor facility [*threshold –28.8 dB, ratio 4.64:1, slow attack, fast release*] (Track 30.68). The *Gain* plugin was intentionally inserted before the *GateComp*, so the compression effect increases as the snare rises in level. On the succeeding *Filteroid* a non-resonant HPF [*185 Hz*] and a LPF [*17.8 kHz*] contain the frequency extremes of the snare (Track 30.69).

Main hats

Inserts: Logic *BitCrusher,* SoundHack +*chebyshev,* t.c. electronic *EQSat.*
Sends: Ambiance Reverb *(*t.c. electronic *ClassicVerb).*

The main hats are the only hats playing during the lead-in section and, as importantly, during the second part of the first break. They also play along other hats during the main sections. The clean sound of the raw track (Track 30.70) could work in the context of many similar productions, but I personally felt that a slightly more dirty sound would better suit the aggressive nature of this track. To achieve this dirty sound, two plugins were used. First, the *BitCrusher* reduces the bit resolution to 4 bits, which adds a noticeable amount of noise. In addition, 13 dB of drive causes clip distortion that adds healthy grit (Track 30.71). Then, the +*chebyshev* adds some harmonics that enhance the effect (Track 30.72). The *BitCrusher* and +*chebyshev* made the hats appear brittle and therefore detached from the mix. To ease this excess of highs, a high-shelving filter [*5 kHz, −1.5 dB*] was engaged on the *EQSat* (Track 30.73). The level of the main hats was automated with respect to the different sections in which they play. It might be worth noting that during the lead-in section the hats are fairly visible as they play an important rhythmical role. They are also sent to the ambiance reverb, which moves them back from the front of the mix (Track 30.74).

Hats 2

Inserts: Logic *BitCrusher.*
Sends: Ambiance Reverb (t.c. electronic *ClassicVerb).*

Hats 2 (Track 30.75) are the only hats playing during the straight section, but they play along other hats tracks during the first part of the second main section. As with the main hats, Hats 2 were treated with the *BitCrusher* [*3 bits, 6 dB drive*], which distorts them quite noticeably (Track 30.76). They were also sent to the ambiance reverb, which sends them back in the depth field (Track 30.77).

Hats 4 Hits

Four hi-hats hits mark the border between the two parts of the first main section. These hits were left untreated (Track 30.78).

Hats Backbeat

Sends: Ambiance Reverb (t.c. electronic *ClassicVerb).*

Hats Backbeat (Track 30.79) plays along with the main hats during the main sections. This track was untreated, but was sent to the ambiance reverb for depth positioning (Track 30.80).

Crash 1, Crash 2 and Crash 3

Donna Pomini involves three crash tracks. None of them were treated or sent to the reverb, but the levels of Crash 2 and 3 were automated between various hits.

WoodZest

Inserts: Logic *Channel EQ.*

WoodZest is part of the buildup to the first main section and plays for most of the production from that point onward (Tracks 30.81–30.83). It was only treated with a HPF [*240 Hz, 24 dB/oct*] from the *Channel EQ* (Track 30.84). Its level was automated, for example, to progressively rise during the ocean section.

PaperLoop

Inserts: Smartelectronix *Cyanide 2,* SoundHack *+chebyshev,* Sonnox *Oxford EQ.*
Sends: Ambiance Reverb *(t.c. electronic ClassicVerb),* Loop Delay *(PSP 84).*

PaperLoop plays during the straight and the second main sections. The raw track (Track 30.85) was distorted using *Cyanide* (Track 30.86). Then additional distortion was produced by the *+chebyshev* (Track 30.87). The *Oxford EQ* was employed to tune this track into the frequency spectrum – a low-shelving filter [*114 Hz, –6.27 dB*] attenuates some lows (Track 30.88).

PaperLoop is sent at low level to the ambiance reverb in order to move slightly backward in the mix (Track 30.89). It is also sent to PSP *84,* with the send level automated to open during the last few hits before the ocean section. The delay time was set to quarter note, –4.89 dB feedback gain, a resonant LPF in the feedback loop and gentle modulation of the delay time (Track 30.90).

Bat Zips, Bat Zaps, Bat Hit1, Bat Hit2, Bat Hit3, Bat Iron and Bat Wiper

Inserts: Smartelectronix *Cyanide 2.*
Sends (for some): Ambiance Reverb *(t.c. electronic ClassicVerb).*

Bat Zips, Bat Zaps, Bat Hit1, Bat Hit2, Bat Hit3, Bat Iron and Bat Wiper (Tracks 30.91–30.98) were all processed using the same *Cyanide* configuration (Track 30.99). This produced a very distinctive sound, which adds some aggressiveness. Bat Zips, Bat Zaps and Bat Iron were sent to the ambiance reverb in order to shift them backward in the mix (Track 30.100).

Beat Group

Inserts: PSP *VintageWarmer.*

Although entitled 'Beat Group', only four beat elements were routed to this audio group: Kick 2, Kick WoodKnock, Rvs Snare and Hats Backbeat. On the group track the *VintageWarmer* was loaded and configured for parallel compression [*mix 32.6% dry, drive 16.5 dB, knee 0, speed 51.5%, release 0.25*]. The idea was to create a thicker sound layer and to draw some compression effect (Tracks 30.101–30.102).

Sound FX

Splutter

Inserts: Logic *BitCrusher,* Logic *PlatinumVerb.*

Splutter appears once, right before the introduction of the lead during the first break (Track 30.103). It was first processed using the *BitCrusher,* but the bit reduction facility was not utilized and the plugin was used purely for distortion and downsampling purposes [*24 bits, drive 22.5 dB, downsampling 2x, clip level –7.8 dB*] (Track 30.104). The *BitCrusher* is followed by the *PlatinumVerb,* which was inserted [*73% dry, 49% wet*] to send Splutter backward and add some effect to the dry sound (Track 30.105).

Laser Gun

Inserts: Logic *Channel EQ.*

Laser Gun appears right before the drop to the main section (Track 30.106). It was only treated by the *Channel EQ*, where a HPF [*240 Hz, 24 dB/oct*] rolls off dispensable lows (Track 30.107).

ChemiWind

*Sends: Ambiance Reverb (*t.c. electronic *ClassicVerb).*

ChemiWind gels the transition between the straight and the ocean sections (Track 30.108). It was only sent at high level to the ambiance reverb, which adds some dimension to it and generates a transition tail (Track 30.109).

FallBall

*Sends: Ambiance Reverb (*t.c. electronic *ClassicVerb).*

FallBall creates some tension toward the end of the first main section and the pitch dive leads into the straight section (Track 30.110). It also appears briefly between the two parts of the second main section. This track was only sent at high level to the ambiance reverb, mostly in order to place it behind the beat and the leads (Track 30.111).

SciFi Waves

SciFi Waves (Track 30.112) appear right before the first break, and also make a short appearance early in the first main section. This track was not treated, but its level was automated to rise before its appearance in the main section.

Bass

JoviBass

Two bass tracks were included on the multitrack. JoviBass (Track 30.113) plays up until the first break. This track was not treated.

ZugBass

In contrast to JoviBass, which called for no treatment, ZugBass (Track 30.114), which plays from the first main section onward, was hopeless. Neither its frequency content nor its dynamics provided workable material and despite quite some time and effort, it sounded wimpy, undefined and flat. So ZugBass was excluded from the mix and I have copied its source MIDI track into Logic and synthesized a new bass using the Access *Virus.* This new bass was called ZugBassR.

ZugBassR

Inserts: Sonnox *Oxford EQ.*

ZugBassR (Track 30.115) was equalized using the *Oxford EQ*, where a HPF [*44.3 Hz, 36 dB/oct*] sets the low-frequency boundary of the bass; this was done to prevent excess of subs that occur on the low notes (Track 30.116). In addition, there is also a parametric filter [*267 Hz, –4.8, Q 2.83*] that combats an excess of low-mids (Track 30.117).

Vocal

Vox

Inserts: Logic *Vocal Transformer,* Logic *Channel EQ,* Logic *Guitar Amp Pro.*

The vocal on *Donna Pomini* (Track 30.118) called for a distinctive effect – just having a normal speaking voice would be mundane and characterless. Also, what's said, and the way in which it is said, suggest an abstract and somewhat dreamy message. A nice vocal effect would improve this mood. The first plugin in the inserts chain is the *Vocal Transformer* – it does not fulfill its original purpose but adds a unique effect, which is a blend between a robotic hollowness and a wired distortion (Track 30.119). Next comes the *Channel EQ*, where a HPF [*450 Hz, 24 dB/oct*] rolls off an excess of lows that really contributed nothing to the speaking voice (Track 30.120). The most dominant contributor to the final vocal effect is the *Guitar Amp Pro*, where in addition to the guitar amp simulation both the tremolo and the spring reverb are enabled (Track 30.121). The vocal track was sent to the ambiance reverb so as to shift it backward in the depth field and to give it some stereo size (Track 30.122).

The squeaking loop that appears in the second part of the main break and in the main sections has its origins in a saturated feedback loop of an automated delay on the words 'each other'. This loop was already bounced on the vocal track on the multitrack.

Other elements

Gidi FM

*Sends: Ambiance Reverb (*t.c. electronic *ClassicVerb).*

Gidi FM plays during the lead-in section (Track 30.123) and also provides the FM riser during the breaks buildup (Track 30.124). This track was automated in level between and within the various sections. It is sent to the ambiance reverb, which pans it backward and increases its stereo size (Track 30.125).

Wacko

Inserts: Sonnox *Oxford EQ.*
*Sends: Ambiance Reverb (*t.c. electronic *ClassicVerb).*

Wacko opens the production and also plays during the first main section and the straight (Track 30.126). Only a stereo version with imprinted delays was included on the multitrack.

This track was automated in three principal ways: its level is automated, for example, to drop twice during the intro section; while being panned center up until the straight section, it then shifts to the right [*+21*] to reduce its masking interaction with Straight Bass; for the same reason, the *Oxford EQ* is enabled at the beginning of the straight section and a HPF [*149 Hz, 18 dB/oct*] rolls off some colliding lows (Track 30.127). It was also sent to the ambiance reverb in order to blend it into the mix space (Track 30.128).

Wacko Filtered

Wacko Filtered plays for the first four bars of the production. On the multitrack this track already contained the filter and the level automation (Track 30.129). It was not treated.

Wacko Support

*Sends: Ambiance Reverb (*t.c. electronic *ClassicVerb).*

Wacko Support plays for two bars during the intro (Track 30.130). This track was mixed at a low level underneath Wacko and its level was automated to rise progressively. It was sent at a high level to the ambiance reverb, which sends it deep to the back of the sound stage (Track 30.131).

Drops

*Sends: Ambiance Reverb (*t.c. electronic *ClassicVerb).*

Drops is one of the main elements playing during the lead-in section. The multitrack included both a version with no delay (Track 30.132) and a stereo version with a delay (Track 30.133). From the early mix stages to the final ones, the stereo version presented no issues, so it is the one used in the final mix. Drops is sent to the ambiance reverb, just to gel it to the mix ambiance (Track 30.134)

Drops Noise

*Sends: Ambiance Reverb (*t.c. electronic *ClassicVerb).*

Just like with Drops, between the with and without delay versions of Drops Noise the former was used (Tracks 30.135–30.136). It is also sent to the ambiance reverb (Track 30.137).

SinSeqRoot, SinSeq3rd, SinSeq5th

Inserts: PSP *Nitro.*
*Sends: Ambiance Reverb (*t.c. electronic *ClassicVerb).*

SinSeqRoot, SinSeq3rd and SinSeq5th all constitute a sequence of sine beats (Tracks 30.138–30.141). Each of the individual tracks was treated with the *Nitro* that was configured for an autopan effect. The modulation frequency on each of the three tracks is different, which creates an unpredictable panning relationship between each track (Track 30.142). To detach the sequence from the front of the mix, each track was also sent to the ambiance reverb (Track 30.143).

SinSeq group

Inserts: PSP 608 MultiDelay.

The three SinSeq tracks are all routed to an audio group. The inserted *608* is set for 35% dry and 65% wet and two taps are operational – one with a delay of 372 ms and the other with 743 ms. The feedback control on the second tap is set to 32.5%. This addition of delays softens the SinSeq and makes it somewhat more dreamy and distant (Track 30.144). The level of the SinSeq group was automated.

LeadSeq

Inserts: UAD Fairchild.
Sends: Ambiance Reverb (t.c. electronic ClassicVerb).

LeadSeq (Track 30.145) is introduced during the second half of the first break and is one of the most important elements in the production. This lead occupies a huge amount of space on the frequency spectrum and as it clearly involves unison it also takes up quite some space on the stereo panorama. It is somewhat surprising that this lead fits nicely into the mix without any frequency or stereo treatment. The *Fairchild* inserted on this track was employed to condense the level of this lead and tuck on some more weight – essentially making the lead even bigger (Track 30.146). The level of this track was automated, for example, it is made louder on the first main section. It is also sent at low level to the ambiance reverb, which gels it into the mix ambiance and shifts it slightly backward (Track 30.147).

LoFi Saw

Sends: Ambiance Reverb (t.c. electronic ClassicVerb).

LoFi Saw accompanies LeadSeq during the first main section and the first part of the second main section. Between the with and the without delay versions on the multitrack I chose to use the one with delay (Tracks 30.148–30.149). This track was untreated, but it was sent at a high level to the ambiance reverb. This not only places it behind LeadSeq, but also fills the ambiance during the most powerful moments of the production, making it denser and richer (Track 30.150).

Straight Bass

Inserts: UAD Cambridge EQ.
Sends: Ambiance Reverb (t.c. electronic ClassicVerb).

Straight Bass (Track 30.151) plays during the straight and ocean sections. The *Cambridge* inserted on this track only has one parametric filter engaged [*90.8 Hz, −2.4 dB, Q 1.66*] that eases an excess of lows around 90 Hz that muddled the mix (Track 30.152). This track was also sent to the ambiance reverb in order to send it slightly backward and gel it to the mix ambiance (Track 30.153).

Straight Beep

Straight Beep plays during the straight section only. I wanted the beep to appear distant and to audibly excite this reverb or another. Initially, I used the dry version from the multitrack and sent it at a high level to the ambiance reverb (Tracks 30.154–30.155). However, the multitrack also contained a version with a reverb and auditioning it a later mix stage worked out better (Track 30.156). It is the only track from the multitrack that had a version with a reverb that was used in the final mix.

Leadar

Inserts: PSP *MasterComp.*
Sends: Ambiance Reverb (t.c. electronic *ClassicVerb).*

Leadar plays from the beginning of the ocean section until the end of the production. Between the with and without delay versions on the multitrack I have chosen the former (Tracks 30.157–30.158). This track was compressed using the *MasterComp* [*threshold –12.93 dB, ratio 1:30:1, RMS sensing, hard knee, attack 10 μs, release 100 ms*], which thickens it and makes it bigger (Track 30.159). It was also sent to the ambiance reverb, which gels it to the sound stage but also makes its stereo appearance richer and bigger (Track 30.160).

OceanPad

Inserts: PSP *MasterQ.*
Sends: Ambiance Reverb (t.c. electronic *ClassicVerb).*

OceanPad plays during the ocean section (Track 30.161). This track was compressed with the *MasterQ*, where a high-shelving filter [*6.41 kHz, –4.15 dB, Q 0.5*] rectifies an excess of brittle highs (Track 30.162). The ocean section is also some kind of a break that I felt needed to have an extra dreamy and spacious sound. Ocean Pad was sent at an extremely high level to the ambiance reverb, which makes it appear distant and facilitates the required dreamy sensation (Track 30.163).

OceanSup

Inserts: PSP *84.*
Sends: Ambiance Reverb (t.c. electronic *ClassicVerb).*

OceanSup is a complementary track to OceanPad – while OceanPad is more concerned with the soundscape, OceanSup gives a clearer idea of the melody being played (Tracks 30.164–30.165). The raw track (Track 30.166) was first treated with the PSP *84* [*30 ms delay on the left channel only, no feedback*], which combats this track's presence in the busy center – the delay converts its central image to a sound bound to the stereo extremes (Track 30.167). It was also sent at a high level to the ambiance reverb to help project the wistful ambience that this section demanded. (Track 30.168).

LowHarm, HighHarm

Sends: Ambiance Reverb (t.c. electronic *ClassicVerb).*

Both LowHarm and HighHarm only appear at the end of the ocean section (Tracks 30.169–30.170). Both tracks were sent at a high level to the ambiance reverb in order to place them far in the depth field (Track 30.171). Both tracks were mixed at a low level, which means that rather than being clearly present they blend with the overall atmosphere.

ContraWaves

ContraWaves appear during the eight-outro bars as a support layer to Leadar (Track 30.172). This track was not treated.

31 The Hustle (DnB)

Written and produced by Dan 'Samurai' Havers & Tom 'Dash' Petais
(*www.dc-breaks.co.uk*).
Mixed (in Digital Performer) by Roey Izhaki.
Mastered by Mandy Parnell (*www.mandyparnell.com*) at Electric Mastering, London.

The Hustle is a powerful DnB production. The 41 tracks submitted with this project – mostly the outcome of original programming – constitute a meticulous arrangement that could easily serve as a production master class. Indeed, from a mix engineer point of view, it was exciting to have an insight into how this epic tune was constructed.

Drum 'n' bass productions, as the name suggests, are characterized by dominant beats and basslines. In mixes, the two tend to be loud to an extent that would not be considered sensibly balanced in many other genres. In addition, most mixes also entail a deep and wide sound stage and a distinguished sense of space – all the outcome of evident reverbs. The fast tempo of DnB productions (174 BPM in the case of *The Hustle*) and typical dense arrangements mean that mostly we are trying to find space in the mix for the various elements.

While many DnB productions are heavily based on a massive beat and a grandiose bassline, the qualities of *The Hustle* are distributed between other elements as well. I thought that the mix should reflect this, so the general approach in mixing the various elements was slightly more balanced compared to many traditional DnB mixes.

It is not uncommon for a rough mix to turn into the mix itself, such was the case with this production. By the time I had studied the various tracks, sections and nature of the production, I already had a mix in progress, and there was little point starting all over again.

Track 31.1 is the producer's rough mix, Track 31.2 is the mix-ready version with rough levels and panning but no processors or effects; Track 31.3 is a snapshot of the mix in progress, Track 31.4 is the final unmastered mix and Track 31.5 is the mastered version.

Ambiance reverb

Audio Ease *Altiverb.*

The reverb choice in productions such as *The Hustle* is a crucial aspect of the mix. In such dense arrangements, any instrument that doesn't need to be forward is panned backward – a panning mostly achieved using reverbs. In addition, the typical prominent space of DnB mixes means that the reverbs used dictate much of the mix character. The plan for this mix was to use one main reverb to which many instruments are sent. This reverb is used for front-back panning and contributes to the space attributes of the mix.

As a common practice, a 'draft' reverb was used during the early mix stages, and the choice of the final ambiance reverb was made once enough tracks were sent to that reverb and once the mix progressed enough to hint at the direction the ambiance needed. After a little experimentation, I chose the Audio Ease *Altiverb*, its *Amsterdam Concertgebouw* IR collection (under the *Concert Halls Large* category), and the *Stereo to Stereo far Omnis* IR. This specific IR was chosen as it excelled at depth positioning, it handled transients effortlessly, and as its warm, defined-yet-elegant nature blended superbly into the mix. The tail of this ambiance reverb can be heard on Tracks 31.6–31.7, while Tracks 31.8–31.9 provide a comparison between a mix with and without the ambiance reverb. One issue with the ambiance reverb was that it felt too wide, so the *Altiverb* output was narrowed by the *MOTU Trim* (Tracks 31.10–31.11).

Drums

Attack Kick

Inserts: Sonnox *Oxford EQ.*
Sends: Kick SC.

Attack Kick (Track 31.12) is the main kick track. It was treated with the *Oxford EQ*, where a low-shelving filter [*136 Hz, 8.2 dB*] adds a healthy lows impact (Track 31.13). A steep HPF [*42 Hz, 30 dB/oct*] prevents boominess and filters the excess of very low frequencies, which many domestic systems cannot reproduce anyway (Track 31.14). There is also a parametric filter [*273 Hz, −9 dB, 1.5 Q*] removing a low-mids tap (Track 31.15). Attack Kick is sent to a bus, which sources the side-chain of the sub-bass gate.

Dirt Kick

Inserts: MOTU *Trim.*

The Attack Kick is layered with Dirt Kick (Track 31.16), which adds some oomph and noise that contributes some definition (Tracks 31.17–31.18). Despite the fact that both are samples, the Attack Kick and the Dirt Kick were out of phase with each other, so the phase of the Dirt Kick was inverted using the MOTU Trim plugin (Tracks 31.19–31.22), otherwise it is untreated.

Kick Triggers

Virtual Instrument: MOTU *Model 12*

Still unhappy with the sound of the kick, I decided to add another layer of samples. This was achieved by exporting the Kick's MIDI track from the original Logic session, importing it into Digital Performer, and routing it to the MOTU *Model 12* Drum Module. The kick061 was chosen from more than a hundred kick samples that ship with *Model 12* (Track 31.23). In addition, I programmed a MIDI sequence with a kick on each downbeat and sent it to the same *Model 12*, this time triggering the kick091 sample. This additional trigger was mixed so it only adds a gentle accent to the downbeats (Track 31.25). Finally, I added sub-bass to the kick by gating a 50 Hz sine wave with respect to the original kick track (Track 31.26).

Thump Kick

There are only six Thump Kick (Track 31.27) hits in the arrangement, three in each break. This track was untreated.

Snare

Inserts: MOTU *Masterworks EQ,* Sonnox *Oxford Dynamics.*
Sends: Snare Reverb (UAD Plate 140).

One problem with the main Snare track (Track 31.28) was that it was too bright and did not sit well in the frequency spectrum – it seemed to end up on higher frequencies than it should have done. To correct this, a high-shelving filter [*4 kHz, −1.6 dB*] on the *Masterworks EQ* was employed to soften the highs (Track 31.29). Following this, the snare sat better in the frequency spectrum but it lost some presence. In order to return to the limelight, the compressor [*threshold −10.2 dB, 2:1 ratio, hard knee, attack 28.6 ms, release 130 ms*] of the *Oxford Dynamics* was employed to add some snap (Track 31.30).

I also experimented with adding a reverb to the snare using the *Plate 140* (Track 31.31). It is the nature of reverbs to soften sounds, so in order to achieve some thrust I gated the reverb using the MOTU *dynamics* (Track 31.32). Since the reverb smears the snare image and sends it slightly backward, it is only mixed during the intro, where the arrangement is sparser and more relaxed.

Sub Snare

Sub Snare (Track 31.33), which was left untreated, is complementary to the Snare track that was layered quietly underneath it (Tracks 31.34–31.35).

Pend Snare

Inserts: MOTU *Gate,* MOTU *Parametric EQ,* Sonnox *Oxford Dynamics.*

Pend Snare (Track 31.36) was one track that I felt contributed little to the overall snare sound, so instead of layering it with the Snare track, it was mute-automated to only appear when the snare plays eight-notes, and therefore accenting the momentary intensity of the beat (Tracks 31.40–31.41). The raw track involved a very long decay, so to make it punchier it was gated using the MOTU *Gate* (Track 31.37). The gate is followed by the *Parametric EQ*, where a high-shelving filter [*1.3 kHz, 4.7 dB*] brightens what was a dull sound (Track 31.38). To add even more punch, the *Oxford Dynamics* compressor [*threshold –18 dB, ratio 3.17:1, attack 10.6 ms, release 5 ms, soft knee 20 dB span*] accents the snare's attack (Track 31.39).

Hi Hats

Inserts: MOTU *Autopan,* MOTU *Masterworks EQ,* MOTU *Dynamics.*
Sends: Ambiance Reverb (Audioease *Altiverb*).

In addition to the kick and snare, the main beat also consists of Hi Hats, Comp Break and Tamb. The Hi Hats (Track 31.42) go through the *AutoPan*, which adds movement. As the Hi Hats shift between left and right, their perceived level rise and fall (partly due to masking), which adds another element of cyclic variation. In order to make the effect not too obvious, the depth on the *AutoPan* was set to 43%; the modulation is synced to a bar (Track 31.43). The *Autopan* is followed by the *Masterworks EQ*, where a high-shelving filter [*3.5 kHz, 2.8 dB*] adds definition and spark (Track 31.44). Although it could be argued, I felt that brightening the hi hats further (Track 31.45) would detach them from the mix and cause exaggerated frequency separation. Following the *Masterworks EQ* is the MOTU *Dynamics*, whose role is explained in the next section. One of the problems with the Hi Hats was that they sounded too in front and as such felt a bit like a fly buzzing in the front of the mix. Sending them to the ambiance reverb shifted them backward and rectified this buzzing sound (Track 31.46).

Comp Break

Inserts: MOTU *Parametric EQ,* MOTU *Dynamics.*

Comp Break (Track 31.47) is yet another important component of the beat. Keeping the kick and snare in this loop served little purpose since each drum has respective tracks already. Using the *Parametric EQ*, a HPF [*725 Hz*] filters much of the kick and snare, while also clearing some low-mids space for other tracks (Track 31.48). As part of tuning this track to the frequency spectrum and in order to give it more presence, a high-shelving filter [*5 kHz, 5.7 dB*] was also applied (Track 31.49). The Comp Break was panned off-center to the left [*< 41*] so as to ease masking interaction with the kick and snare.

Later on in the mix I wanted the leads to stand out more. One way to emphasize them was by ducking other instruments in relation to them – as if they are so powerful that they bring down other instruments. Ducking the whole beat would also attenuate the important kick and snare, so instead only the Hi Hats and Comp Break were ducked. This was achieved by inserting the MOTU *dynamics* compressor on the two tracks with each side-chain set to the same bus. The bus was sourced from a send on the Dina lead track – a send that was pre-fader, so level alterations to the lead would not affect the ducking. The major setting on the compressors was the release, as it determined how quickly the Hi Hats and the Comp Break recover after being ducked. The final settings were 560 ms for the Comp Break and 480 ms for the Hi Hats (Tracks 31.50–31.51).

Tamb

Inserts: MOTU *Parametric EQ.*
Sends: Ambiance Reverb (Audioease *Altiverb).*

The Tamb (Track 31.52) is introduced in the arrangement after the first break. It is mixed, so it blends with the Hi Hats and Comp Break rather than being a distinctive sound (Tracks 31.56–31.57). A HPF [*1.3 kHz*] on the MOTU *Parametric EQ* rolls off dispensable frequency content, which again clears space on the low-mids (Track 31.53). In addition, a parametric filter [*5.3 kHz, 3.3 dB*] fills some missing frequency space (Track 31.54). The Tamb is also sent to the ambiance reverb, which blends it into the depth field (Track 31.55). The Tamb was panned slightly off-center to the right [*12>*], a position in which it best escaped masking without shifting too far (from the beat) to the extreme. It is not ducked since it is masked anyway by the leads, at the moments when it would have been ducked.

Break Beat

Inserts: MOTU *Delay,* SoundHack *+chebyshev.*

The drums during the break are based on four tracks: Break Beat, Hip Hop Beat, Claps and Half Time. Break Beat (Track 31.58) and Hip Hop Beat are two competing tracks and I wanted to maintain separation between them – had the two been blended around the center, the details of both would be lost and the two would combine unimpressively. A break, by nature, involves a departure from the main theme, and this applies to the way we mix it as well. So while Hip Hop Beat was panned center, Break Beat goes through the MOTU *Delay,* which only delays the right channel by a quarter note (Track 31.59). This creates an engaging effect that separates the two competing tracks. Also, this means that during the break some intensity shifts to the extremes and gives an overall wider stereo impression compared to the sections before and after it. The Break Beat delay is followed by the *+chebyshev,* which adds a touch of distortion (Track 31.60).

Hip Hop Beat

Inserts: Sonnox *Oxford Transient Modulator.*

The raw Hip Hop Beat track (Track 31.61) could benefit from more vibrant dynamics, so the *Transient Modulator* [*ratio 0.12*] was employed to accent the transients on this loop (Track 31.62).

Hip Hop Beat distorted layer

Inserts: UAD *Preflex.*

The raw Hip Hop Beat track was duplicated and the new track was distorted using the *Preflex.* The distorted track is layered with the original Hip Hop Beat and automated to progressively rise in level (Tracks 31.63–31.64).

Claps

Inserts: PSP *MasterQ.*

Claps (Track 31.65) were only treated with the *MasterQ,* where a HPF [*1.8 kHz*] rolls off frequency content that contributes little to their sound (Track 31.66).

Half Time

Inserts: MOTU *Masterworks EQ,* Sonnox *Oxford Dynamics.*
Sends: Ambiance Reverb (Audioease *Altiverb), Closing Delay* (MOTU *Delay).*

Half Time (Track 31.67) is a percussive loop that comes and goes throughout the production. The untreated track contained excessive lows and low-mids noise that contributed
nothing but reduced clarity. So it is first filtered by a HPF [*214 Hz, 24 dB/oct*] from the
Masterworks EQ (Track 31.68). Also, a parametric filter [*1.2 kHz, 6.4 dB, Q of 0.85*] adds
some flesh to what I found as thin (Track 31.69). Following the equalizer, an *Oxford
Dynamics* was inserted, where the compressor [*threshold –30 dB, ratio 2:1, attack 2.2 ms,
release 52 ms*] condenses the levels of the percussive sounds, adding some powerful
impact (Track 31.70). Half Time is also sent to the ambiance reverb in order to place it
backward in the depth field (Track 31.71). The level of this track was automated, where
it becomes louder [*2.7 dB*] during the second break. In addition, a send to a ping-pong
delay is automated to take effect during the last bar of the outro. The MOTU *Delay* was
set to quarter note delays with a feedback LPF at 3 kHz (Track 31.72). This delay appears
to escape to the left speaker earlier than normal with this effect, and although this could
have been corrected, I decided to keep it.

Bongos 1 and Bongos 2

Inserts: MOTU *Masterworks EQ.*
Sends: Bongos Reverb (UAD *DreamVerb).*

Bongos 1 and Bongos 2 (Track 31.73) were both treated with a high-shelving filter [*4 kHz,
6.8 dB*] using the *Masterworks EQ* (Track 31.74). Mostly this was done in order to increase
their definition. It was hard not to imagine the bongos with a nice, clear reverb. But the
ambiance reverb was not fit for the job – the bongos either sounded too dry or too far
away (Track 31.75). Also, in the specific case of the bongos I was after brilliance rather
than the native warmth of the ambiance reverb. So the bongos were given their own
reverb – the UAD DreamVerb and its Church Choir preset (Track 31.76). This preset was
only modified by a tiny pull of the high-mids on the pre-reverb EQ.

I did not want the two bongo tracks to be too far apart, yet I did not want them panned to
the busy center. So each track is panned slightly off-center toward a different extreme,
just enough so the two can be distinguished on the horizontal plane. The level of Bongos
1 is automated to balance some hits that jumped out of the mix.

Pen Drums

Inserts: UAD *DM-1L Delay Modulator,* t.c. electronic *Filteroid.*

On the rough mix submitted by the artist, Pen Drums (Track 31.77) were treated with
heavy flanging. To recreate this effect I used the *DM-1L* (Track 31.78), which was followed
by the *Filteroid* (Track 31.79). The *Filteroid* was set with a resonant LPF, with its cut-off
frequency slightly modulated (synced to a quarter note).

In the original arrangement, the Pen Drums appear a few times throughout the production.
Problem was that only during the second break could they be clearly heard – in all other
places they were poorly defined (Track 31.80). This led me to mute all the instances of
these drums except those in the second break (Track 31.81).

Crash

Inserts: MOTU *Masterworks EQ.*

The Crash (Track 31.82) is only treated with a high-shelving pull [*4 kHz, –2.8 dB*] using the *Masterworks EQ* (Track 31.83). This was done in order to tune it into the frequency spectrum that was too bright.

Beat group

Main Group Inserts: PSP *Nitro (automated).*
Parallel Group Inserts: MOTU *Dynamic.*

Parallel compression was applied on the beat. All the beat tracks are routed to a bus, which feeds two auxiliary tracks. One auxiliary is the main beat group and is only processed for two bars during the second theme section with a flanger from the *Nitro* (Tracks 31.84–31.85). The same group is automated so the beat during the intro and breaks is 1.7 dB quieter than during the theme sections. The second auxiliary is the compressed layer, processed by the compressor [*threshold –28 dB, ratio 1.3:1, attack 7.3 ms, release 10 ms*] of the MOTO *Dynamics* (Tracks 31.86–31.89).

Motif elements

Five tracks constitute the bassline and leads: Sick as Funk, Bass, Focus Bass, Dina and Worm.

Sick as Funk

Inserts: MOTU *Masterworks EQ,* t.c. electronic *Chorus Delay,* MOTU *Trim,* MOTU *Quan Jr.*

Sick as Funk was the most problematic track in this production – it choked (Track 31.90). This created a sudden drop in level, which sounded more erroneous than intentional. The first processor Sick as Funk goes through is the *Masterworks EQ,* where a low-shelving filter [*107 Hz, – 12 dB*] reduces some excess of lows (Track 31.91), in addition to a high-shelving filter [*4.3 Hz, 5.6 dB*] that brightens up what was a dull sound (Track 31.92).

The main plugin employed to conceal the choking sound is the t.c. electronic *Chorus Delay,* inserted after the EQ. The *Two-Track* preset was loaded on the *Chorus Delay,* and its 40 ms of delay and 20% feedback were sufficient to fill the quick level drop (Track 31.93). Another plugin that completed the concealment is the *Quan Jr,* which reduces the bit resolution to 4 bits (Track 31.94). This adds a noticeable distortion that also promotes the aggressive nature of the track. The *Quan Jr* only offers a reduction to whole bit values, as opposed to some bit-reduction plugins that offer further bit divisions like 4.5 (while a half bit might not make theoretical sense, it is perfectly possible to have fractional bit reduction in the floating-point domain). When a bit-reduction process only offers whole values, it is possible to fine-tune the effect by boosting or attenuating the signal prior to the bit reduction. And so, the *Quan Jr* is preceded by the MOTU *Trim* plugin, which boosts the signal by 2.4 dB. The bit reduction introduced some high-frequency noise, which was eliminated by a LPF [*13.2 kHz, 12 dB/oct*] from another *Masterworks EQ* that succeeded the *Quan Jr* (Track 31.95). On the same EQ, a parametric filter [*4 kHz, 5.6 dB*] adds some presence and definition (Track 31.96). Sick as Funk is panned slightly off-center to the left [*<11*].

Bass

The Bass track (Track 31.97) is layered with the Sick as Funk track to complement it. It was left untreated and panned hard center (Tracks 31.98–31.99).

Focus Bass

Inserts: Sonnox *Oxford EQ,* MOTU *Dynamics.*

Focus Bass (Track 31.100) was treated with the *Oxford EQ*, where a parametric filter [*131 Hz, −8.4 dB, Q 16*] treated some resonance that occurred on the last note of a rising sequence (Track 31.101). Also, a low-shelving filter [*89.4 Hz, −6.3 dB*] reduces some overemphasized lows (Track 31.102). Then a high-shelving EQ pulls 4 dB at 9.9 kHz to reduce high-frequency noise (Track 31.103). The EQ is followed by the MOTU *Dynamics* where a compressor is set to duck the bass with relation to the Dina track. The release was set to a rather long 3 seconds, so that around 5 dB of gain reduction recovers just before the next Dina hit (Tracks 31.104–31.107). Focus Bass was panned slightly off-center to the left [*<12*].

Dina

Inserts: Sonnox *Oxford Dynamic.*
Sends: Ducking Side-Chain

Dina (Track 31.108) is the main lead track. It is only treated with the compressor [*threshold −22.5 dB, ratio 4.26:1, hard knee, attack 52 ms, release 5 ms*] on the *Oxford Dynamics*, which was employed to add attack (Track 31.109). The Dina lead is the sole track to duck other tracks like the Focus Bass and Hi Hats. It is panned slightly off-center to the left [*<16*].

Worm

Worm (Track 31.110) complements the Dina track. It could have been panned to the same position as Dina (Track 31.111), but in order to open up the sound it was panned off-center to the right [*26>*](Track 31.112). Otherwise, it is untreated.

Lead group

Main Group Inserts: Sonnox *Oxford EQ.*
Parallel Group Inserts: MOTU *PreAmp-1,* MOTU *Trim,* MOTU *Masterworks EQ.*

In order to introduce some extra power and aggression, parallel distortion was applied on the five tracks constituting the bassline and leads. This was achieved by sending the five tracks to a bus, and feeding that bus into two aux tracks. One aux track is simply an audio group, and is only treated with the *Oxford EQ*, where a parametric filter [*125 Hz, −2.88 dB,*

Q 1.87] tunes the leads to the frequency spectrum by reducing some excess content around the 125 Hz area (Track 31.113). The other aux contributes the distorted layer, where the *PreAmp-1* severely distorts the signal (Tracks 31.115–31.116). The output of the *PreAmp-1* involves distinct stereo spread that was unwanted. So it is followed by the MOTU *Trim*, which converts the stereo signal into mono (Track 31.117). The distortion added a generous amount of high-frequency noise, so the *Trim* plugin is followed by *Masterworks EQ*, where a 12 dB/oct LPF rolls off at 14.1 kHz, and a high-shelving filter attenuates 3.6 dB at 4 kHz (Track 31.118).

Initially, the distorted layer was mixed 8.8 dB below the untreated audio group, but this resulted in saturated lows that lacked tightness (Track 31.119). So the balance between the two was adjusted, ending up with the distorted layer 14.3 dB below the audio group (Track 31.120). This means that the distorted layer ended up being felt more than clearly heard, but it still contributes some extra power (Track 31.121).

The level of the lead group was automated. First, the level rises just before the second break (Tracks 31.122–31.123). Then, during the second theme section the level is ridden to accent single note hits (Tracks 31.124–31.125).

Hemorrhage

Inserts: MOTU *Delay*, MOTU *MS Decoder*, MOTU *Dynamics*.

Hemorrhage (Track 31.126) plays between the main motif lines and during the second break. The original mono track would compete with the beat and other main elements had it been panned center, and would create stereo imbalance had it been panned to either extremes. So to make it stereophonic, the Haas trick was applied using 20 ms of delay on the right channel only using the MOTU *Delay* (Track 31.127). But the new stereophonic track was too wide, so the MOTU *MS Decoder* narrows its stereo width; this worked well despite combfiltering that could occur between the delayed and non-delayed channels (Track 31.128). Then, the MOTU *Dynamics* is used to duck Hemorrhage with relation to Dina (Tracks 31.129–31.130). The level of this track was also automated to dive by 3.6 dB during the second break. Hemorrhage is panned off-center to the right [*24>*].

Jumpy

Inserts: MOTU *Masterworks EQ*, MOTU *Delay*, MOTU *Dynamics*.
Sends: Ambiance Reverb (Audioease Altiverb).

Jumpy is a complementary layer for Hemorrhage – it was mixed only to blend with Hemorrhage, not to compete with it, thus it is not easily discerned (Tracks 31.131–31.132). It is first treated with the *Masterworks EQ* where a HPF [*263 Hz, 24 dB/oct*] rolls off dispensable lows. Also, a parametric filter [*4.4 dB, 1.86 kHz, 0.64 Q*] adds some presence, and a high-shelving filter [*8.4 dB, 9.16 kHz*] pulls some high frequencies as part of frequency tuning (Tracks 31.133–31.134). Jumpy was a stereo track that did not blend well with the stereo panorama. So the MOTU *Delay* was used with 8th-note delay on the right channel only – no delay on the left channel. This widens the stereo image to the extremes and adds some rhythmical effect (Track 31.135). Like Hemorrhage, Jumpy is also ducked in relation to Dina using the MOTU *Dynamics*. Jumpy is also sent to the ambiance reverb, which pans it slightly backward (Track 31.136).

Sick Warble

Inserts: MOTU *Masterworks EQ,* MOTU *Delay,* MOTU *Dynamics.*
Sends: Ambiance Reverb (Audioease *Altiverb).*

Sick Warble is a variation of Jumpy. It only plays in the first theme section, where eight bars of Hemorrhage with Jumpy are followed by eight bars of Hemorrhage with Sick Warble. Like Jumpy, it is a complementary layer for Hemorrhage and was mixed to only blend underneath it (Tracks 31.137–31.138). On the *Masterworks EQ,* a HPF [*346.7 Hz, 18 db/oct*] rolls off some dispensable lows (Tracks 31.139–31.140) and a parametric filter [*7.6 dB, 3.24 kHz, 0.83 Q*] boosts some presence (Track 31.141). The MOTU *Delay* also adds 8th-note delay, only this time the non-delayed signal is panned hard right and the delayed signal is panned hard left (Track 31.142). The MOTU *Dynamics* ducks Sick Warble in relation to Dina. In addition, Sick Warble is also sent to the ambiance reverb for backward panning (Track 31.143).

Pads

Atoms 1

Inserts: Sonnox *Oxford EQ,* MOTU *Trim.*
Sends: Ambiance Reverb (Audioease *Altiverb).*

Atoms 1 (Track 31.144) plays during the introduction and during the second break. First it goes through the *Oxford EQ* where a parametric filter [*1.3 kHz, −5.6 dB, 2.83*] pulls some edge to soften the sound (Track 31.145). In addition, a high-shelving filter [*5.2 kHz, 4 dB*] adds some brilliance and definition (Track 31.146). Atoms 1 contributes to the main melodic line during the introduction. Therefore, I found the autopan effect imprinted on the raw stereo track disturbing. To combat this, the stereo track was summed to mono using the MOTU *Trim* (Track 31.147). It is sent to the ambiance reverb to fabricate some atmosphere and to pan the pad backward from the beat. The ambiance reverb also compensates for the mono summing by adding some stereo width (Track 31.148).

Atoms 2

Atoms 2 (Track 31.149) was one track I found annoying, even after various treatments that I applied to it. I found its sharp character to contribution little to the overall mood of the production and decided to exclude it from the final mix.

Korg 174

Inserts: PSP *MasterQ,* MOTU *Delay.*
Sends: Ambiance Reverb (Audioease *Altiverb).*

The Korg 174 (Track 31.150) only plays during the intro. Four bands on the *MasterQ* were employed to tune this track into the frequency spectrum: a HPF [*145 Hz, 12 dB/oct*], a low-shelving filter [*1 kHz, −8.2 dB*], a parametric filter [*2.27 kHz, −9.47 dB, 1.1 Q*] and a high-shelving filter [*9.5 kHz, −2.87 dB*] (Track 31.151). The original stereo track had a limited stereo width and I wanted this pad to stretch to the extremes. So the MOTU *Delay* was inserted to produce the Haas trick, with 30 ms delay on the right channel only

(Track 31.152). This track is also sent to the ambiance reverb in order to detach it from the front of the mix and to blend it into the mix space (Track 31.153).

Horns and brass

Curtis Rev

Inserts: MOTU *Tremolo,* MOTU *Parametric EQ,* MOTU *Dynamics.*
Sends: Ambiance Reverb (Audioease *Altiverb).*

Curtis Rev (Track 31.154) is introduced in the intro, then plays again during the theme sections. It was first treated with the *Tremolo* [*bar-synced, 72% depth*] in order to tighten it to the rhythm and to enhance its come-and-go nature (Track 31.155). The *Tremolo* was followed by the *Parametric EQ* where a HPF [*396 Hz*] removes muddiness (Track 31.156) and a high-shelving filter [*2.7 kHz, 4.5 dB*] adds some brilliance and definition (Track 31.157). The MOTU *Dynamics* ducks the track in relation to Dina. This track is sent to the ambiance reverb for spatial positioning (Track 31.158).

Curtis Rev was automated in two principal ways: First, it was muted during the first theme section, which clears some space in the mix. Then, it was automated to rise in level during the second theme section, including a rise after the first eight bars of each Hemorrhage phrase.

Horns

Inserts: UAD *DM-1 Delay Modulator,* t.c. electronic *EQSat,* t.c. electronic *Filteroid.*

Horns (Track 31.159) is introduced during the intro and plays during the two breaks. The raw track included open–close hi hats, which fortunately responded well to the applied equalization (by that I mean hardly limiting the processing). The original mono track is first treated with a chorus from the *DM-1.* The chorus turns this track into stereo, with the dry/wet control determining the stereo width (ending up at 27.5% dry). The chorus also adds some spatial dimension that sends the horns backward (Track 31.160). Next, a low-shelving filter [*192 Hz, –5 dB*] on the *EQSat* reduces some lows (Track 31.161) and a high-shelving filter [*7 kHz, 3 dB*] adds some brilliance and tunes this track into the mix spectrum (Track 31.162).

Horns is automated in two ways: First, its level is automated to rise during the intro, then it drops with the first break and starts rising again. Second, during the first part of the second break, a resonant LPF from the *Filteroid* sweeps from 668 Hz up to 20 kHz (Track 31.163).

Hip Hornz 2

Inserts: MOTU *Masterworks EQ,* UAD *DM-1 Delay Modulator.*
Sends: Ambiance Reverb (Audioease *Altiverb).*

Hip Hornz 2 (Track 31.164) was mixed to appear distant, somewhat like distant ship horns. First, this track goes through the *Masterworks EQ,* where a HPF [*263 Hz, 18 dB/oct*] removes an excess of lows that resulted in muddiness and lack of definition (Track 31.165); in addition, a high-shelving filter [*1.3 kHz, 7.2 dB*] adds some missing highs and extra definition (Track 31.166). Just like with Horns, the same chorus from the *DM-1* creates some stereo width and sends the track backward (Track 31.167). The track was sent at a high level to the ambiance reverb, which places it deep within the mix space and adds a nice warm reverb tail (Track 31.168).

Hip Hornz 3

Inserts: MOTU *Parametric EQ.*
Sends: Ambiance Reverb (Audioease *Altiverb*).

Hip Hornz 3 (Track 31.169) is something of an answer to a question asked by Hip Hornz 2 – I wanted them to appear closer, so as to create a contrast between the two. The track is only processed with the *Parametric EQ*, where a HPF [*289 Hz*] rolls of an excess of lows (Track 31.170). It is sent to the ambiance reverb but with less level than Hip Hornz 2, thus it appears closer (Track 31.171).

Curtis Verb

Inserts: MOTU *Masterworks EQ,* MOTU *Dynamics.*
Sends: Ambiance Reverb (Audioease *Altiverb*).

Curtis Verb (Track 31.172), essentially a burst of two trumpet-like notes with reverb, was first treated with *Masterworks EQ*, where a high-shelving filter [*4 kHz, −4 dB*] tunes it to the frequency spectrum by softening its highs (Track 31.173). One issue with this track was that it appears too in front. It could be sent backward with more reverb, but as discussed below, the reverb added to this track plays a unique roll that restricted changes to the reverb send level. So the job of sending this track backward was assigned to the *MOTU Dynamics*, where a compressor [*−15 dB threshold, 2:1 ratio, 0.1 ms attack, 100 ms release*] contains its attack (Track 31.174).

Originally, a HPF on the *Masterworks EQ* was employed to roll off the lows, but after sending this track to the ambiance reverb (Track 31.175), disabling the HPF resulted in a metallic thunder that I liked (Track 31.176). This effect, which was discovered by accident, is kept in the final mix.

Risers

Riser and Reversed Rise

Sends: Ambiance Reverb (Audioease *Altiverb*).

Riser (Track 31.177) and Reversed Rise (Track 31.178) mostly play together, but at certain points play alone. Neither tracks are treated, but Riser is sent to the ambiance reverb with a generous send level (Track 31.179).

Bent Phase

Inserts: MOTU *Masterworks EQ.*
Sends: Ambiance Reverb (Audioease *Altiverb).*

Bent Phase (Track 31.180) was treated with the MOTU *Masterworks EQ*, where a HPF [*525 HZ, 18 dB/oct*] removes dispensable low-mids and lows (Track 31.181). It is sent to the ambiance reverb for depth positioning (Track 31.182).

Hi Synth

Inserts: MOTU *PreAmp-1*, MOTU *Dynamics*, MOTU *Trim.*
Sends: Ambiance Reverb (Audioease *Altiverb).*

One problematic track was the Hi Synth (Track 31.183) because it sounded synthetic and foreign to the arrangement. At certain points this track was a candidate for exclusion, but it ended up lightly buried after some treatment: First, it was distorted using the *PreAmp-1*, which adds some grit (Track 31.184). Second, it was compressed using the MOTU *Dynamics* (Track 31.185) – the untreated track rises both in level and pitch and I found that the rises in pitch is sufficient for the buildup, so the compressor contains the level rise. Finally, it goes through the MOTU *Trim*, which inverses the phase of the right channel to produce the out-of-speakers effect (Track 31.186) – neither center nor side panning seemed ideal for this mono track. It is also sent to the ambiance reverb, which blends it into the depth field (Track 31.187).

ES2 Rise

Sends: Ambiance Reverb (Audioease *Altiverb).*

ES2 Rise (Track 31.188) appears once in the arrangement toward the end of the first break. It is untreated, but sent to the ambiance reverb. Apart from blending it with the depth field, some of this track's definition is courtesy of the reverb and its tail (Track 31.189).

Sine Rise

Inserts: MOTU *Masterworks EQ,* MOTU *Delay.*
Sends: Ambiance Reverb (Audioease *Altiverb).*

Sine Rise (Track 31.190) only plays once toward the end of the second break along with Riser, Reversed Rise and Bent Phase, and thus its role is somewhat redundant. A choice had to be made as to which tracks take the spotlight: Riser and Reversed Rise or Sine Rise. Since Riser and Reversed Rise are richer, Sine Rise ended up being layered underneath them. Sine Rise first goes through the *Masterworks EQ* where a parametric filter [*−6 dB, 4 kHz, 0.83 Q*] tunes it into the frequency spectrum (Track 31.191). The originally narrow stereo width was made wider by the MOTU *Delay* which delays the right channel only by 80 ms (Track 31.192). It is sent to the ambiance reverb, which sends it quite deep into the depth field (Track 31.193).

Strings

Curtis Strings

Inserts: MOTU *Masterworks EQ.*
Sends: Ambiance Reverb (Audioease *Altiverb).*

Curtis Strings (Track 31.194) are only treated with the *Masterworks EQ*, where a HPF [*100 Hz, 18 dB/oct*] rolls off dispensable lows (Track 31.195). This track is also sent to the ambiance reverb (Track 31.196) and its level was automated between the various sections.

Strings Finale

Sends: Ambiance Reverb (Audioease *Altiverb).*

Strings Finale (Track 31.197) concludes the production. This track was untreated but to add some drama its imprinted rise in level was enforced by level automation (Track 31.198). It was also sent to the ambiance reverb (Track 31.199).

Appendix A
Notes to Frequencies Chart

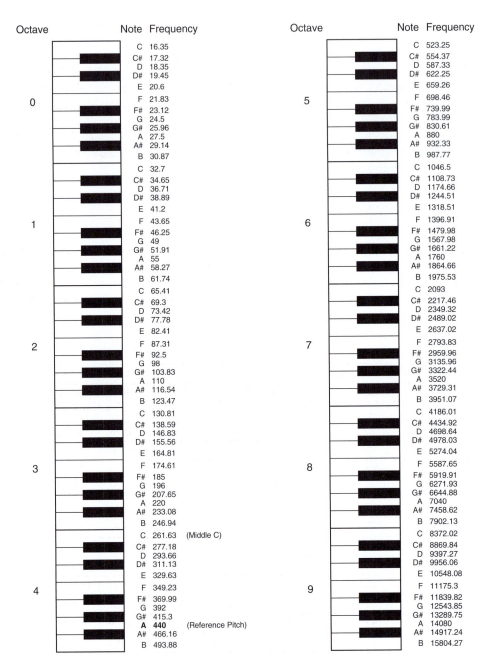

Octave	Note	Frequency
	C	16.35
	C#	17.32
	D	18.35
	D#	19.45
	E	20.6
0	F	21.83
	F#	23.12
	G	24.5
	G#	25.96
	A	27.5
	A#	29.14
	B	30.87
	C	32.7
	C#	34.65
	D	36.71
	D#	38.89
	E	41.2
1	F	43.65
	F#	46.25
	G	49
	G#	51.91
	A	55
	A#	58.27
	B	61.74
	C	65.41
	C#	69.3
	D	73.42
	D#	77.78
	E	82.41
2	F	87.31
	F#	92.5
	G	98
	G#	103.83
	A	110
	A#	116.54
	B	123.47
	C	130.81
	C#	138.59
	D	146.83
	D#	155.56
	E	164.81
3	F	174.61
	F#	185
	G	196
	G#	207.65
	A	220
	A#	233.08
	B	246.94
	C	261.63 (Middle C)
	C#	277.18
	D	293.66
	D#	311.13
	E	329.63
4	F	349.23
	F#	369.99
	G	392
	G#	415.3
	A	**440** (Reference Pitch)
	A#	466.16
	B	493.88

Octave	Note	Frequency
	C	523.25
	C#	554.37
	D	587.33
	D#	622.25
	E	659.26
5	F	698.46
	F#	739.99
	G	783.99
	G#	830.61
	A	880
	A#	932.33
	B	987.77
	C	1046.5
	C#	1108.73
	D	1174.66
	D#	1244.51
	E	1318.51
6	F	1396.91
	F#	1479.98
	G	1567.98
	G#	1661.22
	A	1760
	A#	1864.66
	B	1975.53
	C	2093
	C#	2217.46
	D	2349.32
	D#	2489.02
	E	2637.02
7	F	2793.83
	F#	2959.96
	G	3135.96
	G#	3322.44
	A	3520
	A#	3729.31
	B	3951.07
	C	4186.01
	C#	4434.92
	D	4698.64
	D#	4978.03
	E	5274.04
8	F	5587.65
	F#	5919.91
	G	6271.93
	G#	6644.88
	A	7040
	A#	7458.62
	B	7902.13
	C	8372.02
	C#	8869.84
	D	9397.27
	D#	9956.06
	E	10548.08
9	F	11175.3
	F#	11839.82
	G	12543.85
	G#	13289.75
	A	14080
	A#	14917.24
	B	15804.27

Appendix B
Delay Time Chart

Delay times can be easily calculated. For time signatures involving quarter-note beats (4/4, 3/4, etc.), a quarter-note delay in seconds is calculated using 60/BPM. For example, at 120 BPM, a quarter-note delay would be 60/120, or 0.5 seconds, which is 500 ms. Milliseconds are far more common in delay units, using the calculation 60 000/BPM would give the result in milliseconds. A half-note delay is twice as long as a quarter-note delay for the same BPM, and an eighth-note delay is half the quarter-note delay. For example, a half-note delay at 120 BPM would be 1000 ms, and an eighth-note delay would be 250 ms.

Simple note-values

The following table includes delay times for various simple note-values at 80–190 BPM. For faster BPM figures, simply halve the delay time of half the destination BPM. For example, a quarter-note delay at 240 BPM is 250 ms (half of 500 ms). For slower BPM figures, simply double the delay time of twice the destination BPM. For example, a quarter-note delay in 60 BPM is 1000 ms (twice 500 ms). The delay time of a whole-note in common time (4/4) is also the length of each bar.

BPM/note value	1/1	1/2	1/4	1/8	1/16
80	3000.00	1500.00	750.00	375.00	187.50
81	2962.96	1481.48	740.74	370.37	185.19
82	2926.83	1463.41	731.71	365.85	182.93
83	2891.57	1445.78	722.89	361.45	180.72
84	2857.14	1428.57	714.29	357.14	178.57
85	2823.53	1411.76	705.88	352.94	176.47
86	2790.70	1395.35	697.67	348.84	174.42
87	2758.62	1379.31	689.66	344.83	172.41
88	2727.27	1363.64	681.82	340.91	170.45
89	2696.63	1348.31	674.16	337.08	168.54
90	2666.67	1333.33	666.67	333.33	166.67
91	2637.36	1318.68	659.34	329.67	164.84
92	2608.70	1304.35	652.17	326.09	163.04
93	2580.65	1290.32	645.16	322.58	161.29

BPM/note value	1/1	1/2	1/4	1/8	1/16
94	2553.19	1276.60	638.30	319.15	159.57
95	2526.32	1263.16	631.58	315.79	157.89
96	2500.00	1250.00	625.00	312.50	156.25
97	2474.23	1237.11	618.56	309.28	154.64
98	2448.98	1224.49	612.24	306.12	153.06
99	2424.24	1212.12	606.06	303.03	151.52
100	2400.00	1200.00	600.00	300.00	150.00
101	2376.24	1188.12	594.06	297.03	148.51
102	2352.94	1176.47	588.24	294.12	147.06
103	2330.10	1165.05	582.52	291.26	145.63
104	2307.69	1153.85	576.92	288.46	144.23
105	2285.71	1142.86	571.43	285.71	142.86
106	2264.15	1132.08	566.04	283.02	141.51
107	2242.99	1121.50	560.75	280.37	140.19
108	2222.22	1111.11	555.56	277.78	138.89
109	2201.83	1100.92	550.46	275.23	137.61
110	2181.82	1090.91	545.45	272.73	136.36
111	2162.16	1081.08	540.54	270.27	135.14
112	2142.86	1071.43	535.71	267.86	133.93
113	2123.89	1061.95	530.97	265.49	132.74
114	2105.26	1052.63	526.32	263.16	131.58
115	2086.96	1043.48	521.74	260.87	130.43
116	2068.97	1034.48	517.24	258.62	129.31
117	2051.28	1025.64	512.82	256.41	128.21
118	2033.90	1016.95	508.47	254.24	127.12
119	2016.81	1008.40	504.20	252.10	126.05
120	2000.00	1000.00	500.00	250.00	125.00
121	1983.47	991.74	495.87	247.93	123.97
122	1967.21	983.61	491.80	245.90	122.95
123	1951.22	975.61	487.80	243.90	121.95
124	1935.48	967.74	483.87	241.94	120.97
125	1920.00	960.00	480.00	240.00	120.00
126	1904.76	952.38	476.19	238.10	119.05
127	1889.76	944.88	472.44	236.22	118.11
128	1875.00	937.50	468.75	234.38	117.19
129	1860.47	930.23	465.12	232.56	116.28
130	1846.15	923.08	461.54	230.77	115.38
131	1832.06	916.03	458.02	229.01	114.50
132	1818.18	909.09	454.55	227.27	113.64
133	1804.51	902.26	451.13	225.56	112.78

BPM/note value	1/1	1/2	1/4	1/8	1/16
134	1791.04	895.52	447.76	223.88	111.94
135	1777.78	888.89	444.44	222.22	111.11
136	1764.71	882.35	441.18	220.59	110.29
137	1751.82	875.91	437.96	218.98	109.49
138	1739.13	869.57	434.78	217.39	108.70
139	1726.62	863.31	431.65	215.83	107.91
140	1714.29	857.14	428.57	214.29	107.14
141	1702.13	851.06	425.53	212.77	106.38
142	1690.14	845.07	422.54	211.27	105.63
143	1678.32	839.16	419.58	209.79	104.90
144	1666.67	833.33	416.67	208.33	104.17
145	1655.17	827.59	413.79	206.90	103.45
146	1643.84	821.92	410.96	205.48	102.74
147	1632.65	816.33	408.16	204.08	102.04
148	1621.62	810.81	405.41	202.70	101.35
149	1610.74	805.37	402.68	201.34	100.67
150	1600.00	800.00	400.00	200.00	100.00
151	1589.40	794.70	397.35	198.68	99.34
152	1578.95	789.47	394.74	197.37	98.68
153	1568.63	784.31	392.16	196.08	98.04
154	1558.44	779.22	389.61	194.81	97.40
155	1548.39	774.19	387.10	193.55	96.77
156	1538.46	769.23	384.62	192.31	96.15
157	1528.66	764.33	382.17	191.08	95.54
158	1518.99	759.49	379.75	189.87	94.94
159	1509.43	754.72	377.36	188.68	94.34
160	1500.00	750.00	375.00	187.50	93.75
161	1490.68	745.34	372.67	186.34	93.17
162	1481.48	740.74	370.37	185.19	92.59
163	1472.39	736.20	368.10	184.05	92.02
164	1463.41	731.71	365.85	182.93	91.46
165	1454.55	727.27	363.64	181.82	90.91
166	1445.78	722.89	361.45	180.72	90.36
167	1437.13	718.56	359.28	179.64	89.82
168	1428.57	714.29	357.14	178.57	89.29
169	1420.12	710.06	355.03	177.51	88.76
170	1411.76	705.88	352.94	176.47	88.24
171	1403.51	701.75	350.88	175.44	87.72
172	1395.35	697.67	348.84	174.42	87.21
173	1387.28	693.64	346.82	173.41	86.71

BPM/note value	1/1	1/2	1/4	1/8	1/16
174	1379.31	689.66	344.83	172.41	86.21
175	1371.43	685.71	342.86	171.43	85.71
176	1363.64	681.82	340.91	170.45	85.23
177	1355.93	677.97	338.98	169.49	84.75
178	1348.31	674.16	337.08	168.54	84.27
179	1340.78	670.39	335.20	167.60	83.80
180	1333.33	666.67	333.33	166.67	83.33
181	1325.97	662.98	331.49	165.75	82.87
182	1318.68	659.34	329.67	164.84	82.42
183	1311.48	655.74	327.87	163.93	81.97
184	1304.35	652.17	326.09	163.04	81.52
185	1297.30	648.65	324.32	162.16	81.08
186	1290.32	645.16	322.58	161.29	80.65
187	1283.42	641.71	320.86	160.43	80.21
188	1276.60	638.30	319.15	159.57	79.79
189	1269.84	634.92	317.46	158.73	79.37
190	1263.16	631.58	315.79	157.89	78.95

Triplets

To calculate the delay time of a triplet note, simply multiply the delay time of the simple note-value by 2/3 (0.6667). The following chart displays the delay time for triplet delays:

BPM/note value	1/1T	1/2T	1/4T	1/8T	1/16T
80	2000.00	1000.00	500.00	250.00	125.00
81	1975.31	987.65	493.83	246.91	123.46
82	1951.22	975.61	487.80	243.90	121.95
83	1927.71	963.86	481.93	240.96	120.48
84	1904.76	952.38	476.19	238.10	119.05
85	1882.35	941.18	470.59	235.29	117.65
86	1860.47	930.23	465.12	232.56	116.28
87	1839.08	919.54	459.77	229.89	114.94
88	1818.18	909.09	454.55	227.27	113.64
89	1797.75	898.88	449.44	224.72	112.36
90	1777.78	888.89	444.44	222.22	111.11
91	1758.24	879.12	439.56	219.78	109.89
92	1739.13	869.57	434.78	217.39	108.70
93	1720.43	860.22	430.11	215.05	107.53
94	1702.13	851.06	425.53	212.77	106.38

BPM/note value	1/1T	1/2T	1/4T	1/8T	1/16T
95	1684.21	842.11	421.05	210.53	105.26
96	1666.67	833.33	416.67	208.33	104.17
97	1649.48	824.74	412.37	206.19	103.09
98	1632.65	816.33	408.16	204.08	102.04
99	1616.16	808.08	404.04	202.02	101.01
100	1600.00	800.00	400.00	200.00	100.00
101	1584.16	792.08	396.04	198.02	99.01
102	1568.63	784.31	392.16	196.08	98.04
103	1553.40	776.70	388.35	194.17	97.09
104	1538.46	769.23	384.62	192.31	96.15
105	1523.81	761.90	380.95	190.48	95.24
106	1509.43	754.72	377.36	188.68	94.34
107	1495.33	747.66	373.83	186.92	93.46
108	1481.48	740.74	370.37	185.19	92.59
109	1467.89	733.94	366.97	183.49	91.74
110	1454.55	727.27	363.64	181.82	90.91
111	1441.44	720.72	360.36	180.18	90.09
112	1428.57	714.29	357.14	178.57	89.29
113	1415.93	707.96	353.98	176.99	88.50
114	1403.51	701.75	350.88	175.44	87.72
115	1391.30	695.65	347.83	173.91	86.96
116	1379.31	689.66	344.83	172.41	86.21
117	1367.52	683.76	341.88	170.94	85.47
118	1355.93	677.97	338.98	169.49	84.75
119	1344.54	672.27	336.13	168.07	84.03
120	1333.33	666.67	333.33	166.67	83.33
121	1322.31	661.16	330.58	165.29	82.64
122	1311.48	655.74	327.87	163.93	81.97
123	1300.81	650.41	325.20	162.60	81.30
124	1290.32	645.16	322.58	161.29	80.65
125	1280.00	640.00	320.00	160.00	80.00
126	1269.84	634.92	317.46	158.73	79.37
127	1259.84	629.92	314.96	157.48	78.74
128	1250.00	625.00	312.50	156.25	78.13
129	1240.31	620.16	310.08	155.04	77.52
130	1230.77	615.38	307.69	153.85	76.92
131	1221.37	610.69	305.34	152.67	76.34
132	1212.12	606.06	303.03	151.52	75.76
133	1203.01	601.50	300.75	150.38	75.19
134	1194.03	597.01	298.51	149.25	74.63

BPM/note value	1/1T	1/2T	1/4T	1/8T	1/16T
135	1185.19	592.59	296.30	148.15	74.07
136	1176.47	588.24	294.12	147.06	73.53
137	1167.88	583.94	291.97	145.99	72.99
138	1159.42	579.71	289.86	144.93	72.46
139	1151.08	575.54	287.77	143.88	71.94
140	1142.86	571.43	285.71	142.86	71.43
141	1134.75	567.38	283.69	141.84	70.92
142	1126.76	563.38	281.69	140.85	70.42
143	1118.88	559.44	279.72	139.86	69.93
144	1111.11	555.56	277.78	138.89	69.44
145	1103.45	551.72	275.86	137.93	68.97
146	1095.89	547.95	273.97	136.99	68.49
147	1088.44	544.22	272.11	136.05	68.03
148	1081.08	540.54	270.27	135.14	67.57
149	1073.83	536.91	268.46	134.23	67.11
150	1066.67	533.33	266.67	133.33	66.67
151	1059.60	529.80	264.90	132.45	66.23
152	1052.63	526.32	263.16	131.58	65.79
153	1045.75	522.88	261.44	130.72	65.36
154	1038.96	519.48	259.74	129.87	64.94
155	1032.26	516.13	258.06	129.03	64.52
156	1025.64	512.82	256.41	128.21	64.10
157	1019.11	509.55	254.78	127.39	63.69
158	1012.66	506.33	253.16	126.58	63.29
159	1006.29	503.14	251.57	125.79	62.89
160	1000.00	500.00	250.00	125.00	62.50
161	993.79	496.89	248.45	124.22	62.11
162	987.65	493.83	246.91	123.46	61.73
163	981.60	490.80	245.40	122.70	61.35
164	975.61	487.80	243.90	121.95	60.98
165	969.70	484.85	242.42	121.21	60.61
166	963.86	481.93	240.96	120.48	60.24
167	958.08	479.04	239.52	119.76	59.88
168	952.38	476.19	238.10	119.05	59.52
169	946.75	473.37	236.69	118.34	59.17
170	941.18	470.59	235.29	117.65	58.82
171	935.67	467.84	233.92	116.96	58.48
172	930.23	465.12	232.56	116.28	58.14
173	924.86	462.43	231.21	115.61	57.80
174	919.54	459.77	229.89	114.94	57.47

BPM/note value	1/1T	1/2T	1/4T	1/8T	1/16T
175	914.29	457.14	228.57	114.29	57.14
176	909.09	454.55	227.27	113.64	56.82
177	903.95	451.98	225.99	112.99	56.50
178	898.88	449.44	224.72	112.36	56.18
179	893.85	446.93	223.46	111.73	55.87
180	888.89	444.44	222.22	111.11	55.56
181	883.98	441.99	220.99	110.50	55.25
182	879.12	439.56	219.78	109.89	54.95
183	874.32	437.16	218.58	109.29	54.64
184	869.57	434.78	217.39	108.70	54.35
185	864.86	432.43	216.22	108.11	54.05
186	860.22	430.11	215.05	107.53	53.76
187	855.61	427.81	213.90	106.95	53.48
188	851.06	425.53	212.77	106.38	53.19
189	846.56	423.28	211.64	105.82	52.91
190	842.11	421.05	210.53	105.26	52.63

Dotted notes

A dotted-note duration is achieved by adding half of the note duration to the full-note duration. So to calculate the delay time of a dotted note all we need to do is multiply the delay time of the simple note value by 1.5. The following chart displays the delay time for dotted-note delays:

BPM/note value	1/1D	1/2D	1/4D	1/8D	1/16D
80	4500.00	2250.00	1125.00	562.50	281.25
81	4444.44	2222.22	1111.11	555.56	277.78
82	4390.24	2195.12	1097.56	548.78	274.39
83	4337.35	2168.67	1084.34	542.17	271.08
84	4285.71	2142.86	1071.43	535.71	267.86
85	4235.29	2117.65	1058.82	529.41	264.71
86	4186.05	2093.02	1046.51	523.26	261.63
87	4137.93	2068.97	1034.48	517.24	258.62
88	4090.91	2045.45	1022.73	511.36	255.68
89	4044.94	2022.47	1011.24	505.62	252.81
90	4000.00	2000.00	1000.00	500.00	250.00
91	3956.04	1978.02	989.01	494.51	247.25

BPM/note value	1/1D	1/2D	1/4D	1/8D	1/16D
92	3913.04	1956.52	978.26	489.13	244.57
93	3870.97	1935.48	967.74	483.87	241.94
94	3829.79	1914.89	957.45	478.72	239.36
95	3789.47	1894.74	947.37	473.68	236.84
96	3750.00	1875.00	937.50	468.75	234.38
97	3711.34	1855.67	927.84	463.92	231.96
98	3673.47	1836.73	918.37	459.18	229.59
99	3636.36	1818.18	909.09	454.55	227.27
100	3600.00	1800.00	900.00	450.00	225.00
101	3564.36	1782.18	891.09	445.54	222.77
102	3529.41	1764.71	882.35	441.18	220.59
103	3495.15	1747.57	873.79	436.89	218.45
104	3461.54	1730.77	865.38	432.69	216.35
105	3428.57	1714.29	857.14	428.57	214.29
106	3396.23	1698.11	849.06	424.53	212.26
107	3364.49	1682.24	841.12	420.56	210.28
108	3333.33	1666.67	833.33	416.67	208.33
109	3302.75	1651.38	825.69	412.84	206.42
110	3272.73	1636.36	818.18	409.09	204.55
111	3243.24	1621.62	810.81	405.41	202.70
112	3214.29	1607.14	803.57	401.79	200.89
113	3185.84	1592.92	796.46	398.23	199.12
114	3157.89	1578.95	789.47	394.74	197.37
115	3130.43	1565.22	782.61	391.30	195.65
116	3103.45	1551.72	775.86	387.93	193.97
117	3076.92	1538.46	769.23	384.62	192.31
118	3050.85	1525.42	762.71	381.36	190.68
119	3025.21	1512.61	756.30	378.15	189.08
120	3000.00	1500.00	750.00	375.00	187.50
121	2975.21	1487.60	743.80	371.90	185.95
122	2950.82	1475.41	737.70	368.85	184.43
123	2926.83	1463.41	731.71	365.85	182.93
124	2903.23	1451.61	725.81	362.90	181.45
125	2880.00	1440.00	720.00	360.00	180.00
126	2857.14	1428.57	714.29	357.14	178.57
127	2834.65	1417.32	708.66	354.33	177.17
128	2812.50	1406.25	703.13	351.56	175.78
129	2790.70	1395.35	697.67	348.84	174.42
130	2769.23	1384.62	692.31	346.15	173.08
131	2748.09	1374.05	687.02	343.51	171.76

BPM/note value	1/1D	1/2D	1/4D	1/8D	1/16D
132	2727.27	1363.64	681.82	340.91	170.45
133	2706.77	1353.38	676.69	338.35	169.17
134	2686.57	1343.28	671.64	335.82	167.91
135	2666.67	1333.33	666.67	333.33	166.67
136	2647.06	1323.53	661.76	330.88	165.44
137	2627.74	1313.87	656.93	328.47	164.23
138	2608.70	1304.35	652.17	326.09	163.04
139	2589.93	1294.96	647.48	323.74	161.87
140	2571.43	1285.71	642.86	321.43	160.71
141	2553.19	1276.60	638.30	319.15	159.57
142	2535.21	1267.61	633.80	316.90	158.45
143	2517.48	1258.74	629.37	314.69	157.34
144	2500.00	1250.00	625.00	312.50	156.25
145	2482.76	1241.38	620.69	310.34	155.17
146	2465.75	1232.88	616.44	308.22	154.11
147	2448.98	1224.49	612.24	306.12	153.06
148	2432.43	1216.22	608.11	304.05	152.03
149	2416.11	1208.05	604.03	302.01	151.01
150	2400.00	1200.00	600.00	300.00	150.00
151	2384.11	1192.05	596.03	298.01	149.01
152	2368.42	1184.21	592.11	296.05	148.03
153	2352.94	1176.47	588.24	294.12	147.06
154	2337.66	1168.83	584.42	292.21	146.10
155	2322.58	1161.29	580.65	290.32	145.16
156	2307.69	1153.85	576.92	288.46	144.23
157	2292.99	1146.50	573.25	286.62	143.31
158	2278.48	1139.24	569.62	284.81	142.41
159	2264.15	1132.08	566.04	283.02	141.51
160	2250.00	1125.00	562.50	281.25	140.63
161	2236.02	1118.01	559.01	279.50	139.75
162	2222.22	1111.11	555.56	277.78	138.89
163	2208.59	1104.29	552.15	276.07	138.04
164	2195.12	1097.56	548.78	274.39	137.20
165	2181.82	1090.91	545.45	272.73	136.36
166	2168.67	1084.34	542.17	271.08	135.54
167	2155.69	1077.84	538.92	269.46	134.73
168	2142.86	1071.43	535.71	267.86	133.93
169	2130.18	1065.09	532.54	266.27	133.14
170	2117.65	1058.82	529.41	264.71	132.35
171	2105.26	1052.63	526.32	263.16	131.58

BPM/note value	1/1D	1/2D	1/4D	1/8D	1/16D
172	2093.02	1046.51	523.26	261.63	130.81
173	2080.92	1040.46	520.23	260.12	130.06
174	2068.97	1034.48	517.24	258.62	129.31
175	2057.14	1028.57	514.29	257.14	128.57
176	2045.45	1022.73	511.36	255.68	127.84
177	2033.90	1016.95	508.47	254.24	127.12
178	2022.47	1011.24	505.62	252.81	126.40
179	2011.17	1005.59	502.79	251.40	125.70
180	2000.00	1000.00	500.00	250.00	125.00
181	1988.95	994.48	497.24	248.62	124.31
182	1978.02	989.01	494.51	247.25	123.63
183	1967.21	983.61	491.80	245.90	122.95
184	1956.52	978.26	489.13	244.57	122.28
185	1945.95	972.97	486.49	243.24	121.62
186	1935.48	967.74	483.87	241.94	120.97
187	1925.13	962.57	481.28	240.64	120.32
188	1914.89	957.45	478.72	239.36	119.68
189	1904.76	952.38	476.19	238.10	119.05
190	1894.74	947.37	473.68	236.84	118.42

Index

Note: entries in *italics* are album titles